Jean-Baptiste Hiriart-Urruty
Claude Lemaréchal

Convex Analysis and Minimization Algorithms II

Advanced Theory
and Bundle Methods

With 64 Figures

Springer-Verlag
Berlin Heidelberg New York
London Paris Tokyo
Hong Kong Barcelona
Budapest

Jean-Baptiste Hiriart-Urruty
Département de Mathématiques
Université Paul Sabatier
118, route de Narbonne
F-31062 Toulouse, France

Claude Lemaréchal
INRIA, Rocquencourt
Domaine de Voluceau
B.P. 105
F-78153 Le Chesnay, France

Mathematics Subject Classification (1991): 65-01, 65K, 49M, 49M27, 49-01, 93B60, 90C

ISBN 3-540-56852-2 Springer-Verlag Berlin Heidelberg New York
ISBN 0-387-56852-2 Springer-Verlag New York Berlin Heidelberg

This work is subject to copyright. All rights are reserved, whether the whole or part of the material is concerned, specifically the rights of translation, reprinting, reuse of illustrations, recitation, broadcasting, reproduction on microfilm or in any other way, and storage in data banks. Duplication of this publication or parts thereof is permitted only under the provisions of the German Copyright Law of September 9, 1965, in its current version, and permission for use must always be obtained from Springer-Verlag. Violations are liable for prosecution under the German Copyright Law.

© Springer-Verlag Berlin Heidelberg 1993
Printed in the United States of America

Typesetting: Camera-ready copy produced from the authors' output file using a Springer T$_E$X macro package
41/3140-5 4 3 2 1 0   Printed on acid-free paper

Grundlehren der mathematischen Wissenschaften 306

A Series of Comprehensive Studies in Mathematics

Editors

M. Artin S. S. Chern J. Coates J. M. Fröhlich
H. Hironaka F. Hirzebruch L. Hörmander
C. C. Moore J. K. Moser M. Nagata W. Schmidt
D. S. Scott Ya. G. Sinai J. Tits M. Waldschmidt
S. Watanabe

Managing Editors

M. Berger B. Eckmann S. R. S. Varadhan

Table of Contents Part II

Introduction XV

IX. Inner Construction of the Subdifferential 1
   1 The Elementary Mechanism 2
   2 Convergence Properties 9
      2.1 Convergence 9
      2.2 Speed of Convergence 15
   3 Putting the Mechanism in Perspective 24
      3.1 Bundling as a Substitute for Steepest Descent 24
      3.2 Bundling as an Emergency Device for Descent Methods 27
      3.3 Bundling as a Separation Algorithm 29

X. Conjugacy in Convex Analysis 35
   1 The Convex Conjugate of a Function 37
      1.1 Definition and First Examples 37
      1.2 Interpretations 40
      1.3 First Properties 42
      1.4 Subdifferentials of Extended-Valued Functions 47
      1.5 Convexification and Subdifferentiability 49
   2 Calculus Rules on the Conjugacy Operation 54
      2.1 Image of a Function Under a Linear Mapping 54
      2.2 Pre-Composition with an Affine Mapping 56
      2.3 Sum of Two Functions 61
      2.4 Infima and Suprema 65
      2.5 Post-Composition with an Increasing Convex Function 69
      2.6 A Glimpse of Biconjugate Calculus 71
   3 Various Examples 72
      3.1 The Cramer Transformation 72
      3.2 Some Results on the Euclidean Distance to a Closed Set 73
      3.3 The Conjugate of Convex Partially Quadratic Functions 75
      3.4 Polyhedral Functions 76
   4 Differentiability of a Conjugate Function 79
      4.1 First-Order Differentiability 79
      4.2 Towards Second-Order Differentiability 82

Table of Contents Part II

XI. Approximate Subdifferentials of Convex Functions 91
1 The Approximate Subdifferential 92
  1.1 Definition, First Properties and Examples 92
  1.2 Characterization via the Conjugate Function 95
  1.3 Some Useful Properties 98
2 The Approximate Directional Derivative 102
  2.1 The Support Function of the Approximate Subdifferential 102
  2.2 Properties of the Approximate Difference Quotient 106
  2.3 Behaviour of f'_ε and T_ε as Functions of ε 110
3 Calculus Rules on the Approximate Subdifferential 113
  3.1 Sum of Functions 113
  3.2 Pre-Composition with an Affine Mapping 116
  3.3 Image and Marginal Functions 118
  3.4 A Study of the Infimal Convolution 119
  3.5 Maximum of Functions 123
  3.6 Post-Composition with an Increasing Convex Function 125
4 The Approximate Subdifferential as a Multifunction 127
  4.1 Continuity Properties of the Approximate Subdifferential 127
  4.2 Transportation of Approximate Subgradients 129

XII. Abstract Duality for Practitioners 137
1 The Problem and the General Approach 137
  1.1 The Rules of the Game 137
  1.2 Examples 141
2 The Necessary Theory 147
  2.1 Preliminary Results: The Dual Problem 147
  2.2 First Properties of the Dual Problem 150
  2.3 Primal-Dual Optimality Characterizations 154
  2.4 Existence of Dual Solutions 157
3 Illustrations 161
  3.1 The Minimax Point of View 161
  3.2 Inequality Constraints 162
  3.3 Dualization of Linear Programs 165
  3.4 Dualization of Quadratic Programs 166
  3.5 Steepest-Descent Directions 168
4 Classical Dual Algorithms 170
  4.1 Subgradient Optimization 171
  4.2 The Cutting-Plane Algorithm 174
5 Putting the Method in Perspective 178
  5.1 The Primal Function 178
  5.2 Augmented Lagrangians 181
  5.3 The Dualization Scheme in Various Situations 185
  5.4 Fenchel's Duality 190

XIII. Methods of ε-Descent 195

1 Introduction. Identifying the Approximate Subdifferential 195
  1.1 The Problem and Its Solution 195
  1.2 The Line-Search Function 199
  1.3 The Schematic Algorithm 203
2 A Direct Implementation: Algorithm of ε-Descent 206
  2.1 Iterating the Line-Search 206
  2.2 Stopping the Line-Search 209
  2.3 The ε-Descent Algorithm and Its Convergence 212
3 Putting the Algorithm in Perspective 216
  3.1 A Pure Separation Form 216
  3.2 A Totally Static Minimization Algorithm 219

XIV. Dynamic Construction of Approximate Subdifferentials: Dual Form of Bundle Methods 223

1 Introduction: The Bundle of Information 223
  1.1 Motivation 223
  1.2 Constructing the Bundle of Information 227
2 Computing the Direction 233
  2.1 The Quadratic Program 233
  2.2 Minimality Conditions 236
  2.3 Directional Derivatives Estimates 241
  2.4 The Role of the Cutting-Plane Function 244
3 The Implementable Algorithm 248
  3.1 Derivation of the Line-Search 248
  3.2 The Implementable Line-Search and Its Convergence 250
  3.3 Derivation of the Descent Algorithm 254
  3.4 The Implementable Algorithm and Its Convergence 257
4 Numerical Illustrations 263
  4.1 Typical Behaviour 263
  4.2 The Role of ε 266
  4.3 A Variant with Infinite ε: Conjugate Subgradients 268
  4.4 The Role of the Stopping Criterion 269
  4.5 The Role of Other Parameters 271
  4.6 General Conclusions 273

XV. Acceleration of the Cutting-Plane Algorithm: Primal Forms of Bundle Methods 275

1 Accelerating the Cutting-Plane Algorithm 275
  1.1 Instability of Cutting Planes 276
  1.2 Stabilizing Devices: Leading Principles 279
  1.3 A Digression: Step-Control Strategies 283
2 A Variety of Stabilized Algorithms 285
  2.1 The Trust-Region Point of View 286

- 2.2 The Penalization Point of View 289
- 2.3 The Relaxation Point of View 292
- 2.4 A Possible Dual Point of View 295
- 2.5 Conclusion 299
- 3 A Class of Primal Bundle Algorithms 301
  - 3.1 The General Method 301
  - 3.2 Convergence 307
  - 3.3 Appropriate Stepsize Values 314
- 4 Bundle Methods as Regularizations 317
  - 4.1 Basic Properties of the Moreau-Yosida Regularization . 317
  - 4.2 Minimizing the Moreau-Yosida Regularization 322
  - 4.3 Computing the Moreau-Yosida Regularization 326

Bibliographical Comments 331

References 337

Index 345

Table of Contents Part I

Introduction XV

I. Convex Functions of One Real Variable 1
  1 Basic Definitions and Examples 1
    1.1 First Definitions of a Convex Function 2
    1.2 Inequalities with More Than Two Points 6
    1.3 Modern Definition of Convexity 8
  2 First Properties 9
    2.1 Stability Under Functional Operations 9
    2.2 Limits of Convex Functions 11
    2.3 Behaviour at Infinity 14
  3 Continuity Properties 16
    3.1 Continuity on the Interior of the Domain 16
    3.2 Lower Semi-Continuity: Closed Convex Functions 17
    3.3 Properties of Closed Convex Functions 19
  4 First-Order Differentiation 20
    4.1 One-Sided Differentiability of Convex Functions 21
    4.2 Basic Properties of Subderivatives 24
    4.3 Calculus Rules 27
  5 Second-Order Differentiation 29
    5.1 The Second Derivative of a Convex Function 30
    5.2 One-Sided Second Derivatives 32
    5.3 How to Recognize a Convex Function 33
  6 First Steps into the Theory of Conjugate Functions 36
    6.1 Basic Properties of the Conjugate 38
    6.2 Differentiation of the Conjugate 40
    6.3 Calculus Rules with Conjugacy 43

II. Introduction to Optimization Algorithms 47
  1 Generalities 47
    1.1 The Problem 47
    1.2 General Structure of Optimization Schemes 50
    1.3 General Structure of Optimization Algorithms 52
  2 Defining the Direction 54

2.1 Descent and Steepest-Descent Directions 54
 2.2 First-Order Methods 56
 2.3 Newtonian Methods 61
 2.4 Conjugate-Gradient Methods 65
 3 Line-Searches 70
 3.1 General Structure of a Line-Search 71
 3.2 Designing the Test (0), (R), (L) 74
 3.3 The Wolfe Line-Search 77
 3.4 Updating the Trial Stepsize 81

III. Convex Sets 87
 1 Generalities 87
 1.1 Definition and First Examples 87
 1.2 Convexity-Preserving Operations on Sets 90
 1.3 Convex Combinations and Convex Hulls 94
 1.4 Closed Convex Sets and Hulls 99
 2 Convex Sets Attached to a Convex Set 102
 2.1 The Relative Interior 102
 2.2 The Asymptotic Cone 108
 2.3 Extreme Points 110
 2.4 Exposed Faces 113
 3 Projection onto Closed Convex Sets 116
 3.1 The Projection Operator 116
 3.2 Projection onto a Closed Convex Cone 118
 4 Separation and Applications 121
 4.1 Separation Between Convex Sets 121
 4.2 First Consequences of the Separation Properties 124
 4.3 The Lemma of Minkowski-Farkas 129
 5 Conical Approximations of Convex Sets 132
 5.1 Convenient Definitions of Tangent Cones 133
 5.2 The Tangent and Normal Cones to a Convex Set 136
 5.3 Some Properties of Tangent and Normal Cones 139

IV. Convex Functions of Several Variables 143
 1 Basic Definitions and Examples 143
 1.1 The Definitions of a Convex Function 143
 1.2 Special Convex Functions: Affinity and Closedness 147
 1.3 First Examples 152
 2 Functional Operations Preserving Convexity 157
 2.1 Operations Preserving Closedness 158
 2.2 Dilations and Perspectives of a Function 160
 2.3 Infimal Convolution 162
 2.4 Image of a Function Under a Linear Mapping 166
 2.5 Convex Hull and Closed Convex Hull of a Function 169

    3 Local and Global Behaviour of a Convex Function 173
       3.1 Continuity Properties 173
       3.2 Behaviour at Infinity 178
    4 First- and Second-Order Differentiation 183
       4.1 Differentiable Convex Functions 183
       4.2 Nondifferentiable Convex Functions 188
       4.3 Second-Order Differentiation 190

V. Sublinearity and Support Functions 195

    1 Sublinear Functions 197
       1.1 Definitions and First Properties 197
       1.2 Some Examples 201
       1.3 The Convex Cone of All Closed Sublinear Functions 206
    2 The Support Function of a Nonempty Set 208
       2.1 Definitions, Interpretations 208
       2.2 Basic Properties 211
       2.3 Examples 215
    3 The Isomorphism Between Closed Convex Sets
       and Closed Sublinear Functions 218
       3.1 The Fundamental Correspondence 218
       3.2 Example: Norms and Their Duals, Polarity 220
       3.3 Calculus with Support Functions 225
       3.4 Example: Support Functions of Closed Convex Polyhedra 234

VI. Subdifferentials of Finite Convex Functions 237

    1 The Subdifferential: Definitions and Interpretations 238
       1.1 First Definition: Directional Derivatives 238
       1.2 Second Definition: Minorization by Affine Functions 241
       1.3 Geometric Constructions and Interpretations 243
       1.4 A Constructive Approach to the Existence of a Subgradient . . . 247
    2 Local Properties of the Subdifferential 249
       2.1 First-Order Developments 249
       2.2 Minimality Conditions 253
       2.3 Mean-Value Theorems 256
    3 First Examples 258
    4 Calculus Rules with Subdifferentials 261
       4.1 Positive Combinations of Functions 261
       4.2 Pre-Composition with an Affine Mapping 263
       4.3 Post-Composition with an Increasing Convex Function
           of Several Variables 264
       4.4 Supremum of Convex Functions 266
       4.5 Image of a Function Under a Linear Mapping 272
    5 Further Examples 275
       5.1 Largest Eigenvalue of a Symmetric Matrix 275

- 5.2 Nested Optimization 277
- 5.3 Best Approximation of a Continuous Function
  on a Compact Interval 278
- 6 The Subdifferential as a Multifunction 279
  - 6.1 Monotonicity Properties of the Subdifferential 280
  - 6.2 Continuity Properties of the Subdifferential 282
  - 6.3 Subdifferentials and Limits of Gradients 284

VII. Constrained Convex Minimization Problems: Minimality Conditions, Elements of Duality Theory 291

- 1 Abstract Minimality Conditions 292
  - 1.1 A Geometric Characterization 293
  - 1.2 Conceptual Exact Penalty 298
- 2 Minimality Conditions Involving Constraints Explicitly 301
  - 2.1 Expressing the Normal and Tangent Cones in Terms
    of the Constraint-Functions 303
  - 2.2 Constraint Qualification Conditions 307
  - 2.3 The Strong Slater Assumption 311
  - 2.4 Tackling the Minimization Problem with Its Data Directly 314
- 3 Properties and Interpretations of the Multipliers 317
  - 3.1 Multipliers as a Means to Eliminate Constraints:
    the Lagrange Function 317
  - 3.2 Multipliers and Exact Penalty 320
  - 3.3 Multipliers as Sensitivity Parameters with Respect
    to Perturbations 323
- 4 Minimality Conditions and Saddle-Points 327
  - 4.1 Saddle-Points: Definitions and First Properties 327
  - 4.2 Mini-Maximization Problems 330
  - 4.3 An Existence Result 333
  - 4.4 Saddle-Points of Lagrange Functions 336
  - 4.5 A First Step into Duality Theory 338

VIII. Descent Theory for Convex Minimization: The Case of Complete Information 343

- 1 Descent Directions and Steepest-Descent Schemes 343
  - 1.1 Basic Definitions 343
  - 1.2 Solving the Direction-Finding Problem 347
  - 1.3 Some Particular Cases 351
  - 1.4 Conclusion 355
- 2 Illustration. The Finite Minimax Problem 356
  - 2.1 The Steepest-Descent Method for Finite Minimax Problems 357
  - 2.2 Non-Convergence of the Steepest-Descent Method 363
  - 2.3 Connection with Nonlinear Programming 366

3 The Practical Value of Descent Schemes 371
3.1 Large Minimax Problems 371
3.2 Infinite Minimax Problems 373
3.3 Smooth but Stiff Functions 374
3.4 The Steepest-Descent Trajectory 377
3.5 Conclusion 383

Appendix: Notations 385
1 Some Facts About Optimization 385
2 The Set of Extended Real Numbers 388
3 Linear and Bilinear Algebra 390
4 Differentiation in a Euclidean Space 393
5 Set-Valued Analysis 396
6 A Bird's Eye View of Measure Theory and Integration 399

Bibliographical Comments 401

References 407

Index 415

Introduction

In this second part of "Convex Analysis and Minimization Algorithms", the dichotomous aspect implied by the title is again continuously present. Chapter IX introduces the bundling mechanism to construct the subdifferential of a convex function at a given point. It reveals implementation difficulties which serve as a motivation for another concept: the approximate subdifferential. However, a convenient study of this latter set necessitates the so-called Legendre-Fenchel transform, or conjugacy correspondence, which is therefore the subject of Chap. X. This parent of the celebrated Fourier transform is being used more and more, in its natural context of variational problems, as well as in broader fields from natural sciences. Then we can study the approximate subdifferential in Chap. XI, where our point of view is definitely oriented towards computational utilizations.

Chapter XII, lying on the fringe between theory and applications, makes a break in our development. Its subject, Lagrange relaxation, is probably the most spectacular application of convex analysis; it can be of great help in a huge number of practical optimization problems, but it requires a high level of sophistication. We expose this delicate theory in a setting as applied as possible, so that the major part of the chapter should be accessible to the non-specialist. This chapter is quite voluminous but only its Sections 1 to 3 are essential to understand the subject. Section 4 concerns dual algorithms and its interest is numerical only. Section 5 definitely deviates towards theory and relates the approach with other chapters of the book, especially Chap. X.

The last three chapters can then be entirely devoted to bundle methods, and to the several ways of arriving at them. Chapter XIII does with the approximate subdifferential the job that was done by Chap. IX with the ordinary subdifferential. This allows the development of bundle methods in dual form (Chap. XIV). Finally, Chap. XV gives these same methods in their primal form, which is probably the one having the more promising future.

A reader mainly interested by the convex analysis part of the book can skip the numerical Chapters IX, XIII, XIV. Even if he jumps directly to Chap. XV he will get the necessary comprehension of bundle methods, and more generally of algorithms for nonsmooth optimization. This might even be a good idea, in the sense that Chap. XV is probably much easier to read than the other three. However, we do not recommend this for a reader with professional algorithmic motivations, and this for several reasons:

– Chapter XV relies exclusively on convex analysis, and not on the general algorithmic principles making up Chap. II in the first part. This is dangerous because, as explained in §VIII.3.3, there is no clearcut division between smooth objective func-

tions and merely convex ones. A sound algorithm for convex minimization should therefore not ignore its parents; Chap. XIV, precisely, gives an account of bundle methods on the basis of smooth minimization.
– The bundling mechanism of Chap. IX, and the dual bundle method of Chap. XIV, can be readily extended to nonconvex objective functions. Chapter XV is of no help to understand how this extension can be made.
– More generally, this reader would miss many interesting features of bundle methods, which are important if one wants to do research on the subject. Chapter XV is only a tree, which should not hide the other chapters making up the forest.

We have tried as much as possible to avoid any prospective development, contenting ourselves with well-established material. This explains why the present book ignores many recent works in convex analysis and minimization algorithms, in particular those of the last decade or two, concerning the second-order differentiation of a nonsmooth function. In fact, these are at present just isolated trials towards solving a challenging question, and a well-organized theory is still hard to foresee. It is not clear which of these works will be of real value in the future: they rather belong to speculative science; but yet, "there still are so many beautiful things to write in C major" (S. Prokofiev).

We recall that references to theorems or equations of another chapter are preceded by the chapter number (in roman numerals). The letter A refers to Appendix A of the first part, in which our main system of notation is overviewed. In order to be reasonably self-contained, we give below a short glossary of symbols and terms appearing throughout.

"$\mapsto$" denotes a function or mapping, as in $x \mapsto f(x)$; when $f(x)$ is a set instead of a singleton, we have a *multifunction* and we rather use the notation $x \longmapsto f(x)$.

$\langle \cdot, \cdot \rangle$, $\|\cdot\|$ and $\|\|\cdot\|\|$ are respectively the scalar product, the associated norm and an arbitrary norm (here and below the space of interest is most generally the Euclidean space $\mathbb{R}^n$). Sometimes, $\langle s, x \rangle$ will be the standard *dot-product* $s^\top x = \sum_{i=1}^n s_i x_i$.

$B(0, 1) := \{x : \|x\| \leqslant 1\}$ is the (closed) *unit ball* and its boundary is the *unit sphere*.

cl f is the *closure*, or lower semi-continuous hull, of the function f; if f is convex and lower semi-continuous on the whole of $\mathbb{R}^n$, we say that f is *closed convex*.

$Df(x)$, $D_- f(x)$ and $D_+ f(x)$ are respectively the derivative, the left- and the right-derivative of a univariate function f at a point x; and the *directional derivative* of f (multivariate and convex) at x along d is

$$f'(x, d) := \lim_{t \downarrow 0} \frac{f(x + td) - f(x)}{t}.$$

$\delta_{ij} = 0$ if $i \neq j$, 1 if $i = j$ is the symbol of Kronecker.

$\Delta_k := \{(\alpha_1, \ldots, \alpha_k) : \sum_{i=1}^k \alpha_k = 1, \ \alpha_i \geqslant 0 \text{ for } i = 1, \ldots, k\}$ is the *unit simplex* of $\mathbb{R}^k$. Its elements $\alpha_1, \ldots, \alpha_k$ are called *convex multipliers*. An element of the form $\sum_{i=1}^k \alpha_i x_i$, with $\alpha \in \Delta_k$, is a *convex combination* of the x_i's.

dom $f := \{x : f(x) < +\infty\}$ and epi $f := \{(x, r) : f(x) \leqslant r\}$ are respectively the *domain* and *epigraph* of a function $f : \mathbb{R}^n \to \mathbb{R} \cup \{+\infty\}$.

The function $d \mapsto f'_\infty(d) := \lim_{t \to +\infty} 1/t [f(x + td) - f(x)]$ does not depend on $x \in \text{dom } f$ and is the *asymptotic function* of f ($\in \overline{\text{Conv}} \, \mathbb{R}^n$).

$H^-_{s,r} := \{x : \langle s, x \rangle \leqslant r\}$ is a *half-space*, defined by $s \neq 0$ and $r \in \mathbb{R}$; replacing the inequality by an equality, we obtain the corresponding (affine) *hyperplane*.

$I_S(x)$ is the *indicator function* of the set S at the point x; its value is 0 on S, $+\infty$ elsewhere.

If K is a cone, its *polar* K° is the set of those s such that $\langle s, x \rangle \leqslant 0$ for all $x \in K$.

If $f(x) \leqslant g(x)$ for all x, we say that the function f *minorizes* the function g. A sequence $\{x_k\}$ is *minimizing* for the function $f : X \to \mathbb{R}$ if $f(x_k) \to \bar{f} := \inf_X f$.

$\mathbb{R}^+ = [0, +\infty[$ is the set of *nonnegative* numbers; $(\mathbb{R}^+)^n$ is the nonnegative *orthant* of $\mathbb{R}^n$; a sequence $\{t_k\}$ is decreasing if $k' > k \Rightarrow t_{k'} \leqslant t_k$.

Outer semi-continuity of a multifunction F is what is usually called upper semi-continuity elsewhere. Suppose $F(x)$ is nonempty and included in a fixed compact set for x in a neighborhood of a given x^* (i.e. F is *locally bounded* near x^*); we say that F is outer semi-continuous at x^* if, for arbitrary $\varepsilon > 0$,

$$F(x) \subset F(x^*) + B(0, \varepsilon) \quad \text{for } x \text{ close enough to } x^*.$$

The *perspective-function* $\tilde{f}$ associated with a (convex) function f is obtained by the projective construction $\tilde{f}(x, u) := uf(x/u)$, for $u > 0$.

ri C and rbd C are respectively the *relative interior* and relative boundary of a (convex) set C; these are the interior and boundary for the topology relative to the affine hull aff C of C.

$S_r(f) := \{x : f(x) \leqslant r\}$ is the *sublevel-set* of the function f at the level $r \in \mathbb{R}$.

$\sigma_S(d) := \sup_{s \in S} \langle s, d \rangle$ is the *support function* of the set S at d and $\sigma_S(d) + \sigma_S(-d)$ is its *breadth* in the direction d; $d_S(x)$ is the distance from x to S; $p_S(x)$ is the projection of x onto S, usually considered for S closed convex; when, in addition, S contains the origin, its *gauge* is the function

$$\gamma_S(x) := \inf\{\lambda > 0 : x \in \lambda S\}.$$

(U1) is a black box characterizing a function f, useful for minimization algorithms: at any given x, (U1) computes the value $f(x)$ and a subgradient denoted by $s(x) \in \partial f(x)$.

IX. Inner Construction of the Subdifferential: Towards Implementing Descent Methods

Prerequisites. Chapters VI and VIII (essential); Chap. II (recommended).

Introduction. Chapter VIII has explained how to generalize the steepest-descent method of Chap. II from smooth to merely convex functions; let us briefly recall here how it works. At a given point x (the current iterate of a steepest-descent algorithm), we want to find the Euclidean steepest-descent direction, i.e. the solution of

$$\left| \begin{array}{l} \min f'(x, d) \\ \frac{1}{2}\|d\|^2 = 1, \end{array} \right.$$

knowing that more general normalizations can also be considered. A slight modification of the above problem is much more tractable, just by changing the normalization constraint into an inequality. Exploiting positive homogeneity and using some duality theory, one obtains the classical projection problem $\hat{s} := \operatorname{argmin}_{s \in \partial f(x)} 1/2 \|s\|^2$. The essential results of §VIII.1 are then as follows:

- either $\hat{s} = 0$, which means that x minimizes f;
- or $\hat{s} \neq 0$, in which case the Euclidean (non-normalized) steepest-descent direction is $-\hat{s}$.

This allows the derivation of a steepest-descent algorithm, in which the direction is computed as above, and some line-search can be made along this direction.

We have also seen the major drawbacks of this type of algorithm:

(a) It is subject to zigzags, and need not converge to a minimum point.
(b) The projection problem can be solved only in special situations, basically when the subdifferential is a convex polyhedron, or an ellipsoid.
(c) To compute any kind of descent direction, steepest or not, an explicit description of the whole subdifferential is needed, one way or another. In many applications, this is too demanding.

Our main motivation in the present chapter is to overcome (c), and we will develop a technique which will eliminate (b) at the same time. As in the previous chapters, we assume a finite-valued objective function:

$$\boxed{f : \mathbb{R}^n \to \mathbb{R} \text{ is convex.}}$$

IX. Inner Construction of the Subdifferential

For the sake of simplicity, we limit the study to steepest-descent directions in the sense of the Euclidean norm $\|\cdot\| = \|\cdot\|$; adapting our development to other normings is indeed straightforward and only results in additional technicalities. In fact, our situation is fundamentally characterized by the information concerning the objective function f: we need only the value $f(x)$ and a subgradient $s(x) \in \partial f(x)$, computed in a black box of the type (U1) in Fig. II.1.2.1, at any $x \in \mathbb{R}^n$ that pleases us. Such a computation is possible in theory because $\text{dom}\,\partial f = \mathbb{R}^n$; and it is also natural from the point of view of applications, for reasons given in §VIII.3.

The idea is to construct the subdifferential of f at x, or at least an approximation of it, good enough to mimic the computation of the steepest-descent direction. The approximation in question is obtained by a sequence of compact convex polyhedra, and this takes care of (b) above. On the other hand, our approximation scheme does not solve (a): for this, we need additional material from convex analysis; the question is therefore deferred to Chap. XIII and the chapters that follow it. Furthermore, our approximation mechanism is not quite implementable; this new difficulty, having the same remedy as needed for (a), will also be addressed in the same chapters.

The present chapter is fundamental. It introduces the key ingredients for constructing efficient algorithms for the minimization of convex functions, as well as their extensions to the nonconvex case.

1 The Elementary Mechanism

In view of the poor information available, our very first problem is twofold: (i) how can one compute a descent direction at a given x, or (ii) how can one realize that x is optimal? Let us illustrate this by a simple specific example.

For $n = 2$ take the maximum of three functions f_1, f_2, f_3:

$$x = (\xi, \eta) \mapsto f(x) = \max\{\xi + 2\eta, \xi - 2\eta, -\xi\} \tag{1.1}$$

which is minimal at $x = (0, 0)$. Figure 1.1 shows a level-set of f and the three half-lines of kinks. Suppose that the black box computing $s(x)$ produces the gradient of one of the active functions. For example, the following scheme takes the largest active index:

> First, set $s(x) = \nabla f_1(x) = (1, 2)$;
> if $f_2(x) \geq f_1(x)$, then set $s(x) = \nabla f_2(x) = (1, -2)$;
> if $f_3(x) \geq \max\{f_1(x), f_2(x)\}$, then set $s(x) = \nabla f_3(x) = (-1, 0)$.

(i) When x is as indicated in the picture, $s(x) = \nabla f_2(x)$ and $-s(x)$ points in a direction of increasing f. This is confirmed by calculations: with the usual dot-product for $\langle \cdot, \cdot \rangle$ in $\mathbb{R}^2$ (which is implicitly assumed in the picture),

$$\langle -s(x), \nabla f_1(x) \rangle = 3, \quad \text{which implies} \quad f'(x, -s(x)) \geq 3.$$

Thus the choice $d = -s(x)$ is bad. Of course, it would not help to take $d = s(x)$: a rather odd idea since we already know that

1 The Elementary Mechanism

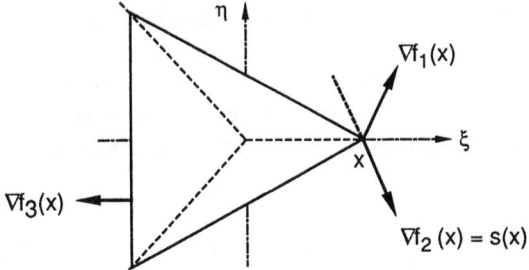

Fig. 1.1. Getting a descent direction is difficult

$$f'(x, s(x)) \geq \|s(x)\|^2 > 0.$$

Finally note that the trouble would be just the same if, instead of taking the largest possible index, the black box (U1) chose the smallest one, say, yielding $\nabla f_1(x)$.

(ii) When $x = 0$, $s(x) = \nabla f_3(x)$ still does not tell us that x is optimal. It would be extremely demanding of (U1) to ask that it select the only subgradient of interest here, namely $s = 0$!

It is the aim of the present chapter to solve this double problem. From now on, x is a fixed point in $\mathbb{R}^n$ and we want to answer the double question: is x optimal? and if not, find a descent direction issuing from x.

Remark 1.1 The difficulty is inherent to the combination of the two words: *descent* direction and *convex* function. In fact, if we wanted an *ascent* direction, it would suffice to take the direction of *any* subgradient, say $d = s(x)$: indeed

$$f'(x, d) \geq \langle s(x), d \rangle = \|s(x)\|^2.$$

This confirms our Remark VII.1.1.3: maximizing a convex function is a completely different problem; here, at least one issue is trivial, namely that of increasing f (barring the very unlikely event $s(x) = 0$). □

Because f is finite everywhere, the multifunction ∂f is locally bounded (Proposition VI.6.2.2): we can write

$$\exists M > 0, \exists \eta > 0 \quad \text{such that} \quad \|s(y)\| \leq M \text{ whenever } \|y - x\| \leq \eta. \tag{1.2}$$

We will make extensive use of the theory from Chap. VIII, more specifically §VIII.1. In particular, let $S [= \partial f(x)]$ be a closed convex polyhedron and consider the problem

$$\min \{\sigma_S(d) : \|d\| \leq 1\}. \quad [\text{i.e. } \min \{f'(x, d) : \|d\| \leq 1\}]$$

We recall that a simple solution can be obtained as follows: compute the unique solution $\hat{s}$ of

$$\min \{\tfrac{1}{2}\|s\|^2 : s \in S\};$$

then take $d = -\hat{s}$.

The problem of finding a descent direction issuing from the given x will be solved by an iterative process. A sequence of "trial" directions d_k will be constructed, together with a sequence of designated subgradients $\hat{s}_k$. These two sequences will be such that: either (i) $\sigma_S(d_k) = f'(x, d_k) < 0$ for some k, or (ii) $k \to \infty$ and $\hat{s}_k \to 0$, indicating that x is optimal. For this process to work, the boundedness property (1.2) is absolutely essential.

(a) The Key Idea. We start the process with the computation of $s(x) \in \partial f(x)$ coming from the black box (U1). We call it s_1, we form the polyhedron $S_1 := \{s_1\}$, and we set $k = 1$.

Recursively, suppose that $k \geq 1$ subgradients

$$s_1, \ldots, s_k \quad \text{with} \quad s_j \in \partial f(x) \text{ for } j = 1, \ldots, k$$

have been computed. They generate the compact convex polyhedron

$$S_k := \operatorname{co}\{s_1, \ldots, s_k\}. \tag{1.3}$$

Then we compute the trial direction d_k. Because we want

$$0 > f'(x, d_k) = \sigma_{\partial f(x)}(d_k) \geq \max\{\langle s_j, d_k \rangle : j = 1, \ldots, k\},$$

we must certainly take d_k so that

$$\langle s_j, d_k \rangle < 0 \quad \text{for } j = 1, \ldots, k \tag{1.4}$$

or equivalently

$$\langle s, d_k \rangle < 0 \quad \text{for all } s \in S_k \tag{1.5}$$

or again

$$\sigma_{S_k}(d_k) < 0. \tag{1.6}$$

Only then, has d_k a chance of being a descent direction.

The precise computation of d_k satisfying (1.4) will be specified later. Suppose for the moment that this computation is done; then the question is whether or not d_k is a descent direction. The best way to check is to compare $f(x + td_k)$ with $f(x)$ for $t \downarrow 0$. This is nothing but a *line-search*, which works schematically as follows.

Algorithm 1.2 (Line-Search for Mere Descent) The direction d ($= d_k$) is given. Start from some $t > 0$, for example $t = 1$.

STEP 1. Compute $f(x+td)$. If $f(x+td) < f(x)$ stop. This line-search is successful, d is downhill.

STEP 2. Take a smaller $t > 0$, for example $1/2\,t$, and loop to Step 1. □

Compared to the line-searches of §II.3, the present one is rather special, indeed. No extrapolations are made: t is never too small. Our aim is not to find a new iterate $x_+ = x + td$, but just to check the (unknown) sign of $f'(x, d)$. Only the case $t \downarrow 0$ is interesting for our present concern: it indicates that $f'(x, d) \geq 0$, a new direction must therefore be computed.

Then the key lies in the following simple result.

1 The Elementary Mechanism

Proposition 1.3 *For any $x \in \mathbb{R}^n$, $d \neq 0$ and $t > 0$ we have*
$$f(x+td) \geq f(x) \implies \langle s(x+td), d \rangle \geq 0.$$

PROOF. Straightforward: because $s(x+td) \in \partial f(x+td)$, the subgradient inequality written at x gives
$$f(x) \geq f(x+td) + \langle s(x+td), x - x - td \rangle.$$
□

This property, which relies upon convexity, has an obvious importance for our problem. At Step 1 of the Line-search 1.2, we are entitled to use from the black box (U1) not only $f(x+td)$ but also $s(x+td)$. Now, suppose in Algorithm 1.2 that $t \downarrow 0$, and suppose that we extract a cluster point s^* from the corresponding $\{s(x+td)\}_t$. Because the mapping $\partial f(\cdot)$ is outer semi-continuous (§VI.6.2), such an s^* exists and lies in $\partial f(x)$. Furthermore, letting $t \downarrow 0$ in Proposition 1.3 shows that
$$\langle s^*, d_k \rangle \geq 0.$$

In other words: not only have we observed that d_k was not a descent direction; but we have also explained why: we have *singled out* an additional subgradient s^* which does not satisfy (1.5); therefore $s^* \notin S_k$. We can call s_{k+1} this additional subgradient, and recompute a new direction satisfying (1.4) (with k increased by 1). This new direction d_{k+1} will certainly be different from d_k, so the process can be safely repeated.

Remark 1.4 When doing this, we are exploiting the characterization of $\partial f(x)$ detailed in §VI.6.3(b). First of all, Proposition 1.3 goes along with Lemma VI.6.3.4: s^* is produced by a directional sequence $\{x + td\}_{t \downarrow 0}$, and therefore lies in the face of $\partial f(x)$ exposed by d. Algorithm 1.2 is one instance of the process called there Π, constructing the subgradient $s^* = s_\Pi(d_k)$. In view of Theorem VI.6.3.6, we know that, if we take "sufficiently many" such directional sequences, then we shall be able to make up the whole set $\partial f(x)$. More precisely (see also Proposition V.3.1.5), the whole boundary of $\partial f(x)$ is described by the collection of subgradients $s_\Pi(d)$, with d describing the unit sphere of $\mathbb{R}^n$.

The game is not over, though: it is prohibited to run Algorithm 1.2 infinitely many times, one time per normalized direction – not even mentioning the fact that each run takes an infinite computing time, with t tending to 0. Our main concern is therefore to select the directions d_k carefully, so as to obtain a sufficiently rich approximation of $\partial f(x)$ for some reasonable value of k. □

(b) Choosing the Sequence of Directions. To specify how each d_k should be computed, (1.4) leaves room for infinitely many possibilities. As observed in Remark 1.4 above, this computation is crucial if we do not want to exhaust a whole neighborhood of x. To guide our choice, let us observe that the above mechanism achieves two things at the same time:

– It performs a sequence of line-searches along the trial directions $d_1, d_2, \ldots, d_k$; the hope is really to obtain a descent direction d, having $\sigma_{\partial f(x)}(d) < 0$.

– It builds up a sequence $\{S_k\}$ of compact convex polyhedra defined by (1.3), having the property that
$$\{s(x)\} = S_1 \subset S_k \subset S_{k+1} \subset \partial f(x);$$
and S_k is hoped to "tend" to $\partial f(x)$, in the sense that (1.6) should eventually imply $\sigma_{\partial f(x)}(d_k) < 0$.

Accordingly, the aim of the game is double as well:

– We want d_k to make the number $\sigma_{S_k}(d_k)$ as negative as possible: this will increase the chances of having $\sigma_{\partial f(x)}(d_k)$ negative as well.
– We also want the sequence of polyhedra $\{S_k\}$ to fill up as much room as possible within $\partial f(x)$.

Fortunately, the above two requirements are not antagonistic but they really go together: the sensible idea is to take d_k *minimizing* $\sigma_{S_k}(\cdot)$. It certainly is good for our first requirement, but also for the second. In fact, if the process is not going to terminate with the present d_k, we know in advance that $\langle s_{k+1}, d_k \rangle$ will be nonnegative. Hence, making the present $\sigma_{S_k}(d_k)$ minimal, i.e. as negative as possible, is a good way of making sure that the next s_{k+1} will be as remote as possible from S_k.

When we decide to minimize the (finite) sublinear function σ_{S_k}, we place ourselves in the framework of Chap. VIII. We know that a norming is necessary and, as agreed in the introduction, we just choose the Euclidean norm. In summary, d_k is computed as the solution of

$$\left| \begin{array}{l} \min r \\ \langle s_j, d \rangle \leq r \quad \text{for } j = 1, \ldots, k, \\ \tfrac{1}{2} \|d\|^2 = 1. \end{array} \right. \tag{1.7}$$

Once again, the theory of §VIII.1 tells us that it is a good idea to replace the normalization by an inequality, so as to solve instead the more tractable problem

$$\min \{ \tfrac{1}{2} \|d\|^2 \; : \; -d \in S_k \}. \tag{1.8}$$

In summary: with S_k on hand, the best we have to do is to pretend that S_k coincides with $\partial f(x)$, and compute the "pretended steepest-descent direction" accordingly. If S_k is [close to] $\partial f(x)$, d_k will be [close to] the steepest-descent direction. On the other hand, a failure (d_k not downhill) will indicate that S_k is too poor an approximation of $\partial f(x)$; this is going to be corrected by an enrichment of it.

Remark 1.5 It is worth mentioning that, with d_k solving (1.8), $-\|d_k\|^2$ is an estimate of $f'(x, d_k)$: there holds (see Remark VIII.1.3.7)

$$-\|d_k\|^2 = \sigma_{S_k}(d_k) \leq \sigma_{\partial f(x)}(d_k) = f'(x, d_k)$$

with equality if $S_k = \partial f(x)$. This estimate is therefore an *underestimate*, supposedly improved at each enrichment of S_k. □

(c) The Bundling Process The complete algorithm is now well-defined:

Algorithm 1.6 (Quest for a Descent Direction) Given $x \in \mathbb{R}^n$, compute $f(x)$ and $s_1 = s(x) \in \partial f(x)$. Set $k = 1$.

STEP 1. With S_k defined by (1.3), solve (1.8) to obtain d_k. If $d_k = 0$ stop: x is optimal.
STEP 2. Execute the line-search 1.2 with its two possible exits:
STEP 3. If the line-search has produced $t > 0$ with $f(x + td_k) < f(x)$, then stop: d_k is a descent direction.
STEP 4. If the line-search has produced $t \downarrow 0$ and $s_{k+1} \in \partial f(x)$ such that

$$\langle s_{k+1}, d_k \rangle \geq 0, \tag{1.9}$$

then replace k by $k + 1$ and loop to Step 1. □

Two questions must be addressed concerning this algorithm.

(i) Does it solve our problem, i.e. does it really stop at some iteration k? This will be the subject of §2.
(ii) How can s_{k+1} be actually computed at Step 4? (a normal computer cannot extract a cluster point).

Remark 1.7 Another question, which is not raised by the algorithm itself but rather by §VIII.2.2, is the following: if we plug Algorithm 1.6 into a steepest-descent scheme such as VIII.1.1.7 to minimize f, the result will probably converge to a wrong point. A descent direction d_k produced by Algorithm 1.6 is "at best" the steepest-descent direction, which is known to generate zigzags. In view of (ii) above, we have laid down a non-implementable computation, which results in a non-convergent method!

Chapter XIII and its successors will be devoted to remedying this double drawback. For the moment, we just mention once more that the ambition of a minimization method should not be limited to iterating along steepest-descent directions, which are already bad enough in the smooth case. The remedy, precisely, will also result in improvements of the steepest-descent idea. □

The scheme described by Algorithm 1.6 will be referred to as a *bundling* mechanism: the pieces of information obtained from f are bundled together, so as to construct the subdifferential. Looping from Step 4 to Step 1 will be called a *null-step*. The difference between a normal line-search and this procedure is that the latter does not use the new point $x + td$ as such, but only via the information returned by (U1): the next line-search will start from the same x, which is not updated.

Example 1.8 Let us give a geometric illustration of the key assessing the introduction of s_{k+1} into S_k. Take again the function f of (1.1) and, for given x, consider the one-dimensional function $t \mapsto q(t) := f(x - ts(x))$.

If $q(t) < q(0)$ for some $t > 0$, no comment. Suppose, therefore, that $q(t) \geq q(0)$ for all $t \geq 0$. This certainly implies that $x = (\xi, \eta)$ is a kink; say $\eta = 0$, as in Fig. 1.1. More importantly, the function that is active at x – here f_2, say, with $s(x) = \nabla f_2(x)$ – is certainly active at *no* $t > 0$. Otherwise, we would have for $t > 0$ small enough

$$q(t) = f_2(x - ts(x)) = q(0) - t\langle \nabla f_2(x), s(x)\rangle = q(0) - t\|\nabla f_2(x)\|^2 < q(0).$$

In other words: the black box (U1) is *forced* to produce some other gradient – here $\nabla f_1(x)$ – at any $x - ts(x)$, no matter how small $t > 0$ is. The discontinuous nature of $s(\cdot)$ is here turned into an advantage. In the particular example of the picture, once this new gradient is picked, our knowledge of $\partial f(x)$ is complete and we can iterate. □

Remark 1.9 Note in Example 1.8 that, f_1 being affine, $s(x - ts(x))$ is already in $\partial f(x)$ for $t > 0$. It is not necessary to let $t \downarrow 0$, the argument (ii) above is eliminated. This is a nice feature of piecewise affine functions.

It is interesting to consider this example with §VIII.3.4 in mind. Specifically, observe in the picture that an arbitrary $s \in \partial f(x)$ is usually not in $\partial f(x - ts)$, no matter how close the positive t is to 0. There is, however, exactly one $\hat{s} \in \partial f(x)$ satisfying $s \in \partial f(x - t\hat{s})$ for $t > 0$ small enough; and this $\hat{s}$ is precisely the Euclidean steepest-descent direction at x. The same observation can be made in the example of §VIII.2.2, even though the geometry is slightly different. Indeed, we have here an "explanation" of the uniqueness result in Theorem VIII.3.4.1, combined with the stability result VIII.3.4.5(i). □

Needless to say, the bundling mechanism is not bound to the Euclidean norm $\|\cdot\|$. Choosing an arbitrary $\|\|\cdot\|\|$ would amount to solving

$$\left|\begin{array}{l} \min r \\ \langle s_j, d\rangle \leqslant r \quad \text{for } j = 1, \ldots, k, \\ \|\|d\|\| = 1 \end{array}\right.$$

instead of (1.7), or

$$\min\{\|\|d\|\|^* : -d \in S_k\}$$

instead of (1.8). This would preserve the key property that $\sigma_{S_k}(d_k)$ is as negative as possible, in some different sense. As an example, we recall from Proposition VIII.1.3.4 that, if a quadratic norm $\langle d, Qd\rangle^{1/2}$ is chosen, the essential calculations become

$$\hat{s}_k = \operatorname{argmin}\{\tfrac{1}{2}\langle s, Q^{-1}s\rangle : s \in S_k\}, \quad d_k = -Q^{-1}\hat{s}_k,$$

which gives the directional derivative estimate of Remark 1.5:

$$f'(x, d_k) \geqslant \langle \hat{s}_k, d_k\rangle = -\langle \hat{s}_k, Q^{-1}\hat{s}_k\rangle = -\langle d_k, Qd_k\rangle.$$

Among all possible norms, could there be one (possibly depending on k) such that the bundling Algorithm 1.6 converges as fast as possible? This is a fascinating question indeed, which is important because the quality of algorithms for nonsmooth optimization depends directly on it. Unfortunately, no clear answer can be given with our present knowledge. Worse: the very meaning of this question depends on a proper definition of "as fast as possible", and this is not quite clear either, in the context of nonsmooth optimization.

On the other hand, Remark 1.9 above reveals a certain property of "stability" of the steepest-descent direction, which is special to the Euclidean norming.

2 Convergence Properties

In this section, we study the convergence properties of the schematic bundling process described as Algorithm 1.6. Its convergence and speed of convergence will be analyzed by techniques which, once again, serve as a basis for virtually all minimization methods to come.

2.1 Convergence

We will use the notation S_k of (1.3); and, to make the writing less cumbersome, we will denote by

$$\text{Proj } v/A := p_{\overline{co}\, A}(v)$$

the Euclidean projection of a vector v onto the closed convex hull of a set A.

From the point of view of its convergence, the process of §1 can be reduced to the following essentials. It consists of generating two sequences of $\mathbb{R}^n$: $\{s_k\}$ (the subgradients) and $\{d_k\}$ (the directions, or the projections). Knowing that

$$-d_k = \hat{s}_k := \text{Proj } 0/S_k \in \partial f(x) \quad \text{for } k = 1, 2, \ldots$$

our sequences satisfy the fundamental properties:

$$\langle s, \hat{s}_k \rangle \geq \|\hat{s}_k\|^2 \quad \text{for all } s \in S_k \tag{2.1.1}$$

(minimality conditions for the projection) and

$$\langle s_{k+1}, \hat{s}_k \rangle \leq 0 \tag{2.1.2}$$

which is (1.9). Furthermore, we know from the boundedness property (1.2) that

$$\|\hat{s}_k\| \leq \|s_k\| \leq M \quad \text{for } k = 1, 2, \ldots \tag{2.1.3}$$

All our convergence theory is based on the property that $\hat{s}_k \to 0$ if Algorithm 1.6 loops forever.

Before giving a proof, we illustrate this last property in Fig. 2.1.1. Suppose $\partial f(x)$ lies entirely in the upper half-space. Start from $s_1 = (-1, 0)$, say (we assume $M = 1$). Given s_k somewhere as indicated in the picture, s_{k+1} has to lie below the dashed line. "At worst", this s_{k+1} has norm 1 and satisfies equality in (2.1.2). The picture does suggest that, when the operation is repeated, the ordinate of s_{k+1} must tend to 0, implying $\hat{s}_k \to 0$.

The property $\hat{s}_k \to 0$ results from the combination of the inequalities (2.1.1), (2.1.2), (2.1.3): they ensure that the triangle formed by 0, $\hat{s}_k$ and s_{k+1} is not too elongated. Taking (2.1.3) for granted, a further examination of Fig. 2.1.1 reveals what exactly (2.1.1) and (2.1.2) are good for: the important thing is that the projection of the segment $[s_1, s_{k+1}]$ along the vector $\hat{s}_k$ has a length which is not "infinitely smaller" than $\|\hat{s}_k\|$.

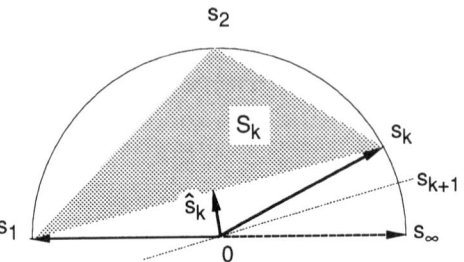

Fig. 2.1.1. Why the bundling process converges

(a) A Qualitative Result. Establishing mere convergence of $\hat{s}_k$ is easy:

Lemma 2.1.1 *Let $m > 0$ be fixed. Consider two sequences $\{s_k\}$ and $\{\hat{s}_k\}$, satisfying for $k = 1, 2, \ldots$*

$$\langle s_j - s_{k+1}, \hat{s}_k \rangle \geq m \|\hat{s}_k\|^2 \quad \text{for } j = 1, \ldots, k. \tag{2.1.4}$$

If, in addition, $\{s_k\}$ is bounded, then $\hat{s}_k \to 0$ as $k \to +\infty$.

PROOF. Using the Cauchy-Schwarz inequality in (2.1.4) gives

$$\|\hat{s}_k\| \leq \tfrac{1}{m} \|s_j - s_{k+1}\| \quad \text{for all } k \geq 1 \text{ and } j = 1, \ldots, k.$$

Suppose that there is a subsequence $\{s_k\}_{k \in N_1}$, $N_1 \subset \mathbb{N}$ such that

$$0 < \delta \leq \|\hat{s}_k\| \leq \tfrac{1}{m} \|s_j - s_{k+1}\| \quad \text{for all } k \in N_1 \text{ and } j = 1, \ldots, k.$$

Extract from the (bounded) corresponding subsequence $\{s_{k+1}\}$ a convergent subsequence and take j preceding $k+1$ in that last subsequence to obtain a contradiction. □

Corollary 2.1.2 *Consider the bundling Algorithm 1.6. If x is not optimal, there must be some integer k with $f'(x, d_k) < 0$.*

PROOF. For all $k \geq 1$ we have from (2.1.1), (2.1.2)

$$\langle s_j, \hat{s}_k \rangle \geq \|\hat{s}_k\|^2 \quad \text{for } j = 1, \ldots, k \tag{2.1.5}$$

$$\langle s_{k+1}, \hat{s}_k \rangle \leq 0, \tag{2.1.6}$$

so Lemma 2.1.1 applies (with $m = 1$). Being in the compact convex set $\partial f(x)$, $\{s_k\}$ cannot tend to 0 if $0 \notin \partial f(x)$. Therefore Algorithm 1.6 cannot loop forever. Since it cannot stop in Step 1 either, the only possibility left is a stop in Step 3. □

Figure 2.1.2 is another illustration of this convergence property. Along the first $d_1 = -s_1$, the s_2 produced by the line-search has to be in the dashed area. In the situation displayed in the picture, d_2 will separate $\partial f(x)$ from the origin, i.e. d_2 will be a descent direction (§VIII.1.1).

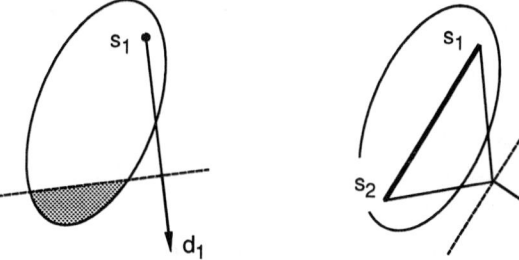

Fig. 2.1.2. Two successive projections

Lemma 2.1.1 also provides a stopping criterion for the case of an optimal x. It suffices to insert in Step 1 of Algorithm 1.6 the test

$$\text{if } \|d_k\| \leq \delta \quad \text{stop,} \qquad (2.1.7)$$

where $\delta > 0$ is some pre-specified tolerance. As explained in Remark VIII.1.3.7, $-d_k$ can be viewed as an approximation of "the" gradient of f at x_k; (2.1.7) thus appears as the equivalent of the "classical" stopping test (II.1.2.1). When (2.1.7) is inserted, Algorithm 1.6 stops in any case: it either produces a descent direction, or signals (approximate) optimality of x.

There are several ways of interpreting (2.1.7) to obtain some information on the quality of x_k:

– First we have by the Cauchy-Schwarz inequality:

$$f(y) \geq f(x) + \langle \hat{s}_k, y - x \rangle \geq f(x) - \delta \|y - x\| \quad \text{for all } y \in \mathbb{R}^n. \qquad (2.1.8)$$

If x and a minimum of f are a priori known to lie in some ball $B(0, R)$, this gives an estimate of the type

$$f(x) \leq \inf f + 2\delta R.$$

– Second x minimizes the perturbed function

$$y \mapsto f_1(y) := f(y) + \delta \|y - x\|;$$

this can be seen either from (2.1.8) or from the formula $\partial f_1(x) = \partial f(x) + B(0, \delta)$.

– Finally consider another perturbation:

$$y \mapsto f_2(y) := f(y) + \tfrac{1}{2}c\|y - x\|^2,$$

for some fixed $c > 0$; then $B\left(x, \tfrac{1}{2c}\delta\right)$ contains all the minima of f. Indeed, for any $s \in \partial f(x)$,

$$f_2(y) - f(x) \geq \langle s, y - x \rangle + \tfrac{1}{2}c\|y - x\|^2 \geq \|y - x\| \left[\tfrac{1}{2}c\|y - x\| - \|s\|\right];$$

thus, the property $f_2(y) \leq f_2(x) = f(x)$ implies $\|y - x\| \leq \tfrac{1}{2c}\|s\|$.

Remark 2.1.3 The number m in Lemma 2.1.1 is not used in Corollary 2.1.2. It will appear as a tolerance for numerical implementations of the bundling Algorithm 1.6. In fact, it will be convenient to set

$$m = m'' - m' \quad \text{with} \quad 0 < m' < m'' < 1. \qquad (2.1.9)$$

– The tolerance m' plays the role of 0 and will actually be essential for facilitating the line-search: with $m' > 0$, the test

$$\langle s_{k+1}, d_k \rangle \geqslant -m' \|d_k\|^2$$

is easier to meet than $\langle s_{k+1}, d_k \rangle \geqslant 0$.
– The tolerance m'' plays the role of 1 and will be helpful for the quadratic program computing $\hat{s}_k$: with $m'' < 1$, the test

$$\langle s_j, \hat{s}_k \rangle \geqslant m'' \|\hat{s}_k\|^2 \quad \text{for } j = 1, 2, \ldots, k$$

is easier to meet than (2.1.1).
– Finally $m'' - m'$ must be positive, to allow application of Lemma 2.1.1. Altogether, (2.1.9) appears as the convenient choice. In practice, values such as $m' = 0.1$, $m'' = 0.9$ are reasonable.

For an illustration, see Fig. 2.1.3, where the dashed area represents the current S_k of Fig. 2.1.1. Everything is all right if $\hat{s}_k$ lies above D'' and s_{k+1} lies below D', so that the thick segment has a definitely positive length. □

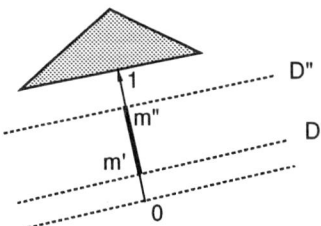

Fig. 2.1.3. Tolerances for convergence

(b) A Quantitative Result and its Consequences. The next result is a more accurate alternative to Lemma 2.1.1; it will be useful for numerical implementations, and also when we study speeds of convergence. Here, s and $\hat{s}$ play the role of s_{k+1} and $\hat{s}_k$ respectively; m' is the m' of Remark 2.1.3; as for z, it is an alternative to $\hat{s}_{k+1}$.

Lemma 2.1.4 *Let s and $\hat{s}$ ($\hat{s} \neq 0$) be two vectors satisfying the relation*

$$\langle s, \hat{s} \rangle \leqslant m' \|\hat{s}\|^2 \tag{2.1.10}$$

for some $m' < 1$; let $M \geqslant \max\{\|s\|, \|\hat{s}\|\}$ be an upper bound for their norm; call z the projection of the origin onto the segment $[s, \hat{s}]$ and set

$$\mu := (1 - m')^2 \frac{\|\hat{s}\|^2}{M^2 - m'^2 \|\hat{s}\|^2} > 0. \tag{2.1.11}$$

Then the following inequality holds:

$$\|z\|^2 \leqslant \frac{1}{1+\mu} \|\hat{s}\|^2. \tag{2.1.12}$$

Furthermore, if equality holds in (2.1.10) and $\|s\| = M$, equality holds in (2.1.12).

PROOF. Develop

$$\|\alpha\hat{s} + (1-\alpha)s\|^2 = \alpha^2\|\hat{s}\|^2 + 2\alpha(1-\alpha)\langle s, \hat{s}\rangle + (1-\alpha)^2\|s\|^2$$

and consider the function

$$\mathbb{R} \ni \alpha \mapsto \varphi(\alpha) := \alpha^2\|\hat{s}\|^2 + 2\alpha(1-\alpha)m'\|\hat{s}\|^2 + (1-\alpha)^2 M^2.$$

By assumption, $\|\alpha s + (1-\alpha)\hat{s}\|^2 \leq \varphi(\alpha)$ for $\alpha \in [0, 1]$, so we have the bound

$$\|z\|^2 \leq \min\{\varphi(\alpha) : \alpha \in [0, 1]\}$$

which is exact if (2.1.10) holds as an equality and $\|s\| = M$.

Now observe that

$$\varphi'(0) = 2m'\|\hat{s}\|^2 - 2M^2 \leq 2(m' - 1)M^2 < 0$$

$$\varphi'(1) = 2(1 - m')\|\hat{s}\|^2 \geq 0,$$

so the minimum of φ on $[0, 1]$ is actually unconstrained. Then the minimal value is given by straightforward calculations:

$$\varphi_{\min} = \frac{M^2 - m'^2\|\hat{s}\|^2}{M^2 - 2m'\|\hat{s}\|^2 + \|\hat{s}\|^2}\|\hat{s}\|^2$$

(which, of course, is nonnegative). Write this equality as $\varphi_{\min} = c\|\hat{s}\|^2$. This defines

$$\mu \ [=\tfrac{1}{c} - 1] \ := \frac{M^2 - 2m'\|\hat{s}\|^2 + \|\hat{s}\|^2}{M^2 - m'^2\|\hat{s}\|^2} - 1$$

and another straightforward calculation shows that this μ is just (2.1.11). It is nonnegative because its denominator is

$$M^2 - m'^2\|\hat{s}\|^2 \geq M^2 - m'^2 M^2 > 0. \qquad \square$$

A first use of this additional convergence result appears when considering numerical implementations. In the bundling Algorithm 1.6, S_k is defined by k vectors. Since k has no a priori upper bound, an infinite memory is required for the algorithm to work. Furthermore, the computing time necessary to solve (1.8) is also potentially unbounded. Let us define, however, the following extension of Algorithm 1.6.

Algorithm 2.1.5 (Economic Bundling for a Descent Direction) Given $x \in \mathbb{R}^n$, compute $f(x)$ and $s_1 = s(x) \in \partial f(x)$; set $S'_1 := \{s_1\}$, $k = 1$. The tolerances $\delta > 0$ and $m' < 1$ are also given.

STEP 1 (direction-finding and stopping criterion). Compute $\hat{s}_k := \operatorname{Proj} 0/S'_k$. Stop if $\|\hat{s}_k\| \leq \delta$: x is nearly optimal. Otherwise set $d_k := -\hat{s}_k$.

STEP 2 (line-search). Execute the line-search 1.2 with its two possible exits:

STEP 3 (descent direction found). If the line-search has produced a stepsize $t > 0$ with $f(x + td_k) < f(x)$, then stop: d_k is a descent direction.

STEP 4 (null-step, update S'). Here, the line-search has produced $t \downarrow 0$ and $s_{k+1} \in \partial f(x)$ such that
$$\langle s_{k+1}, d_k \rangle \geqslant -m' \|d_k\|^2.$$
Take for S'_{k+1} any compact convex set included in $\partial f(x)$ but containing at least $\hat{s}_k$ and s_{k+1}.

STEP 5 (loop). Replace k by $k+1$ and loop to Step 1. □

The idea exploited in this scheme should be clear, but let us demonstrate the mechanism in some detail. Suppose for example that we do not want to store more than 10 generators to characterize the *bundle of information* S'_k. Until the 10th iteration, Algorithm 1.6 can be reproduced normally: the successive subgradients s_k can be accumulated one after the other, S'_k is just S_k. After the 10th iteration has been completed, however, there is no room to store s_{11}. Then we can carry out the following operations:

- Extract from $\{1, 2, \ldots, 10\}$ a subset containing at most 8 indices – call K_{10} this subset, let it contain $k' \leqslant 8$ indices.
- Delete all those s_j such that $j \notin K_{10}$. After renumbering, the subgradients that are still present can be denoted by $\tilde{s}_1, \ldots, \tilde{s}_{k'}$. We have here a "compression" of the bundle.
- Append $\hat{s}_{10}$ and s_{11} to the above list of subgradients and define S'_{10} to be the convex hull of $\tilde{s}_1, \ldots, \tilde{s}_{k'}, \hat{s}_{10}, s_{11}$.
- Then loop to perform the 11th iteration.

Algorithm 1.6 is then continued: $s_{12}, s_{13}, \ldots$ are appended, until the next compression occurs, at some future k ($\leqslant 19$, depending on the size of K_{10}).

This is just an example, other compression mechanisms are possible. An important observation is that, at subsequent iterations, the polyhedron S'_k ($k > 10$) will be generated by vectors of several origins: (i) some will be original subgradients, computed by the black box (U1) during previous iterations; (ii) some will be projections $\hat{s}_\ell$ for some $10 \leqslant \ell < k$, i.e. convex combinations of those in (i); (iii) some will be "projections of projections" (after two compressions at least have been made), i.e. convex combinations of those in (i) and (ii); and so on. The important thing to understand is that all these generators are in $\partial f(x)$ anyway: we have $S'_k \subset \partial f(x)$ for all k.

Remark 2.1.6 Returning to the notation of §VIII.2.1, suppose that solving the quadratic problem (VIII.2.1.10) at the 10th iteration has produced $\alpha \in \Delta \subset \mathbb{R}^{10}$. Then choose K_{10}. If K_{10} happens to contain each $j \leqslant 10$ such that $\alpha_j > 0$, it is not necessary to append subsequently $\hat{s}_{10}$: it is already in co$\{\tilde{s}_1, \ldots, \tilde{s}_{k'}\}$.

A possible strategy for defining the compressed set K_{10} is to delete first all those s_j such that $\alpha_j = 0$, if any (one is enough). If there is none, then call for a more stringent deletion rule; for example delete the two subgradients with largest norm. □

Lemma 2.1.4 suggests that the choice of K_{10} above is completely arbitrary – including $K_{10} := \emptyset$ – leaving convergence unaffected. This is confirmed by the next result.

2 Convergence Properties

Theorem 2.1.7 *Algorithm 2.1.5 stops for some finite iteration index, producing either a downhill d_k or an $\hat{s}_k \in \partial f(x)$ of norm less than δ.*

PROOF. Suppose for contradiction that the algorithm runs forever. Because of Step 4, we have at each iteration

$$\|\hat{s}_{k+1}\|^2 \leqslant \|\operatorname{Proj} 0/\{\hat{s}_k, s_{k+1}\}\|^2 \leqslant \tfrac{1}{1+\mu_k}\|\hat{s}_k\|^2, \qquad (2.1.13)$$

where μ_k is defined by (2.1.11) with $\hat{s}$ replaced by $\hat{s}_k$. If the algorithm runs forever, the stop of Step 1 never operates and we have

$$\|\hat{s}_k\| > \delta \quad \text{for } k = 1, 2, \ldots$$

Set $r := m'^2\|\hat{s}_k\|^2/M^2 \in [0, 1[$ and observe that the function $r \mapsto r/(1-r)$ is increasing on $[0, 1[$. This implies

$$\mu_k > (1-m')^2 \frac{\delta^2}{M^2 - m'^2\delta^2} =: \mu > 0 \quad \text{for } k = 1, 2, \ldots$$

and we obtain with (2.1.13) the contradiction

$$0 < \delta \leqslant \|\hat{s}_{k+1}\|^2 \leqslant \tfrac{1}{1+\mu}\|\hat{s}_k\|^2 \quad \text{for } k = 1, 2, \ldots \qquad \square$$

2.2 Speed of Convergence

As demonstrated by Fig. 2.1.1, the efficiency of a bundling algorithm such as 2.1.5 directly depends on how fast $\|\hat{s}_k\|$ decreases at each iteration. In fact, the algorithm must stop at the latest when $\|\hat{s}_k\|$ has reached its minimal value, which is either δ or the distance from the origin to $\partial f(x)$.

Here Lemma 2.1.4 comes in again. It was used first to control the storage needed by the algorithm; a second use is now to measure its speed. Examining (2.1.11), (2.1.12) with $\hat{s} = \hat{s}_k$ and $z = \hat{s}_{k+1}$, we see that $\mu = \mu_k$ tends to 0 with $\hat{s}_k$, and it is impossible to bound $\|\hat{s}_k\|^2$ by an exponential function of k: a linear rate of convergence is impossible to obtain. Indeed the majorization expressed by (2.1.12) becomes fairly weak when k grows. The following general result quantifies this.

Lemma 2.2.1 *Let $\{\delta_k\}$ be a sequence of nonnegative numbers satisfying, with $\lambda > 0$ fixed,*

$$\delta_{k+1} = \frac{\lambda}{\lambda + \delta_k}\delta_k \quad [resp. \leqslant, \ resp. \geqslant] \quad for \ k = 1, 2, \ldots \qquad (2.2.1)$$

Then we have for $k = 1, 2, \ldots$

$$\delta_k = \frac{\lambda\delta_1}{k\delta_1 + \lambda - \delta_1} \quad [resp. \leqslant, \ resp. \geqslant]. \qquad (2.2.2)$$

PROOF. We prove the "="-part by induction on k. First, (2.2.2) obviously holds for $k = 1$. Assume that the equality in (2.2.2) holds for some k. Plug this value of δ_k into (2.2.1) and work out the algebra to obtain after some manipulations

$$\delta_{k+1} = \frac{\lambda \delta_1}{k\delta_1 + \lambda} = \frac{\lambda \delta_1}{(k+1)\delta_1 + \lambda - \delta_1}.$$

The recurrence is established. The " $\leqslant$ "- and " $\geqslant$ "-parts follow since the function $x \mapsto \lambda x/(\lambda + x)$ is increasing on $\mathbb{R}^+$. □

In other words, a nonnegative sequence yielding equality in (2.2.1) tends to 0, with an asymptotic speed of $1/k$. This is very slow indeed; it is fair to say that an algorithm behaving so slowly must be considered as not convergent at all. We now proceed to relate the bundling algorithm to the above analysis.

(a) Worst Possible Behaviour. Our analysis so far allows an immediate bound on the speed of convergence of the bundling algorithm:

Corollary 2.2.2 *There holds at each iteration k of Algorithm 2.1.5*

$$\|\hat{s}_k\| \leqslant \frac{1}{(1-m')\sqrt{k}} M . \tag{2.2.3}$$

PROOF. First check that (2.2.3) holds for $k = 1$. Then use Lemma 2.1.4: we clearly have in (2.1.11)

$$\mu \geqslant (1 - m')^2 \frac{\|\hat{s}_k\|^2}{M^2} ;$$

plug this majorization into (2.1.12) to obtain

$$\|\hat{s}_{k+1}\|^2 \ [\leqslant \|z\|^2] \ \leqslant \frac{\lambda}{\lambda + \|\hat{s}_k\|^2} \|\hat{s}_k\|^2$$

with $\lambda := M^2/(1 - m')^2$. The " $\leqslant$ "-part in Lemma 2.2.1 applies (with $\delta_k = \|\hat{s}_k\|^2$ and observing that $\lambda > M \geqslant \|\hat{s}_1\|^2$); (2.2.3) follows by taking square roots. □

Thus a bundling algorithm must reach its stopping criterion after not more than $[M/(1-m')\delta]^2$ iterations – and we already mentioned that this is a prohibitive number. Suppose that the estimate (2.2.3) is sharp; then it may take like a million iterations, just to obtain a subgradient of norm $10^{-3} M$. Now, if we examine the proof of Theorem 2.1.7, we see that the only upper bound that is possibly not sharp is the first inequality in (2.1.13). From the second part of Lemma 2.1.4, we see that the equality

$$\| \operatorname{Proj} 0 / \{\hat{s}_k, s_{k+1}\} \|^2 = \frac{\|s_{k+1}\|^2}{\|s_{k+1}\|^2 + \|\hat{s}_k\|^2} \|\hat{s}_k\|^2$$

does hold at each iteration for which $\langle \hat{s}_k, s_{k+1} \rangle = 0$.

The poor speed resulting from Corollary 2.2.2 is therefore the real one if:

(i) the generated subgradients do not cluster to 0 (say all have roughly the same norm M),
(ii) they are all approximately orthogonal to the corresponding $\hat{s}_k$ (so (2.1.10) becomes roughly an equality), and
(iii) a cheap form of Algorithm 2.1.5 is used (say, S'_{k+1} is systematically taken as the segment $[\hat{s}_k, s_{k+1}]$, or a compact convex polyhedron hardly richer).

The following example illustrates our point.

Example 2.2.3 Take the function
$$\mathbb{R}^2 \ni x = (\xi, \eta) \mapsto f(x) = \max\{\xi, -\xi, \eta\}$$
and let $x := (0, 0)$ be the reference point in Algorithm 2.1.5. Suppose $s_1 = (0, 1)$ (why not!). Then the first line-search is done along $(0, -1)$. Suppose it produces $s_2 = (1, 0)$, so $d_2 = (-1, -1)$ (no compression is allowed at this stage). Then s_3 is certainly $(-1, 0)$ (see Fig. 2.2.1) and our knowledge of $\partial f(0)$ is complete.

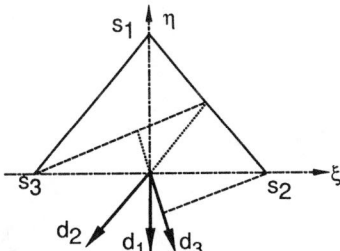

Fig. 2.2.1. Compression gives slow convergence

From then on, suppose that the "minimal" bundle $\{\hat{s}_k, s_{k+1}\}$ is taken at each iteration. Calling $(\hat{\sigma}, \hat{\tau}) \in \mathbb{R}^2$ the generic projection $\hat{s}$ (note: $\hat{\tau} > 0$), the generic line-search yields the next subgradient $s = (-\hat{\sigma}/|\hat{\sigma}|, 0)$. Now observe the symmetry in Fig. 2.2.1: the projections of s_2 and s_3 onto $\hat{s}$ are opposite, so we have
$$\langle s, \hat{s} \rangle = -\|\hat{s}\|^2.$$

Thus (2.1.10) holds as an equality with $m' := -1$, and we conclude from the last part of Lemma 2.1.4 that
$$\|\hat{s}_{k+1}\|^2 = \frac{1 - \|\hat{s}_k\|^2}{1 + 3\|\hat{s}_k\|^2} \|\hat{s}_k\|^2 = \frac{\lambda_k}{\lambda_k + 4\|\hat{s}_k\|^2} \|\hat{s}_k\|^2 \quad \text{for } k = 2, 3, \ldots$$

(here $\lambda_k = 1 - \|\hat{s}_k\|^2 \simeq 1$). With Lemma 2.2.1 in mind, it is clear that $\|\hat{s}_k\|^2 \simeq 1/(4k)$.
□

(b) Best Possible Behaviour. We are naturally led to the next question: might the bound established in Corollary 2.2.2 be unduly pessimistic? More specifically: since the only non-sharp majorization in our analysis is $\|\hat{s}_{k+1}\| \leq \|z\|$, where z denotes Proj $0/[\hat{s}_k, s_{k+1}]$, do we have $\|\hat{s}_{k+1}\| \ll \|z\|$? The answer is no; the bundling algorithm does suffer a sublinear convergence rate, even if no subgradient is ever discarded:

Counter-example 2.2.4 To reach the stopping criterion $\|\hat{s}_k\| \leq \delta$, Corollary 2.2.2 tells us that an order of $(M/\delta)^2$ iterations are *sufficient*, no matter how $\{s_k\}$ is generated. As a partial converse to this result, we claim that a particular sequence $\{s_k\}$ can be constructed, such that an order of M/δ iterations are indeed *necessary*.

To substantiate our claim, consider again the example of Fig. 2.1.1, with $M = 1$; but place the first iterate close to 0, say at $(-\varepsilon, 0)$, instead of $(-1, 0)$ (see the left part of Fig. 2.2.2); then, place all the subsequent iterates as before, namely take each s_{k+1} of norm 1 and orthogonal to $\hat{s}_k$. Look at the left part of Fig. 2.2.2 to see that $\hat{s}_k$ is, as before, the projection of the origin onto the line-segment $[s_1, s_k]$; but the new thing is that $\|s_1\| \ll \|s_k\|$ already for $k = 2$: the majorization obtained in Lemma 2.1.4 (which is sharp, knowing that $\hat{s}$ plays now the role of s_1), becomes weak from the very beginning.

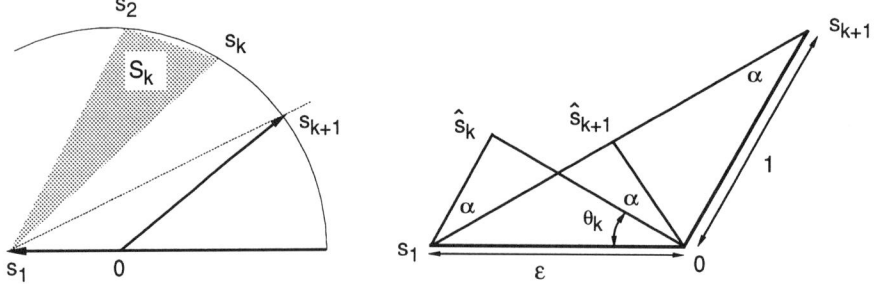

Fig. 2.2.2. Perverse effect of a short initial subgradient

These observations can be quantified, with the help of some trigonometry: the angle θ_k being as drawn in the right part of Fig. 2.2.2, we have

$$\|\hat{s}_k\| = \varepsilon \cos \theta_k, \quad \theta_1 = 0. \tag{2.2.4}$$

Draw the angle $\alpha = \theta_{k+1} - \theta_k$ and observe that $\sin \alpha = \|\hat{s}_{k+1}\| = \varepsilon \cos \theta_{k+1}$ to obtain

$$\varepsilon \cos \theta_{k+1} = \sin \theta_{k+1} \cos \theta_k - \cos \theta_{k+1} \sin \theta_k.$$

Thus, the variable $u_k := \tan \theta_k$ satisfies the recurrence formula

$$u_{k+1} = u_k + \varepsilon \sqrt{1 + u_k^2}, \quad u_1 = 0, \tag{2.2.5}$$

which shows in particular that $\{u_k\}$ is increasing.

Now take $\varepsilon = \delta\sqrt{2}$. In view of (2.2.4), the event $\|\hat{s}_k\| \leq \delta$ happens when $\theta_k \geq \pi/4$. Our claim thus amounts to showing that the process (2.2.5) needs $1/\delta \simeq 1/\varepsilon$ iterations to reach the value $u_k \geq 1$. This, however, can be established: we have

$$u_{k+1} < u_k + \varepsilon\sqrt{2} \quad \text{if} \quad u_k < 1$$

which, together with the value $u_1 = 0$, shows that

$$u_{k+1} < k\varepsilon\sqrt{2} \quad \text{if} \quad u_2 < \cdots < u_k < 1.$$

Thus the event $u_{k+1} \geq 1$ requires $k \geq 1/(\varepsilon\sqrt{2}) = 1/(2\delta)$; our claim is proved. □

The above counter-example is slightly artificial, in that s_1 is "good" (i.e. small), and the following subgradients are as nasty as possible. The example can be modified, however, so that all the subgradients generated by the black box (U1) have comparable norms. For this, add a third coordinate to the space, and start with two preliminary iterations: the first subgradient, say s_0, is $(-\varepsilon, 0, 1)$, so $\hat{s}_0 = s_0 = (-\varepsilon, 0, 1)$. For the next subgradient, take $s_1 = (-\varepsilon, 0, -1)$, which obviously satisfies (2.1.2); it is easy to see that $\hat{s}_1$ is then $(-\varepsilon, 0, 0)$. From then on, generate the subgradients $s_2 = (0, 1, 0)$ and so on as before, all having their third coordinate null. Draw a picture if necessary to see that the situation is then exactly the same as in 2.2.4, with $\hat{s}_1 = (-\varepsilon, 0, 0)$ replacing the would-be $s_1 = (-\varepsilon, 0)$: the third coordinate no longer plays a role.

In this variant, a wide angle $(\langle s_1, \hat{s}_0 \rangle \ll 0)$ has the same effect as a small initial subgradient. The message from our counter-examples is that, in a bundling algorithm, an exceptionally good event (small subgradient, or wide angle) must subsequently be paid for by a slower convergence.

Among other things, our Counter-example 2.2.4 suggests that the choice of $S'_{k+1} \subset S_{k+1}$ plays a minor role. A situation worth mentioning is one in which this choice is strictly irrelevant: when all the gradients are mutually orthogonal. Remembering Remark II.2.4.6, assume

$$\langle s_i, s_j \rangle = 0 \quad \text{for all } i, j = 1, \ldots, k \text{ with } i \neq j. \tag{2.2.6}$$

Then the projection of the origin onto $\mathrm{co}\{s_1, \ldots, s_k\}$ and onto $[\hat{s}_{k-1}, s_k]$ coincide.

This remark has some important consequences. Naturally, (2.2.6) cannot hold forever in a finite-dimensional space. Instead of $\mathbb{R}^n$, consider therefore the space ℓ_2 of square-summable sequences $s = (s^1, s^2, \ldots)$. Let s_k be the kth vector of the canonical Hilbertian basis of this space:

$$s_k := (0, \ldots, 0, 1, 0, \ldots) \quad \text{with the "1" in position } k$$

so that, of course, $\langle s_k, s_{k'} \rangle = \delta_{kk'}$. Then the projection is particularly simple to compute:

$$\hat{s}_k := \mathrm{Proj}\, 0/\{s_1, \ldots, s_k\} = \frac{1}{k}\sum_{j=1}^{k} s_j.$$

This is because

$$\|\hat{s}_k\|^2 = \frac{1}{k^2} \sum_{j=1}^{k} \|s_j\|^2 = \frac{1}{k} = \langle \hat{s}_k, s_j \rangle \quad \text{for } j = 1, \ldots, k, \tag{2.2.7}$$

which characterizes the projection. It is also clear that

$$\langle \hat{s}_k, s_{k+1} \rangle = 0 \quad \text{for } k = 1, 2, \ldots$$

Thus, the above sequence $\{s_k\}$ could well be generated by the bundling Algorithm 1.6. In fact, such would be the case with

$$f(x) = f(\xi^1, \ldots, \xi^k, \ldots) := \max\{\xi^1, \ldots, \xi^k, \ldots\}$$

if (U1) computed the smallest index giving the max, and if the algorithm was initialized on the reference point $x = 0$. In this case, the majorization (2.2.3) would be definitely sharp, as predicted by our theory and confirmed by (2.2.7). It may be argued that *numerical algorithms*, intended for implementation on a computer, are not supposed to solve infinite-dimensional problems. Yet, if n is large, say $n \geqslant 100$, then (2.2.6) may hold till $k = 100$; in this case, it requires 100 iterations to reduce the initial $\|s\|$ by a mere factor of 10.

Remark 2.2.5 Let us conclude. When $k \to +\infty$, $\|\hat{s}_k\|^2$ in Algorithm 2.1.5 is at least as "good" as $1/k$, even if the bundle S'_k is kept minimal, just containing two subgradients. On the other hand, $\|\hat{s}_k\|^2$ can be as bad as $1/k^2$, at least if S'_k is kept minimal, or also when the dimension of the space is really large. In particular, a speed of convergence of the form

$$\|\hat{s}_k\|^2 \simeq q^k \quad \text{when} \quad k \to +\infty, \tag{2.2.8}$$

with a rate $q < 1$ independent of the particular function f, cannot hold for our algorithm – at least when the space is infinite-dimensional. There is actually a reason for that: according to *complexity theory*, no algorithm can enjoy the "good" behaviour (2.2.8); or if it does, the rate q in (2.2.8) must tend to 1 when the dimension n of the space tends to infinity; in fact q must be "at best" of the form $q \simeq 1 - 1/n$.

Actually, complexity theory also tells us that the estimate (2.2.3) cannot be improved, unless n is taken explicitly into account; a brief and informal account of this complexity result is as follows. Consider an arbitrary algorithm, having called our black box (U1) at k points $x_1, \ldots, x_k$, and denote by $s_1, \ldots, s_k$ the output from (U1). Then, an estimate of the type

$$\|\operatorname{Proj} 0/\{s_1, \ldots, s_k\}\|^2 = o(1/k) \quad \text{for } k \to +\infty \tag{2.2.9}$$

is impossible to prove *independently* of n and of (U1). In particular: no matter how the algorithm selects its kth point x_k, an s_k can be produced (like in our Counter-example 2.2.4) so that (2.2.8) does not hold independently of n. Of course, it is a particularly nasty black box that does the trick: it uses its own output $\{(f(x_i), s_i)\}_{i<k}$ to concoct its kth answer $(f(x_k), s_k)$. The corresponding convex function f is not fixed a priori, but is recursively constructed, depending on the sequence $\{x_k\}$ chosen by the algorithm.

This result gives rise to an additional and frustrating observation: comparing the behaviour (2.2.9) with (2.2.3), we see that the bundling algorithm, although apparently very weak, is "optimal" in some sense. □

(c) Practical Behaviour. Our analysis in (a) and (b) above predicts a certain (mediocre) convergence speed of Algorithm 2.1.5; it also indicates that the choice of S'_{k+1} has little influence on this speed. Unfortunately, this is blatantly contradicted empirically. Some numerical experiments will give an idea of what is observed in practice.

Test-problem 2.2.6 (TR48) Given a symmetric $n \times n$ matrix A, and two vectors d and s in $\mathbb{R}^n$, consider the piecewise affine objective function

$$f(x) := \sum_{i=1}^{n} [s_i x_i + d_i \max \{a_{i1} - x_1, a_{i2} - x_2, \ldots, a_{in} - x_n\}] \,. \qquad (2.2.10)$$

To minimize this f originates in the problem of minimizing the cost of transporting goods between pairs of locations (after the duality transformation of §VII.4 is performed on the mathematical formulation of this problem). Formulation (2.2.10) is not a particularly good one for obtaining effective solutions of *transportation problems*; but it is an excellent academic example to illustrate some of our points.

Needless to say, the black box (U1) computing $f(x)$ and $s(x)$ works as follows: for each i, select *some* j – call it $j(i)$ – such that $a_{ij(i)} - x_{j(i)}$ is maximal. Then form

$$f(x) = \sum_{i=1}^{n} [s_i x_i + d_i (a_{ij(i)} - x_{j(i)})]$$

and compute $s(x)$ accordingly.

In what follows, the standard dot-product $\langle x, y \rangle = x^\top y$ on $\mathbb{R}^n$ will be used. Call u the vector of $\mathbb{R}^n$ whose coordinates are $1, 1, \ldots, 1$. Because

$$f(x + ru) = f(x) + r \sum_{i=1}^{n} (s_i - d_i) \quad \text{for all } r \in \mathbb{R},$$

we assume that $u^\top (s-d) = 0$; otherwise, f would have no minimum. This assumption implies

$$s^\top u = 0 \quad \text{for all } x \in \mathbb{R}^n \text{ and } s \in \partial f(x)$$

(easy exercise). Thus, fixing the starting point at $x = 0$, we are actually faced with a problem posed in $\mathbb{R}^{n-1}$, the subspace orthogonal to u.

We have selected a particular instance of the above problem, in which $n = 48$ and the minimal value is $\bar{f} = -638565$. The data are organized in such a way that, at a minimum point $\bar{x}$, there are for each i an average of 2–3 terms that yield the max in (2.2.10). As it happens, the number of possible outputs $s(\bar{x})$ for the black box (U1) is of the order 10^{11}. □

Apply Algorithm 2.1.5 to this example, the starting point x being a minimum $\bar{x}$. Then no stop can occur at Step 3 and the algorithm loops between Step 4 and Step 1 until a small enough $\|\hat{s}_k\|$ is found.

We have tested two forms of Algorithm 2.1.5:

– The first form is the most expensive possible, in which no subgradient is ever deleted: at each iteration, $S'_{k+1} = S_{k+1}$.

– The second form is the cheapest possible: all subgradients are systematically deleted and $S'_{k+1} = [\hat{s}_k, s_{k+1}]$ at each iteration.

Figure 2.2.3 displays the respective evolutions of $\|\hat{s}_k\|$, as a function of k. We believe that it illustrates well enough how pessimistic (2.1.13) can be. The same phenomenon is observed with the test-problem MAXQUAD of VIII.3.3.3: with the expensive form, the squared norm of the subgradient is divided by 10^5 in 3 iterations; with the cheap form, it is divided by 10^2, and then stagnates there.

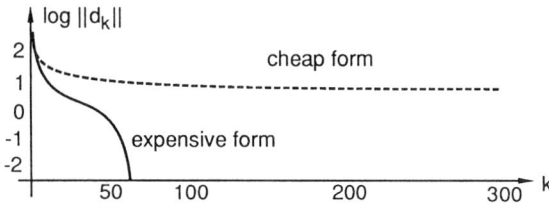

Fig. 2.2.3. Efficiency of storing subgradients in a bundling algorithm; TR48

In view of Remark 2.2.5, one could expect that the cheap and expensive forms tend to equalize when the dimension n increases. To check this, we introduce another example:

Test-problem 2.2.7 (TSP) Consider a complete graph, i.e. a set of n nodes ("cities") and $m = 1/2\,n(n-1)$ arcs ("routes", linking each pair of cities); let a cost ("length") be associated with each arc. The *traveling salesman problem* is that of constructing a path visiting all the cities exactly once, and having the shortest length. This is not simple, but a closely related problem is to minimize an ordinary piecewise affine function associated with the graph, say

$$\mathbb{R}^n \ni x \mapsto f(x) := \max \left\{ s_i^\top x + b_i \,:\, i \in T \right\}, \qquad (2.2.11)$$

where x is a vector of "dummy costs", associated with the nodes. We do not give a precise statement of this last problem; let us just say that the basis is once again *duality*, introduced in a fashion to be seen in Chap. XII.

Here the index-set T (the set of "1-trees" in the graph) is finite but intractable: $|T| \simeq n^n$. However, finding for given x an index $k \in T$ realizing the max in (2.2.11) is a reasonable task, which can be performed in $O(m) = O(n^2)$ operations. Furthermore, the resulting s_k is an integer vector.

This example will be used with several datasets corresponding to several values of n, and referred to as TSPn. □

Take first TSP442, having thus $n = 442$ variables; the minimal value of f turns out to be -50505. Just as with TR48, we start from an optimal $\bar{x}$ and we run the two forms of Algorithm 2.1.5. With this larger test-problem, Fig. 2.2.4 displays an evolution similar to that of TR48; but at least the cheap variant is no longer "non-convergent": the two curves can be drawn on the same picture.

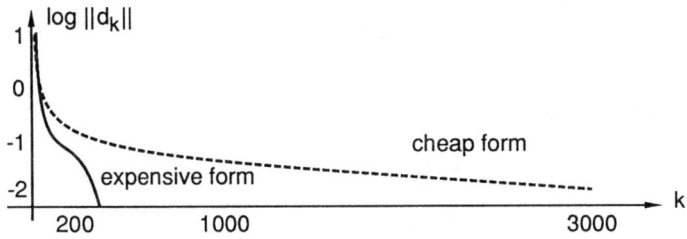

Fig. 2.2.4. Efficiency of storing subgradients in a bundling algorithm; TSP442

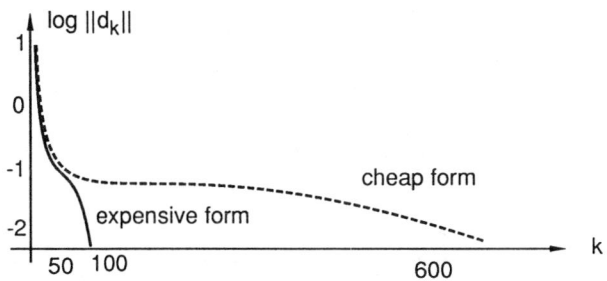

Fig. 2.2.5. Efficiency of storing subgradients in a bundling algorithm; TSP1173

Figure 2.2.5 is a further illustration, using another dataset called TSP1173. A new feature is now the unexpectedly small number of iterations, which can be partly explained by the special structure of the subgradients s_k:

(i) All their coordinates have very small values, say in the range $\{-1, 0, 1\}$, with exceptionally some at ± 2; as a result, it becomes easier to form the 0-vector with them.
(ii) More importantly, most of their components are systematically 0, for all x near the optimal reference point: actually, less than 100 of them ever become nonzero; it follows that the sequence $\{\hat{s}_k\}$ evolves in a space of dimension not more than 100 (instead of 1173). In TSP442, this space has dimension 150; and we recall that in TR48, it has dimension exactly 47.

Remark 2.2.8 The snaky shape of the curves in Figs. 2.2.3, 2.2.5 (at least for the expensive variant) is probably due to piecewise affine character of f. Roughly speaking, $\|\hat{s}_k\|$ decreases fast at the beginning, and then slows down, simply because $1/k$ follows this pattern; then comes the structure of $\partial f(\bar{x})$, which is finitely generated: less and less room is left for the next s_{k+1}; the combinatorial aspect of the problem somehow vanishes, and $\|\hat{s}_k\|$ decreases more rapidly. □

Let us conclude this section with a comment similar to Remark II.2.4.4: numerical analysis is difficult, in that theory and practice often contradict each other. More precisely, a theoretical framework establishing specific properties of an algorithm may not be relevant for the actual behaviour of that algorithm. Here, it is difficult to find an appropriate theoretical framework explaining why the bundling algorithm works so much better in its expensive form. It is also difficult to explain why its actual behaviour contradicts so much the theory.

3 Putting the Mechanism in Perspective

In this section, we look at the bundling algorithm from some different points of view, and we relate it to various topics.

3.1 Bundling as a Substitute for Steepest Descent

The first obvious way of looking at the mechanism of §2 is to consider it as one single execution of Step 2 in the steepest-descent Algorithm VIII.1.1.7. After all, the original motivation of this mechanism lay precisely there. (From this point of view, it is an unfortunate notation to use the same index k in Algorithm 1.6 and in Algorithm VIII.1.1.7; we have good reasons for doing so, which will appear later in this book: indeed, we will see that one should not distinguish the outer iteration of a descent algorithm, from the inner iteration used to find the descent direction). Somehow, we have reached our aim of making implementable the steepest descent scheme; the bundling Algorithms 1.6 or 2.1.5 merely requires:

(i) the projection of the origin onto a compact convex polyhedron: a perfectly solvable convex quadratic problem;
(ii) the computation of an arbitrary subgradient of the objective-function at each given x; this can be done in many instances, §VIII.3 was devoted to this point; anyway, it can be done whenever the whole subdifferential can be computed!

On the other hand, some other difficulties are not resolved yet:

(iii) The descent direction d_k, found via Corollary 2.1.2, is not steepest; this point, however, is not too serious because of (iv) below.
(iv) As already mentioned, the resulting descent algorithm is likely not to minimize f, since the direction d_k suffers the same deficiencies as a steepest one (non-continuity as a function of x, short-sightedness, zigzags, ...).
(v) We have introduced the new problem of letting $t \downarrow 0$ before Algorithm 2.1.5 can iterate.
(vi) The applications mentioned in §VIII.3.3 are still not covered: for such problems, our algorithm trivially terminates after the first iteration. We end up with the mere gradient method, which is among the worst possible ideas; remember Table VIII.3.3.2.

At this point, the reader may feel that arguments (iv) – (vi) are due to the same single cause: all this is theory and, for proper implementations, some small parameter ε must come into play; Remark VIII.2.3.4 made an allusion to it already. Such is indeed the case and this will motivate subsequent developments: Chap. XI for theory, Chap. XIII for a first implementation; Chap. XIV will globalize the approach and enlarge this ε so as to definitely escape from the steepest-descent concept.

Here we give some comments on (i) and (ii), and we start with an important remark.

3 Putting the Mechanism in Perspective

Remark 3.1.1 As already mentioned in Remark 1.4, each cycle of the bundling mechanism generates a subgradient s_{k+1} lying in the face of $\partial f(x)$ exposed by the direction d_k.

This s_{k+1} is interesting for d_k itself: not only is d_k uphill (because $\langle s_{k+1}, d_k \rangle \geq 0$), but we can say more. In terms of the descent property of d_k, the subgradient s_{k+1} is the worst possible; or, reverting the argument, s_{k+1} is the best possible in terms of the useful information concerning d_k. Figure 2.1.2 shows how to interpret this property geometrically: s_{k+1} (there s_2), known to lie in the dashed area, is even at the bottom tip of it. □

In light of this remark, we see that the subgradients generated by the bundling mechanism are fairly special ones. Their special character is also enhanced by the black box (U1): the way $s(x)$ is computed in practice makes it likely that it will lie on the boundary of $\partial f(x)$, or even at an extreme point. For example, in the case of a minimax problem (§VIII.2), each $s(x)$, and hence each s_{k+1}, is certainly the gradient at x of some differentiable active function.

In summary, the bundling mechanism 1.6 can be viewed as a *filter*, which extracts some carefully selected subgradients from the whole set $\partial f(x)$. What is even more interesting is that it has a tendency to select the most interesting ones. Let us admit that the directions d_k are not completely random – they are probably closer and closer to directions of interest, namely descent directions. Then the subgradients s_{k+1} generated by the algorithm are also likely to "flirt" with the subgradients of interest, namely those near the face of $\partial f(x)$ exposed by (steepest) descent directions. In other words, our algorithm can be interpreted as *selecting* those subgradients that are *important* for defining *descent* directions, disregarding the "hidden side of the moon"; see Fig. 3.1.1, where the visible side of the moon is the set of faces of $\partial f(x)$ that are exposed by descent directions.

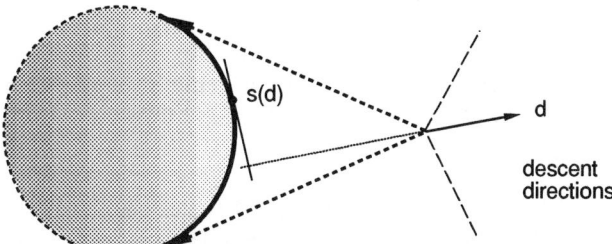

Fig. 3.1.1. The visible side of the moon

According to these observations, the bundling mechanism can also be viewed as a constructive way of computing the steepest-descent direction, playing its role even when a full description of the subdifferential is available. For an illustration, consider a case in which the subdifferential

$$\partial f(x) = \text{co}\{s^1, s^2, \ldots, s^m\} \tag{3.1.1}$$

is a known compact convex polyhedron, with m possibly large. When Algorithm 1.6 is applied to this example, the full set (3.1.1) is replaced by a growing sequence of internal approximations S_k. Correspondingly, there is a sequence of approximate steepest-descent directions d_k, with its sequence of approximate derivatives $-\|d_k\|^2$ (see Remark 1.5).

Now, it is reasonable to admit that each subgradient s_{k+1} generated by the algorithm is extracted from the list $\{s^1, \ldots, s^m\}$: first, (U1) is likely to choose $s(\cdot)$ in that list; and second, each s_{k+1} is in the face of $\partial f(x)$ exposed by d_k; remember Remark 1.4. Then, projecting onto S_k is a simpler problem than projecting onto (3.1.1). Except for its stopping test, the bundling algorithm really provides a constructive way to compute $\hat{s} = \text{Proj}\, 0/\partial f(x)$.

Of course, this interpretation is of limited value when a descent direction does exist: Algorithm 1.6 stops much too early, namely as soon as

$$0 > f'(x, d_k) \geq -\|\hat{s}_k\|^2$$

(knowing that $d_k = -\hat{s}_k$). A real algorithm computing $\hat{s}$ should stop much later, namely when $f'(x, d_k)$ is "negative enough":

$$f'(x, d_k) = -\|\hat{s}_k\|^2,$$

since the latter characterizes $\hat{s}_k$ as the projection of the origin onto $\partial f(x)$.

On the other hand, when x is optimal, the problem of computing $\hat{s} = 0$ becomes that of finding the correct convex multipliers, satisfying $\sum_{j=1}^m \alpha_j s^j = 0$ in (3.1.1). Then the bundling Algorithm 1.6 is a valid alternative to direct methods for projection. In fact, if (3.1.1) is complex enough, Algorithm 1.6 can be a very efficient approach: it consists of replacing the single convex quadratic problem

$$\min_{\alpha \in \Delta_m} \tfrac{1}{2} \left\| \sum_{j=1}^m \alpha_j s^j \right\|^2$$

by a sequence of similar problems with m replaced by $1, 2, \ldots, k \leq m$. From our discussion above, all of these problems are probably "strictly simpler" than the original one, insofar as their generators $s^1, \ldots, s^k$ are extracted from the list $\{s^1, \ldots, s^m\}$ in (3.1.1).

Example 3.1.2 The test-problem TR48 of 2.2.6 illustrates our point in a particularly spectacular way. We have said that, at an optimum $\bar{x}$, there are an order of 10^{11} active functions; and there are just as many possible values of $s(\bar{x})$, as computed by (U1). On the other hand, it is (probably) necessary to have 48 of them to obtain the correct convex multipliers α_j in $\sum_j \alpha_j s^j = 0$ (this is Carathéodory's theorem in $\mathbb{R}^{48}$, taking into account that all subgradients are actually in $\mathbb{R}^{47}$ because of degeneracy).

Figure 2.2.3 reveals a remarkable phenomenon: in the expensive variant of Algorithm 2.1.5, $\|d_{65}\|$ is zero within roundoff error. In other words, the bundling mechanism has managed to extract only 65 "interesting" subgradients: to obtain a correct set of multipliers, it has made only $17 (= 65 - 48)$ "mistakes", among the 10^{11} possible ones. □

To conclude, we add that all intermediate black boxes (U1) are conceivable: in this chapter, we considered only the simplest one, computing just one subgradient; Chap. VIII required the most sophisticated one, computing the full set (3.1.1). When (U1) returns richer information, Algorithm 2.1.5 will converge faster, but possibly at the price of a more expensive iteration; among other things, each individual projection will be more difficult. The bundling approach is open to a proper balance between simplicity (one subgradient) and rapidity (all of them).

3.2 Bundling as an Emergency Device for Descent Methods

Another, more subtle, way of looking at the bundling mechanism comes from (vi) in the introduction to §3.1. The rather delicate material contained in this Section 3.2 is an introduction to the numerical Chaps XIII and XIV. Consider an "ordinary" method such as those in Chap. II for minimizing a smooth function. Within one iteration, consider the line-search: given the current x and d, and setting $q(t) := f(x+td)$, we want to find a convenient stepsize $t > 0$, satisfying in particular $q(t) < q(0)$.

Anyone familiar with optimization algorithms has experienced line-searches which "do not work", just because q doggedly refuses to decrease. A frequent reason is that the $s(x)$ computed by (U1) is not the gradient of f at x – because of some programming mistake. Another reason, more fundamental, is ill-conditioning: although $q'(0) < 0$, the property $q(t) < q(0)$ may hold only for values of $t > 0$ so small that the computer cannot find any; remember Fig. II.1.3.1. In this situation, any reasonable line-search can only produce a sequence of trial stepsizes tending to 0. In the terminology of §II.3, $t_L \equiv 0$, $t_R \downarrow 0$: the minimization algorithm is stuck at the present iteration. Normally there is not much to do, then – except checking (U1)!

It is precisely in this situation that the mechanism of §2 can come into play. As stated, Algorithm 1.2 can be viewed as the beginning of a "normal" line-search, of the type of those in §II.3. More precisely, compare the line-search 1.2 and Fig. II.3.3.1 with $m = 0$. As long as t_L does not become positive, i.e. as long as $q(t) \geq q(0)$, both line-searches do the same thing: they perform interpolations aimed at pulling t towards the region having $q(t) < q(0)$ (if it is not void). It is only when some $t_L > 0$ is found, that the difference occurs. Then the line-search 1.2 becomes pointless, simply because the bundling Algorithm 1.6 stops: the *local* problem of finding a descent direction is solved.

If, however, the question is considered from the higher point of view of really minimizing f, it is certainly not a good idea to stop the line-search 1.2 as we do, when $q(t) < q(0)$. Rather, it is time to call t_L the present t, and to start a second phase in the line-search: our aim is now to find a convenient t, satisfying some criterion (0) in the terminology of §II.3.2. In other words: knowing that the line-search 1.2 is just a slave of Algorithm 1.6, it should be a good idea to complement it with some mechanism coming from §II.3.2.

The result would produce a compound algorithm, which can be outlined as follows:

Algorithm 3.2.1 (Compound Descent Scheme) Given the current iterate x:

STEP 1. Compute a direction d according to the principles of §II.2.

STEP 2. Do a line-search along d, according to the principles of §II.3. This line-search has two possible exits:
- Either a *descent step* is found ($t_L > 0$), and a stopping criterion of the type of those in §II.3.2 can be fulfilled. Then loop normally to Step 1 with x appropriately updated.
- Or no descent step can be found ($t_L \equiv 0$, $t_R \downarrow 0$); then make a *null-step*: do not update x but switch to

STEP 3. Follow the rules of Algorithm 1.6 by projecting the origin onto the current approximation of $\partial f(x)$. Loop to Step 2. □

We have thus grafted the bundling mechanism on a descent scheme as described in §II.2. Algorithm 3.2.1 can be interpreted as a way to rescue a good "smooth algorithm" from some difficult situations, in which ill-conditioning provokes zigzags.

Remark 3.2.2 We do not specify how to compute the direction in Step 1. Let us just observe that, if it is the gradient method of Definition II.2.2.2 that is chosen, then Algorithm 3.2.1 just becomes the method we started from (§2.1), simulating the steepest descent. It is in Chap. XIII that we will start studying approaches making possible harmonious combination of Steps 1 and 2.

Our interpretation of Algorithm 1.6, as an anti-zigzagging mechanism for a descent method, suggests a way in which one might suppress the difficulty (v) of the introduction to §3.1. Instead of letting $t = t_R \downarrow 0$, it is sensible in Algorithm 3.2.1 to switch to Step 3 when t_R becomes small. To make this kind of decision, we will see that the following considerations are relevant:

- The expected decrease of q in the current direction d should be estimated. If it is small, d can be relinquished.
- Allowing such a "nonzero null-step" implies appending to S_k some vectors which are not in $\partial f(x)$. One should therefore estimate how close $s(x + td)$ is to $\partial f(x)$; this makes sense because of the outer semi-continuity of ∂f.

Thus, the expression "small stepsize" can be given an appropriate meaning, in terms of objective-value differences, and of subgradient proximity. □

An interesting point here is to see how Step 2 could be implemented, along the lines of §II.3. The issue would be to inject into the test (0), (R), (L) a provision for the switch to Step 3, when t is small enough – whatever this means. In anticipation of Chap. XIV, and without giving too much detail, let us mention the starting ideas:

- First, the initial derivative $q'(0)$ is needed. Although it exists (being the directional derivative $f'(x, d)$ along the current direction d) it is not known. Nevertheless, Remark 1.5 gives a substitute: we have $-\|d\|^2 \leq q'(0)$, with equality if the current polyhedron S approximates the subdifferential $\partial f(x)$ well enough.
- Then the descent-test, declaring that t is not too large, is that of (II.3.2.1):

$$q(t) \leq q(0) - mt\|d\|^2 \qquad (3.2.1)$$

for a given coefficient $m \in \,]0, 1[$.

– Now, when switching to Step 3 in Algorithm 3.2.1, a new subgradient s must be found so as to recompute the direction. Ideally, it should satisfy $\langle s, d\rangle \geqslant 0$; but this can be relaxed, thanks to Lemma 2.1.4: everything will be all right if the new subgradient merely satisfies

$$\langle s, d\rangle \geqslant -m'\|d\|^2 \tag{3.2.2}$$

for some $m' < 1$.

At this point, we are bound to realize from (3.2.1), (3.2.2) that *Wolfe's* criterion (II.3.2.4) suggests itself as the appropriate one for declaring a descent step. The provision for a null-step will operate when $-\|d\|^2$ is a too optimistic underestimate of the real $f'(x, d)$. Note that the value $-\|d\|^2$, in (3.2.1) and (3.2.2), should not be taken too literally: it does not depend positively homogeneously on d. What actually counts is $\langle \hat{s}, d\rangle$, where $\hat{s}$ is the "gradient", in the interpretation of Remark VIII.1.3.7.

Remark 3.2.3 The need for a null-step appears when (3.2.1) is never satisfied, despite repeated interpolations. If the coefficients m and m' are properly chosen, namely $m < m'$, then we have

$$\langle s(x+td), d\rangle \geqslant \frac{q(t) - q(0)}{t} > -m\|d\|^2 > -m'\|d\|^2 ;$$

here the first inequality expresses that $s(x+td) \in \partial f(x+td)$, and the second is the negation of (3.2.1). We see that (3.2.2) holds automatically in this case; one more manifestation of the second part in Remark 3.1.1. On the other hand, if the black box (U1) is bugged, the line-search may well generate a sequence $t_R \downarrow 0$ which never satisfies (3.2.1) nor (3.2.2). Then, even our rescuing mechanism is helpless. The same phenomenon may happen with a nonconvex f (but a fairly pathological one, indeed). There is a moral: a totally arbitrary function, such as the output of a wrong (U1), cannot be properly minimized! □

3.3 Bundling as a Separation Algorithm

In its purest form, the mechanism of §1 can be thought of independently of any convex function to be minimized. What it actually does is: given a compact convex set $S = \partial f(x)$ and a point 0, find an affine hyperplane strictly separating $\{0\}$ and S, if there is one; or conclude that $0 \in S$.

By definition, a (strictly) separating hyperplane is a pair $(d, r) \in \mathbb{R}^n \times \mathbb{R}$ such that (see §III.4.1):

$$\langle d, s\rangle < r \text{ for all } s \in S \quad \text{and} \quad r < \langle d, s'\rangle \text{ for all } s' \in \{0\}.$$

In simpler terms, it is a $d \in \mathbb{R}^n$ such that

$$\langle d, s\rangle < 0 \quad \text{for all } s \in S,$$

or equivalently, $\sigma_S(d) < 0$ because S is compact. When S is actually a subdifferential of a convex function, a separating hyperplane gives a descent direction, while a non-separating hyperplane corresponds to a d such that neither d nor $-d$ is downhill.

The above separation problem can be more or less trivial, depending on how S is characterized.

– If
$$S := \{s \in \mathbb{R}^n : c(s) \leq 0\}$$
is defined by constraints (several constraints c_j can be included in c, simply by setting $c = \max_j c_j$), then answering the question is easy: it amounts to computing $c(0)$ and a subgradient d of c at 0. In fact
$$0 \notin S \iff c(0) > 0.$$
In case $c(0) > 0$, then any $d \in \partial c(0)$ defines a separating hyperplane. The reason is that the property
$$s \in S \implies 0 \geq c(s) > c(s) - c(0) \geq \langle d, s \rangle$$
gives the equation of our hyperplane:
$$H_{d,-c(0)} = \{s \in \mathbb{R}^n : \langle s, d \rangle = -c(0)\}.$$

– If S is a compact convex polyhedron characterized as a convex hull, the mere question "$0 \in S$?" is already more difficult to answer. From the preceding sections, a suggestion is to project 0 onto S. If the projection is 0, the answer is yes. If it is nonzero, a separating hyperplane is found.

Our bundling mechanism works in another situation, which we now explain. Algorithm 1.6 can be considered as made up of two parts: Step 2 and the rest. Step 2, i.e. Algorithm 1.2, represents the problem to be solved: it is in charge of providing the rest with information concerning the function f to minimize, or the set $\partial f(x)$ to separate. By contrast, the rest of Algorithm 1.6 is the "decision maker", which decides whether to stop, or what other direction to try, etc. It might even decide what norm to use for the projection.

If we make a parallel with the hill-climbing problem of §II.1.3, we see that Step 2 is one more black box (the concept of black box is very important in numerical analysis!). Given d, it computes $s \in S$ yielding $\sigma_S(d)$, or detects that $\sigma_S(d) < 0$. In other words, our present mechanism is adapted to situations in which S is not known explicitly, but in which the only available information is *pointwise* and concerns the support function of S. In a word, the main task of our black box of Step 2 is to solve the problem
$$\max \{\langle d, s \rangle : s \in S\} \qquad (3.3.1)$$
and to return an optimal s. Remark 3.1.1 makes it clear that this is exactly what is done when extracting a cluster point of $\{s(x + td)\}$ for $t \downarrow 0$.

With this in mind, Algorithm 1.6 can be read as follows; we deliberately combine the two possible exits from the line-search: the following algorithm is supposed to run forever.

Algorithm 3.3.1 (Elementary Separation Algorithm) To initialize the algorithm, solve (3.3.1) with $d_0 = 0$ and obtain $s_1 \in S$. Set $k = 1$.

STEP 1. Compute $d_k = -\operatorname{Proj} 0/\{s_1, \ldots, s_k\}$.

STEP 2. Solve (3.3.1) with $d = d_k$ to obtain $s_{k+1} \in S$ such that
$$\langle s_{k+1}, d_k \rangle = \sigma_S(d_k).$$
Replace k by $k+1$ and loop to Step 1. □

The black box involved in this form of algorithm is, somehow, an "intelligent (U1)". Its input is no longer x but rather the couple (x, d); its output is no longer an arbitrary $s(x) \in S = \partial f(x)$, but a somewhat particular $s_d(x)$, lying in the face of S exposed by d. The initialization is artificial: with $d = 0$, $s_0(x)$ is now an arbitrary subgradient.

The next idea that comes to mind is: how about computing not only a separating hyperplane, but a "best" one, separating S as much as possible from 0? The set of directions defining a separating hyperplane is an open cone, characterized by the equation $\sigma_S(d) < 0$. Minimizing $\sigma_S(d)$ can be interpreted as seeking a ray which is central in this cone, i.e. as remote as possible from its boundary. This must be understood among normalized directions only, because positive homogeneity of σ_S and properties of cones make the picture invariant by homothety. Finally, the theory of §VIII.1.2 tells us that we are exactly in the situation of §3.1: what we want to do, really, is to project the origin onto S.

Remark 3.3.2 It is interesting to recall that the above set of separating hyperplanes (the cone of descent directions!) has a geometric characterization. When S does not contain the origin, σ_S is not minimal at 0. In view of Theorem VI.1.3.4, our set of separating hyperplanes is therefore given by the interior of the polar [cone S]° of the cone generated by S. Incidentally, note also from the compactness of S that the conical hull cone S is closed (Proposition III.1.4.7). □

When a mere descent direction was sought, the bundling Algorithm 1.6 was stopped as soon as $\sigma_S(d_k)$ became negative. Here, to compute a steepest-descent direction, Algorithm 3.3.1 must continue until
$$\sigma_S(d_k) = -\|d_k\|^2,$$
which characterizes the projection of the origin onto S.

In the spirit of Lemma 2.1.1, it is not difficult to prove that the present process does converge.

Theorem 3.3.3 *When $k \to +\infty$ in Algorithm* 3.3.1,
$$d_k \to \hat{d} := -\operatorname{Proj} 0/S.$$

PROOF. It is convenient to set $\hat{s}_k := -d_k$, $\hat{s} := -\hat{d}$. Then we have
$$\langle \hat{s}_k, s_i \rangle \geq \|\hat{s}_k\|^2 \quad \text{for } i = 1, \ldots, k \text{ and } k = 1, 2, \ldots$$
and, since $-\hat{d} \in S$,
$$\langle \hat{s}_k, s_{k+1} \rangle \leq \langle \hat{s}_k, \hat{s} \rangle \quad [\text{i.e. } \sigma_S(d_k) = \langle d_k, s_{k+1} \rangle \geq \langle d_k, -\hat{d} \rangle].$$

We therefore deduce

$$\begin{aligned}\langle \hat{s}_k, s_i - s_{k+1}\rangle &\geq \langle \hat{s}_k, \hat{s}_k - \hat{s}\rangle = \\ &= \langle \hat{s}, \hat{s}_k - \hat{s}\rangle + \langle \hat{s}_k - \hat{s}, \hat{s}_k - \hat{s}\rangle \\ &\geq \|\hat{s}_k - \hat{s}\|^2.\end{aligned}$$

Finish as in the proof of Lemma 2.1.1. The inequality

$$\langle \hat{s}_k, s_i - s_{k+1}\rangle \geq \|\hat{s}_k - \hat{s}\|^2 \geq \delta > 0 \quad \text{for } i = 1, \ldots, k \text{ and } k = 1, 2, \ldots$$

infinitely often would lead to a contradiction: we could extract a further subsequence such that $s_i - s_{k+1} \to 0$. □

Some comments can be made concerning the speed of convergence. When $0 \in S$, Algorithms 1.6 and 3.3.1 both construct a sequence $\{d_k\}$ tending to 0. The latter gets from its black box information of better quality, though. Algorithm 1.6 receives at each iteration an s_{k+1} such that

$$\langle s_{k+1}, d_k\rangle \geq 0,$$

while the black box for Algorithm 3.3.1 is "optimal": among all possible such s_{k+1}, it selects one having the *largest* possible such scalar product.

Then a natural question arises: is this reflected by a faster convergence of Algorithm 3.3.1? A complete answer is unknown but we mention the following partial result.

Proposition 3.3.4 *Suppose $0 \in \operatorname{ri} S$. Then $d_k = 0$ for some finite index k in Algorithm 3.3.1.*

PROOF. Remember that $-d_k \in S$. As always,

$$\langle d_k, s_i\rangle \leq -\|d_k\|^2 \quad \text{for all } k \text{ and } i \leq k. \tag{3.3.2}$$

Assume $d_k \neq 0$ (otherwise there is nothing to prove) and, for $\delta > 0$, define $s(\delta) := \delta d_k / \|d_k\| \in B(0, \delta)$. Because $0 \in S$, the affine and linear hulls of S coincide, and $s(\delta) \in \operatorname{aff} S$. Then, by Definition III.2.1.1 of the relative interior, our assumption implies that $s(\delta) \in S$ if δ is small enough. In this case

$$\langle d_k, s_{k+1}\rangle = \sigma_S(d_k) \geq \langle s(\delta), d_k\rangle = \delta \|d_k\|$$

and we deduce by subtraction from (3.3.2)

$$\langle d_k, s_{k+1} - s_i\rangle \geq \|d_k\|^2 + \delta\|d_k\| \geq \delta\|d_k\|.$$

From the Cauchy-Schwarz inequality,

$$\|s_{k+1} - s_i\| \geq \delta > 0 \quad \text{for all } k \text{ and } i \leq k,$$

which implies that the compact sequence $\{s_k\}$ is actually finite. □

Of course, nothing of this sort holds for Algorithm 1.6: observe in Fig. 2.1.1 that $\partial f(x)$ may contain some points with negative ordinates, without changing the sequence $\{s_k\}$. The assumption for the above result is interesting, to the extent that S is a subdifferential and that we are checking optimality of our given x: remember from the end of §VI.2.2 that, if $x = \bar{x}$ actually minimizes f, the property $0 \in \text{ri}\,\partial f(\bar{x})$ is natural.

We illustrate this by the following numerical experiments. Take a convex compact set S containing the origin and consider two forms of Algorithm 3.3.1, differing in their Step 2, i.e. in their black box.

– The first form is Algorithm 3.3.1 itself, in which Step 2 does solve (3.3.1) at each iteration.
– In the second form, Step 2 merely computes an s_{k+1} such that $\langle s_{k+1}, d_k \rangle = 0$ (note: since $0 \in S$, $\sigma_S(d) \geq 0$ for all d). This second form is therefore pretty much like Algorithm 1.6. It is even ideal for illustrating Lemma 2.1.4 with equality in (2.1.10).

Just how these two forms can be constructed will be seen in §XIII.3. Here, it suffices to say that the test-problem TR48 is used; the set S in question is a certain enlargement of the subdifferential at a non-optimal x; this enlargement contains the origin; Algorithm 3.3.1 is implemented via Algorithm 1.6, with two variants of the line-search 1.2.

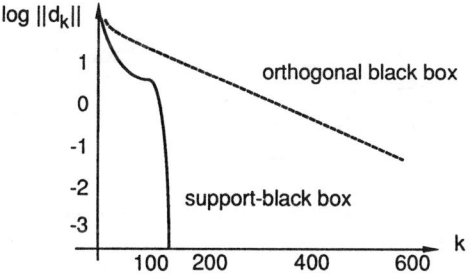

Fig. 3.3.1. Two forms of Algorithm 3.3.1 with TR48

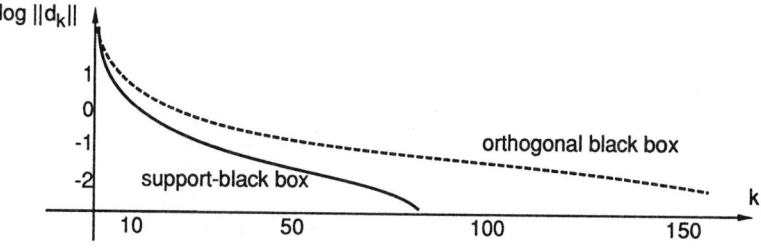

Fig. 3.3.2. Two forms of Algorithm 3.3.1 with MAXQUAD

Figure 3.3.1 shows the two corresponding speeds of convergence of $\{d_k\}$ to 0. In this test, 0 is actually an interior point of $S \subset \mathbb{R}^{48}$. This explains the very fast decrease of $\|d_k\|$ at the end of the first variant (finite convergence, cf. Remark 2.2.8). With another test, based on MAXQUAD of VIII.3.4.2, 0 is a boundary point of $S \subset \mathbb{R}^{10}$. The results are plotted in Fig. 3.3.2: $\|d_k\|$ decreases much more regularly. Finally, Figs. 3.3.3 and 3.3.4 give another illustration, using the test-problem TSP of 2.2.7. In all these pictures, note the same qualitative behaviour of $\|d_k\|$.

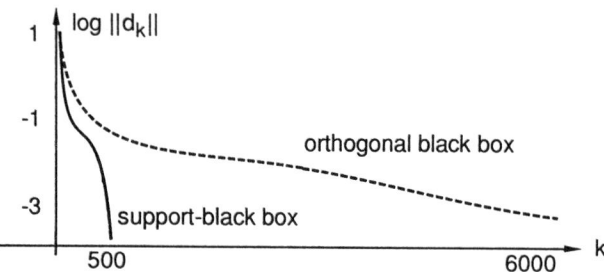

Fig. 3.3.3. Two forms of Algorithm 3.3.1 with TSP120

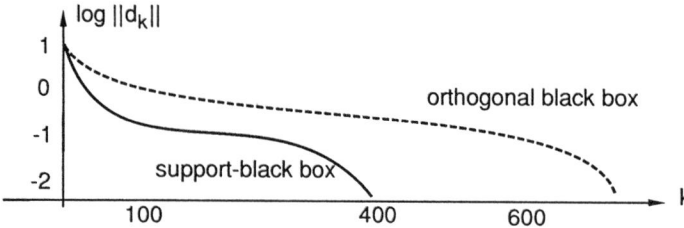

Fig. 3.3.4. Two forms of Algorithm 3.3.1 with TSP442

X. Conjugacy in Convex Analysis

Prerequisites. Definitions, properties and operations concerning convex sets (Chap. III) and convex functions (Chap. IV); definition of sublinear functions and associated convex sets (Chap. V); definitions of the subdifferential of a finite convex function (§VI.1). One-dimensional conjugacy (§I.6) can be helpful but is not necessary.

Introduction. In classical real analysis, the gradient of a differentiable function $f : \mathbb{R}^n \to \mathbb{R}$ plays a key role – to say the least. Considering this gradient as a mapping $x \mapsto s(x) = \nabla f(x)$ from (some subset X of) $\mathbb{R}^n$ to (some subset S of) $\mathbb{R}^n$, an interesting object is then its inverse: to a given $s \in S$, associate the $x \in X$ such that $s = \nabla f(x)$. This question may be meaningless: not all mappings are invertible! but could for example be considered locally, taking for $X \times S$ a neighborhood of some $(x_0, s_0 = \nabla f(x_0))$, with $\nabla^2 f$ continuous and invertible at x_0 (use the local inverse theorem).

Let us skip for the moment all such technicalities. Geometrically, we want to find x such that the hyperplane in $\mathbb{R}^n \times \mathbb{R}$ defined by the given $(s, -1)$, and passing through $(x, f(x))$, is tangent to gr f at x; the problem is meaningful when this x exists and is unique. Its construction is rather involved but analytically, an amazing fact is that the new mapping $x(\cdot) = (\nabla f)^{-1}$ thus defined is itself a *gradient* mapping: say $x(\cdot) = \nabla h$, with $h : S \subset \mathbb{R}^n \to \mathbb{R}$. Even more surprising: this function h has a simple expression, namely

$$S \ni s \mapsto h(s) = \langle s, x(s) \rangle - f(x(s)) . \tag{0.1}$$

To explain this, do a formal a posteriori calculation in (0.1): a differential ds induces the differentials dx and dh, which are linked by the relation

$$dh = \langle s, dx \rangle + \langle ds, x \rangle - \langle \nabla f(x), dx \rangle = \langle s, dx \rangle + \langle ds, x \rangle - \langle s, dx \rangle = \langle ds, x \rangle$$

and this *defines* x as the gradient of h with respect to s.

Remark 0.1 When ∇f is invertible, the function

$$s \mapsto h(s) = \langle s, (\nabla f)^{-1}(s) \rangle - f\big((\nabla f)^{-1}(s)\big)$$

is called the *Legendre transform* (relative to f). It should never be forgotten that, from the above motivation itself, it is not really the function h which is primarily interesting, but rather its gradient ∇h. □

The gradients of f and of h are inverse to each other by definition, so they establish a reciprocal correspondence:

$$s = \nabla f(x) \quad \Longleftrightarrow \quad x = \nabla h(s). \tag{0.2}$$

In particular, applying the Legendre transform to h, we have to get back f. This symmetry appears in the expression of h itself: (0.1) tells us that, for x and s related by (0.2),

$$f(x) + h(s) = \langle s, x \rangle.$$

Once again, the above statements are rather formal, insofar as they implicitly assume that $(\nabla f)^{-1}$ is well-defined. Convex analysis, however, provides a nice framework to give this last operation a precise meaning.

First of all, observe that the mapping $x \mapsto \nabla f(x)$ is now replaced by a *set-valued* mapping $x \mapsto \partial f(x)$ – see Chap. VI. To invert it is to find x such that $\partial f(x)$ contains a given s; and we can accept a nonunique such x: a set-valued $(\partial f)^{-1}$ will be obtained, but the price has been already paid anyway.

Second, the construction of $x(s)$ is now much simpler: $s \in \partial f(x)$ means that $0 \in \partial f(x) - \{s\}$ and, thanks to convexity, the last property means that x minimizes $f - \langle s, \cdot \rangle$ over $\mathbb{R}^n$. In other words, to find $x(s)$, we have to solve

$$\inf \{ f(x) - \langle s, x \rangle : x \in \mathbb{R}^n \}; \tag{0.3}$$

and the Legendre transform – in the classical sense of the term – is well-defined when this problem has a unique solution.

Let us sum up: if f is convex, (0.3) is a possible way of defining the Legendre transform when it exists unambiguously. It is easy to see that the latter holds when f satisfies three properties:

- differentiability – so that there is something to invert;
- strict convexity – to have uniqueness in (0.3);
- $\nabla f(\mathbb{R}^n) = \mathbb{R}^n$ – so that (0.3) does have a solution for all $s \in \mathbb{R}^n$; this latter property essentially means that, when $\|x\| \to \infty$, $f(x)$ increases faster than any linear function: f is *1-coercive*.

In all other cases, there is no well-defined Legendre transform; but then, the transformation implied by (0.3) can be taken as a new definition, generalizing the initial inversion of ∇f. We can even extend this definition to *nonconvex* f, namely to any function such that (0.3) is meaningful! Finally, an important observation is that the infimal value in (0.3) is a concave, and even closed, function of s; this is Proposition IV.2.1.2: the infimand is affine in s, f has little importance, $\mathbb{R}^n$ is nothing more than an index set.

The concept of conjugacy in convex analysis results from all the observations above. It is often useful to simplify algebraic calculations; it plays an important role in deriving duality schemes for convex minimization problems; it is also a basic operation for formulating variational principles in optimization (convex or not), with applications in other areas of applied mathematics, such as probability, statistics, nonlinear elasticity, economics, etc.

1 The Convex Conjugate of a Function

1.1 Definition and First Examples

As suggested by (0.3), conjugating a function f essentially amounts to minimizing a perturbation of it. There are two degenerate situations we want to avoid:

– the result is $+\infty$ for some s; observe that this is the case if and only if the result is $+\infty$ for all s;
– the result is $-\infty$ for all s.

Towards this end, we assume throughout that $f : \mathbb{R}^n \to \mathbb{R} \cup \{+\infty\}$ (not necessarily convex) satisfies

$$\boxed{f \not\equiv +\infty, \text{ and there is an affine function minorizing } f \text{ on } \mathbb{R}^n.} \quad (1.1.1)$$

Note in particular that this implies $f(x) > -\infty$ for all x. As usual, we use the notation $\operatorname{dom} f := \{x : f(x) < +\infty\} \neq \emptyset$. We know from Proposition IV.1.2.1 that (1.1.1) holds for example if $f \in \operatorname{Conv} \mathbb{R}^n$.

Definition 1.1.1 The *conjugate* of a function f satisfying (1.1.1) is the function f^* defined by

$$\mathbb{R}^n \ni s \mapsto f^*(s) := \sup\{\langle s, x\rangle - f(x) : x \in \operatorname{dom} f\}. \quad (1.1.2)$$

For simplicity, we may also let x run over the whole space instead of $\operatorname{dom} f$.

The mapping $f \mapsto f^*$ will often be called the *conjugacy* operation, or the *Legendre-Fenchel* transform. □

A very first observation is that a conjugate function is associated with a scalar product on $\mathbb{R}^n$. Of course, note also with relation to (0.3) that

$$f^*(s) = -\inf\{f(x) - \langle s, x\rangle : x \in \operatorname{dom} f\}.$$

As an immediate consequence of (1.1.2), we have for all $(x, s) \in \operatorname{dom} f \times \mathbb{R}^n$

$$f^*(s) + f(x) \geq \langle s, x\rangle. \quad (1.1.3)$$

Furthermore, this inequality is obviously true if $x \notin \operatorname{dom} f$: it does hold for all $(x, s) \in \mathbb{R}^n \times \mathbb{R}^n$ and is called *Fenchel's inequality*. Another observation is that $f^*(s) > -\infty$ for all $s \in \mathbb{R}^n$; also, if $x \mapsto \langle s_0, x\rangle + r_0$ is an affine function smaller than f, we have

$$-f^*(s_0) = \inf_x [f(x) - \langle s_0, x\rangle] \geq r_0, \quad \text{i.e.} \quad f^*(s_0) \leq -r_0 < +\infty.$$

In a word, f^* satisfies (1.1.1).

Thus, $\operatorname{dom} f^*$ – the set where f^* is finite – is the set of slopes of all the possible affine functions minorizing f over $\mathbb{R}^n$. Likewise, any $s \in \operatorname{dom} f$ is the slope of an affine function smaller than f^*.

Theorem 1.1.2 *For f satisfying (1.1.1), the conjugate f^* is a closed convex function: $f^* \in \overline{\mathrm{Conv}}\, \mathbb{R}^n$.*

PROOF. See Example IV.2.1.3. □

Example 1.1.3 (Convex Quadratic Functions) Let Q be a symmetric positive definite linear operator on $\mathbb{R}^n$, $b \in \mathbb{R}^n$ and consider

$$f(x) := \tfrac{1}{2}\langle x, Qx\rangle + \langle b, x\rangle \quad \text{for all } x \in \mathbb{R}^n. \tag{1.1.4}$$

A straightforward calculation gives the optimal $x = Q^{-1}(s - b)$ in the defining problem (1.1.2), and the resulting f^* is

$$f^*(s) = \tfrac{1}{2}\langle s - b, Q^{-1}(s - b)\rangle \quad \text{for all } s \in \mathbb{R}^n.$$

In particular, the function $1/2 \,\|\cdot\|^2$ is its own conjugate.

Needless to say, the Legendre transformation is present here, in a particularly simple setting: ∇f is the affine mapping $x \mapsto Qx + b$, its inverse is the affine mapping $s \mapsto Q^{-1}(s - b)$, the gradient of f^*. What we have done is a parametrization of $\mathbb{R}^n$ via the change of variable $s = \nabla f(x)$; with respect to the new variable, f is given by

$$f(x(s)) = \tfrac{1}{2}\langle Q^{-1}(s - b), s - b\rangle + \langle b, Q^{-1}(s - b)\rangle.$$

Adding $f^*(s)$ on both sides gives

$$f(x(s)) + f^*(s) = \langle Q^{-1}(s - b), s\rangle = \langle x(s), s\rangle,$$

which illustrates (0.1).

Taking $b = 0$ and applying Fenchel's inequality (1.1.3) gives

$$\langle x, Qx\rangle + \langle s, Q^{-1}s\rangle \geqslant 2\langle s, x\rangle \quad \text{for all } (x, s),$$

a generalization of the well-known inequality (obtained for $Q = cI$, $c > 0$):

$$c\|x\|^2 + \tfrac{1}{c}\|s\|^2 \geqslant 2\langle s, x\rangle \quad \text{for all } c > 0 \text{ and } x, s \in \mathbb{R}^n. \quad \square$$

When Q in the above example is merely positive semi-definite, a meaning can still be given to $(\nabla f)^{-1}$ provided that two problems are taken care of: first, s must be restricted to $\nabla f(\mathbb{R}^n)$, which is the affine manifold $b + \mathrm{Im}\, Q$; second, $\nabla f(x + y) = \nabla f(x)$ for all $y \in \mathrm{Ker}\, Q$, so $(\nabla f)^{-1}(s)$ can be defined only up to the subspace $\mathrm{Ker}\, Q$. Using the definition of the conjugate function, we obtain the following:

Example 1.1.4 (Convex Degenerate Quadratic Functions) Take the convex quadratic function (1.1.4), but with Q symmetric positive semi-definite. The supremum in (1.1.2) is finite only if $s - b \in (\mathrm{Ker}\, Q)^\perp$, i.e. $s - b \in \mathrm{Im}\, Q$; it is attained at an x such that $s - \nabla f(x) = 0$ (optimality condition VI.2.2.1). In a word, we obtain

$$f^*(s) = \begin{cases} +\infty & \text{if } s \notin b + \mathrm{Im}\, Q, \\ \tfrac{1}{2}\langle x, s - b\rangle & \text{otherwise}, \end{cases}$$

where x is any element satisfying $Qx + b = s$. This formulation can be condensed to

$$f^*(Qx + b) = \tfrac{1}{2}\langle x, Qx\rangle \quad \text{for all } x \in \mathbb{R}^n,$$

which is one more illustration of (0.1): add $f(x)$ to both sides and obtain

$$f(x) + f^*(Qx + b) = \langle x, Qx + b\rangle = \langle x, \nabla f(x)\rangle.$$

It is also interesting to express f^* in terms of a *pseudo-inverse*: for example, Q^- denoting the Moore-Penrose pseudo-inverse of Q (see §A.3.4), f^* can be written

$$f^*(s) = \begin{cases} +\infty & \text{if } s \notin b + \operatorname{Im} Q, \\ \tfrac{1}{2}\langle s - b, Q^-(s - b)\rangle & \text{if } s \in b + \operatorname{Im} Q. \end{cases} \qquad \square$$

In the above example, take $b = 0$ and let $Q = p_H$ be the orthogonal projection onto a subspace H of $\mathbb{R}^n$. Then $\operatorname{Im} Q = H$ and $Q^- = Q$, so

$$(p_H)^*(s) = \begin{cases} +\infty & \text{if } s \notin H, \\ \tfrac{1}{2}\|s\|^2 & \text{if } s \in H. \end{cases}$$

Another interesting example is when Q is a rank-one operator, i.e. $Q = uu^\top$, with $0 \neq u \in \mathbb{R}^n$ (we assume the usual dot-product for $\langle \cdot, \cdot\rangle$). Then $\operatorname{Im} Q = \mathbb{R}u$ and, for $x \in \operatorname{Im} Q$, $Qx = \|u\|^2 x$. Therefore,

$$\left(uu^\top\right)^*(s) = \begin{cases} \frac{1}{2\|u\|^2} s & \text{if } s \text{ and } u \text{ are collinear}, \\ +\infty & \text{otherwise}. \end{cases}$$

Example 1.1.5 Let I_C be the indicator function of a nonempty set $C \subset \mathbb{R}^n$. Then

$$(\mathrm{I}_C)^*(s) = \sup_{x \in \operatorname{dom} \mathrm{I}_C} [\langle s, x\rangle - \mathrm{I}_C(x)] = \sup_{x \in C} \langle s, x\rangle$$

is just the support function of C. If C is a closed convex cone, we conclude from Example V.2.3.1 that $(\mathrm{I}_C)^*$ is the indicator of its polar C°. If C is a subspace, $(\mathrm{I}_C)^*$ is the indicator of its orthogonal $C^\perp$.

If $C = \mathbb{R}^n$, $C^\circ = \{0\}$; $\mathrm{I}_C \equiv 0$ and the conjugate of I_C is 0 at 0, $+\infty$ elsewhere (this is the indicator of $\{0\}$, or the support of $\mathbb{R}^n$). A similar example is the nonconvex function $x \mapsto f(x) = \|x\|^{1/2}$, where $\|\cdot\|$ is some norm. Then $f^*(0) = 0$; but if $s \neq 0$, take x of the form $x = ts$ to realize that

$$f^*(s) \geq \sup_{t \geq 0}\left[t\|s\|^2 - \sqrt{t}\,\|s\|\right] = +\infty.$$

In other words, f^* is still the indicator function of $\{0\}$. The conjugacy operation has ignored the difference between f and the zero function, simply because f increases at infinity more slowly than any linear function. $\qquad \square$

1.2 Interpretations

Geometrically, the computation of f^* can be illustrated in the graph-space $\mathbb{R}^n \times \mathbb{R}$. For given $s \in \mathbb{R}^n$, consider the family of affine functions $x \mapsto \langle s, x \rangle - r$, parametrized by $r \in \mathbb{R}$. They correspond to affine hyperplanes $H_{s,r}$ orthogonal to $(s, -1) \in \mathbb{R}^{n+1}$; see Fig. 1.2.1. From (1.1.1), $H_{s,r}$ is below gr f whenever (s, r) is properly chosen, namely $s \in \text{dom } f^*$ and r is large enough. To construct $f^*(s)$, we lift $H_{s,r}$ as much as possible subject to supporting gr f. Then, admitting that there is contact at some $(x, f(x))$, we write

$$\langle s, x \rangle - r = f(x) \quad \text{or rather} \quad r = \langle s, x \rangle - f(x),$$

to see that $r = f^*(s)$. This means that the best $H_{s,r}$ intersects the vertical axis $\{0\} \times \mathbb{R}$ at the altitude $-f^*(s)$.

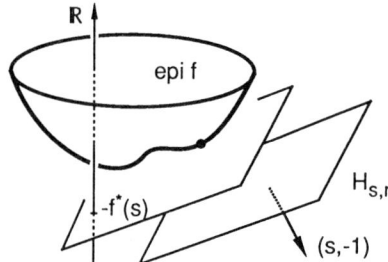

Fig. 1.2.1. Computation of f^* in the graph-space

Naturally, the horizontal hyperplanes $H_{0,r}$ correspond to minimizing f:

$$-f^*(0) = \inf \{f(x) : x \in \mathbb{R}^n\}.$$

The picture illustrates another definition of f^*: being normal to $H_{s,r}$, the vector $(s, -1)$ is also normal to gr f (more precisely to epi f) at the contact when it exists.

Proposition 1.2.1 *There holds for all $x \in \mathbb{R}^n$*

$$f^*(s) = \sigma_{\text{epi } f}(s, -1) = \sup\{\langle s, x \rangle - r : (x, r) \in \text{epi } f\}. \tag{1.2.1}$$

It follows that the support function of epi f has the expression

$$\sigma_{\text{epi } f}(s, -u) = \begin{cases} uf^*(\frac{1}{u}s) & \text{if } u > 0, \\ \sigma_{\text{epi } f}(s, 0) = \sigma_{\text{dom } f}(s) & \text{if } u = 0, \\ +\infty & \text{if } u < 0. \end{cases} \tag{1.2.2}$$

PROOF. In (1.2.1), the right-most term can be written

$$\sup_{x} \sup_{r \geq f(x)} [\langle s, x \rangle - r] = \sup_{x} [\langle s, x \rangle - f(x)]$$

and the first equality is established. As for (1.2.2), the case $u < 0$ is trivial; when $u > 0$, use the positive homogeneity of support functions to get

$$\sigma_{\text{epi } f}(s, -u) = u\sigma_{\text{epi } f}\left(\tfrac{1}{u}s, -1\right) = uf^*\left(\tfrac{1}{u}s\right) ;$$

finally, for $u = 0$, we have by definition

$$\sigma_{\text{epi } f}(s, 0) = \sup\{\langle s, x\rangle : (x, r) \in \text{epi } f \text{ for some } r \in \mathbb{R}\},$$

and we recognize $\sigma_{\text{dom } f}(s)$. □

Assume $f \in \overline{\text{Conv}}\,\mathbb{R}^n$. This result, illustrated in Fig. 1.2.1, confirms that the contact-set between the optimal hyperplane and epi f is the face of epi f exposed by the given $(s, -1)$. From (1.2.2), we see that $\sigma_{\text{epi } f}$ and the perspective-function of f^* (§IV.2.2) coincide for $u \neq 0$ – up to the change of variable $u \mapsto -u$. As a closed function, $\sigma_{\text{epi } f}$ therefore coincides (still up to the change of sign in u) with the closure of the perspective of f^*. As for $u = 0$, we obtain a relation which will be used on several occasions:

Proposition 1.2.2 *For $f \in \overline{\text{Conv}}\,\mathbb{R}^n$,*

$$\sigma_{\text{dom } f}(s) = \sigma_{\text{epi } f}(s, 0) = (f^*)'_\infty(s) \quad \text{for all } s \in \mathbb{R}^n. \tag{1.2.3}$$

PROOF. Use direct calculations; or see Proposition IV.2.2.2 and the calculations in Example IV.3.2.4. □

The additional variable u introduced in Proposition 1.2.1 gives a second geometric construction: Fig. 1.2.1 plays the role of Fig. V.2.1.1, knowing that $\mathbb{R}^n$ is replaced here by $\mathbb{R}^{n+1}$. Here again, note in passing that the closed convex hull of epi f gives the same f^*. Now, playing the same game as we did for Interpretation V.2.1.6, f^* can be looked at from the point of view of projective geometry, $\mathbb{R}^n$ being identified with $\mathbb{R}^n \times \{-1\}$.

Consider the set epi $f \times \{-1\} \subset \mathbb{R}^n \times \mathbb{R} \times \mathbb{R}$, i.e. the copy of epi f, translated down vertically by one unit. It generates a cone $K_f \subset \mathbb{R}^n \times \mathbb{R} \times \mathbb{R}$:

$$K_f := \{t(x, r, -1) : t > 0, \ (x, r) \in \text{epi } f\}. \tag{1.2.4}$$

Now take the polar cone $(K_f)^\circ \subset \mathbb{R}^n \times \mathbb{R} \times \mathbb{R}$. We know from Interpretation V.2.1.6 that it is the epigraph (in $\mathbb{R}^{n+1} \times \mathbb{R}!$) of the support function of epi f. In view of Proposition 1.2.1, its intersection with $\mathbb{R}^n \times \{-1\} \times \mathbb{R}$ is therefore epi f^*. A short calculation confirms this:

$$\begin{aligned}(K_f)^\circ &= \{(s, \alpha, \beta) \in \mathbb{R}^n \times \mathbb{R} \times \mathbb{R} : t\langle s, x\rangle + t\alpha r - t\beta \leq 0 \\ &\qquad \text{for all } (x, r) \in \text{epi } f \text{ and } t > 0\} \\ &= \{(s, \alpha, \beta) : \langle s, x\rangle + \alpha r \leq \beta \text{ for all } (x, r) \in \text{epi } f\}.\end{aligned}$$

Imposing $\alpha = -1$, we just obtain the translated epigraph of the function described in (1.2.1).

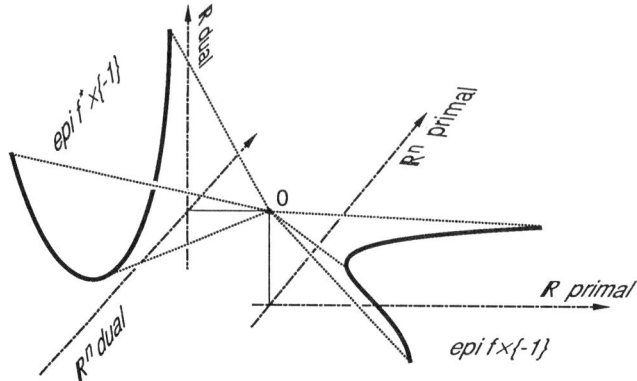

Fig. 1.2.2. A projective view of conjugacy

Figure 1.2.2 illustrates the construction, with $n = 1$ and epi f drawn horizontally; this epi f plays the role of S in Fig. V.2.1.2. Note once more the symmetry: K_f [resp. $(K_f)°$] is defined in such a way that its intersection with the hyperplane $\mathbb{R}^n \times \mathbb{R} \times \{-1\}$ [resp. $\mathbb{R}^n \times \{-1\} \times \mathbb{R}$] is just epi f [resp. epi f^*].

Finally, we mention a simple economic interpretation. Suppose $\mathbb{R}^n$ is a set of goods and its dual $(\mathbb{R}^n)^*$ a set of prices: to produce the goods x costs $f(x)$, and to sell x brings an income $\langle s, x \rangle$. The net benefit associated with x is then $\langle s, x \rangle - f(x)$, whose supremal value $f^*(s)$ is the best possible profit, resulting from the given set of unit prices s. Incidentally, this last interpretation opens the way to *nonlinear* conjugacy, in which the selling price would be a nonlinear (but concave) function of x.

Remark 1.2.3 This last interpretation confirms the warning already made on several occasions: $\mathbb{R}^n$ should not be confused with its dual; the arguments of f and of f^* are not comparable, an expression like $x + s$ is meaningless (until an isomorphism is established between the Euclidean space $\mathbb{R}^n$ and its dual).

On the other hand, f-values and f^*-values are comparable, indeed: for example, they can be added to each other – which is just done by Fenchel's inequality (1.1.3)! This is explicitly due to the particular value "-1" in (1.2.1), which goes together with the "-1" of (1.2.4). □

1.3 First Properties

Some properties of the conjugacy operation $f \mapsto f^*$ come directly from its definition.

(a) Elementary Calculus Rules Direct arguments prove easily a first result:

Proposition 1.3.1 *The functions f, f_j appearing below are assumed to satisfy* (1.1.1).

(i) *The conjugate of the function $g(x) := f(x) + \alpha$ is $g^*(s) = f^*(s) - \alpha$.*

(ii) *With $\alpha > 0$, the conjugate of the function $g(x) := \alpha f(x)$ is $g^*(s) = \alpha f^*(s/\alpha)$.*
(iii) *With $\alpha \neq 0$, the conjugate of the function $g(x) := f(\alpha x)$ is $g^*(s) = f^*(s/\alpha)$.*
(iv) *More generally: if A is an invertible linear operator, $(f \circ A)^* = f^* \circ (A^{-1})^*$.*
(v) *The conjugate of the function $g(x) := f(x - x_0)$ is $g^*(s) = f^*(s) + \langle s, x_0 \rangle$.*
(vi) *The conjugate of the function $g(x) := f(x) + \langle s_0, x \rangle$ is $g^*(s) = f^*(s - s_0)$.*
(vii) *If $f_1 \leq f_2$, then $f_1^* \geq f_2^*$.*
(viii) *"Convexity" of the conjugation: if $\mathrm{dom}\, f_1 \cap \mathrm{dom}\, f_2 \neq \emptyset$ and $\alpha \in\,]0, 1[$,*

$$[\alpha f_1 + (1-\alpha) f_2]^* \leq \alpha f_1^* + (1-\alpha) f_2^* \, ;$$

(ix) *The Legendre-Fenchel transform preserves decomposition: with*

$$\mathbb{R}^n := \mathbb{R}^{n_1} \times \cdots \times \mathbb{R}^{n_m} \ni x \mapsto f(x) := \sum_{j=1}^{m} f_j(x_j)$$

and assuming that $\mathbb{R}^n$ has the scalar product of a product-space,

$$f^*(s_1, \ldots, s_m) = \sum_{j=1}^{m} f_j^*(s_j) \, . \qquad \square$$

Among these results, (iv) deserves comment: it gives the effect of a change of variables on the conjugate function; this is of interest for example when the scalar product is put in the form $\langle x, y \rangle = (Ax)^\top Ay$, with A invertible. Using Example 1.1.3, an illustration of (vii) is: the only function f satisfying $f = f^*$ is $1/2\, \|\cdot\|^2$ (start from Fenchel's inequality); and also, for symmetric positive definite Q and P:

$$Q \leq P \quad \Longrightarrow \quad P^{-1} \leq Q^{-1} \, ;$$

and an illustration of (viii) is:

$$[\alpha Q + (1-\alpha) P]^{-1} \leq \alpha Q^{-1} + (1-\alpha) P^{-1} \, .$$

Our next result expresses how the conjugate is transformed, when the starting function is restricted to a subspace.

Proposition 1.3.2 *Let f satisfy (1.1.1), let H be a subspace of $\mathbb{R}^n$, and call p_H the operator of orthogonal projection onto H. Suppose that there is a point in H where f is finite. Then $f + I_H$ satisfies (1.1.1) and its conjugate is*

$$(f + I_H)^* = (f \circ p_H)^* \circ p_H \, . \tag{1.3.1}$$

PROOF. When y describes $\mathbb{R}^n$, $p_H y$ describes H so we can write, knowing that p_H is symmetric:

$$\begin{aligned}
(f + I_H)^*(s) &:= \sup\{\langle s, x \rangle - f(x) : x \in H\} \\
&= \sup\{\langle s, p_H y \rangle - f(p_H y) : y \in \mathbb{R}^n\} \\
&= \sup\{\langle p_H s, y \rangle - f(p_H y) : y \in \mathbb{R}^n\} \, . \qquad \square
\end{aligned}$$

When conjugating $f + I_H$, we disregard f outside H, i.e. we consider f as a function defined on the space H; this is reflected in the "$f \circ p_H$"-part of (1.3.1). We then obtain a "partial" conjugate, say $\hat{f}^* \in \overline{\text{Conv}}\, H$; the last "$\circ p_H$" of (1.3.1) says that, to recover the whole conjugate $(f + I_H)^*$, which is a function of $\overline{\text{Conv}}\, \mathbb{R}^n$, we just have to translate horizontally the graph of $\hat{f}^*$ along $H^\perp$.

Remark 1.3.3 Thus, if a subspace H and a function g (standing for $f + I_H$) are such that $\text{dom}\, g \subset H$, then $g^*(\cdot + s) = g^*$ for all $s \in H^\perp$. It is interesting to note that this property has a converse: if, for some $x_0 \in \text{dom}\, g$, we have $g(x_0 + y) = g(x_0)$ for all $y \in H$, then $\text{dom}\, g^* \subset H^\perp$. The proof is immediate: take x_0 as above and, for $s \notin H^\perp$, take $y_0 \in H$ such that $\langle s, y_0 \rangle = \alpha \neq 0$; there holds $\langle s, \lambda y_0 \rangle - g(x_0 + \lambda y_0) = \lambda \alpha - g(x_0)$ for all $\lambda \in \mathbb{R}$; hence

$$\begin{aligned} g^*(s) &\geq \sup_\lambda [\langle s, x_0 + \lambda y_0 \rangle - g(x_0 + \lambda y_0)] \\ &= \langle s, x_0 \rangle + \sup_\lambda [\lambda \alpha - g(x_0)] = +\infty. \end{aligned}$$
□

The above formulae can be considered from a somewhat opposite point of view: suppose that $\mathbb{R}^n$, the space on which our function f is defined, is embedded in some larger space $\mathbb{R}^{n+p}$. Various corresponding extensions of f to the whole of $\mathbb{R}^{n+p}$ are then possible. One is to set $f := +\infty$ outside $\mathbb{R}^n$ (which is often relevant when minimizing f), the other is the horizontal translation

$$f(x + y) = f(x) \quad \text{for all } y \in \mathbb{R}^p;$$

these two possibilities are, in a sense, dual to each other (see Proposition 1.3.4 below). Such a duality, however, holds only if the extended scalar product preserves the structure of $\mathbb{R}^n \times \mathbb{R}^p$ as a product-space: so is the case of the decomposition $H + H^\perp$ appearing in Proposition 1.3.2, because

$$\langle s, x \rangle = \langle p_H s, p_H x \rangle + \langle p_{H^\perp} s, p_{H^\perp} x \rangle.$$

Considering affine manifolds instead of subspaces, we mention the following useful result:

Proposition 1.3.4 *For f satisfying (1.1.1), let a subspace V contain the subspace parallel to $\text{aff dom}\, f$ and set $U := V^\perp$. For any $z \in \text{aff dom}\, f$ and any $s \in \mathbb{R}^n$ decomposed as $s = s_U + s_V$, there holds*

$$f^*(s) = \langle s_U, z \rangle + f^*(s_V).$$

PROOF. In (1.1.2), the variable x can range through $z + V \supset \text{aff dom}\, f$:

$$\begin{aligned} f^*(s) &= \sup_{v \in V}[\langle s_U + s_V, z + v \rangle - f(z + v)] \\ &= \langle s_U, z \rangle + \sup_{v \in V}[\langle s_V, z + v \rangle - f(z + v)] \\ &= \langle s_U, z \rangle + f^*(s_V). \end{aligned}$$
□

(b) The Biconjugate of a Function What happens if we take the conjugate of f^* again? Remember that f^* satisfies automatically (1.1.1) if f does. We can therefore compute the *biconjugate* function of f: for all $x \in \mathbb{R}^n$,

$$f^{**}(x) := (f^*)^*(x) = \sup\{\langle s, x \rangle - f^*(s) \,:\, s \in \mathbb{R}^n\}. \tag{1.3.2}$$

This operation appears as fundamental. The function f^{**} thus defined is the "close-convexification" of f, in the sense that its epigraph is the closed convex hull of epi f:

Theorem 1.3.5 *For f satisfying (1.1.1), the function f^{**} of (1.3.2) is the pointwise supremum of all the affine functions on $\mathbb{R}^n$ majorized by f. In other words*

$$\operatorname{epi} f^{**} = \overline{\operatorname{co}} (\operatorname{epi} f). \tag{1.3.3}$$

PROOF. Call $\Sigma \subset \mathbb{R}^n \times \mathbb{R}$ the set of pairs (s, r) defining affine functions $x \mapsto \langle s, x \rangle - r$ majorized by f:

$$\begin{aligned}
(s, r) \in \Sigma &\iff f(x) \geq \langle s, x \rangle - r \quad \text{for all } x \in \mathbb{R}^n \\
&\iff r \geq \sup \{\langle s, x \rangle - f(x) : x \in \mathbb{R}^n\} \\
&\iff r \geq f^*(s) \quad (\text{and } s \in \operatorname{dom} f^*!).
\end{aligned}$$

Then we obtain, for $x \in \mathbb{R}^n$,

$$\begin{aligned}
\sup_{(s,r) \in \Sigma}[\langle s, x \rangle - r] &= \sup\{\langle s, x \rangle - r : s \in \operatorname{dom} f^*, -r \leq -f^*(s)\} \\
&= \sup\{\langle s, x \rangle - f^*(s) : s \in \operatorname{dom} f^*\} = f^{**}(x).
\end{aligned}$$

Geometrically, the epigraphs of the affine functions associated with $(s, r) \in \Sigma$ are the (non-vertical) closed half-spaces containing $\operatorname{epi} f$. From §IV.2.5, the epigraph of their supremum is the closed convex hull of $\operatorname{epi} f$, and this proves (1.3.3). □

Note: the biconjugate of an $f \in \operatorname{Conv} \mathbb{R}^n$ is not exactly f itself but its *closure*: $f^{**} = \operatorname{cl} f$. Thanks to Theorem 1.3.5, the general notation

$$\overline{\operatorname{co}} f := f^{**} \tag{1.3.4}$$

can be – and will be – used for a function simply satisfying (1.1.1); it reminds one more directly that f^{**} is the closed convex hull of f:

$$f^{**}(x) = \sup_{r,s} \left\{ \langle s, x \rangle - r : \langle s, y \rangle - r \leq f(y) \text{ for all } y \in \mathbb{R}^n \right\}. \tag{1.3.5}$$

Corollary 1.3.6 *If g is a function satisfying $\overline{\operatorname{co}} f \leq g \leq f$, then $g^* = f^*$. The function f is equal to its biconjugate f^{**} if and only if $f \in \overline{\operatorname{Conv}} \mathbb{R}^n$.*

PROOF. Immediate. □

Thus, the conjugacy operation defines an *involution* on the set of closed convex functions. When applied to strictly convex quadratic functions, it corresponds to the inversion of symmetric positive definite operators (Example 1.1.3 with $b = 0$). When applied to indicators of closed convex cones, it corresponds to the polarity correspondence (Example 1.1.5 – note also in this example that the biconjugate of the square root of a norm is the zero-function). For general $f \in \overline{\operatorname{Conv}} \mathbb{R}^n$, it has a geometric counterpart, also based on polarity, which is the correspondence illustrated in Fig. 1.3.1. Of course, this involution property implies a lot of symmetry, already alluded to, for pairs of conjugate functions.

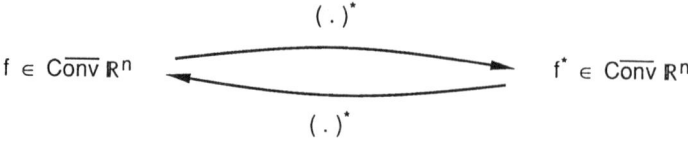

Fig. 1.3.1. The *-involution

(c) Conjugacy and Coercivity A basic question in (1.1.2) is whether the supremum is going to be $+\infty$. This depends only on the behaviour of f at infinity, so we extend to non-convex situations the concepts seen in Definition IV.3.2.6:

Definition 1.3.7 A function f satisfying (1.1.1) is said to be *0-coercive* [resp. *1-coercive*] when

$$\lim_{\|x\|\to+\infty} f(x) = +\infty \quad \left[\text{resp.}\quad \lim_{\|x\|\to+\infty} \frac{f(x)}{\|x\|} = +\infty\right]. \qquad \square$$

Proposition 1.3.8 *If f satisfying (1.1.1) is 1-coercive, then $f^*(s) < +\infty$ for all $s \in \mathbb{R}^n$.*

PROOF. For given s, the 1-coercivity of f implies the existence of a number R such that

$$\|x\| \geqslant R \implies f(x) \geqslant \|x\|(\|s\| + 1),$$

so that we have in (1.1.2)

$$\langle s, x \rangle - f(x) \leqslant -\|x\| \quad \text{for all } x \text{ such that } \|x\| \geqslant R,$$

hence

$$\sup\{\langle s, x \rangle - f(x) : \|x\| \geqslant R\} \leqslant -R.$$

On the other hand, (1.1.1) implies an upper bound

$$\sup\{\langle s, x \rangle - f(x) : \|x\| \leqslant R\} \leqslant M. \qquad \square$$

For a converse to this property, we have the following:

Proposition 1.3.9 *Let f satisfy (1.1.1). Then*
(i) $x_0 \in \mathrm{int\,dom}\, f \implies f^* - \langle x_0, \cdot \rangle$ *is 0-coercive;*
(ii) *in particular, if f is finite over $\mathbb{R}^n$, then f^* is 1-coercive.*

PROOF. We know from (1.2.3) that $\sigma_{\mathrm{dom}\, f} = (f^*)'_\infty$ so, using Theorem V.2.2.3(iii), $x_0 \in \mathrm{int\,dom}\, f \subset \mathrm{int}(\mathrm{co\,dom}\, f)$ implies

$$(f^*)'_\infty(s) - \langle x_0, s \rangle > 0 \quad \text{for all } s \neq 0.$$

By virtue of Proposition IV.3.2.5, this means exactly that $f^* - \langle x_0, \cdot \rangle$ has compact sublevel-sets; (i) is proved.

Then, as demonstrated in Definition IV.3.2.6, 0-coercivity of $f^* - \langle x_0, \cdot \rangle$ for all x_0 means 1-coercivity of f^*. $\qquad \square$

Piecing together, we see that the 1-coercivity of a function f implies that f^* is finite everywhere, and this in turn implies 1-coercivity of $\overline{\mathrm{co}}\, f$.

1 The Convex Conjugate of a Function

Remark 1.3.10 If we assume in particular $f \in \overline{\text{Conv}}\,\mathbb{R}^n$, (i) and (ii) become equivalences:

$$x_0 \in \text{int dom } f \iff f^* - \langle x_0, \cdot \rangle \text{ is 0-coercive},$$

$$\text{dom } f = \mathbb{R}^n \iff f^* \text{ is 1-coercive}. \qquad \square$$

1.4 Subdifferentials of Extended-Valued Functions

For a function f satisfying (1.1.1), consider the following set:

$$\partial f(x) := \{ s \in \mathbb{R}^n : f(y) \geq f(x) + \langle s, y - x \rangle \text{ for all } y \in \mathbb{R}^n \}. \tag{1.4.1}$$

When f happens to be convex and finite-valued, this is just the subdifferential of f at x, defined in VI.1.2.1; but (1.4.1) can be used in a much more general framework. We therefore keep the terminology *subdifferential* for the set of (1.4.1), and *subgradients* for its elements. Note here that $\partial f(x)$ is empty if $x \notin \text{dom } f$: take $y \in \text{dom } f$ in (1.4.1).

Theorem 1.4.1 *For f satisfying (1.1.1) and ∂f defined by (1.4.1), $s \in \partial f(x)$ if and only if*

$$f^*(s) + f(x) - \langle s, x \rangle = 0 \quad (\text{or } \leq 0). \tag{1.4.2}$$

PROOF. To say that s is in the set (1.4.1) is to say that

$$\langle s, y \rangle - f(y) \leq \langle s, x \rangle - f(x) \quad \text{for all } y \in \text{dom } f,$$

i.e.

$$f^*(s) \leq \langle s, x \rangle - f(x);$$

but this is indeed an equality, in view of Fenchel's inequality (1.1.3). $\qquad \square$

As before, $\partial f(x)$ is closed and convex: it is a sublevel-set of the closed convex function $f^* - \langle \cdot, x \rangle$, namely at the level $-f(x)$. A subgradient of f at x is the slope of an affine function minorizing f and coinciding with f at x; $\partial f(x)$ can therefore be empty: for example if epi f has a vertical tangent hyperplane at $(x, f(x))$; or also if f is not convex-like near x.

Theorem 1.4.2 *Let $f \in \text{Conv }\mathbb{R}^n$. Then $\partial f(x) \neq \emptyset$ whenever $x \in \text{ri dom } f$.*

PROOF. This is Proposition IV.1.2.1. $\qquad \square$

When $\partial f(x) \neq \emptyset$, we obtain a particular relationship at x between f and its convexified version $\overline{\text{co}}\, f$:

Proposition 1.4.3 *For f satisfying (1.1.1), the following properties hold:*

$$\partial f(x) \neq \emptyset \implies (\overline{\text{co}}\, f)(x) = f(x); \tag{1.4.3}$$

$$\overline{\text{co}}\, f \leq g \leq f \text{ and } g(x) = f(x) \implies \partial g(x) = \partial f(x); \tag{1.4.4}$$

$$s \in \partial f(x) \implies x \in \partial f^*(s). \tag{1.4.5}$$

PROOF. Let s be a subgradient of f at x. From the definition (1.4.1) itself, the function $y \mapsto \ell_s(y) := f(x) + \langle s, y - x \rangle$ is affine and minorizes f, hence $\ell_s \leqslant \overline{\text{co}}\, f \leqslant f$; because $\ell_s(x) = f(x)$, this implies (1.4.3).

Now, $s \in \partial f(x)$ if and only if

$$f^*(s) + f(x) - \langle s, x \rangle = 0.$$

From our assumption, $f^* = g^* = (\overline{\text{co}}\, f)^*$ (see Corollary 1.3.6) and $g(x) = f(x)$; the above equality can therefore be written

$$g^*(s) + g(x) - \langle s, x \rangle = 0$$

which expresses exactly that $s \in \partial g(x)$, and (1.4.4) is proved.

Finally, we know that $f^{**} = \overline{\text{co}}\, f \leqslant f$; so, when s satisfies (1.4.2), we have

$$f^*(s) + f^{**}(x) - \langle s, x \rangle = f^*(s) + (\overline{\text{co}}\, f)(x) - \langle s, x \rangle \leqslant 0,$$

which means $x \in \partial f^*(s)$: we have just proved (1.4.5). □

Among the consequences of (1.4.3), we note the following sufficiency condition for convexity: if $\partial f(x)$ is nonempty for all $x \in \mathbb{R}^n$, then f is convex and finite-valued on $\mathbb{R}^n$. Another consequence is important:

Corollary 1.4.4 *If $f \in \overline{\text{Conv}}\, \mathbb{R}^n$, the following equivalences hold:*

$$f(x) + f^*(s) - \langle s, x \rangle = 0 \ (or \leqslant 0) \quad \Longleftrightarrow \quad s \in \partial f(x) \quad \Longleftrightarrow \quad x \in \partial f^*(s).$$

PROOF. This is a rewriting of Theorem 1.4.1, taking into account (1.4.5) and the symmetric role played by f and f^* when $f \in \overline{\text{Conv}}\, \mathbb{R}^n$. □

If $s \in \partial f(x)$ (which in particular implies $x \in \text{dom}\, f$), the property $s \in \text{dom}\, f^*$ comes immediately; beware that, conversely, $\text{dom}\, f^*$ is not entirely covered by $\partial f(\mathbb{R}^n)$: take $f(x) = \exp x$, for which $f^*(0) = 0$; but 0 is only in the closure of $f'(\mathbb{R})$.

Even though it is attached to some designated x, the concept of subdifferential, as defined by (1.4.1), is *global*, in the sense that it uses the values of f on the whole of $\mathbb{R}^n$. For example,

$$\inf \{f(x) : x \in \mathbb{R}^n\} = -f^*(0)$$

and obviously

$$x \text{ minimizes } f \text{ satisfying (1.1.1)} \quad \Longleftrightarrow \quad 0 \in \partial f(x).$$

Then a consequence of Corollary 1.4.4 is:

$$\underset{x \in \mathbb{R}^n}{\text{Argmin}}\, f(x) = \partial f^*(0) \quad \text{if} \quad f \in \overline{\text{Conv}}\, \mathbb{R}^n. \tag{1.4.6}$$

1.5 Convexification and Subdifferentiability

For a function f satisfying (1.1.1), the biconjugate defined in §1.3(b) is important in optimization. First of all, minimizing f or $\overline{\text{co}} f$ is almost the same problem:

– As a consequence of Corollary 1.3.6, we have the equality in $\mathbb{R} \cup \{-\infty\}$

$$\inf \{f(x) : x \in \mathbb{R}^n\} = \inf \{(\overline{\text{co}} f)(x) : x \in \mathbb{R}^n\}.$$

– We also have from Proposition 1.4.3 that the minimizers of f minimize $\overline{\text{co}} f$ as well; and since the latter form a closed convex set,

$$\overline{\text{co}}(\text{Argmin } f) \subset \text{Argmin}(\overline{\text{co}} f). \qquad (1.5.1)$$

This inclusion is strict: think of $f(x) = |x|^{1/2}$. Equality holds under appropriate assumptions, as can be seen from our analysis below (see Remark 1.5.7).

The next results relate the smoothness of f and of $\overline{\text{co}} f$.

Proposition 1.5.1 *Suppose f satisfying (1.1.1) is Gâteaux differentiable at x, and has at x a nonempty subdifferential in the sense of (1.4.1). Then*

$$\partial f(x) = \{\nabla f(x)\} = \partial(\overline{\text{co}} f)(x).$$

PROOF. Let $s \in \partial f(x)$: for all $d \in \mathbb{R}^n$ and $t > 0$,

$$\frac{f(x + td) - f(x)}{t} \geqslant \langle s, d \rangle.$$

Let $t \downarrow 0$ to obtain $\langle \nabla f(x), d \rangle \geqslant \langle s, d \rangle$ for all d, and this implies $\nabla f(x) = s$. The second equality then follows using (1.4.3) and (1.4.4). □

For a given x to minimize a differentiable f, a necessary condition is "$\nabla f(x) = 0$". A natural question is then: what additional condition ensures that a stationary point is a global minimum of f? Clearly, the missing condition has to be global: it involves the behaviour of f on the whole space. The conjugacy correspondence turns out to be an appropriate tool for obtaining such a condition.

Corollary 1.5.2 *Let f be Gâteaux differentiable on $\mathbb{R}^n$. Then, x is a global minimum of f on $\mathbb{R}^n$ if and only if*

(i) $\nabla f(x) = 0$ *and*
(ii) $(\overline{\text{co}} f)(x) = f(x)$.

In such a case, $\overline{\text{co}} f$ is differentiable at x and $\nabla(\overline{\text{co}} f)(x) = 0$.

PROOF. Let x minimize f; then $\partial f(x)$ is nonempty, hence (ii) follows from (1.4.3); furthermore, $\partial(\overline{\text{co}} f)(x)$ reduces to $\{0\}$ (Proposition 1.5.1).

Conversely, let x satisfy (i), (ii); because the differentiable function f is finite everywhere, the convex function $\overline{\text{co}} f$ is such. Then, by virtue of Proposition 1.5.1, $0 \in \partial(\overline{\text{co}} f)(x)$; we therefore obtain immediately that x minimizes f on $\mathbb{R}^n$. □

Thus, the global property (ii) is just what is missing in the local property (i) for a stationary point to be a global minimum. One concludes for example that differentiable functions whose stationary points are global minima are those f for which

$$\nabla f(x) = 0 \implies (\overline{\text{co }} f)(x) = f(x).$$

We turn now to a characterization of the subdifferential of $\overline{\text{co }} f$ in terms of that of f, which needs the following crucial assumption:

f satisfies (1.1.1), and is lower semi-continuous and 1-coercive. (1.5.2)

Lemma 1.5.3 *For f satisfying (1.5.2), co(epi f) is a closed set.*

PROOF. Take a sequence $\{x_k, r_k\}$ in co(epi f) converging to (x, r) for $k \to +\infty$; in order to establish $(x, r) \in$ co(epi f), we will prove that $(x, \rho) \in$ co(epi f) for some $\rho \leqslant r$.

By definition of a convex hull in $\mathbb{R}^{n+1}$, there are $n + 2$ sequences $\{x_k^i, r_k^i\}$ with $f(x_k^i) \leqslant r_k^i$, and a sequence $\{\alpha_k\}$ in the unit simplex Δ_{n+2} such that

$$(x_k, r_k) = \sum_{i=1}^{n+2} \alpha_k^i \left(x_k^i, r_k^i\right).$$

[*Step 1*] For each i, the sequence $\{\alpha_k^i r_k^i\}$ is bounded: in fact, because of (1.5.2), f is bounded from below, say by μ; then we write

$$r_k \geqslant \alpha_k^i r_k^i + \sum_{j \neq i} \alpha_k^j f(x_k^j) \geqslant \alpha_k^i r_k^i + \left(1 - \alpha_k^i\right)\mu,$$

and our claim is true because $\{r_k\}$ and $\{\alpha_k^i\}$ are bounded.

The sequences $\{r_k^i\}$ are also bounded from below by μ. Now, if some $\{r_k^i\}$ is not bounded from above, go to Key Step; otherwise go to Last Step.

[*Key Step*] We proceed to prove that, if $\{r_k^i\}$ is not bounded from above for some index i, the corresponding sequence can be omitted from the convex combination. Assume without loss of generality that 1 is such an index and extract a subsequence if necessary, so $r_k^1 \to +\infty$. Then

$$\frac{f(x_k^1)}{\|x_k^1\|} \leqslant \frac{r_k^1}{\|x_k^1\|} \to +\infty;$$

this is clear if $\{x_k^1\}$ is bounded; if not, it is a consequence of 1-coercivity. Remembering Step 1, we thus have the key properties:

$$\alpha_k^1 \to 0, \quad \|\alpha_k^1 x_k^1\| = \frac{\alpha_k^1 r_k^1}{r_k^1/\|x_k^1\|} \to 0.$$

Now, for each k, define $\beta_k \in \Delta_{n+1}$ by

$$\beta_k^i := \frac{\alpha_k^{i+1}}{1 - \alpha_k^1} \quad \text{for } i = 1, \ldots, n+1.$$

We have

$$\sum_{i=1}^{n+1} \beta_k^i x_k^i = \frac{1}{1 - \alpha_k^1}(x_k - \alpha_k^1 x_k^1) \to x,$$

$$\mu \leq \sum_{i=1}^{n+1} \beta_k^i r_k^i = \frac{1}{1 - \alpha_k^1}(r_k - \alpha_k^1 r_k^1) \leq \frac{1}{1 - \alpha_k^1}(r_k - \alpha_k^1 \mu) \to r.$$

Let us summarize this key step: starting from the $n + 2$ sequences of triples $\{\alpha_k^i, x_k^i, r_k^i\}$, we have eliminated one having $\{r_k^i\}$ unbounded, and thus obtained $\ell = n + 1$ sequences of triples $\{\beta_k^i, x_k^i, r_k^i\}$ satisfying

$$\beta_k \in \Delta_\ell, \quad \sum_{i=1}^{\ell} \beta_k^i x_k^i \to x, \quad \sum_{i=1}^{\ell} \beta_k^i r_k^i \to \rho \leq r. \quad (1.5.3)$$

Execute this procedure as many times as necessary, to end up with $\ell \geq 1$ sequences of triples satisfying (1.5.3), and *all* having $\{r_k^i\}$ bounded.

[*Last Step*] At this point of the proof, each $\{r_k^i\}$ is bounded; from coercivity of f, each $\{x_k^i\}$ is bounded as well. Extracting subsequences if necessary, we are therefore in the following situation: there are sequences of triples $\{\beta_k^i, x_k^i, r_k^i\}, i = 1, \ldots, \ell \leq n + 2$, satisfying

$$\beta_k \in \Delta_\ell, \quad \beta_k \to \beta \in \Delta_\ell;$$
$$f(x_k^i) \leq r_k^i, \quad x_k^i \to x^i, \quad r_k^i \to r^i \quad \text{for } i = 1, \ldots, \ell;$$
$$\sum_{i=1}^{\ell} \beta_k^i x_k^i \to x, \quad \sum_{i=1}^{\ell} \beta_k^i r_k^i \to \sum_{i=1}^{\ell} \beta^i r^i = \rho \leq r.$$

Because f is lower semi-continuous, $f(x^i) \leq r^i$ for $i = 1, \ldots, \ell$ and the definition (IV.2.5.3) of co f gives

$$(\text{co } f)(x) \leq \sum_{i=1}^{\ell} \beta^i f(x^i) \leq \sum_{i=1}^{\ell} \beta^i r^i = \rho \leq r.$$

In a word, $(x, r) \in \text{epi co } f$. □

This closedness property has important consequences:

Proposition 1.5.4 *Let f satisfy* (1.5.2). *Then*

(i) co $f = \overline{\text{co}} f$ *(hence* co $f \in \overline{\text{Conv}} \, \mathbb{R}^n$*)*.
(ii) *For any $x \in \text{dom co } f = \text{co dom } f$, there are $x_j \in \text{dom } f$ and convex multipliers α_j for $j = 1, \ldots, n+1$ such that*

$$x = \sum_{j=1}^{n+1} \alpha_j x_j \quad \text{and} \quad (\text{co } f)(x) = \sum_{j=1}^{n+1} \alpha_j f(x_j).$$

PROOF. By definition, $\overline{\text{co}}\, f$ is the function whose epigraph is the closure of co epi f; but the latter set is already closed. Since the relations

$$\text{co epi } f \subset \text{epi co } f \subset \text{epi } \overline{\text{co}}\, f = \overline{\text{co}}\, \text{epi } f$$

are always true, (i) is proved.

Using the definition co $f = f_1$ of Proposition IV.2.5.1, and knowing that co epi f is closed, we infer that the point $(x, \text{co } f(x))$ is on the boundary of co epi f. Then, invoking Proposition III.4.2.3, we can describe $(x, \text{co } f(x))$ as a convex combination of $n+1$ elements in epi f:

$$(x, \text{co } f(x)) = \sum_{j=1}^{n+1} \alpha_j (x_j, r_j),$$

with $f(x_j) \leqslant r_j$. Actually each r_j has to be $f(x_j)$ if the corresponding α_j is positive, simply because

$$\sum_{j=1}^{n+1} \alpha_j \left(x_j, f(x_j)\right) \in \text{co epi } f,$$

and, again from the definition of co $f(x)$:

$$\sum_{j=1}^{n+1} \alpha_j r_j = \text{co } f(x) \leqslant \sum_{j=1}^{n+1} \alpha_j f(x_j). \qquad \square$$

The combinations exhibited in (ii) are useful for an explicit calculation of co f and $\partial(\text{co } f)$, given in the next two results.

Theorem 1.5.5 *Let f satisfy* (1.1.1). *For a given $x \in \text{co dom } f$, suppose there exists a family $\{x_j, \alpha_j\}$ as described in Proposition 1.5.4(ii); set*

$$J := \{j \,:\, \alpha_j > 0\}.$$

Then

(i) $f(x_j) = (\text{co } f)(x_j)$ *for all $j \in J$,*
(ii) co f *is affine on the polyhedron* $P := \text{co}\{x_j \,:\, j \in J\}$.

PROOF. [(i)] The function co f is convex and minorizes f:

$$(\text{co } f)(x) \leqslant \sum_{j=1}^{n+1} \alpha_j (\text{co } f)(x_j) \leqslant \sum_{j=1}^{n+1} \alpha_j f(x_j) = (\text{co } f)(x);$$

following an argument already seen on several occasions, this implies

$$\alpha_j (\text{co } f)(x_j) = \alpha_j f(x_j) \quad \text{for } j = 1, \ldots, n+1$$

and (i) is proved.

[(ii)] Consider the affine function (here $\{\beta_j\}$ is a set of convex multipliers)
$$P \ni x' = \sum_{j \in J} \beta_j x_j \mapsto \ell(x') := \sum_{j \in J} \beta_j f(x_j).$$
In view of (i), the convexity of co f implies that co $f \leq \ell$ on P, with equality at the given x.

Now take $x' \neq x$ in P. The affine line passing through x and x' cuts rbd P at two points y_1 and y_2, both different from x (since the latter is in ri P). Then write (see Fig. 1.5.1)
$$x = \beta x' + (1 - \beta) y_1 \quad \text{with} \quad \beta \in \,]0, 1[\,.$$
By convexity,
$$\begin{aligned}(\text{co } f)(x) &\leq \beta (\text{co } f)(x') + (1 - \beta)(\text{co } f)(y_1) \\ &\leq \beta \ell(x') + (1 - \beta) \ell(y_1) \quad &&[\text{because co } f \leq \ell\,] \\ &= \ell(x). &&[\text{because } \ell \text{ is affine}]\end{aligned}$$
Since $(\text{co } f)(x) = \ell(x)$, this is actually a chain of equalities:
$$\beta [(\text{co } f)(x') - \ell(x')] + (1 - \beta)[(\text{co } f)(y_1) - \ell(y_1)] = 0.$$
Once again, we have two nonnegative numbers having a zero convex combination; they are both zero and (ii) is proved. □

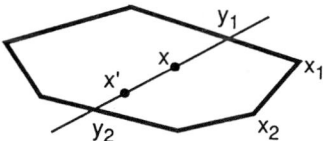

Fig. 1.5.1. Affinity of a convex hull

Theorem 1.5.6 *Under the hypotheses and notations of Theorem 1.5.5,*
$$\partial (\text{co } f)(x) = \bigcap_{j \in J} \partial f(x_j); \tag{1.5.4}$$
$$\forall s \in \partial (\text{co } f)(x), \quad \langle s, x \rangle - (\text{co } f)(x) = \langle s, x_j \rangle - f(x_j) \text{ for all } j \in J.$$

PROOF. A subgradient of co f at x is an s characterized by
$$\begin{aligned}&(\text{co } f)^*(s) + (\text{co } f)(x) - \langle s, x \rangle = 0 &&[\text{Theorem 1.4.1}] \\ \Longleftrightarrow\ & f^*(s) + \sum_{j \in J} \alpha_j f(x_j) - \langle s, \sum_{j \in J} \alpha_j x_j \rangle = 0 &&[(\text{co } f)^* = f^*] \\ \Longleftrightarrow\ & \sum_{j \in J} \alpha_j [f^*(s) + f(x_j) - \langle s, x_j \rangle] = 0 \\ \Longleftrightarrow\ & f^*(s) + f(x_j) - \langle s, x_j \rangle = 0 \quad \text{for all } j \in J. &&[\text{Fenchel (1.1.3)}]\end{aligned}$$

The last line means precisely that $s \in \partial f(x_j)$ for all $j \in J$; furthermore, we can write
$$f(x_j) - \langle s, x_j \rangle = -f^*(s) = -(\text{co } f)^*(s) = (\text{co } f)(x) - \langle s, x \rangle. \quad \square$$

Note in the right-hand side of (1.5.4) that each $\partial f(x_j)$ could be replaced by $\partial (\text{co } f)(x_j)$: this is due to (1.4.4).

Remark 1.5.7 As a supplement to (1.5.1), the above result implies the following: for f satisfying (1.5.2), Argmin f is a compact set (trivial), whose convex hull coincides with Argmin(co f). Indeed, use (1.5.4): an $x \in $ Argmin(co f) is characterized as being a convex combination of points $\{x_j\}_{j \in J}$ such that $0 \in \partial f(x_j)$, i.e. these x_j minimize f. □

Corollary 1.5.8 *Let the function $f : \mathbb{R}^n \to \mathbb{R}$ be lower semi-continuous, Gâteaux differentiable, and 1-coercive. Then co $f = \overline{co}\, f$ is continuously differentiable on $\mathbb{R}^n$; furthermore,*

$$\nabla(\mathrm{co}\, f)(x) = \nabla f(x_j) \quad \text{for all } j \in J, \tag{1.5.5}$$

where we have used the notation of Theorem 1.5.5.

PROOF. Our f satisfies (1.5.2), so Proposition 1.5.4(i) directly gives co $f = \overline{co}\, f$. The latter function is finite everywhere, hence it has a nonempty subdifferential at any x (Theorem 1.4.2): by virtue of (1.5.4), all the $\partial f(x_j)$'s are nonempty. Together with Proposition 1.5.1, we obtain that all these $\partial f(x_j)$'s are actually the same singleton, described by (1.5.5). Finally, the continuity of $\nabla(\mathrm{co}\, f)$ follows from Theorem VI.6.2.4. □

It is worth mentioning that, even if we impose more regularity on f (say C^∞), co f as a rule is not C^2.

2 Calculus Rules on the Conjugacy Operation

The function f, whose conjugate is to be computed, is often obtained from some other functions f_i, whose conjugates are known. In this section, we develop a set of calculus rules expressing f^* in terms of the $(f_i)^*$ (some rudimentary such rules were already given in Proposition 1.3.1).

2.1 Image of a Function Under a Linear Mapping

Given a function $g : \mathbb{R}^m \to \mathbb{R} \cup \{+\infty\}$ satisfying (1.1.1), and a linear mapping $A : \mathbb{R}^m \to \mathbb{R}^n$, we recall that the image of g under A is the function defined by

$$\mathbb{R}^n \ni x \mapsto (Ag)(x) := \inf\{g(y) : Ay = x\}. \tag{2.1.1}$$

Let g^* be associated with a scalar product $\langle \cdot, \cdot \rangle_m$ in $\mathbb{R}^m$; we denote by $\langle \cdot, \cdot \rangle_n$ the scalar product in $\mathbb{R}^n$, with the help of which we want to define $(Ag)^*$. To make sure that Ag satisfies (1.1.1), some additional assumption is needed: among all the affine functions minorizing g, there is one with slope in $(\mathrm{Ker}\, A)^\perp = \mathrm{Im}\, A^*$.

Theorem 2.1.1 *With the above notation, assume that $\mathrm{Im}\, A^* \cap \mathrm{dom}\, g^* \neq \emptyset$. Then Ag satisfies (1.1.1); its conjugate is*

$$(Ag)^* = g^* \circ A^*.$$

2 Calculus Rules on the Conjugacy Operation

PROOF. First, it is clear that $Ag \not\equiv +\infty$ (take $x = Ay$, with $y \in \text{dom } g$). On the other hand, our assumption implies the existence of some $p_0 = A^*s_0$ such that $g^*(p_0) < +\infty$; with Fenchel's inequality (1.1.3), we have for all $y \in \mathbb{R}^m$:

$$g(y) \geq \langle A^*s_0, y \rangle_m - g^*(p_0) = \langle s_0, Ay \rangle_n - g^*(p_0).$$

For each $x \in \mathbb{R}^n$, take the infimum over those y satisfying $Ay = x$: the affine function $\langle s_0, \cdot \rangle - g^*(p_0)$ minorizes Ag. Altogether, Ag satisfies (1.1.1).

Then we have for $s \in \mathbb{R}^n$

$$\begin{aligned}(Ag)^*(s) &= \sup_{x \in \mathbb{R}^n}[\langle s, x \rangle - \inf_{Ay=x} g(y)] \\ &= \sup_{x \in \mathbb{R}^n, Ay=x}[\langle s, x \rangle - g(y)] \\ &= \sup_{y \in \mathbb{R}^m}[\langle s, Ay \rangle - g(y)] = g^*(A^*s).\end{aligned}$$
□

For example, when $m = n$ and A is invertible, Ag reduces to $g \circ A^{-1}$, whose conjugate is therefore $g^* \circ A^*$, given by Proposition 1.3.1(iv).

As a first application of Theorem 2.1.1, it is straightforward to compute the conjugate of a marginal function:

$$f(x) := \inf\{g(x, z) : z \in \mathbb{R}^p\}, \tag{2.1.2}$$

where g operates on the product-space $\mathbb{R}^n \times \mathbb{R}^p$. Indeed, just call A the projection onto $\mathbb{R}^n$: $A(x, z) = x \in \mathbb{R}^n$, so f is clearly Ag. We obtain:

Corollary 2.1.2 *With $g : \mathbb{R}^n \times \mathbb{R}^p =: \mathbb{R}^m \to \mathbb{R} \cup \{+\infty\}$ not identically $+\infty$, let g^* be associated with a scalar product preserving the structure of $\mathbb{R}^m$ as a product space:*

$$\langle \cdot, \cdot \rangle_m = \langle \cdot, \cdot \rangle_n + \langle \cdot, \cdot \rangle_p,$$

and suppose that there is $s_0 \in \mathbb{R}^n$ such that $(s_0, 0) \in \text{dom } g^*$. Then the conjugate of f defined by (2.1.2) is

$$f^*(s) = g^*(s, 0) \quad \text{for all } s \in \mathbb{R}^n.$$

PROOF. It suffices to observe that, A being the projection defined above, there holds for all $y_1 = (x_1, z_1) \in \mathbb{R}^m$ and $x_2 \in \mathbb{R}^n$,

$$\langle Ay_1, x_2 \rangle_n = \langle x_1, x_2 \rangle_n = \langle x_1, x_2 \rangle_n + \langle z_1, 0 \rangle_p = \langle y_1, (x_2, 0) \rangle_m,$$

which *defines* the adjoint $A^*x = (x, 0)$ for all $x \in \mathbb{R}^n$. Then apply Theorem 2.1.1.
□

More will be said on this operation in §2.4. Here we consider another application: the infimal convolution which, to f_1 and f_2 defined on $\mathbb{R}^n$, associates (see §IV.2.3)

$$(f_1 \downarrow\!\!\!+ f_2)(x) := \inf\{f_1(x_1) + f_2(x_2) : x_1 + x_2 = x\}. \tag{2.1.3}$$

To use Theorem 2.1.1, we take $m = 2n$ and

$$g(x_1, x_2) := f_1(x_1) + f_2(x_2) \quad \text{and} \quad A(x_1, x_2) := x_1 + x_2.$$

Corollary 2.1.3 *Let two functions f_1 and f_2 from $\mathbb{R}^n$ to $\mathbb{R} \cup \{+\infty\}$, not identically $+\infty$, satisfy $\operatorname{dom} f_1^* \cap \operatorname{dom} f_2^* \neq \emptyset$. Then $f_1 \mathbin{\dot{+}} f_2$ satisfies (1.1.1), and $(f_1 \mathbin{\dot{+}} f_2)^* = f_1^* + f_2^*$.*

PROOF. Equipping $\mathbb{R}^n \times \mathbb{R}^n$ with the scalar product $\langle \cdot, \cdot \rangle + \langle \cdot, \cdot \rangle$, we obtain $g^*(s_1, s_2) = f_1^*(s_1) + f_2^*(s_2)$ (Proposition 1.3.1(ix)) and $A^*(s) = (s, s)$. Then apply the definitions. □

Example 2.1.4 As sketched in Proposition I.2.2.4, a function can be regularized if we take its infimal convolution with one of the kernels $1/2 \, c \| \cdot \|^2$ or $c \| \cdot \|$. Then Corollary 2.1.3 gives, with the help of Proposition 1.3.1(iii) and Example 1.1.3:

$$\left(f \mathbin{\dot{+}} \tfrac{1}{2} c \| \cdot \|^2\right)^* = f^* + \tfrac{1}{2c} \| \cdot \|^2,$$
$$\left(f \mathbin{\dot{+}} c \| \cdot \|\right)^* = f^* + \tfrac{1}{c} I_{B(0,c)}.$$

In particular, if f is the indicator of a nonempty set $C \subset \mathbb{R}^n$, the above formulae yield

$$\left(\tfrac{1}{2} d_C^2\right)^* = \sigma_C + \tfrac{1}{2} \| \cdot \|^2, \quad d_C^* = \sigma_C + I_{B(0,1)}. \qquad \square$$

2.2 Pre-Composition with an Affine Mapping

In view of the symmetry of the conjugacy operation, Theorem 2.1.1 suggests that the conjugate of $g \circ A$, when A is a linear mapping, is the image-function $A^* g^*$. In particular, a condition was needed in §2.1 to prevent $Ag(x) = -\infty$ for some x. Likewise, a condition will be needed here to prevent $g \circ A \equiv +\infty$. The symmetry is not quite perfect, though: the composition of a closed convex function with a linear mapping is still a closed convex function; but an image-function need not be closed, and therefore cannot be a conjugate function.

We use notation similar to that of §2.1, but we find it convenient to distinguish between the linear and affine cases.

Theorem 2.2.1 *With $g \in \overline{\operatorname{Conv}} \mathbb{R}^m$ and A_0 linear from $\mathbb{R}^n$ to $\mathbb{R}^m$, define $A(x) := A_0 x + y_0 \in \mathbb{R}^m$ and suppose that $A(\mathbb{R}^n) \cap \operatorname{dom} g \neq \emptyset$. Then $g \circ A \in \overline{\operatorname{Conv}} \mathbb{R}^n$ and its conjugate is the closure of the convex function*

$$\mathbb{R}^n \ni s \mapsto \inf_p \{g^*(p) - \langle y_0, p \rangle_m \ : \ A_0^* p = s\}. \tag{2.2.1}$$

PROOF. We start with the linear case ($y_0 = 0$): suppose that $h \in \overline{\operatorname{Conv}} \mathbb{R}^n$ satisfies $\operatorname{Im} A_0 \cap \operatorname{dom} h \neq \emptyset$. Then Theorem 2.1.1 applied to $g := h^*$ and $A := A_0^*$ gives $(A_0^* h^*)^* = h \circ A_0$; conjugating both sides, we see that the conjugate of $h \circ A_0$ is the closure of the image-function $A_0^* h^*$.

In the affine case, consider the function $h := g(\cdot + y_0) \in \overline{\operatorname{Conv}} \mathbb{R}^m$; its conjugate is given by Proposition 1.3.1(v): $h^* = g^* - \langle y_0, \cdot \rangle_m$. Furthermore, it is clear that

$$(g \circ A)(x) = g(A_0 x + y_0) = h(A_0 x) = (h \circ A_0)(x),$$

so (2.2.1) follows from the linear case. □

2 Calculus Rules on the Conjugacy Operation

Thus, the conjugation of a composition by a linear (or affine) mapping is not quite straightforward, as it requires a closure operation. A natural question is therefore: when does (2.2.1) define a closed function of s? In other words, when is an image-function closed? Also, when is the infimum in (2.2.1) attained at some p? We start with a technical result.

Lemma 2.2.2 *Let $g \in \overline{\mathrm{Conv}}\,\mathbb{R}^m$ be such that $0 \in \mathrm{dom}\,g$ and let A_0 be linear from $\mathbb{R}^n$ to $\mathbb{R}^m$. Make the following assumption:*

$$\mathrm{Im}\,A_0 \cap \mathrm{ri}\,\mathrm{dom}\,g \neq \emptyset \quad \text{i.e.} \quad 0 \in \mathrm{ri}\,\mathrm{dom}\,g - \mathrm{Im}\,A_0 \;[= \mathrm{ri}(\mathrm{dom}\,g - \mathrm{Im}\,A_0)].$$

Then $(g \circ A_0)^* = A_0^* g^*$ and, for every $s \in \mathrm{dom}(g \circ A_0)^*$, the problem

$$\inf_p \{g^*(p) \,:\, A_0^* p = s\} \tag{2.2.2}$$

has an optimal solution $\bar{p}$, which therefore satisfies $g^*(\bar{p}) = (g \circ A_0)^*(s) = A_0^* g^*(s)$.

PROOF. To prove $(g \circ A_0)^* = A_0^* g^*$, we have to prove that $A_0^* g^*$ is a closed function, i.e. that its sublevel-sets are closed (Definition IV.1.2.3). Thus, for given $r \in \mathbb{R}$, take a sequence $\{s_k\}$ such that

$$(A_0^* g^*)(s_k) \leqslant r \quad \text{and} \quad s_k \to s.$$

Take also $\delta_k \downarrow 0$; from the definition of the image-function, we can find $p_k \in \mathbb{R}^m$ such that

$$g^*(p_k) \leqslant r + \delta_k \quad \text{and} \quad A_0^* p_k = s_k.$$

Let q_k be the orthogonal projection of p_k onto the subspace $V := \mathrm{lin}\,\mathrm{dom}\,g - \mathrm{Im}\,A_0$. Because V contains $\mathrm{lin}\,\mathrm{dom}\,g$, Proposition 1.3.4 (with $z = 0$) gives $g^*(p_k) = g^*(q_k)$. Furthermore, $V^\perp = (\mathrm{lin}\,\mathrm{dom}\,g)^\perp \cap \mathrm{Ker}\,A_0^*$; in particular, $q_k - p_k \in \mathrm{Ker}\,A_0^*$. In summary, we have singled out $q_k \in V$ such that

$$g^*(q_k) \leqslant r + \delta_k \quad \text{and} \quad A_0^* q_k = s_k \quad \text{for all } k. \tag{2.2.3}$$

Suppose we can bound q_k. Extracting a subsequence if necessary, we will have $q_k \to \bar{q}$ and, passing to the limit, we will obtain (since g^* is l.s.c)

$$g^*(\bar{q}) \leqslant \liminf g^*(q_k) \leqslant r \quad \text{and} \quad A_0^* \bar{q} = s.$$

The required closedness property $A_0^* g^*(\bar{q}) \leqslant r$ will follow by definition. Furthermore, this $\bar{q}$ will be a solution of (2.2.2) in the particular case $s_k \equiv s$ and $r = (A_0^* g^*)(s)$. In this case, $\{q_k\}$ will be actually a minimizing sequence of (2.2.2).

To prove boundedness of q_k, use the assumption: for some $\varepsilon > 0$, $B_m(0, \varepsilon) \cap V$ is included in $\mathrm{dom}\,g - \mathrm{Im}\,A$. Thus, for arbitrary $z \in B_m(0, \varepsilon) \cap V$, we can find $y \in \mathrm{dom}\,g$ and $x \in \mathbb{R}^n$ such that $z = y - A_0 x$. Then

$$\begin{aligned}
\langle q_k, z \rangle_m &= \langle q_k, y \rangle_m - \langle A_0^* q_k, x \rangle_n \\
&\leqslant g(y) + g^*(q_k) - \langle A_0^* q_k, x \rangle_n && \text{[Fenchel (1.1.3)]} \\
&\leqslant g(y) + r + \delta_k - \langle s_k, x \rangle_n. && \text{[(2.2.3)]}
\end{aligned}$$

We conclude

$$\sup\{\langle q_k, z\rangle : k = 1, 2, \ldots\} \text{ is bounded for any } z \in B_m(0, \varepsilon) \cap V,$$

which implies that q_k is bounded; this is Proposition V.2.1.3 in the vector space V. □

Adapting this result to the general case is now a matter of translation:

Theorem 2.2.3 *With $g \in \overline{\text{Conv}}\,\mathbb{R}^m$ and A_0 linear from $\mathbb{R}^n$ to $\mathbb{R}^m$, define $A(x) := A_0 x + y_0 \in \mathbb{R}^m$. Make the following assumption:*

$$A(\mathbb{R}^n) \cap \text{ri dom}\, g \neq \emptyset. \tag{2.2.4}$$

Then, for every $s \in \text{dom}(g \circ A_0)^*$, the following minimization problem has a solution:

$$\min_p \{g^*(p) - \langle p, y_0\rangle : A_0^* p = s\} = (g \circ A)^*(s). \tag{2.2.5}$$

PROOF. By assumption, we can choose $\bar{x} \in \mathbb{R}^n$ such that $\bar{y} := A(\bar{x}) \in \text{ri dom}\, g$. Consider the function $\bar{g} := g(\bar{y} + \cdot) \in \overline{\text{Conv}}\,\mathbb{R}^m$. Observing that

$$(g \circ A)(x) = \bar{g}(A(x) - \bar{y}) = (\bar{g} \circ A_0)(x - \bar{x}),$$

we obtain from the calculus rule 1.3.1(v)

$$(g \circ A)^* = (\bar{g} \circ A_0)^* - \langle \cdot, \bar{x}\rangle.$$

Then Lemma 2.2.2 allows the computation of this conjugate. We have 0 in the domain of $\bar{g}$, and even in its relative interior:

$$\text{ri dom}\,\bar{g} = \text{ri dom}\, g - \{\bar{y}\} \ni 0 \in \text{Im}\, A_0.$$

We can therefore write: for all $s \in \text{dom}(\bar{g} \circ A_0)^* [= \text{dom}(g \circ A)^*]$,

$$(\bar{g} \circ A_0)^*(s) = \min_p\{\bar{g}^*(p) : A_0^* p = s\},$$

or also

$$\begin{aligned}(g \circ A)^*(s) - \langle s, \bar{x}\rangle &= \min_p\{g^*(p) - \langle p, \bar{y}\rangle : A_0^* p = s\} \\ &= -\langle s, \bar{x}\rangle + \min_p\{g^*(p) - \langle p, y_0\rangle : A_0^* p = s\}.\end{aligned}$$ □

An example will be given in Remark 2.2.4 below, to show that the calculus rule (2.2.5) does need a *qualification assumption* such as (2.2.4). First of all, we certainly need to avoid $g \circ A \equiv +\infty$, i.e.

$$A(\mathbb{R}^n) \cap \text{dom}\, g \neq \emptyset; \tag{2.2.Q.i}$$

but this is not sufficient, unless g is a polyhedral function (this situation has essentially been treated in §V.3.4).

A "comfortable" sharpening of (2.2.Q.i) is

$$A(\mathbb{R}^n) \cap \text{int dom } g \neq \emptyset \quad \text{i.e. } 0 \in \text{int dom } g - A(\mathbb{R}^n), \qquad (2.2.\text{Q.ii})$$

but it is fairly restrictive, implying in particular that dom g is full-dimensional. More tolerant is

$$0 \in \text{int}[\text{dom } g - A(\mathbb{R}^n)], \qquad (2.2.\text{Q.iii})$$

which is rather common. Use various results from Chap. V, in particular Theorem V.2.2.3(iii), to see that (2.2.Q.iii) means

$$\sigma_{\text{dom } g}(p) + \langle p, y_0 \rangle > 0 \quad \text{for all nonzero } p \in \text{Ker } A_0^*.$$

Knowing that $\sigma_{\text{dom } g} = (g^*)'_\infty$ (Proposition 1.2.2), this condition has already been alluded to at the end of §IV.3.

Naturally, our assumption (2.2.4) is a further weakening; actually, it is only a slight generalization of (2.2.Q.iii). It is interesting to note that, if (2.2.4) is replaced by (2.2.Q.iii), the solution-set in problem (2.2.5) becomes bounded; to see this, read again the proof of Lemma 2.2.2, in which V becomes $\mathbb{R}^m$.

Remark 2.2.4 Condition (2.2.4) is rather natural: when it does not hold, almost all information on g is ignored by A, and pathological things may happen when conjugating. Consider the following example with $m = n = 2$, $A(\xi, \eta) := (\xi, 0)$,

$$\mathbb{R}^2 \ni (\xi, \eta) \mapsto g(\xi, \eta) := \begin{cases} \eta \log \eta & \text{for } \eta > 0, \\ 0 & \text{for } \eta = 0 \\ +\infty & \text{elsewhere}, \end{cases}$$

and the scalar product

$$\langle s, x \rangle = \langle (\rho, \tau), (\xi, \eta) \rangle = (\rho + \tau)\xi + (\rho + 2\tau)\eta.$$

Easy calculations give:

$$g^*(\rho, \tau) = \begin{cases} \exp(\tau - 1) & \text{if } \rho + \tau = 0, \\ +\infty & \text{otherwise}, \end{cases}$$

and $A^*(\rho, \tau) = (\rho + \tau)(2, -1)$.

Note: taking the canonical basis of $\mathbb{R}^2$ (which is not orthonormal for $\langle \cdot, \cdot \rangle$!), we can define
$$A := \begin{bmatrix} 1 & 0 \\ 0 & 0 \end{bmatrix}, M := \begin{bmatrix} 1 & 1 \\ 1 & 2 \end{bmatrix} \text{ so that}$$

$$\langle s, x \rangle = s^\top M x \quad \text{and} \quad A^* = \begin{bmatrix} 2 & 2 \\ -1 & -1 \end{bmatrix} = M^{-1} A^\top M.$$

Thus dom $g \cap \text{Im } A = \mathbb{R} \times \{0\}$; none of the assumptions in Theorem 2.2.3 are satisfied; dom $g^* \cap \text{Im } A^* = \{0\}$,

$$A^* g^*(0) = \inf \{\exp(\tau - 1) : \rho + \tau = 0\} = 0,$$

and the infimum is not attained "at finite distance". Note also that $A^* g^*$ is closed (how could it not be, its domain being a singleton!) and is the conjugate of $g \circ A \equiv 0$.

The trouble illustrated by this example is that composition with A extracts from g only its nasty behaviour, namely its vertical tangent plane for $\eta = 0$. □

The message of the above counter-example is that conjugating brutally $g \circ A$, with a scalar product defined on the whole of $\mathbb{R}^m$, is clumsy; this is the first line in Fig. 2.2.1. Indeed, the restriction of g to Im A, considered as a Euclidean space, is a closed convex function by itself, and this is the relevant function to consider: see the second line of Fig. 2.2.1 (but a scalar product is then needed in Im A). Alternatively, if we insist on working in the whole environment space $\mathbb{R}^m$, the relevant function is rather $g + I_{\text{Im } A}$ (third line in Fig. 2.2.1, which requires the conjugate of a sum). All three constructions give the same $g \circ A$, but the intermediate conjugates are quite different.

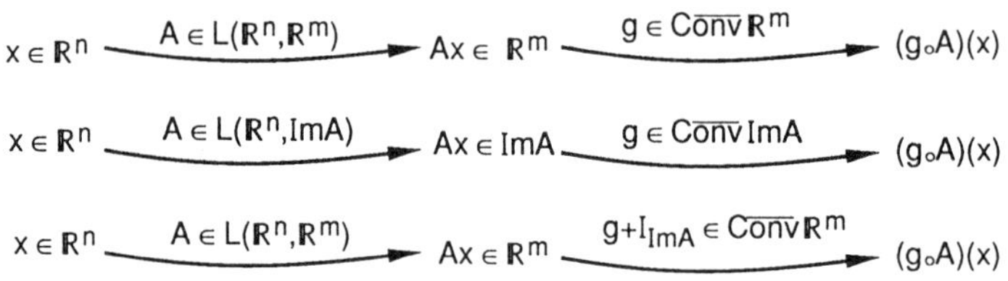

$x \in \mathbb{R}^n \xrightarrow{A \in L(\mathbb{R}^n, \mathbb{R}^m)} Ax \in \mathbb{R}^m \xrightarrow{g \in \overline{\text{Conv}}\, \mathbb{R}^m} (g \circ A)(x)$

$x \in \mathbb{R}^n \xrightarrow{A \in L(\mathbb{R}^n, \text{Im}A)} Ax \in \text{Im}A \xrightarrow{g \in \overline{\text{Conv}}\, \text{Im}A} (g \circ A)(x)$

$x \in \mathbb{R}^n \xrightarrow{A \in L(\mathbb{R}^n, \mathbb{R}^m)} Ax \in \mathbb{R}^m \xrightarrow{g + I_{\text{Im}A} \in \overline{\text{Conv}}\, \mathbb{R}^m} (g \circ A)(x)$

Fig. 2.2.1. Three possible expressions for $g \circ A$

Corollary 2.2.5 *Let $g \in \overline{\text{Conv}}\, \mathbb{R}^m$ and let A be linear from $\mathbb{R}^n$ to $\mathbb{R}^m$ with $\text{Im } A \cap \text{dom } g \neq \emptyset$. Then*

$$(g \circ A)^* = A^*(g + I_{\text{Im } A})^* \tag{2.2.6}$$

and, for every $s \in \text{dom}(g \circ A)^*$, the problem

$$\inf_p \{(g + I_{\text{Im } A})^*(p) : A^* p = s\}$$

has an optimal solution $\bar{p}$, which therefore satisfies

$$(g + I_{\text{Im } A})^*(\bar{p}) = A^*(g + I_{\text{Im } A})^*(s) = (g \circ A)^*(s). \tag{2.2.7}$$

PROOF. The closed convex functions $g \circ A$ and $(g + I_{\text{Im } A}) \circ A$ are identical on $\mathbb{R}^n$, so they have the same conjugate. Also, $g + I_{\text{Im } A}$ is a closed convex function whose domain is included in Im A, hence

$$\text{ri dom}(g + I_{\text{Im } A}) \cap \text{Im } A = \text{ri dom}(g + I_{\text{Im } A}) \neq \emptyset.$$

The result follows from Theorem 2.2.3. □

We leave it as an exercise to study how Example 2.2.4 is modified by this result. To make it more interesting, one can also take $g(\xi, \eta) + 1/2\xi^2$ instead of g. Theorem 2.2.3 and Corollary 2.2.5 are two different ways of stating essentially the same thing; the former requires that a qualification assumption such as (2.2.4) be checked; the latter requires that the conjugate of the additional function $g + I_{\text{Im } A}$ be computed. Either result may be simpler to use, depending on the particular problem under consideration.

Remark 2.2.6 We know from Proposition 1.3.2 that $(g + I_{\text{Im } A})^* = (g \circ P_A)^* \circ P_A$, where $P_A : \mathbb{R}^m \to \mathbb{R}^m$ is the orthogonal projection onto the subspace Im A (beware that, by contrast to A, P_A operates on $\mathbb{R}^m$). It follows that (2.2.6) – (2.2.7) can be replaced respectively by:

$$(g \circ A)^* = A^*[(g \circ P_A)^* \circ P_A],$$

2 Calculus Rules on the Conjugacy Operation   61

$$\inf \{(g \circ P_A)^*(P_A p) \ : \ A^* p = s\},$$
$$(g \circ P_A)^*(P_A \bar{p}) = A^*[(g \circ P_A)^* \circ P_A](s) = (g \circ A)^*(s).$$

When the qualification assumption allows the application of Theorem 2.2.3, we are bound to realize that

$$A^* g^* = A^*(g + I_{\text{Im } A})^* = A^*[(g \circ P_A)^* \circ P_A]. \tag{2.2.8}$$

Indeed, Theorem 2.2.3 actually applies also to the couple (P_A, g) in this case: $(g \circ P_A)^*$ can be developed further, to obtain (since P_A is symmetric)

$$[(g \circ P_A)^* \circ P_A](p) = [(P_A g^*) \circ P_A](p) = \inf \{g^*(q) \ : \ P_A q = P_A p\}.$$

Now, to say that $P_A q = P_A p$, i.e. that $P_A(q-p) = 0$, is to say that $q - p \in (\text{Im } A)^\perp = \text{Ker } A^*$, i.e. $A^* p = A^* q$. Altogether, we deduce

$$A^*(g + I_{\text{Im } A})^*(s) = \inf_{p,q}\{g^*(q) \ : \ A^* p = A^* q = s\} = A^* g^*(s)$$

and (2.2.8) does hold. □

To conclude this subsection, consider an example with $n = 1$: given a function $g \in \overline{\text{Conv}} \, \mathbb{R}^m$, fix $x_0 \in \text{dom } g$, $0 \neq d \in \mathbb{R}^m$, and set

$$\psi(t) := g(x_0 + td) \quad \text{for all } t \in \mathbb{R}.$$

This ψ is the composition of g with the affine mapping which, to $t \in \mathbb{R}$, assigns $x_0 + td \in \mathbb{R}^m$. It is an exercise to apply Theorem 2.2.1 and obtain the conjugate:

$$\psi^*(\alpha) = \min \{g^*(s) - \langle s, x_0 \rangle \ : \ \langle s, d \rangle = \alpha\}$$

whenever, for example, x_0 and d are such that $x_0 + td \in \text{ri dom } g$ for some t – this is (2.2.4). This example can be further particularized to various functions mentioned in §1 (quadratic, indicator, ...).

2.3 Sum of Two Functions

The formula for conjugating a sum will supplement Proposition 1.3.1(ii) to obtain the conjugate of a positive combination of closed convex functions. Summing two functions is a simple operation (at least it preserves closedness); but Corollary 2.1.3 shows that a sum is the conjugate of something rather involved: an inf-convolution. Likewise, compare the simplicity of the composition $g \circ A$ with the complexity of its dual counterpart Ag; as seen in §2.2, difficulties are therefore encountered when conjugating a function of the type $g \circ A$. The same kind of difficulties must be expected when conjugating a sum, and the development in this section is quite parallel to that of §2.2.

Theorem 2.3.1 *Let g_1 and g_2 be in $\overline{\text{Conv}} \, \mathbb{R}^n$ and assume that $\text{dom } g_1 \cap \text{dom } g_2 \neq \emptyset$. The conjugate $(g_1 + g_2)^*$ of their sum is the closure of the convex function $g_1^* \triangledown g_2^*$.*

PROOF. Call $f_i^* := g_i$, for $i = 1, 2$; apply Corollary 2.1.3: $(g_1^* \mathbin{\dot\triangledown} g_2^*)^* = g_1 + g_2$; then take the conjugate again. □

The above calculus rule is very useful in the following framework: suppose we want to compute an inf-convolution, say $h = f \mathbin{\dot\triangledown} g$ with f and g in $\overline{\text{Conv}}\,\mathbb{R}^n$. Compute f^* and g^*; if their sum happens to be the conjugate of some known function in $\overline{\text{Conv}}\,\mathbb{R}^n$, this function has to be the closure of h.

Just as in the previous section, it is of interest to ask whether the closure operation is necessary, and whether the inf-convolution is exact.

Theorem 2.3.2 *The assumptions are those of Theorem 2.3.1. Assume in addition that*

$$\text{the relative interiors of } \mathrm{dom}\, g_1 \text{ and } \mathrm{dom}\, g_2 \text{ intersect} \qquad (2.3.1)$$
$$\text{or equivalently: } 0 \in \mathrm{ri}(\mathrm{dom}\, g_1 - \mathrm{dom}\, g_2)\,.$$

Then $(g_1 + g_2)^* = g_1^* \mathbin{\dot\triangledown} g_2^*$ and, for every $s \in \mathrm{dom}(g_1 + g_2)^*$, the problem

$$\inf\{g_1^*(p) + g_2^*(q) \,:\, p + q = s\}$$

has an optimal solution $(\bar p, \bar q)$, which therefore satisfies

$$g_1^*(\bar p) + g_2^*(\bar q) = (g_1^* \mathbin{\dot\triangledown} g_2^*)(s) = (g_1 + g_2)^*(s)\,.$$

PROOF. Define $g \in \overline{\text{Conv}}(\mathbb{R}^n \times \mathbb{R}^n)$ by $g(x_1, x_2) := g_1(x_1) + g_2(x_2)$ and the linear operator $A : \mathbb{R}^n \to \mathbb{R}^n \times \mathbb{R}^n$ by $Ax := (x, x)$. Then $g_1 + g_2 = g \circ A$, and we proceed to use Theorem 2.2.3. As seen in Proposition 1.3.1(ix), $g^*(p, q) = g_1^*(p) + g_2^*(q)$ and straightforward calculation shows that $A^*(p, q) = p + q$. Thus, if we can apply Theorem 2.2.3, we can write

$$\begin{aligned}(g_1 + g_2)^*(s) &= (g \circ A)^*(s) = (A^* g^*)(s) \\ &= \inf_{p,q}\{g_1^*(p) + g_2^*(q) \,:\, p + q = s\} = (g_1^* \mathbin{\dot\triangledown} g_2^*)(s)\end{aligned}$$

and the above minimization problem does have an optimal solution.

To check (2.2.4), we note that $\mathrm{dom}\, g = \mathrm{dom}\, g_1 \times \mathrm{dom}\, g_2$, and $\mathrm{Im}\, A$ is the diagonal set $\Delta := \{(s, s) : s \in \mathbb{R}^n\}$. We have

$$(x, x) \in \mathrm{ri}\,\mathrm{dom}\, g_1 \times \mathrm{ri}\,\mathrm{dom}\, g_2 = \mathrm{ri}(\mathrm{dom}\, g_1 \times \mathrm{dom}\, g_2)$$

(Proposition III.2.1.11), and this just means that $\mathrm{Im}\, A = \Delta$ has a nonempty intersection with $\mathrm{ri}\,\mathrm{dom}\, g$. □

As with Theorem 2.2.3, a qualification assumption – taken here as (2.3.1) playing the role of (2.2.4) – is necessary to ensure the stated properties; but other assumptions exist that do the same thing. First of all,

$$\mathrm{dom}\, g_1 \cap \mathrm{dom}\, g_2 \neq \emptyset \quad \text{i.e.} \quad 0 \in \mathrm{dom}\, g_1 - \mathrm{dom}\, g_2 \qquad (2.3.\mathrm{Q.j})$$

is obviously necessary to have $g_1 + g_2 \not\equiv +\infty$. We saw that this "minimal" condition was sufficient in the *polyhedral* case. Here the results of Theorem 2.3.2 do hold if (2.3.1) is replaced by:

g_1 and g_2 are both polyhedral, or also
g_1 is polyhedral and $\operatorname{dom} g_1 \cap \operatorname{ri} \operatorname{dom} g_2 \neq \emptyset$.

The "comfortable" assumption playing the role of (2.2.Q.ii) is

$$\operatorname{int} \operatorname{dom} g_1 \cap \operatorname{int} \operatorname{dom} g_2 \neq \emptyset, \tag{2.3.Q.jj}$$

which obviously implies (2.3.1). We mention a non-symmetric assumption, particular to a sum:

$$(\operatorname{int} \operatorname{dom} g_1) \cap \operatorname{dom} g_2 \neq \emptyset. \tag{2.3.Q.jj'}$$

Finally, the weakening (2.2.Q.iii) is

$$0 \in \operatorname{int}(\operatorname{dom} g_1 - \operatorname{dom} g_2) \tag{2.3.Q.jjj}$$

which means

$$\sigma_{\operatorname{dom} g_1}(s) + \sigma_{\operatorname{dom} g_2}(-s) > 0 \quad \text{for all } s \neq 0;$$

we leave it as an exercise to prove (2.3.Q.jj') $\Rightarrow$ (2.3.Q.jjj).

The following application of Theorem 2.3.2 is important in optimization: take two functions g_1 and g_2 in $\overline{\operatorname{Conv}} \mathbb{R}^n$ and assume that

$$\mu := \inf \{g_1(x) + g_2(x) : x \in \mathbb{R}^n\}$$

is a finite number. Under some qualification assumption such as (2.3.1),

$$[(g_1 + g_2)^*(0) =] \; -\mu = \min_{s \in \mathbb{R}^n} [g_1^*(s) + g_2^*(-s)], \tag{2.3.2}$$

a relation known as *Fenchel's duality Theorem*. If, in particular, $g_2 = I_C$ is an indicator function, we read (still under some qualification assumption)

$$\inf \{f(x) : x \in C\} = -\min \{f^*(s) + \sigma_C(-s) : s \in \mathbb{R}^n\}.$$

It is appropriate to recall here that conjugate functions are interesting primarily via their arguments and subgradients; thus, it is the *existence* of a minimizing s in the right-hand side of (2.3.2) which is useful, rather than a mere equality between two real numbers.

Remark 2.3.3 Our proof of Theorem 2.3.2 is based on the fact that the sum of two functions can be viewed as the composition of a function with a linear mapping. It is interesting to demonstrate the converse mechanism: suppose we want to prove Theorem 2.2.3, assuming that Theorem 2.3.2 is already proved.

Given $g \in \overline{\operatorname{Conv}} \mathbb{R}^m$ and $A : \mathbb{R}^n \to \mathbb{R}^m$ linear, we then select the two functions with argument $(x, y) \in \mathbb{R}^n \times \mathbb{R}^m$:

$$g_1(x, y) := g(y) \quad \text{and} \quad g_2 := I_{\operatorname{gr} A}.$$

Direct calculations give

$$g_1^*(s, p) = \begin{cases} g^*(p) & \text{if } s = 0, \\ +\infty & \text{if not,} \end{cases} \quad (2.3.3)$$

$$g_2^*(s, p) = \sup_{x,y} \{\langle s, x\rangle_n + \langle p, y\rangle_m : y = Ax\} = \sup_x \langle x, s + A^*p\rangle_n.$$

Hence

$$g_2^*(s, p) = \begin{cases} 0 & \text{if } s + A^*p = 0, \\ +\infty & \text{if not.} \end{cases} \quad (2.3.4)$$

We want to study $(g \circ A)^*$, which is obtained from the following calculation:

$$(g_1 + g_2)^*(s, 0) = \sup_{x,y} \{\langle s, x\rangle_n - g(y) : y = Ax\} = (g \circ A)^*(s),$$

so the whole question is to compute the conjugate of $g_1 + g_2$. If (2.3.1) holds, we have

$$(g \circ A)^*(s) = (g_1^* \mathbin{\dot{\vee}} g_2^*)(s, 0) = \inf_{s_1, s_2, p} \{g_1^*(s_1, p) + g_2^*(s_2, -p) : s_1 + s_2 = s\};$$

but, due to the particular structure appearing in (2.3.3) and (2.3.4), the above minimization problem can be written

$$\inf \{g^*(p) : s - A^*p = 0\} = A^*g^*(s).$$

There remains to check (2.3.1), and this is easy:

$$\text{dom } g_1 = \mathbb{R}^n \times \text{dom } g \quad \text{hence} \quad \text{ri dom } g_1 = \mathbb{R}^n \times \text{ri dom } g;$$
$$\text{dom } g_2 = \text{gr } A \quad \text{hence} \quad \text{ri dom } g_2 = \text{gr } A.$$

Under these conditions, (2.3.1) expresses the existence of an x_0 such that $Ax_0 \in \text{ri dom } g$, and this is exactly (2.2.4), which was our starting hypothesis. □

To conclude this study of the sum, we mention one among the possible results concerning its dual operation: the infimal convolution.

Corollary 2.3.4 *Take f_1 and f_2 in $\overline{\text{Conv}}\,\mathbb{R}^n$, with f_1 0-coercive and f_2 bounded from below. Then the inf-convolution problem (2.1.3) has a nonempty compact set of solutions; furthermore $f_1 \mathbin{\dot{\vee}} f_2 \in \overline{\text{Conv}}\,\mathbb{R}^n$.*

PROOF. Letting μ denote a lower bound for f_2, we have

$$f_1(x_1) + f_2(x - x_1) \geq f_1(x_1) + \mu \quad \text{for all } x_1 \in \mathbb{R}^n,$$

and the first part of the claim follows. For closedness of the infimal convolution, we set $g_i := f_i^*$, $i = 1, 2$; because of 0-coercivity of f_1, $0 \in \text{int dom } g_1$ (Remark 1.3.10), and $g_2(0) \leq -\mu$. Thus, we can apply Theorem 2.3.2 with the qualification assumption (2.3.Q.jj'). □

2.4 Infima and Suprema

A result of the previous sections is that the conjugacy correspondence establishes a symmetry between the sum and the inf-convolution, and also between the image and the pre-composition with a linear mapping. Here we will see that the operations "supremum" and "closed convex hull of the infimum" are likewise symmetric to each other.

Theorem 2.4.1 *Let $\{f_j\}_{j \in J}$ be a collection of functions satisfying (1.1.1) and assume that there is a common affine function minorizing all of them:*

$$\sup \{f_j^*(s) : j \in J\} < +\infty \quad \text{for some } s \in \mathbb{R}^n.$$

Then their infimum $f := \inf_{j \in J} f_j$ satisfies (1.1.1), and its conjugate is the supremum of the f_j^*'s:

$$\left(\inf_{j \in J} f_j\right)^* = \sup_{j \in J} f_j^*. \tag{2.4.1}$$

PROOF. By definition, for all $s \in \mathbb{R}^n$

$$\begin{aligned} f^*(s) &= \sup_x [\langle s, x \rangle - \inf_{j \in J} f_j(x)] \\ &= \sup_x \sup_j [\langle s, x \rangle - f_j(x)] \\ &= \sup_j \sup_x [\langle s, x \rangle - f_j(x)] = \sup_{j \in J} f_j^*(s). \end{aligned}$$ □

This result should be compared to Corollary 2.1.2, which after all expresses the conjugate of the same function, and which is proved in just the same way; compare the above proof with that of Theorem 2.1.1. The only difference (apart from the notation j instead of z) is that the space $\mathbb{R}^p$ is now replaced by an arbitrary set J, with no special structure, in particular no scalar product. In fact, Theorem 2.4.1 supersedes Corollary 2.1.2: the latter just says that, for g defined on a product-space, the easy-to-prove formula

$$g^*(s, 0) = \sup_z g_x^*(s, z)$$

holds, where the subscript x of g_x^* indicates conjugacy with respect to the first variable only. In a word, the conjugate with respect to the *couple* (x, z) is of no use for computing the conjugate at $(\cdot, 0)$. Beware that this is true only when the scalar product considers x and z as two separate variables (i.e. is compatible with the product-space). Otherwise, trouble may occur: see Remark 2.2.4.

Example 2.4.2 The Euclidean distance to a nonempty set $C \subset \mathbb{R}^n$ is an inf-function:

$$x \mapsto d_C(x) = \inf \{f_y(x) : y \in C\} \quad \text{with} \quad f_y(x) := \|x - y\|.$$

Remembering that the conjugate of the norm $\|\cdot\|$ is the indicator of the unit ball (combine (V.2.3.1) and Example 1.1.5), the calculus rule 1.3.1(v) gives

$$(f_y)^*(s) = I_{B(0,1)}(s) + \langle s, y \rangle \tag{2.4.2}$$

and (2.4.1) may be written as

$$d_C^*(s) = I_{B(0,1)}(s) + \sup_{y \in C} \langle s, y \rangle = I_{B(0,1)}(s) + \sigma_C(s),$$

a formula already given in Example 2.1.4. The same exercise can be carried out with the squared distance. □

Example 2.4.3 (Directional Derivative) With $f \in \overline{\text{Conv}}\,\mathbb{R}^n$, let x_0 be a point where $\partial f(x_0)$ is nonempty and consider for $t > 0$ the functions

$$\mathbb{R}^n \ni d \mapsto f_t(d) := \frac{f(x_0 + td) - f(x_0)}{t}. \tag{2.4.3}$$

They are all minorized by $\langle s_0, \cdot \rangle$, with $s_0 \in \partial f(x_0)$; also, the difference quotient is an increasing function of the stepsize; their infimum is therefore obtained for $t \downarrow 0$. This infimum is denoted by $f'(x_0, \cdot)$, the *directional derivative* of f at x_0 (already encountered in Chap. VI, in a finite-valued setting). Their conjugates can be computed directly or using Proposition 1.3.1:

$$s \mapsto (f_t)^*(s) = \frac{f^*(s) + f(x_0) - \langle s, x_0 \rangle}{t}, \tag{2.4.4}$$

so we obtain from (2.4.1)

$$[f'(x_0, \cdot)]^*(s) = \sup_{t>0} \frac{f^*(s) + f(x_0) - \langle s, x_0 \rangle}{t}.$$

Observe that the supremand is always nonnegative, and that it is 0 if and only if $s \in \partial f(x_0)$ (cf. Theorem 1.4.1). In a word:

$$[f'(x_0, \cdot)]^* = I_{\partial f(x_0)}.$$

Taking the conjugate again, we see that the support function of the subdifferential is *the closure of* the directional derivative; a result which must be compared to §VI.1.1. □

As already mentioned, sup-functions are fairly important in optimization, so a logical supplement to (2.4.1) is the conjugate of a supremum.

Theorem 2.4.4 *Let $\{g_j\}_{j \in J}$ be a collection of functions in $\overline{\text{Conv}}\,\mathbb{R}^n$. If their supremum $g := \sup_{j \in J}$ is not identically $+\infty$, it is in $\overline{\text{Conv}}\,\mathbb{R}^n$, and its conjugate is the closed convex hull of the g_i^*'s:*

$$(\sup_{j \in J} g_j)^* = \overline{\text{co}}\left(\inf_{j \in J} g_j^*\right). \tag{2.4.5}$$

PROOF. Call $f_j := g_j^*$, hence $f_j^* = g_j$, and g is nothing but the f^* of (2.4.1). Taking the conjugate of both sides, the result follows from (1.3.4). □

Example 2.4.5 Given, as in Example 2.4.2, a nonempty bounded set C, let

$$\mathbb{R}^n \ni x \mapsto \Delta_C(x) := \sup \{\|x - y\| : y \in C\}$$

be the distance from x to the most remote point of cl C. Using (2.4.2),

$$\inf_{y \in C} (f_y)^*(s) = I_{B(0,1)}(s) + \inf_{y \in C} \langle s, y \rangle,$$

so (2.4.5) gives

$$\Delta_C^* = \overline{\mathrm{co}}\left(I_{B(0,1)} - \sigma_{-C}\right). \qquad \square$$

Example 2.4.6 (Asymptotic Function) With $f \in \overline{\mathrm{Conv}}\,\mathbb{R}^n$, $x_0 \in \mathrm{dom}\,f$ and f_t as defined by (2.4.3), consider

$$f'_\infty(d) := \sup\{f_t(d) : t > 0\} = \lim\{f_t(d) : t \to +\infty\}$$

(see §IV.3.2). Clearly, $f_t \in \overline{\mathrm{Conv}}\,\mathbb{R}^n$ and $f_t(0) = 0$ for all $t > 0$, so we can apply Theorem 2.4.4: $f'_\infty \in \overline{\mathrm{Conv}}\,\mathbb{R}^n$. In view of (2.4.4), the conjugate of f'_∞ is therefore the closed convex hull of the function

$$s \mapsto \inf_{t>0} \frac{f^*(s) + f(x_0) - \langle s, x_0 \rangle}{t}.$$

Since the infimand is in $[0, +\infty]$, the infimum is 0 if $s \in \mathrm{dom}\,f^*$, $+\infty$ if not. In summary, $(f'_\infty)^*$ is the indicator of cl dom f^*. Conjugating again, we obtain

$$f'_\infty = \sigma_{\mathrm{dom}\,f^*},$$

a formula already seen in (1.2.3).

The comparison with Example 2.4.3 is interesting. In both cases we extremize the same function, namely the difference quotient (2.4.3); and in both cases a support function is obtained (neglecting the closure operation). Naturally, the supported set is larger in the present case of maximization. Indeed, dom f^* and the union of all the subdifferentials of f have the same closure. $\qquad \square$

The conjugate of a sup-function appears to be rather difficult to compute. One possibility is to take the supremum of all the affine functions minorizing all the f_j^*'s, i.e. to solve (1.3.5) (a problem with infinitely many constraints, where f stands for $\inf_j f_j^*$). Another possibility, based on (IV.2.5.3), is to solve for $\{\alpha_j, s_j\}$

$$\inf\left\{\sum_{j=1}^{n+1} \alpha_j f_j^*(s_j) : \alpha \in \Delta_{n+1},\; s_j \in \mathrm{dom}\,f_j^*,\; \sum_{j=1}^{n+1} \alpha_j s_j = s\right\}; \qquad (2.4.6)$$

but then we still have to take the lower semi-continuous hull of the result. Both possibilities are rather involved, let us mention a situation where the second one takes an easier form.

Theorem 2.4.7 Let $f_1, \ldots, f_p$ be finitely many convex functions from $\mathbb{R}^n$ to $\mathbb{R}$, and let $f := \max_j f_j$; denote by $m := \min\{p, n+1\}$. For every
$$s \in \operatorname{dom} f^* = \operatorname{co} \cup \{\operatorname{dom} f_j^* : j = 1, \ldots, p\},$$
there exist m vectors $s_j \in \operatorname{dom} f_j^*$ and convex multipliers α_j such that
$$s = \sum_j \alpha_j s_j \quad \text{and} \quad f^*(s) = \sum_j \alpha_j f_j^*(s_j). \tag{2.4.7}$$

In other words: an optimal solution exists in the minimization problem (2.4.6), and the optimal value is a closed convex function of s.

PROOF. Set $g := \min_j f_j^*$, so $f^* = \overline{\operatorname{co}}\, g$; observe that $\operatorname{dom} g = \cup_j \operatorname{dom} f_j^*$. Because each f_j^* is closed and 1-coercive, so is g. Then we apply Proposition 1.5.4: first, $\operatorname{co} g = \overline{\operatorname{co}}\, g$; second, for every
$$s \in \operatorname{dom} f^* = \operatorname{dom} \operatorname{co} g = \operatorname{co}\{\cup_j \operatorname{dom} f_j^*\},$$
there are $s_j \in \operatorname{dom} g$ and convex multipliers α_j, $j = 1, \ldots, n+1$ such that
$$s = \sum_{j=1}^{n+1} \alpha_j s_j \quad \text{and} \quad (\operatorname{co} g)(s) = f^*(s) = \sum_{j=1}^{n+1} \alpha_j g^*(s_j).$$

In the last sum, each $g^*(s_j)$ is a certain $f_i^*(s_j)$; furthermore, several s_j's having the same f_i^* can be compressed to a single convex combination (thanks to convexity of each f_i^*, see Proposition IV.2.5.4). Thus, we obtain (2.4.7). □

The framework in which we proved this result may appear very restrictive; however, enlarging it is not easy. For example, extended-valued function f_j create a difficulty: (2.4.7) does not hold with $n = 1$ and
$$f_1(x) = \exp x \quad \text{and} \quad f_2 = I_{\{0\}}.$$

Example 2.4.8 Consider the function $g^+ := \max\{0, g\}$, where $g : \mathbb{R}^n \to \mathbb{R}$ is convex. With $f_1 := g$ and $f_2 \equiv 0$, we have $f_2^* = I_{\{0\}}$ and, according to Theorem 2.4.7: for all $s \in \operatorname{dom}(g^+)^* = \operatorname{co}\{\{0\} \cup \operatorname{dom} g^*\}$,
$$(g^+)^*(s) = \min\{\alpha g^*(s_1) : s_1 \in \operatorname{dom} g^*, \ \alpha \in [0, 1], \ \alpha s_1 = s\}.$$
For $s \neq 0$, the value $\alpha = 0$ is infeasible: we can write
$$(g^+)^*(s) = \min_{0 < \alpha \leq 1} \alpha g^*(\tfrac{1}{\alpha} s) \quad \text{for } s \neq 0.$$
For $s = 0$, the feasible solutions have either $\alpha = 0$ or $s_1 = 0$. Both cases are covered by the condensed formulation
$$(g^+)^*(0) = \min\{g^*(0), 0\},$$
which could also have been obtained by successive applications of the formula $\inf f = -f^*(0)$. □

2.5 Post-Composition with an Increasing Convex Function

For $f \in \overline{\text{Conv}}\,\mathbb{R}^n$ and $g \in \overline{\text{Conv}}\,\mathbb{R}$, the function $g \circ f$ is in $\overline{\text{Conv}}\,\mathbb{R}^n$ if g is increasing (we assume $f(\mathbb{R}^n) \cap \text{dom}\,g \neq \emptyset$ and we set $g(+\infty) = +\infty$). A relevant question is then how to express its conjugate, in terms of f^* and g^*.

We start with some preliminary observations, based on the particular nature of g. The domain of g is an interval unbounded from the left, whose relative interior is int dom $g \neq \emptyset$. Also, dom $g^* \subset \mathbb{R}^+$ and (remember §I.3.2)

$$g(y) = g^{**}(y) = \sup\{py - g^*(p) : p \in \text{ri dom}\,g^*\}.$$

Theorem 2.5.1 *With f and g as above, assume that $f(\mathbb{R}^n) \cap \text{int dom}\,g \neq \emptyset$. For all $s \in \text{dom}(g \circ f)^*$, define the function $\psi_s \in \overline{\text{Conv}}\,\mathbb{R}$ by*

$$\mathbb{R} \ni \alpha \mapsto \psi_s(\alpha) = \begin{cases} \alpha f^*(\frac{1}{\alpha}s) + g^*(\alpha) & \text{if } \alpha > 0, \\ \sigma_{\text{dom}\,f}(s) + g^*(0) & \text{if } \alpha = 0, \\ +\infty & \text{if } \alpha < 0. \end{cases}$$

Then

$$(g \circ f)^*(s) = \min_{\alpha \in \mathbb{R}} \psi_s(\alpha).$$

PROOF. By definition,

$$\begin{aligned} -(g \circ f)^*(s) &= \inf_x [g(f(x)) - \langle s, x \rangle] \\ &= \inf_{x,r}\{g(r) - \langle s, x \rangle : f(x) \leq r\} \quad [g \text{ is increasing}] \\ &= \inf_{x,r}[g(r) - \langle s, x \rangle + I_{\text{epi}\,f}(x, r)]. \end{aligned}$$

Then we must compute the conjugate of a sum; we set

$$\mathbb{R}^n \times \mathbb{R} \ni (x, r) \mapsto f_1(x, r) := g(r) - \langle s, x \rangle$$

and $f_2 := I_{\text{epi}\,f}$. We have

$$\text{dom}\,f_1 = \mathbb{R}^n \times \text{dom}\,g, \quad \text{int dom}\,f_1 = \mathbb{R}^n \times \text{int dom}\,g;$$

so, by assumption:

$$\text{int dom}\,f_1 \cap \text{dom}\,f_2 = (\mathbb{R}^n \times \text{int dom}\,g) \cap \text{epi}\,f \neq \emptyset.$$

Theorem 2.3.2, more precisely Fenchel's duality theorem (2.3.2), can be applied with the qualification condition (2.3.Q.jj'):

$$(g \circ f)^*(s) = \min\{f_1^*(-p, \alpha) + f_2^*(p, -\alpha) : (p, \alpha) \in \mathbb{R}^n \times \mathbb{R}\}.$$

The computation of the above two conjugates is straightforward and gives

$$(g \circ f)^*(s) = \min_{p,\alpha}[g^*(\alpha) + I_{\{-s\}}(-p) + \sigma_{\text{epi}\,f}(p, -\alpha)] = \min_\alpha \psi_s(\alpha),$$

where the second equality comes from (1.2.2). □

Let us take two examples, using the ball-pen function defined on its domain $B(0, 1)$ by $f(x) = 1 - \sqrt{1 - \|x\|^2}$: here $f(\mathbb{R}^n) = [0, 1] \cup \{+\infty\}$.

- With $g(r) = r^+$, we leave it to the reader to compute the relevant conjugates: $\psi_s(\alpha)$ has the unique minimum-point $\bar{\alpha} = 0$ for all s.
- With $g = I_{]-\infty,0]}$, the qualification assumption in the above theorem is no longer satisfied. We have $(g \circ f)^*(s) = 0$ for all s, while $\min_\alpha \psi_s(\alpha) = \|s\| - 1$.

Remark 2.5.2 Under the qualification assumption, ψ_s always attains its minimum at some $\bar{\alpha} \geqslant 0$. As a result, if for example $\sigma_{\text{dom } f}(s) = (f^*)'_\infty(s) = +\infty$, or if $g^*(0) = +\infty$, we are sure that $\bar{\alpha} > 0$. □

Theorem 2.5.1 takes an interesting form in case of positive homogeneity, which can apply to f or g. For the two examples below, we recall that a closed sublinear function is the support of its subdifferential at the origin.

Example 2.5.3 If g is positively homogeneous, say $g = \sigma_\ell$ for some closed interval ℓ, the domain of ψ_s is contained in ℓ, on which the term $g^*(\alpha)$ disappears. The interval ℓ supported by g has to be contained in $\mathbb{R}^+$ (to preserve monotonicity) and there are essentially two instances: Example 2.4.8 was one, in which ℓ was $[0, 1]$; in the other instance, ℓ is unbounded (a half-line), and the following example gives an application.

Consider the set $C := \{x \in \mathbb{R}^n : f(x) \leqslant 0\}$, with $f : \mathbb{R}^n \to \mathbb{R}$ convex; its support function can be expressed in terms of f^*. Indeed, write the indicator of C as the composed function:

$$I_C = I_{]-\infty,0]} \circ f.$$

This places us in the present framework: g is the support of $\mathbb{R}^+$. To satisfy the hypotheses of Theorem 2.5.1, we have only to assume the existence of an x_0 with $f(x_0) < 0$; then the result is: for $0 \neq s \in \text{dom } \sigma_C$, there exists $\bar{\alpha} > 0$ such that

$$\sigma_C(s) = \min_{\alpha > 0} \alpha f^*(\tfrac{1}{\alpha}s) = \bar{\alpha} f^*(s/\bar{\alpha}).$$

Indeed, we are in the case of Remark 2.5.2 because $\text{dom } f = \mathbb{R}^n$. □

Example 2.5.4 In our second example, it is f that is positively homogeneous. Then, for all $s \in \text{dom}(g \circ f)^*$,

$$\psi_s(\alpha) = \begin{cases} g^*(\alpha) & \text{if } \alpha > 0 \text{ and } \tfrac{1}{\alpha}s \in \partial f(0), \\ \sigma_{\text{dom } f}(s) + g^*(0) & \text{if } \alpha = 0, \\ +\infty & \text{if } \alpha < 0. \end{cases}$$

If f is finite everywhere, i.e. $\partial f(0)$ is bounded, we have for all $s \neq 0$

$$(g \circ f)^*(s) = \min\{g^*(\alpha) : \alpha > 0 \text{ and } \tfrac{1}{\alpha}s \in \partial f(0)\}.$$

As an application, let Q be symmetric positive definite and define the function

$$\mathbb{R}^n \ni x \mapsto f(x) := \sqrt{\langle Q^{-1}x, x \rangle},$$

which is the support function of the elliptic set (see Example V.2.3.4)

$$E_Q := \partial f(0) = \{s \in \mathbb{R}^n : \langle s, Qs \rangle \leqslant 1\}.$$

The composition of f with the function $r \mapsto g(r) := 1/2 r^2$ is the quadratic from associated with Q^{-1}, whose conjugate is known (see Example 1.1.3):
$$(g \circ f)(x) = \tfrac{1}{2}\langle Q^{-1}x, x\rangle, \quad (g \circ f)^*(s) = \tfrac{1}{2}\langle s, Qs\rangle.$$
Then
$$\tfrac{1}{2}\langle s, Qs\rangle = \min\{\tfrac{1}{2}\alpha^2 : \alpha \geq 0 \text{ and } s \in \alpha E_Q\}$$
is the half-square of the gauge function γ_{E_Q} (Example V.2.3.4). □

2.6 A Glimpse of Biconjugate Calculus

Taking the biconjugate of a function, i.e. "close-convexifying" it, is an important operation. A relevant question is therefore to derive rules giving the biconjugate of a function obtained from other functions having known biconjugates.

First of all, Proposition 1.3.1 can easily be reproduced to give the following statements, in which the various f's and f_j's are assumed to satisfy (1.1.1).

(j) The biconjugate of $g(\cdot) := f(\cdot) + \alpha$ is $(\overline{\text{co}}\, g)(\cdot) = (\overline{\text{co}}\, f)(\cdot) + \alpha$.
(jj) With $\alpha > 0$, the biconjugate of $g := \alpha f$ is $(\overline{\text{co}}\, g) = \alpha(\overline{\text{co}}\, f)$.
(jjj) With $\alpha \neq 0$, the biconjugate of the function $g(x) := f(\alpha x)$ is $(\overline{\text{co}}\, g)(x) = (\overline{\text{co}}\, f)(\alpha x)$.
(jv) More generally: if A is an invertible linear operator, $\overline{\text{co}}(f \circ A) = (\overline{\text{co}}\, f) \circ A$.
(v) The biconjugate of $g(\cdot) := f(\cdot - x_0)$ is $(\overline{\text{co}}\, g)(\cdot) = (\overline{\text{co}}\, f)(\cdot - x_0)$.
(vj) The biconjugate of $g := f + \langle s_0, \cdot\rangle$ is $(\overline{\text{co}}\, g) = (\overline{\text{co}}\, f) + \langle s_0, \cdot\rangle$.
(vjj) If $f_1 \leq f_2$, then $\overline{\text{co}}\, f_1 \leq \overline{\text{co}}\, f_2$.
(jx) The biconjugate of $f(x) = \sum_{j=1}^m f_j(x_j)$ is $\overline{\text{co}}\, f = \sum_{j=1}^m \overline{\text{co}}\, f_j$.

Note: there is nothing corresponding to (viii). All these results are straightforward, from the very definition of the closed convex hull. Other calculus rules are not easy to derive: for most of the operations involved, only *inequalities* are obtained.

Consider first the sum: let f_1 and f_2 satisfy (1.1.1), as well as
$$\text{dom } f_1 \cap \text{dom } f_2 \neq \emptyset.$$
Clearly, $\overline{\text{co}}\, f_1 + \overline{\text{co}}\, f_2$ is then a function of $\overline{\text{Conv}}\,\mathbb{R}^n$ which minorizes $f_1 + f_2$. Therefore
$$\overline{\text{co}}\, f_1 + \overline{\text{co}}\, f_2 \leq \overline{\text{co}}(f_1 + f_2).$$
This inequality is the only general comparison result one can get, as far as sums are concerned; and closed convexity of f_1, say, does not help: it is only when f_1 is *affine* that equality holds.

Let now $(f_j)_{j \in J}$ be a collection of functions and $f := \sup f_j$. Starting from $f_j \leq f$ for $j \in J$, we obtain immediately $\overline{\text{co}}\, f_j \leq \overline{\text{co}}\, f$ and
$$\sup_{j \in J}(\overline{\text{co}}\, f_j) \leq \overline{\text{co}}\, f.$$
Nothing more accurate can be said in general.

The case of an infimum is hardly more informative:

Proposition 2.6.1 *Let $\{f_j\}_{j \in J}$ be a collection of functions satisfying (1.1.1) and assume that there is a common affine function minorizing all of them. Then*

$$\overline{\mathrm{co}}(\inf_{j \in J} f_j) = \overline{\mathrm{co}}(\inf_{j \in J} \overline{\mathrm{co}}\, f_j) \leqslant \inf_{j \in J} \overline{\mathrm{co}}\, f_j\,.$$

PROOF. The second relation is trivial. As for the first, Theorem 2.4.1 gives $(\inf_j f_j)^* = \sup_j f_j^*$. The left-hand function is in $\overline{\mathrm{Conv}}\,\mathbb{R}^n$, so we can take the conjugate of both sides and apply Theorem 2.4.4. □

So far, the calculus developed in the previous sections has been of little help. The situation improves for the image function under a linear mapping.

Proposition 2.6.2 *Let $g : \mathbb{R}^m \to \mathbb{R} \cup \{+\infty\}$ satisfy (1.1.1) and let A be linear from $\mathbb{R}^n$ to $\mathbb{R}^m$. Assuming $\mathrm{Im}\, A^* \cap \mathrm{ri}\,\mathrm{dom}\, g^* \neq \emptyset$, the equality $\overline{\mathrm{co}}(Ag) = A(\overline{\mathrm{co}}\, g)$ holds. Actually, for each $x \in \mathrm{dom}\,\overline{\mathrm{co}}(Ag)$, there exists y such that*

$$[\overline{\mathrm{co}}(Ag)](x) = (\overline{\mathrm{co}}\, g)(y) \quad \text{and} \quad Ay = x\,.$$

PROOF. Since $\mathrm{Im}\, A^* \cap \mathrm{dom}\, g^* \neq \emptyset$, Theorem 2.1.1 applies: $(Ag)^* = g^* \circ A^*$. Likewise, Theorem 2.2.3 (with g and A replaced by g^* and A^*) allows the computation of the conjugate of the latter. □

The qualification assumption needed in this result is somewhat abstract; a more concrete one is for example 0-coercivity of $\overline{\mathrm{co}}\, g$: then $\mathrm{Im}\, A^* \ni 0 \in \mathrm{int}\,\mathrm{dom}\, g^*$ (see Remark 1.3.10).

The cases of marginal functions and inf-convolutions are immediately taken care of by Proposition 2.6.2; the corresponding statements and proofs are left to the reader.

3 Various Examples

The examples listed below illustrate some applications of the conjugacy operation. It is an obviously useful tool thanks to its convexification ability; note also that the knowledge of f^* fully characterizes $\overline{\mathrm{co}}\, f$, i.e. f in case the latter is already in $\overline{\mathrm{Conv}}\,\mathbb{R}^n$; another fundamental property is the characterization (1.4.2) of the subdifferential. All this concerns convex analysis more or less directly, but the conjugacy operation is instrumental in several areas of mathematics.

3.1 The Cramer Transformation

Let P be a probability measure on $\mathbb{R}^n$ and define its *Laplace transform* $\mathrm{L}P : \mathbb{R}^n \to \,]0, +\infty]$ by

$$\mathbb{R}^n \ni s \mapsto \mathrm{L}P(s) := \int_{\mathbb{R}^n} \exp\langle s, z\rangle\, \mathrm{d}P(z)\,.$$

The so-called *Cramer transform* of P is the function

$$\mathbb{R}^n \ni x \mapsto C_P(x) := \sup\{\langle s, x\rangle - \log L_P(s) : s \in \mathbb{R}^n\}.$$

It is used for example in statistics. We recognize in C_P the conjugate of the function $\log L_P$, which can be shown to be convex (as a consequence of Hölder's inequality) and lower semi-continuous (from Fatou's lemma). We conclude that C_P is a closed convex function, whose conjugate is $\log L_P$.

3.2 Some Results on the Euclidean Distance to a Closed Set

Let S be a nonempty closed set in $\mathbb{R}^n$ and define the function $f := 1/2 \|\cdot\|^2 + I_S$:

$$f(x) := \begin{cases} \frac{1}{2}\|x\|^2 & \text{if } \in S, \\ +\infty & \text{if not}. \end{cases} \qquad (3.2.1)$$

Clearly, f satisfies (1.1.1), and is lower semi-continuous and 1-coercive; it is convex if and only if S is convex; its conjugate turns out to be of special interest. By definition,

$$f^*(s) = \sup\left\{\langle s, x\rangle - \tfrac{1}{2}\|x\|^2 : x \in S\right\},$$

which is the function φ_S of Example IV.2.1.4:

$$f^*(s) = \tfrac{1}{2}\|s\|^2 - \tfrac{1}{2}\min_{x \in S}\|s - x\|^2 = \tfrac{1}{2}\left[\|s\|^2 - d_S^2(s)\right]. \qquad (3.2.2)$$

From Theorem 1.5.4, $\operatorname{co} f = \overline{\operatorname{co}} f$ and, for each $x \in \operatorname{co} S$, there are $x_1, \ldots, x_{n+1}$ in S and convex multipliers $\alpha_1, \ldots, \alpha_{n+1}$ such that

$$x = \sum_{j=1}^{n+1} \alpha_j x_j \quad \text{and} \quad (\operatorname{co} f)(x) = \tfrac{1}{2}\sum_{j=1}^{n+1} \alpha_j \|x_j\|^2.$$

The (nonconvex) function f of (3.2.1) has a subdifferential in the sense of (1.4.1), which is characterized via the projection operation onto S:

$$P_S(x) := \{y \in S : d_S(x) = \|y - x\|\}$$

(a compact-valued multifunction, having the whole space as domain).

Proposition 3.2.1 *Let $x \in S$. With f given by (3.2.1),*

(i) s is a subgradient of f at x if and only if its projection onto S contains x; in other words,

$$\partial f(x) = \{s \in \mathbb{R}^n : x \in P_S(s)\} = \{s \in \mathbb{R}^n : d_S(s) = \|s - x\|\};$$

in particular, $x \in \partial f(x)$;

(ii) $(\operatorname{co} f)(x) = f(x) = \tfrac{1}{2}\|x\|^2$ and $\partial(\operatorname{co} f)(x) = \partial f(x)$.

PROOF. From the expression (3.2.2) of f^*, $s \in \partial f(x)$ if and only if

$$\tfrac{1}{2}\|s\|^2 - \tfrac{1}{2}d_S^2(s) + \tfrac{1}{2}\|x\|^2 + I_S(x) = \langle s, x \rangle.$$

Using $I_S(x) = 0$, this means $d_S(s) = \|s-x\|$, i.e. $x \in P_S(s)$. In particular, $s = x$ is in $\partial f(x)$, just because $P_S(x) = \{x\}$. Hence, $\partial f(x)$ is nonempty and (ii) is a consequence of (1.4.3), (1.4.4). □

Having thus characterized ∂f, we can compute $\partial(\operatorname{co} f)$ using Theorem 1.5.6, and this in turn allows the computation of ∂f^*:

Proposition 3.2.2 *For all $s \in \mathbb{R}^n$,*

$$\partial f^*(s) = \operatorname{co} P_S(s). \tag{3.2.3}$$

PROOF. Because $\operatorname{co} f$ is closed, $x \in \partial f^*(s) \Leftrightarrow s \in \partial(\operatorname{co} f)(x)$, which in turn is expressed by (1.5.4): there are $x_1, \ldots, x_{n+1}$ in S and convex multipliers $\alpha_1, \ldots, \alpha_{n+1}$ such that

$$x = \sum_{j=1}^{n+1} \alpha_j x_j \quad \text{and} \quad s \in \cap\{\partial f(x_j) : \alpha_j > 0\}.$$

It then suffices to apply Proposition 3.2.1(i): $s \in \partial f(x_j) \Leftrightarrow x_j \in P_S(s)$ for $j = 1, \ldots, n+1$. □

This result is interesting in itself. For example, we deduce that the projection onto a closed set is almost everywhere a singleton, just because ∇f^* exists almost everywhere. When S is convex, $P_S(s)$ is for any s a singleton $\{p_S(s)\}$, which thus appears as the gradient of f^* at s:

$$\nabla\left(\tfrac{1}{2}d_S^2\right) = I - p_S \quad \text{if } S \text{ is convex}.$$

Another interesting result is that the converse to this property is true: if d_S^2 (or equivalently f^*) is differentiable, then S is convex and we obtain a convexity criterion:

Theorem 3.2.3 *For a nonempty closed set S, the following statements are equivalent:*
  (i) *S is convex;*
  (ii) *the projection operation P_S is single-valued on $\mathbb{R}^n$;*
  (iii) *the squared distance d_S^2 is differentiable on $\mathbb{R}^n$;*
  (iv) *the function $f = I_S + 1/2\|\cdot\|^2$ is convex.*

PROOF. We know that (i) $\Rightarrow$ (ii); when (ii) holds, (3.2.3) tells us that the function $f^* = 1/2(\|\cdot\|^2 - d_S^2)$ is differentiable, i.e. (iii) holds. When f in (iv) is convex, its domain S is convex. There remains to prove (iii) $\Rightarrow$ (iv).

Thus, assuming (iii), we want to prove that $f = \operatorname{co} f$ ($=\overline{\operatorname{co}} f$). Take first $x \in \operatorname{ri} \operatorname{dom} \operatorname{co} f$, so that there is an $s \in \partial(\operatorname{co} f)(x)$ (Theorem 1.4.2). In view of (1.5.4), there are $x_j \in S$ and positive convex multipliers α_j such that

$$x = \sum_j \alpha_j x_j, \quad (\text{co } f)(x) = \sum_j \alpha_j f(x_j), \quad s \in \cap_j \partial f(x_j).$$

The last property implies that each x_j is the unique element of $\partial f^*(s)$: they are all equal and it follows $(\text{co } f)(x) = f(x)$.

Now use Proposition IV.1.2.5, expressing $\overline{\text{co}} \, f$ outside the relative interior of its domain: for $x \in \text{dom co } f$, there is a sequence $\{x_k\} \subset \text{ri dom co } f$ tending to x such that

$$f(x) \geq (\text{co } f)(x) = \lim_{k \to +\infty} (\text{co } f)(x_k) = \lim f(x_k) \geq f(x),$$

where the last inequality comes from the lower semi-continuity of f. □

3.3 The Conjugate of Convex Partially Quadratic Functions

A simple case in §3.2 is one in which S is a subspace: we essentially obtain a quadratic function "restricted" to a subspace, resembling the f^* obtained in Example 1.1.4. Thus, given a linear symmetric positive semi-definite operator B and a subspace H, it is of interest to compute the conjugate of

$$g(x) := \begin{cases} \frac{1}{2}\langle Bx, x \rangle & \text{if } x \in H, \\ +\infty & \text{otherwise}. \end{cases} \quad (3.3.1)$$

Such a function (closed, convex, homogeneous of degree 2) is said to be *partially quadratic*. Its conjugate turns out to have just the same form, so the set of convex partially quadratic functions is stable with respect to the conjugacy operation (cf. Example 1.1.4).

Proposition 3.3.1 *The function g of (3.3.1) has the conjugate*

$$g^*(s) = \begin{cases} \frac{1}{2}\langle s, (P_H \circ B \circ P_H)^- s \rangle & \text{if } s \in \text{Im } B + H^\perp, \\ +\infty & \text{otherwise}, \end{cases} \quad (3.3.2)$$

where P_H is the operator of orthogonal projection onto H and $(\cdot)^-$ is the Moore-Penrose pseudo-inverse.

PROOF. Writing g as $f + I_H$ with $f := 1/2 \langle B\cdot, \cdot \rangle$, use Proposition 1.3.2: $g^* = (f \circ P_H)^* \circ P_H$; knowing that

$$(f \circ P_H)(x) = \tfrac{1}{2}\langle P_H \circ B \circ P_H x, x \rangle,$$

we obtain from Example 1.1.4 $g^*(s)$ under the form

$$(f \circ P_H)^*(P_H s) = \begin{cases} \frac{1}{2}\langle s, (P_H \circ B \circ P_H)^- s \rangle & \text{if } P_H s \in \text{Im}(P_H \circ B \circ P_H), \\ +\infty & \text{otherwise}. \end{cases}$$

It could be checked directly that $\text{Im}(P_H \circ B \circ P_H) + H^\perp = \text{Im } B + H^\perp$. A simpler argument, however, is obtained via Theorem 2.3.2, which can be applied since dom $f = \mathbb{R}^n$. Thus,

$$g^*(s) = \left(f^* \dotplus I_{H^\perp}\right)(s) = \min\left\{\tfrac{1}{2}\langle p, B^- p\rangle \ : \ p \in \operatorname{Im} B, \ s - p \in H^\perp\right\},$$

which shows that $\operatorname{dom} g^* = \operatorname{Im} B + H^\perp$. □

For example, suppose $H = \operatorname{Im} B$. Calling $A := B^-$ the pseudo-inverse of B (hence $A^- = (B^-)^- = B$), we see that g of (3.3.1) is just f^* of Example 1.1.4 with $b = 0$. Since $\operatorname{Im} B + H^\perp = \mathbb{R}^n$, and using the relations $P_H \circ B = B \circ P_H = B$, we obtain finally that g^* is the f of (1.1.4) – and this is perfectly normal: $g^* = f^{**} = f$.

As another example, take the identity for B; then $\operatorname{Im} B = \mathbb{R}^n$ and Proposition 3.3.1 gives immediately

$$g^*(s) = \tfrac{1}{2}\langle s, (P_H)^- s\rangle = \tfrac{1}{2}\langle s, P_H s\rangle \quad \text{for all } s \in \mathbb{R}^n.$$

This last example fits exactly into the framework of §3.2, and (3.2.2) becomes

$$\langle s, P_H s\rangle = \|s\|^2 - d_H^2(s) = \|s\|^2 - \|s - P_H s\|^2,$$

a classical relation known as Pythagoras' Theorem.

3.4 Polyhedral Functions

For given $(s_i, b_i) \in \mathbb{R}^n \times \mathbb{R}$, $i = 1, \ldots, k$, set

$$f_i(x) := \langle s_i, x\rangle - b_i \quad \text{for } i = 1, \ldots, k$$

and define the piecewise affine function

$$\mathbb{R}^n \ni x \mapsto f(x) := \max\{f_i(x) \ : \ i = 1, \ldots, k\}. \tag{3.4.1}$$

Proposition 3.4.1 *At each $s \in \operatorname{co}\{s_1, \ldots, s_k\} = \operatorname{dom} f^*$, the conjugate of f has the value (Δ_k is the unit simplex)*

$$f^*(s) = \min\left\{\sum_{i=1}^k \alpha_i b_i \ : \ \alpha \in \Delta_k, \ \sum_{i=1}^k \alpha_i s_i = s\right\}. \tag{3.4.2}$$

PROOF. Theorem 2.4.7 is directly applicable: since $f_i^* = I_{\{s_i\}} + b_i$, the variables s_j in the minimization problem (2.4.6) become the given s_j. This problem simplifies to (3.4.2), which at the same time reveals $\operatorname{dom} f^*$. □

In $\mathbb{R}^n \times \mathbb{R}$ (considered here as the dual of the graph-space), each of the k vertical half-lines $\{(s_i, r) \ : \ r \geq b_i\}$, $i = 1, \ldots, k$ is the epigraph of f_i^*, and these half-lines are the "needles" of Fig. IV.2.5.1. Now, the convex hull of these half-lines is a closed convex polyhedron, which is just $\operatorname{epi} f^*$. Needless to say, (3.4.1) is obtained by conjugating again (3.4.2). This double operation has an interesting application in the design of minimization algorithms:

Example 3.4.2 Suppose that the only available information about a function $f \in \overline{\mathrm{Conv}}\,\mathbb{R}^n$ is a finite sampling of function- and subgradient-values, say

$$f(x_i) \text{ and } s_i \in \partial f(x_i) \quad \text{for } i = 1, \ldots, k.$$

By convexity, we therefore know that $\varphi_1 \leqslant f \leqslant \varphi_2$, where

$$\varphi_1(y) := \max\{f(x_i) + \langle s_i, y - x_i \rangle : i = 1, \ldots, k\},$$
$$\varphi_2 := \mathrm{co}\,(\min_i g_i) \quad \text{with} \quad g_i(y) := \mathrm{I}_{\{x_i\}}(y) + f(x_i) \text{ for } i = 1, \ldots, k.$$

The resulting bracket on f is drawn in the left-part of Fig. 3.4.1. It has a counterpart in the dual space: $\varphi_2^* \leqslant f^* \leqslant \varphi_1^*$, illustrated in the right-part of Fig. 3.4.1, where

$$\varphi_1^*(s) = \mathrm{co}\left[\min\{\mathrm{I}_{\{s_i\}}(s) + \langle s_i, x_i\rangle - f(x_i)\}\right],$$
$$\varphi_2^*(s) = \max\{\langle s, x_i\rangle - f(x_i) : i = 1, \ldots, k\}.$$

Note: these formulae can be made more symmetric, with the help of the relations $\langle s_i, x_i \rangle - f(x_i) = f^*(s_i)$.

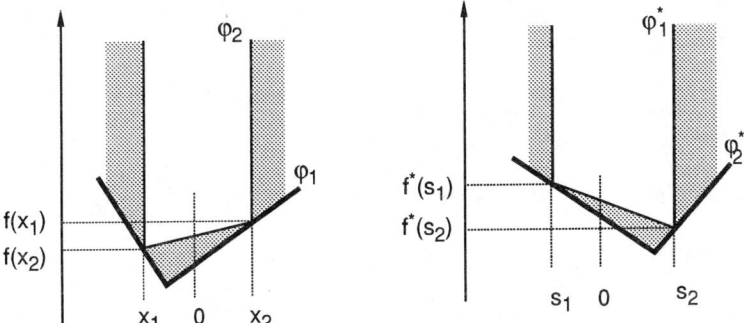

Fig. 3.4.1. Sandwiching a function and its conjugate

In the framework of optimization, the relation $\varphi_1 \leqslant f$ is well-known; it will be studied in more detail in Chapters XII and XV. It is certainly richer than the relation $f \leqslant \varphi_2$, which ignores all the first-order information contained in the subgradients s_i. The interesting point is that the situation is reversed in the dual space: the "cutting-plane" approximation of f^* (namely φ_2^*) is obtained from the poor primal approximation φ_2 and vice versa. □

Because the f of (3.4.1) increases at infinity no faster than linearly, its conjugate f^* has a bounded domain: it is not piecewise affine, but rather polyhedral. A natural idea is then to develop a full calculus for polyhedral functions, just as was done in §3.3 for the quadratic case.

A polyhedral function has the general form $g = f + \mathrm{I}_P$, where P is a closed convex polyhedron and f is defined by (3.4.1). The conjugate of g is given by Theorem 2.3.2: $g^* = f^* \, \dot{\triangledown} \, \sigma_P$, i.e.

$$g^*(s) = \min_{\alpha \in \Delta_k} \left[\sum_{i=1}^{k} \alpha_i b_i + \sigma_P(s - \sum_{i=1}^{k} \alpha_i s_i) \right]. \quad (3.4.3)$$

This formula may take different forms, depending on the particular description of P. When

$$P = \text{co}\{p_1, \ldots, p_m\}$$

is described as a convex hull, σ_P is the maximum of just as many linear functions $\langle p_j, \cdot \rangle$ and (3.4.3) becomes an explicit formula. Dual descriptions of P are more frequent in applications, though.

Example 3.4.3 With the notations above, suppose that P is described as an intersection (assumed nonempty) of half-spaces:

$$H_j := \{x \in \mathbb{R}^n : \langle c_j, x \rangle \leq d_j\} \quad \text{for } j = 1, \ldots, m$$
$$\text{and} \quad P := \cap H_j = \{x \in \mathbb{R}^n : Cx \leq d\}.$$

It is convenient to introduce the following notation: $\mathbb{R}^k$ and $\mathbb{R}^m$ are equipped with their standard dot-products; A [resp. C] is the linear operator which, to $x \in \mathbb{R}^n$, associates the vector whose coordinates are $\langle s_i, x \rangle$ in $\mathbb{R}^k$ [resp. $\langle c_j, x \rangle \in \mathbb{R}^m$]. Then it is not difficult to compute the adjoints of A and C: for $\alpha \in \mathbb{R}^k$ and $\gamma \in \mathbb{R}^m$,

$$A^*\alpha = \sum_{i=1}^{k} \alpha_i s_i, \quad C^*\gamma = \sum_{j=1}^{m} \gamma_j c_j.$$

Then we want to compute the conjugate of the polyhedral function

$$x \mapsto g(x) := \begin{cases} \max_{j=1,\ldots,k} \langle s_j, x \rangle - b_j & \text{if } Cx \leq d, \\ +\infty & \text{if not}, \end{cases} \quad (3.4.4)$$

and (3.4.3) becomes

$$g^*(s) = \min_{\alpha \in \Delta_k} \left[b^\top \alpha + \sigma_P(s - A^*\alpha) \right].$$

To compute σ_P, we have its conjugate I_P as the sum of the indicators of the H_j's; using Theorem 2.3.2 (with the qualification assumption (2.3.Q.j), since all the functions involved are polyhedral),

$$\sigma_P(p) = \min \left\{ \sum_{j=1}^{m} \sigma_{H_j}(s_j) : \sum_{j=1}^{m} s_j = p \right\}.$$

Using Example V.3.4.4 for the support function of H_j, it comes

$$\sigma_P(p) = \min \{ d^\top \gamma : C^*\gamma = p \}.$$

Piecing together, we finally obtain $g^*(s)$ as the optimal value of the problem in the pair of variables (α, γ):

$$\left| \begin{array}{l} \min \left[b^\top \alpha + d^\top \gamma \right] \quad \alpha \in \Delta_k, \ \gamma \in (\mathbb{R}^+)^m, \\ A^* \alpha + C^* \gamma = s. \end{array} \right. \qquad (3.4.5)$$

Observe the nice image-function exhibited by the last constraint, characterized by the linear operator $[A^*|C^*]$. The only difference between the variables α (involving the piecewise affine f) and γ (involving the constraints in the primal space) is that the latter do not have to sum up to 1; they have no a priori bound. Also, note that the constraint of (3.4.4) can be more generally expressed as $Cx - d \in K$ (a closed convex polyhedral cone), which induces in (3.4.5) the constraint $\gamma \in K^\circ$ (another closed convex polyhedral cone, standing for the nonnegative orthant).

Finally, g^* is a closed convex function. Therefore, there is some $s \in \mathbb{R}^n$ such that the feasible domain of (3.4.5) is nonempty, and the minimal value is never $-\infty$. These properties hold provided that g in (3.4.4) satisfies (1.1.1), i.e. that the domain $Cx \leqslant d$ is nonempty. □

4 Differentiability of a Conjugate Function

For $f \in \overline{\mathrm{Conv}}\, \mathbb{R}^n$, we know from Corollary 1.4.4 that

$$s \in \partial f(x) \quad \text{if and only if} \quad x \in \partial f^*(s);$$

and this actually was our very motivation for defining f^*, see the introduction to this chapter. Geometrically, the graph of ∂f and of ∂f^* in $\mathbb{R}^n \times \mathbb{R}^n$ are images of each other under the mapping $(x, s) \mapsto (s, x)$. Knowing that a convex function is differentiable when its subdifferential is a singleton, smoothness properties of f^* correspond to monotonicity properties of ∂f, in the sense of §VI.6.1.

4.1 First-Order Differentiability

Theorem 4.1.1 *Let $f \in \overline{\mathrm{Conv}}\, \mathbb{R}^n$ be strictly convex. Then $\mathrm{int}\,\mathrm{dom}\, f^* \neq \emptyset$ and f^* is continuously differentiable on $\mathrm{int}\,\mathrm{dom}\, f^*$.*

PROOF. For arbitrary $x_0 \in \mathrm{dom}\, f$ and nonzero $d \in \mathbb{R}^n$, consider Example 2.4.6. Strict convexity of f implies that

$$0 < \frac{f(x_0 - td) - f(x_0)}{t} + \frac{f(x_0 + td) - f(x_0)}{t} \quad \text{for all } t > 0,$$

and this inequality extends to the suprema:

$$0 < f'_\infty(-d) + f'_\infty(d).$$

Remembering that $f'_\infty = \sigma_{\mathrm{dom}\, f^*}$ (Proposition 1.2.2), this means

$$\sigma_{\mathrm{dom}\, f^*}(d) + \sigma_{\mathrm{dom}\, f^*}(-d) > 0,$$

80   X. Conjugacy in Convex Analysis

i.e. dom f^* has a positive breadth in every nonzero direction d: its interior is nonempty – Theorem V.2.2.3(iii).

Now, suppose that there is some $s \in \text{int dom } f^*$ such that $\partial f^*(s)$ contains two distinct points x_1 and x_2. Then $s \in \partial f(x_1) \cap \partial f(x_2)$ and, by convex combination of the relations
$$f^*(s) + f(x_i) = \langle s, x_i \rangle \quad \text{for } i = 1, 2,$$
we deduce, using Fenchel's inequality (1.1.3):
$$f^*(s) + \sum_{i=1}^{2} \alpha_i f(x_i) = \langle s, \sum_{i=1}^{2} \alpha_i x_i \rangle \leqslant f^*(s) + f\left(\sum_{i=1}^{2} \alpha_i x_i\right),$$
which implies that f is affine on $[x_1, x_2]$, a contradiction. In other words, ∂f^* is single-valued on int dom f^*, and this means f^* is continuously differentiable there. □

For an illustration, consider
$$\mathbb{R}^n \ni x \mapsto f(x) := \sqrt{1 + \|x\|^2},$$
whose conjugate is
$$\mathbb{R}^n \ni s \mapsto f^*(s) = \begin{cases} -\sqrt{1 - \|s\|^2} & \text{if } \|s\| \leqslant 1, \\ +\infty & \text{otherwise.} \end{cases}$$

Here, f is strictly convex (compute $\nabla^2 f$ to check this), dom f^* is the unit ball, on the interior of which f^* is differentiable, but ∂f^* is empty on the boundary.

Incidentally, observe also that f^* is strictly convex; as a result, f is differentiable. Such is not the case in our next example: with $n = 1$, take
$$f(x) := |x| + \tfrac{1}{2}x^2 = \max\left\{\tfrac{1}{2}x^2 + x, \tfrac{1}{2}x^2 - x\right\}. \tag{4.1.1}$$

Use direct calculations or calculus rules from §2 to compute
$$f^*(s) = \begin{cases} \tfrac{1}{2}(s+1)^2 & \text{for } s \leqslant -1, \\ 0 & \text{for } -1 \leqslant s \leqslant 1, \\ \tfrac{1}{2}(s-1)^2 & \text{for } s \geqslant 1. \end{cases}$$

This example is illustrated by the instructive Fig. 4.1.1. Its left part displays gr f, made up of two parabolas; $f^*(s)$ is obtained by leaning onto the relevant parabola a straight line with slope s. The right part illustrates Theorem 2.4.4: it displays $1/2 s^2 + s$ and $1/2 s^2 - s$, the conjugates of the two functions making up f; epi f^* is then the convex hull of the union of their epigraphs.

Example 4.1.2 We have studied in §IV.1.3(f) the volume of an ellipsoid: on the Euclidean space $(S_n(\mathbb{R}), \langle\!\langle \cdot, \cdot \rangle\!\rangle)$ of symmetric matrices with the standard scalar product of $\mathbb{R}^{n \times n}$, the function
$$f(A) := \begin{cases} -\log(\det A) - n/2 & \text{if } A \text{ is positive definite}, \\ +\infty, & \text{otherwise} \end{cases}$$

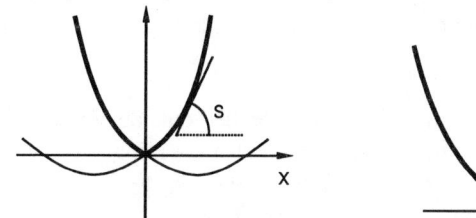
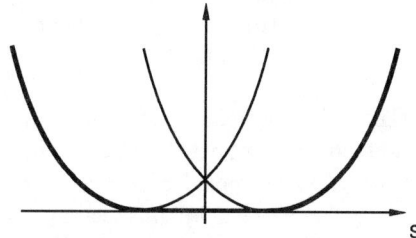

Fig. 4.1.1. Strict convexity corresponds to differentiability of the conjugate

is convex. It is also differentiable on its domain, with gradient $\nabla f(A) = -A^{-1}$ (cf. §A.4.3). Its conjugate can then be computed using the Legendre transform (Remark 0.1):

$$f^*(B) := \begin{cases} -\log(\det(-B)) - n/2 & \text{if } B \text{ is negative definite}, \\ +\infty & \text{otherwise}. \end{cases}$$

Thus f^* is again a differentiable convex function; actually, f is strictly convex. This property, by no means trivial, can be seen either from the Hessian operator $\nabla^2 f(A) = A^{-1} \otimes A^{-1}$, i.e.

$$\frac{\partial^2 f}{\partial a_{ij} \partial a_{k\ell}}(A) = (A^{-1})_{\ell j}(A^{-1})_{ik}$$

or from Theorem IV.4.1.4: ∇f is strictly monotone because

$$\langle\!\langle B^{-1} - A^{-1}, A - B \rangle\!\rangle = \operatorname{tr}(A^{-1}B) + \operatorname{tr}(B^{-1}A) - 2n > 0$$

for all symmetric positive definite matrices $A \neq B$. □

Actually, the strict convexity of f in Example 4.1.2 can be directly established in our present framework. In fact, Theorem 4.1.1 gives a sufficient condition for smoothness of a closed convex function (namely f^*). Apart from possible side-effects on the boundary, this condition is also necessary:

Theorem 4.1.3 *Let $f \in \overline{\operatorname{Conv}} \mathbb{R}^n$ be differentiable on the set $\Omega := \operatorname{int} \operatorname{dom} f$. Then f^* is strictly convex on each convex subset $C \subset \nabla f(\Omega)$.*

PROOF. Let C be a convex set as stated. Suppose that there are two distinct points s_1 and s_2 in C such that f^* is affine on the line-segment $[s_1, s_2]$. Then, setting $s := 1/2\,(s_1 + s_2) \in C \subset \nabla f(\Omega)$, there is $x \in \Omega$ such that $\nabla f(x) = s$, i.e. $x \in \partial f^*(s)$. Using the affine character of f^*, we have

$$0 = f(x) + f^*(s) - \langle s, x \rangle = \tfrac{1}{2} \sum_{i=1}^{2} \left[f(x) + f^*(s_i) - \langle s_i, x \rangle \right]$$

and, in view of Fenchel's inequality (1.1.3), this implies that each term in the bracket is 0: $x \in \partial f^*(s_1) \cap \partial f^*(s_2)$, i.e. $\partial f(x)$ contains the two points s_1 and s_2, a contradiction to the existence of $\nabla f(x)$. □

The strict convexity of f^* cannot in general be extended outside $\nabla f(\Omega)$; and be aware that this set may be substantially smaller than one might initially guess: the function f_5^* in Example I.6.2.4(e) is finite everywhere, but not strictly convex.

We gather the results of this section, with a characterization of the "ideal situation", in which the Legendre transform alluded to in the introduction is well-defined.

Corollary 4.1.4 *Let $f : \mathbb{R}^n \to \mathbb{R}$ be strictly convex, differentiable and 1-coercive. Then*

(i) *f^* is likewise finite-valued on $\mathbb{R}^n$, strictly convex, differentiable and 1-coercive;*
(ii) *the continuous mapping ∇f is one-to-one from $\mathbb{R}^n$ onto $\mathbb{R}^n$, and its inverse is continuous;*
(iii) *$f^*(s) = \langle s, (\nabla f)^{-1}(s) \rangle - f\big((\nabla f)^{-1}(s)\big)$ for all $s \in \mathbb{R}^n$.* □

The simplest such situation occurs when f is a strictly convex quadratic function, as in Example 1.1.3, corresponding to an affine Legendre transform. Another example is $f(x) := \exp(\frac{1}{2} \|x\|^2)$.

4.2 Towards Second-Order Differentiability

Along the same lines as in §4.1, better than strict convexity of f and better than differentiability of f^* correspond to each other. We start with the connection between strong convexity of f and Lipschitz continuity of ∇f^*.

(a) Lipschitz Continuity of the Gradient Mapping. The next result is stated for a finite-valued f, mainly because the functions considered in Chap. VI were such; but this assumption is actually useless.

Theorem 4.2.1 *Assume that $f : \mathbb{R}^n \to \mathbb{R}$ is strongly convex with modulus $c > 0$ on $\mathbb{R}^n$: for all $(x_1, x_2) \in \mathbb{R}^n \times \mathbb{R}^n$ and $\alpha \in \,]0, 1[$,*

$$f(\alpha x_1 + (1-\alpha) x_2) \leq \alpha f(x_1) + (1-\alpha) f(x_2) - 1/2 c\alpha(1-\alpha) \|x_1 - x_2\|^2. \quad (4.2.1)$$

Then $\mathrm{dom}\, f^* = \mathbb{R}^n$ and ∇f^* is Lipschitzian with constant $1/c$ on $\mathbb{R}^n$:

$$\|\nabla f^*(s_1) - \nabla f^*(s_2)\| \leq \tfrac{1}{c} \|s_1 - s_2\| \quad \text{for all } (s_1, s_2) \in \mathbb{R}^n \times \mathbb{R}^n.$$

PROOF. We use the various equivalent definitions of strong convexity (see Theorem VI.6.1.2). Fix x_0 and $s_0 \in \partial f(x_0)$: for all $0 \neq d \in \mathbb{R}^n$ and $t \geq 0$

$$f(x_0 + td) \geq f(x_0) + t \langle s_0, d \rangle + \tfrac{1}{2} ct^2 \|d\|^2,$$

hence $f'_\infty(d) = \sigma_{\mathrm{dom}\,f^*}(d) = +\infty$, i.e. dom $f^* = \mathbb{R}^n$. Also, f is in particular strictly convex, so we know from Theorem 4.1.1 that f^* is differentiable (on $\mathbb{R}^n$). Finally, the strong convexity of f can also be written

$$\langle s_1 - s_2, x_1 - x_2 \rangle \geq c\|x_1 - x_2\|^2,$$

in which we have $s_i \in \partial f(x_i)$, i.e. $x_i = \nabla f^*(s_i)$, for $i = 1, 2$. The rest follows from the Cauchy-Schwarz inequality. □

This result is quite parallel to Theorem 4.1.1: improving the convexity of f from "strict" to "strong" amounts to improving ∇f^* from "continuous" to "Lipschitzian". The analogy can even be extended to Theorem 4.1.3:

Theorem 4.2.2 *Let $f : \mathbb{R}^n \to \mathbb{R}$ be convex and have a gradient-mapping Lipschitzian with constant $L > 0$ on $\mathbb{R}^n$: for all $(x_1, x_2) \in \mathbb{R}^n \times \mathbb{R}^n$,*

$$\|\nabla f(x_1) - \nabla f(x_2)\| \leq L\|x_1 - x_2\|.$$

Then f^* is strongly convex with modulus $1/L$ on each convex subset $C \subset \mathrm{dom}\,\partial f^*$. In particular, there holds for all $(x_1, x_2) \in \mathbb{R}^n \times \mathbb{R}^n$

$$\langle \nabla f(x_1) - \nabla f(x_2), x_1 - x_2 \rangle \geq \tfrac{1}{L}\|\nabla f(x_1) - \nabla f(x_2)\|^2. \tag{4.2.2}$$

PROOF. Let s_1 and s_2 be arbitrary in dom $\partial f^* \subset$ dom f^*; take s and s' on the segment $[s_1, s_2]$. To establish the strong convexity of f^*, we need to minorize the remainder term $f^*(s') - f^*(s) - \langle x, s' - s \rangle$, with $x \in \partial f^*(s)$. For this, we minorize $f^*(s') = \sup_y[\langle s', y \rangle - f(y)]$, i.e. we majorize $f(y)$:

$$\begin{aligned} f(y) &= f(x) + \langle \nabla f(x), y - x \rangle + \int_0^1 \langle \nabla f(x + t(y - x)) - \nabla f(x), y - x \rangle dt \\ &\leq f(x) + \langle \nabla f(x), y - x \rangle + \tfrac{1}{2}L\|y - x\|^2 = \\ &= -f^*(s) + \langle s, y \rangle + \tfrac{1}{2}L\|y - x\|^2 \end{aligned}$$

(we have used the property $\int_0^1 t\,dt = 1/2$, as well as $x \in \partial f^*(s)$, i.e. $\nabla f(x) = s$). In summary, we have

$$f^*(s') \geq f^*(s) + \sup_y \left[\langle s' - s, y \rangle - \tfrac{1}{2}L\|y - x\|^2 \right].$$

Observe that the last supremum is nothing but the value at $s' - s$ of the conjugate of $1/2\,L\|\cdot - x\|^2$. Using the calculus rule 1.3.1, we have therefore proved

$$f^*(s') \geq f^*(s) + \langle s' - s, x \rangle + \tfrac{1}{2L}\|s' - s\|^2 \tag{4.2.3}$$

for all s, s' in $[s_1, s_2]$ and all $x \in \partial f^*(s)$. Replacing s' in (4.2.3) by s_1 and by s_2, and setting $s = \alpha s_1 + (1 - \alpha)s_2$, the strong convexity (4.2.1) for f^* is established by convex combination.

On the other hand, replacing (s, s') by (s_1, s_2) in (4.2.3):
$$f^*(s_2) \geqslant f^*(s_1) + \langle s_2 - s_1, x_1 \rangle + \tfrac{1}{2L}\|s_2 - s_1\|^2 \quad \text{for all } x_1 \in \partial f^*(s_1).$$
Then, replacing (s, s') by (s_2, s_1) and summing:
$$\langle x_1 - x_2, s_1 - s_2 \rangle \geqslant \tfrac{1}{L}\|s_1 - s_2\|^2.$$
In view of the differentiability of f, this is just (4.2.2), which has to hold for all (x_1, x_2) simply because $\operatorname{Im}\partial f^* = \operatorname{dom}\nabla f = \mathbb{R}^n$. □

Remark 4.2.3 For a convex function, the Lipschitz property of the gradient mapping thus appears as equivalently characterized by (4.2.2); a result which is of interest in itself. As an application, let us return to §3.2: for a convex set C, the convex function $\|\cdot\|^2 - d_C^2$ has gradient p_C, a nonexpansive mapping. Therefore
$$\|p_C(x) - p_C(y)\|^2 \leqslant \langle p_C(x) - p_C(y), x - y \rangle$$
for all x and y in $\mathbb{R}^n$. Likewise, $\nabla\left(1/2\, d_S^2\right) = I - p_C$ is also nonexpansive:
$$\|(I - p_C)(x) - (I - p_C)(y)\|^2 \leqslant \langle (I - p_C)(x) - (I - p_C)(y), x - y \rangle. \quad \square$$

Naturally, Corollary 4.1.4 has also its equivalent, namely: if f is strongly convex and has a Lipschitzian gradient-mapping on $\mathbb{R}^n$, then f^* enjoys the same properties. These properties do not leave much room, though: f (and f^*) must be "sandwiched" between two positive definite quadratic functions.

(b) Second-Order Approximations. Some Lipschitz continuity of the gradient is of course necessary for second-order differentiability, but is certainly not sufficient: (4.1.1) is strongly convex but its conjugate is not C^2. More must therefore be assumed to ensure the existence of $\nabla^2 f^*$, and we recall from §I.6.2 that finding minimal assumptions is a fairly complex problem.

In the spirit of the "unilateral" and "conical" character of convex analysis, we proceed as follows. First, a function $\varphi \in \overline{\operatorname{Conv}}\,\mathbb{R}^n$ is said to be *directionally quadratic* if it is positively homogeneous of degree 2:
$$\varphi(tx) = t^2 \varphi(x) \quad \text{for all } x \in \mathbb{R}^n \text{ and } t > 0.$$
If φ is directionally quadratic, immediate consequences of the definition are:
- $\varphi(0) = 0$ (see the beginning of §V.1.1);
- φ^* is directionally quadratic as well (easy to check);
- hence $\varphi^*(0) = 0$, which in turn implies that φ is minimal at 0: a directionally quadratic function is *nonnegative*;
- as a result (see Example V.1.2.7), $\sqrt{\varphi}$ is the support function of a closed convex set containing 0.

Of particular importance are the directionally quadratic functions which are finite everywhere, and/or positive (except at 0); these two properties are dual to each other:

4 Differentiability of a Conjugate Function

Lemma 4.2.4 *For a directionally quadratic function φ, the following properties are equivalent:*

(i) *φ is finite everywhere;*
(ii) *$\nabla\varphi(0)$ exists (and is equal to 0);*
(iii) *there is $C \geq 0$ such that $\varphi(x) \leq 1/2\, C\|x\|^2$ for all $x \in \mathbb{R}^n$;*
(iv) *there is $c > 0$ such that $\varphi^*(s) \geq 1/2\, c\|s\|^2$ for all $s \in \mathbb{R}^n$;*
(v) *$\varphi^*(s) > 0$ for all $s \neq 0$.*

PROOF. [(i) $\Leftrightarrow$ (ii) $\Rightarrow$ (iii)] When (i) holds, φ is continuous; call $1/2\, C \geq 0$ its maximal value on the unit sphere: (iii) holds by positive homogeneity. Furthermore, compute difference quotients to observe that $\varphi'(0, \cdot) \equiv 0$, and (ii) follows from the differentiability criterion IV.4.2.1.

Conversely, existence of $\nabla\varphi(0)$ implies finiteness of φ in a neighborhood of 0 and, by positive homogeneity, on the whole space.

[(iii) $\Rightarrow$ (iv) $\Rightarrow$ (v)] When (iii) holds, (iv) comes from the calculus rule 1.3.1(vii), with for example $0 < c := 1/C$ (or rather $1/(C+1)$, to take care of the case $C = 0$); this implies (v) trivially.

[(v) $\Rightarrow$ (i)] The lower semi-continuous function φ^* is positive on the unit sphere, and has a minimal value $c > 0$ there. Being positively homogeneous of degree 2, φ^* is then 1-coercive; finiteness of its conjugate φ follows from Proposition 1.3.8. □

Definition 4.2.5 We say that the directionally quadratic function φ_s defines a *minorization to second order* of $f \in \overline{\text{Conv}}\, \mathbb{R}^n$ at $x_0 \in \text{dom}\, f$, and associated with $s \in \partial f(x_0)$, when

$$f(x_0 + h) \geq f(x_0) + \langle s, h \rangle + \varphi_s(h) + o(\|h\|^2). \quad (4.2.4)$$

We say that φ_s defines likewise a *majorization* to second order when

$$f(x_0 + h) \leq f(x_0) + \langle s, h \rangle + \varphi_s(h) + o(\|h\|^2). \quad (4.2.5)$$

□

Note in passing that, because $\varphi_s \geq 0$, (4.2.4) could not hold if s were not a subgradient of f at x_0. Whenever $x_0 \in \text{dom}\, \partial f$, the zero function and $I_{\{0\}}$ define trivial minorizations and majorizations to second order respectively, valid for all $s \in \partial f(x_0)$. This observation motivates the particular class of directionally quadratic functions introduced in Lemma 4.2.4.

We proceed to show that the correspondences established in Lemma 4.2.4 are somehow conserved when remainder terms are introduced, as in Definition 4.2.5. First observe that, if (4.2.5) holds with φ_s finite everywhere, then $\nabla f(x_0)$ exists (and is equal to s). The next two results concern the equivalence (iii) $\Leftrightarrow$ (iv).

Proposition 4.2.6 *With the notation of Definition 4.2.5, suppose that there is a directionally quadratic function φ_s satisfying, for some $c > 0$,*

X. Conjugacy in Convex Analysis

$$\varphi_S(h) \geq \tfrac{1}{2}c\|h\|^2 \quad \text{for all } h \in \mathbb{R}^n, \tag{4.2.6}$$

and defining a minorization to second order of $f \in \overline{\text{Conv}}\,\mathbb{R}^n$, *associated with* $s \in \partial f(x_0)$. *Then* φ_S^* *defines a majorization of* f^* *at* s, *associated with* x_0. *This implies in particular* $\nabla f^*(s) = x_0$.

PROOF. [*Preamble*] First note that φ_S^* is finite everywhere: as already mentioned after Definition 4.2.5, differentiability of f^* will follow from the majorization property of f^*.

We take the tilted function

$$\mathbb{R}^n \ni h \mapsto g(h) := f(x_0 + h) - f(x_0) - \langle s, h \rangle,$$

so that the assumption means

$$g(h) \geq \varphi_S(h) + o(\|h\|^2).$$

In view of the relation

$$g^*(p) = f^*(s+p) + f(x_0) - \langle s+p, x_0 \rangle = f^*(s+p) - f^*(s) - \langle p, x_0 \rangle,$$

we have only to prove that φ_S^* defines a majorization of g^*, i.e.

$$g^*(p) \leq \varphi_S^*(p) + o(\|p\|^2) \quad \text{for all } p \in \mathbb{R}^n.$$

[*Step 1*] By assumption, for any $\varepsilon > 0$, there is a $\delta > 0$ such that

$$g(h) \geq \varphi_S(h) - \varepsilon \|h\|^2 \quad \text{whenever} \quad \|h\| \leq \delta;$$

and, in view of (4.2.6), we write:

$$\text{whenever } \|h\| \leq \delta, \quad \begin{vmatrix} g(h) \geq \left(1 - \tfrac{2\varepsilon}{c}\right)\varphi_S(h) & (*) \\ g(h) \geq \left(\tfrac{1}{2}c - \varepsilon\right)\|h\|^2. & (**) \end{vmatrix}$$

Our aim is then to establish that, even though $(*)$ holds locally only, the calculus rule 1.3.1(vii) applies, at least in a neighborhood of 0.

[*Step 2*] For $\|h\| > \delta$, write the convexity of g on $[0, h]$ and use $g(0) = 0$ to obtain with $(**)$

$$\tfrac{\delta}{\|h\|} g(h) \geq g\left(\tfrac{\delta}{\|h\|} h\right) \geq \left(\tfrac{1}{2}c - \varepsilon\right)\delta^2,$$

hence

$$g(h) \geq \left(\tfrac{1}{2}c - \varepsilon\right)\delta \|h\|.$$

Then, if $\|p\| \leq (1/2\,c - \varepsilon)\delta =: \delta'$, we have

$$\langle p, h \rangle - g(h) \leq \left(\tfrac{1}{2}c - \varepsilon\right)\|h\| - g(h) \leq 0 \quad \text{whenever } \|h\| > \delta.$$

Since $g^* \geq 0$, this certainly implies that

4 Differentiability of a Conjugate Function   87

$$\begin{aligned} g^*(p) &= \sup_{\|h\| \leq \delta}[\langle p, h \rangle - g(h)] \\ &\leq \sup_{\|h\| \leq \delta}\left[\langle p, h \rangle - \left(1 - \tfrac{2\varepsilon}{c}\right)\varphi_s(h)\right] \quad \text{[from (*)]} \\ &\leq \left[\left(1 - \tfrac{2\varepsilon}{c}\right)\varphi_s\right]^*(p) \quad \text{for } \|p\| \leq \delta'. \end{aligned}$$

[*Step 3*] Thus, using the calculus rule 1.3.1(ii) and positive homogeneity of φ_s^*, we have for $p \in B(0, \delta')$,

$$\begin{aligned} g^*(p) \leq \tfrac{1}{1-2\varepsilon/c}\varphi_s^*(p) &= \varphi_s^*(p) + \tfrac{2\varepsilon}{c-2\varepsilon}\varphi_s^*(p) \\ &\leq \varphi_s^*(p) + \tfrac{2\varepsilon}{c-2\varepsilon}\tfrac{1}{2c}\|p\|^2. \quad \text{[from (4.2.6)]} \end{aligned}$$

Let us sum up: given $\varepsilon' > 0$, choose ε in Step 1 such that $\frac{\varepsilon}{c(c-2\varepsilon)} \leq \varepsilon'$. This gives $\delta > 0$ and $\delta' > 0$ in Step 2 yielding the required majorization

$$g^*(p) \leq \varphi_s^*(p) + \varepsilon' \|p\|^2 \quad \text{for all } p \in B(0, \delta'). \qquad \Box$$

Proposition 4.2.7 *With the notation of Definition 4.2.5, suppose that there is a directionally quadratic function φ_s satisfying (4.2.6) for some $c > 0$ and defining a majorization to second order of $f \in \overline{\text{Conv}}\,\mathbb{R}^n$, associated with $s \in \partial f(x_0)$. Then φ_s^* defines a minorization to second order of f^* at s, associated with x_0.*

PROOF. We use the proof pattern of Proposition 4.2.6; using in particular the same tilting technique, we assume $x_0 = 0$, $f(0) = 0$ and $s = 0$. For any $\varepsilon > 0$, there is by assumption $\delta > 0$ such that

$$f(h) \leq \varphi_s(h) + \varepsilon \|h\|^2 \leq (1 + 2\varepsilon/c)\varphi_s(h) \quad \text{whenever} \quad \|h\| \leq \delta.$$

It follows that

$$f^*(p) \geq \sup\{\langle p, h \rangle - (1 + 2\varepsilon/c)\varphi_s(h) : \|h\| \leq \delta\} \quad (4.2.7)$$

for all p; we have to show that, for $\|p\|$ small enough, the right-hand side is close to $\varphi_s^*(p)$.

Because of (4.2.6), φ_s^* is finite everywhere and $\nabla\varphi_s^*(0) = 0$ (Lemma 4.2.4). It follows from the outer semi-continuity of $\partial\varphi_s^*$ (Theorem VI.6.2.4) that, for some $\delta' > 0$,

$$\emptyset \neq \partial\varphi_s^*\left(\tfrac{1}{1+2\varepsilon/c}p\right) \subset B(0, \delta) \quad \text{for all } p \in B(0, \delta').$$

Said otherwise: whenever $\|p\| \leq \delta'$, there exists $h_0 \in B(0, \delta)$ such that

$$\varphi_s^*\left(\tfrac{1}{1+2\varepsilon/c}p\right) = \frac{\langle p, h_0 \rangle}{1 + 2\varepsilon/c} - \varphi_s(h_0).$$

Thus, the function $\langle p, \cdot \rangle - (1 + 2\varepsilon/c)\varphi_s$ attains its (unconstrained) maximum at h_0. In summary, provided that $\|p\|$ is small enough, the index h in (4.2.7) can run in the whole space, to give

$$f^*(p) \geq [(1 + 2\varepsilon/c)\varphi_s]^*(p) \quad \text{for all } p \in B(0, \delta').$$

To conclude, mimic Step 3 in the proof of Proposition 4.2.6. □

To show that (4.2.6) cannot be removed in this last result, take
$$\mathbb{R}^2 \ni (\xi, \eta) \mapsto f(\xi, \eta) = \tfrac{1}{2}\xi^2 + \tfrac{1}{4}\eta^4,$$
so that $\varphi_0(\rho, \tau) := 1/2\,\rho^2$ defines a majorization to second order of f at $x_0 = 0$. We leave it as an exercise to compute f^* and φ_0^*, and to realize that the latter is not suitable for a minorization to second order.

Example 4.2.8 Take
$$f(x) = \max\{f_j(x) : j = 1, \ldots, m\},$$
where each $f_j : \mathbb{R}^n \to \mathbb{R}$ is convex and twice differentiable at x_0. Using the notation $J := \{j : f_j(x_0) = f(x_0)\}$ for the active index-set at x_0, and Δ being the associated unit simplex, assume for simplicity that $A_j := \nabla^2 f_j(x_0)$ is positive definite for each $j \in J$. For $\alpha \in \Delta$, define
$$s_\alpha := \sum_{j \in J} \alpha_j \nabla f_j(x_0) \in \partial f(x_0)$$
and
$$\varphi_\alpha(h) := \tfrac{1}{2} \sum_{j \in J} \alpha_j \langle A_j h, h \rangle;$$
this last function is convex quadratic and yields the minorization
$$f(x_0 + h) \geqslant \sum_{j \in J} \alpha_j f_j(x_0 + h) = f(x_0) + \langle s_\alpha, h \rangle + \varphi_\alpha(h) + o(\|h\|^2).$$

Because φ_α is strongly convex, Proposition 4.2.6 tells us that
$$f^*(s_\alpha + p) \leqslant f^*(s_\alpha) + \langle p, x_0 \rangle + \varphi_\alpha^*(p) + o(\|p\|^2),$$
where
$$\varphi_\alpha^*(p) = \tfrac{1}{2} \left\langle \left[\sum_{j \in J} \alpha_j A_j\right]^{-1} p, p \right\rangle.$$

Note that the function $\varphi := \mathrm{co}(\min_{\alpha \in \Delta} \varphi_\alpha)$ also defines a minorization to second order of f at x_0, which is no longer quadratic, but which still satisfies (4.2.6), and is *independent* of α, i.e. of $s \in \partial f(x_0)$. Dually, it corresponds to
$$f^*(s + p) \leqslant f^*(s) + \langle p, x_0 \rangle + \tfrac{1}{2} \max_{j \in J} \langle A_j^{-1} p, p \rangle + o(\|p\|^2)$$
for all $p \in \mathbb{R}^n$, and this majorization is valid at all $s \in \partial f(x_0)$. □

If a directionally quadratic function φ_s defines both a minorization and a majorization of f (at x_0, associated with s), it will be said to define an *approximation* to second order of f. As already stated, existence of such an approximation implies a rather strong smoothness property of f at x_0.

Convex quadratic functions are particular instances of directionally quadratic functions. With Propositions 4.2.6 and 4.2.7 in mind, the next result is straightforward:

4 Differentiability of a Conjugate Function

Corollary 4.2.9 *Let $f \in \overline{\mathrm{Conv}}\,\mathbb{R}^n$ have an approximation at x_0 which is actually quadratic: for $s := \nabla f(x_0)$ and some symmetric positive semi-definite linear operator A,*

$$f(x_0 + h) = f(x_0) + \langle s, h \rangle + \tfrac{1}{2}\langle Ah, h \rangle + o(\|h\|^2).$$

If, in addition, A is positive definite, then f^* has also a quadratic approximation at s, namely

$$f^*(s + p) = f^*(s) + \langle p, x_0 \rangle + \tfrac{1}{2}\langle A^{-1}p, p \rangle + o(\|p\|^2). \qquad \square$$

The global version of Corollary 4.2.9 is:

Corollary 4.2.10 *Let $f : \mathbb{R}^n \to \mathbb{R}$ be convex, twice differentiable and 1-coercive. Assume, moreover, that $\nabla^2 f(x)$ is positive definite for all $x \in \mathbb{R}^n$. Then f^* enjoys the same properties and*

$$\nabla^2 f^*(s) = \left[\nabla^2 f(\nabla f^{-1}(s))\right]^{-1} \quad \text{for all } s \in \mathbb{R}^n.$$

PROOF. The 1-coercivity of f implies that $\mathrm{dom}\, f^* = \mathbb{R}^n$, so $\mathrm{Im}\,\nabla f = \mathrm{dom}\, f^* = \mathbb{R}^n$: apply Corollary 4.2.9 to any $x \in \mathbb{R}^n$ with $A = \nabla^2 f(x)$. $\qquad \square$

The assumptions of Corollary 4.2.10 are certainly satisfied if the Hessian of f is uniformly positive definite, i.e. if

$$\exists c > 0 \quad \text{such that} \quad \langle \nabla^2 f(x)d, d \rangle \geq c\|d\|^2 \text{ for all } (x, d) \in \mathbb{R}^n \times \mathbb{R}^n.$$

Use the 1-dimensional counter-example $f(x) := \exp x$ to realize that 1-coercivity is really more than mere positive definiteness of $\nabla^2 f(x)$ for all $x \in \mathbb{R}^n$.

Remark 4.2.11 To conclude, let us return to Example 4.2.8, which illustrates a fundamental difficulty for a second-order analysis. Under the conditions of that example, f can be approximated to second order in a number of ways.

(a) For $h \in \mathbb{R}^n$, define

$$J_h := \{j \in J : f'(x_0, h) = \langle \nabla f_j(x_0), h \rangle\};$$

in other words, the convex hull of the gradients indexed by J_h forms the face of $\partial f(x_0)$ exposed by h. The function

$$\mathbb{R}^n \ni h \mapsto \varphi(h) := \tfrac{1}{2} \max\{\langle A_j h, h \rangle : j \in J_h\}$$

is directionally quadratic and, for all $d \in \mathbb{R}^n$, we have the estimate

$$f(x_0 + td) = f(x_0) + tf'(x_0, d) + t^2\varphi(d) + o(t^2) \quad \text{for } t > 0.$$

Unfortunately, the remainder term depends on d, and actually φ is *not even continuous* (too bad if we want to approximate a nicely convex function). To see this, just take $m = 2$, f_1 and f_2 quadratic, and x_0 at a kink: $f_1(x_0) = f_2(x_0)$. Then let d_0 be tangent to the kinky surface (if necessary look at Fig. VI.2.1.3, which displays such a surface). For d in the neighborhood of this d_0, $\varphi(d)$ is equal to either $1/2 \langle A_1 d, d \rangle$ or $1/2 \langle A_2 d, d \rangle$, depending on which index prevails at $x_0 + td$ for small $t > 0$.

(b) Another second-order estimate is

$$f(x_0 + h) = f(x_0) + \max_{j \in J} \left[\langle \nabla f_j(x_0), h \rangle + 1/2 \langle \nabla^2 f_j(x_0) h, h \rangle \right] + o(\|h\|^2).$$

It does not suffer the non-uniformity effect of (a) and the approximating function is now nicely convex. However, it patches together the first- and second-order information and is no longer the sum of a sublinear function and a directionally quadratic function. □

XI. Approximate Subdifferentials of Convex Functions

Prerequisites. Sublinear functions and associated convex sets (Chap. V); characterization of the subdifferential via the conjugacy correspondence (§X.1.4); calculus rules on conjugate functions (§X.2); and also: behaviour at infinity of one-dimensional convex functions (§I.2.3).

Introduction. There are two motivations for the concepts introduced in this chapter: a practical one, related with descent methods, and a more theoretical one, in the framework of differential calculus.

– In §VIII.2.2, we have seen that the steepest-descent method is not convergent, essentially because the subdifferential is not a continuous mapping. Furthermore, we have defined Algorithm IX.1.6 which, to find a descent direction, needs to extract limits of subgradients: an impossible task on a computer.
– On the theoretical side, we have seen in Chap. VI the directional derivative of a finite convex function, which supports a convex set: the subdifferential. This latter set was generalized to extended-valued functions in §X.1.4; and infinite-valued directional derivatives have also been seen (Proposition I.4.1.3, Example X.2.4.3). A natural question is then: is the supporting property still true in the extended-valued case? The answer is not quite yes, see below the example illustrated by Fig. 2.1.1.

The two difficulties above are overcome altogether by the so-called ε-subdifferential of f, denoted $\partial_\varepsilon f$, which is a certain perturbation of the subdifferential studied in Chap. VI for finite convex functions. While the two sets are identical for $\varepsilon = 0$, the properties of $\partial_\varepsilon f$ turn out to be substantially different from those of ∂f. We therefore study $\partial_\varepsilon f$ with the help of the relevant tools, essentially the conjugate function f^* (which was of no use in Chap. VI). In return, particularizing our study to the case $\varepsilon = 0$ enables us to generalize the results of Chap. VI to extended-valued functions.

Throughout this chapter, and unless otherwise specified, we therefore have

$$\boxed{f \in \overline{\mathrm{Conv}}\,\mathbb{R}^n \text{ and } \varepsilon \geq 0.}$$

However, keeping in mind that our development has a *practical* importance for numerical optimization, we will often pay special attention to the finite-valued case.

XI. Approximate Subdifferentials of Convex Functions

1 The Approximate Subdifferential

1.1 Definition, First Properties and Examples

Definition 1.1.1 Given $x \in \text{dom } f$, the vector $s \in \mathbb{R}^n$ is called an ε-*subgradient* of f at x when the following property holds:

$$f(y) \geq f(x) + \langle s, y - x \rangle - \varepsilon \quad \text{for all } y \in \mathbb{R}^n. \tag{1.1.1}$$

Of course, s is still an ε-subgradient if (1.1.1) holds only for $y \in \text{dom } f$. The set of all ε-subgradients of f at x is the ε-*subdifferential* (of f at x), denoted by $\partial_\varepsilon f(x)$.

Even though we will rarely take ε-subdifferentials of functions not in $\overline{\text{Conv}}\, \mathbb{R}^n$, it goes without saying that the relation of definition (1.1.1) can be applied to any function finite at x. Also, one could set $\partial_\varepsilon f(x) = \emptyset$ for $x \notin \text{dom } f$. □

It follows immediately from the definition that

$$\partial_\varepsilon f(x) \subset \partial_{\varepsilon'} f(x) \quad \text{whenever} \quad \varepsilon \leq \varepsilon'; \tag{1.1.2}$$

$$\partial f(x) = \partial_0 f(x) = \cap \{\partial_\varepsilon f(x) : \varepsilon > 0\} \ [= \lim_{\varepsilon \downarrow 0} \partial_\varepsilon f(x)]; \tag{1.1.3}$$

$$\partial_{\alpha\varepsilon + (1-\alpha)\varepsilon'} f(x) \supset \alpha \partial_\varepsilon f(x) + (1-\alpha)\partial_{\varepsilon'} f(x) \quad \text{for all } \alpha \in \,]0, 1[. \tag{1.1.4}$$

The last relation means that the graph of the multifunction $\mathbb{R}^+ \ni \varepsilon \mapsto \partial_\varepsilon f(x)$ is a convex set in $\mathbb{R}^{n+1}$; more will be said about this set later (Proposition 1.3.3).

We will continue to use the notation ∂f, rather than $\partial_0 f$, for the *exact* subdifferential – knowing that $\partial_\varepsilon f$ can be called an *approximate* subdifferential when $\varepsilon > 0$.

Figure 1.1.1 gives a geometric illustration of Definition 1.1.1: s, together with $r \in \mathbb{R}$, defines the affine function $y \mapsto a_{s,r}(y) := r + \langle s, y - x \rangle$; we say that s is an ε-subgradient of f at x when it is possible to have simultaneously $r \geq f(x) - \varepsilon$ and $a_{s,r} \leq f$. The condition $r = f(x)$, corresponding to exact subgradients, is thus relaxed by ε; thanks to closed convexity of f, this relaxation makes it possible to find such an $a_{s,r}$:

Fig. 1.1.1. Supporting hyperplanes within ε

Theorem 1.1.2 *For all $x \in \text{dom } f$, $\partial_\varepsilon f(x) \neq \emptyset$ whenever $\varepsilon > 0$.*

PROOF. Theorem III.4.1.1 implies the existence in $\mathbb{R}^n \times \mathbb{R}$ of a hyperplane $H_{s,\alpha}$ separating strictly the point $(x, f(x) - \varepsilon)$ from the closed convex set epi f: for all $y \in \mathbb{R}^n$

$$\langle s, x \rangle + \alpha[f(x) - \varepsilon] < \langle s, y \rangle + \alpha f(y).$$

Taking $y = x$ in this inequality shows that

$$\alpha[f(x) - \varepsilon] < \alpha f(x) < +\infty$$

hence $\alpha > 0$ (the hyperplane is non-vertical). Then $s' := -s/\alpha \in \partial_\varepsilon f(x)$. □

Incidentally, the above proof shows that, among the ε-subgradients, there is one for which strict inequality holds in (1.1.1). A consequence of this result is that, for $\varepsilon > 0$, the domain of the multifunction $x \mapsto \partial_\varepsilon f(x)$ is the convex set dom f. Here is a difference with the case $\varepsilon = 0$: we know that dom ∂f need not be the whole of dom f; it may even be a nonconvex set.

Consider for example the one-dimensional convex function

$$f(x) := \begin{cases} -2\sqrt{x} & \text{if } x \geq 0, \\ +\infty & \text{otherwise}. \end{cases} \tag{1.1.5}$$

Applying the definition, an ε-subgradient of f at $x = 0$ is an $s \in \mathbb{R}$ satisfying

$$sy + 2\sqrt{y} - \varepsilon \leq 0 \quad \text{for all } y \geq 0,$$

and easy calculation shows that the ε-subdifferential of f at 0 is $]-\infty, -1/\varepsilon]$: it is nonempty for all $\varepsilon > 0$, even though $\partial f(0) = \emptyset$. Also, note that it is unbounded, just because 0 is on the boundary of dom f.

For a further illustration, let $f : \mathbb{R} \to \mathbb{R}$ be $f(x) := |x|$. We have

$$\partial_\varepsilon f(x) = \begin{cases} [-1, -1 - \varepsilon/x] & \text{if } x < -\varepsilon/2, \\ [-1, +1] & \text{if } -\varepsilon/2 \leq x \leq \varepsilon/2, \\ [1 - \varepsilon/x, 1] & \text{if } x > \varepsilon/2. \end{cases}$$

The two parts of Fig. 1.1.2 display this set, as a multifunction of ε and x respectively. It is always a segment, reduced to the singleton $\{f'(x)\}$ only for $\varepsilon = 0$ (when $x \neq 0$). This example suggests that the approximate subdifferential is usually a proper enlargement of the exact subdifferential; this will be confirmed by Proposition 1.2.3 below.

Another interesting instance is the indicator function of a nonempty closed convex set:

Definition 1.1.3 The set of ε-*normal* directions to a closed convex set C at $x \in C$, or the ε-normal set for short, is the ε-subdifferential of the indicator function I_C at x:

$$N_{C,\varepsilon}(x) := \partial_\varepsilon I_C(x) = \{s \in \mathbb{R}^n : \langle s, y - x \rangle \leq \varepsilon \text{ for all } y \in C\}. \tag{1.1.6}$$
□

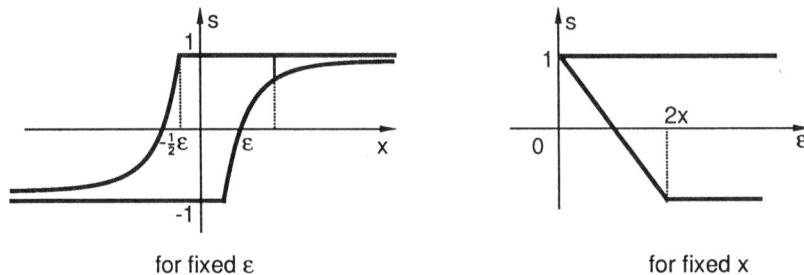

Fig. 1.1.2. Graphs of the approximate subdifferential of $|x|$

The ε-normal set is thus an intersection of half-spaces but is usually *not* a cone; it contains the familiar normal cone $N_C(x)$, to which it reduces when $\varepsilon = 0$. A condensed form of (1.1.6) uses the polar of the set $C - \{x\}$:

$$\partial_\varepsilon I_C(x) = \varepsilon (C - x)^\circ \quad \text{for all } x \in C \text{ and } \varepsilon > 0.$$

These examples raise the question of the boundedness of $\partial_\varepsilon f(x)$.

Theorem 1.1.4 *For $\varepsilon \geqslant 0$, $\partial_\varepsilon f(x)$ is a closed convex set, which is nonempty and bounded if and only if $x \in \operatorname{int} \operatorname{dom} f$.*

PROOF. Closedness and convexity come immediately from the definition (1.1.1).

Now, if $x \in \operatorname{int} \operatorname{dom} f$, then $\partial_\varepsilon f(x)$ contains the nonempty set $\partial f(x)$ (Theorem X.1.4.2). Then let $\delta > 0$ be such that $B(x, \delta) \subset \operatorname{int} \operatorname{dom} f$, and let L be a Lipschitz constant for f on $B(x, \delta)$ (Theorem IV.3.1.2). For $0 \neq s \in \partial_\varepsilon f(x)$, take $y = x + \delta s / \|s\|$:

$$f(x) + L\delta \geqslant f(y) \geqslant f(x) + \delta \langle s, s/\|s\| \rangle - \varepsilon$$

i.e. $\|s\| \leqslant L + \varepsilon/\delta$. Thus, the nonempty $\partial_\varepsilon f(x)$ is also bounded.

Conversely, take any s_1 in the normal cone to $\operatorname{dom} f$ at $x \in \operatorname{dom} f$:

$$\langle s_1, y - x \rangle \leqslant 0 \quad \text{for all } y \in \operatorname{dom} f.$$

If $\partial_\varepsilon f(x) \neq \emptyset$, add this inequality to (1.1.1) to obtain

$$f(y) \geqslant f(x) + \langle s + s_1, y - x \rangle - \varepsilon$$

for all $y \in \operatorname{dom} f$. We conclude $s + s_1 \in \partial_\varepsilon f(x)$: $\partial_\varepsilon f(x)$ contains (actually: is equal to) $\partial_\varepsilon f(x) + N_{\operatorname{dom} f}(x)$, which is unbounded if $x \notin \operatorname{int} \operatorname{dom} f$. □

In the introduction to this chapter, it was mentioned that one motivation for the ε-subdifferential was practical. The next result gives a first explanation: $\partial_\varepsilon f$ can be used to characterize the ε-solutions of a convex minimization problem – but, starting from Chap. XIII, we will see that its role is much more important than that.

Theorem 1.1.5 *The following two properties are equivalent:*

$$0 \in \partial_\varepsilon f(x),$$

$$f(x) \leqslant f(y) + \varepsilon \quad \text{for all } y \in \mathbb{R}^n.$$

PROOF. Apply the definition. □

1.2 Characterization via the Conjugate Function

The conjugate function f^* allows a condensed writing for (1.1.1):

Proposition 1.2.1 *The vector $s \in \mathbb{R}^n$ is an ε-subgradient of f at $x \in \text{dom } f$ if and only if*
$$f^*(s) + f(x) - \langle s, x \rangle \leqslant \varepsilon. \tag{1.2.1}$$

As a result,
$$s \in \partial_\varepsilon f(x) \iff x \in \partial_\varepsilon f^*(s). \tag{1.2.2}$$

PROOF. The definition (1.1.1) of $\partial_\varepsilon f(x)$ can be written
$$f(x) + [\langle s, y \rangle - f(y)] - \langle s, x \rangle \leqslant \varepsilon \quad \text{for all } y \in \text{dom } f$$
which, remembering that $f^*(s)$ is the supremum of the bracket, is equivalent to (1.2.1). This implies that $\partial_\varepsilon f(x) \subset \text{dom } f^*$ and, applying (1.2.1) with f replaced by f^*:
$$x \in \partial_\varepsilon f^*(s) \iff f^{**}(x) + f^*(s) - \langle s, x \rangle \leqslant \varepsilon.$$
The conclusion follows since $\overline{\text{Conv}} \, \mathbb{R}^n \ni f = f^{**}$. □

In contrast with the case $\varepsilon = 0$, note that
$$\partial_\varepsilon f(\mathbb{R}^n) = \text{dom } f^* \quad \text{whenever} \quad \varepsilon > 0.$$

Indeed, the inclusion "$\subset$" comes directly from (1.2.1); conversely, if $s \in \text{dom } f^*$, we know from Theorem 1.1.2 that there exists $x \in \partial_\varepsilon f^*(s)$, i.e. $s \in \partial_\varepsilon f(x)$.

Likewise, for fixed $x \in \text{dom } f$,
$$[\lim_{\varepsilon \to +\infty} \partial_\varepsilon f(x) =] \cup \{\partial_\varepsilon f(x) : \varepsilon > 0\} = \text{dom } f^*. \tag{1.2.3}$$

Here again, the "$\subset$" comes directly from (1.2.1) while, if $s \in \text{dom } f^*$, set
$$\varepsilon_s := f^*(s) + f(x) - \langle s, x \rangle \in [0, +\infty[$$
to see that s is in the corresponding $\partial_{\varepsilon_s} f(x)$.

Example 1.2.2 (Convex Quadratic Functions) Consider the function
$$\mathbb{R}^n \ni x \mapsto f(x) := \tfrac{1}{2} \langle Qx, x \rangle + \langle b, x \rangle$$
where Q is symmetric positive semi-definite with pseudo-inverse Q^-. Using Example X.1.1.4, we see that $\partial_\varepsilon f(x)$ is the set
$$\left\{ s \in b + \text{Im } Q : f(x) + \tfrac{1}{2} \langle s - b, Q^-(s - b) \rangle \leqslant \langle s, x \rangle + \varepsilon \right\}.$$

This set has a nicer expression if we single out $\nabla f(x)$: setting $p = s - b - Qx$, we see that $s - b \in \text{Im } Q$ means $p \in \text{Im } Q$ and, via some algebra,

$$\partial_\varepsilon f(x) = \{\nabla f(x)\} + \{p \in \operatorname{Im} Q : \tfrac{1}{2}\langle p, Q^- p\rangle \leqslant \varepsilon\}. \tag{1.2.4}$$

Another equivalent formula is obtained if we set $y := Q^- p$ (so $p = Qy$):

$$\partial_\varepsilon f(x) = \{\nabla f(x) + Qy : \tfrac{1}{2}\langle Qy, y\rangle \leqslant \varepsilon\}. \tag{1.2.5}$$

To illustrate Theorem 1.1.5, we see from (1.2.4) that

$$x \text{ minimizes } f \text{ within } \varepsilon \iff \begin{cases} \nabla f(x) = Qx + b \in \operatorname{Im} Q & [\text{i.e. } b \in \operatorname{Im} Q], \\ \text{and } \tfrac{1}{2}\langle \nabla f(x), Q^- \nabla f(x)\rangle \leqslant \varepsilon. \end{cases}$$

When Q is invertible and $b = 0$, f defines the norm $\|x\| = \langle Qx, x\rangle^{1/2}$. Its ε-subdifferential is then a neighborhood of the gradient for the metric associated with the dual norm $\|s\|^* = \langle s, Q^{-1}s\rangle^{1/2}$. □

The above example suggests once more that $\partial_\varepsilon f$ is usually a proper enlargement of ∂f; in particular it is "never" reduced to a singleton, in contrast with ∂f, which is "often" reduced to the gradient of f. This is made precise by the next results, which somehow describe two opposite situations.

Proposition 1.2.3 *Let $f \in \overline{\operatorname{Conv}}\, \mathbb{R}^n$ be 1-coercive, i.e. $f(x)/\|x\| \to +\infty$ if $\|x\| \to \infty$. Then, for all $x \in \operatorname{dom} f$,*

$$0 \leqslant \varepsilon < \varepsilon' \implies \partial_\varepsilon f(x) \subset \operatorname{int} \partial_{\varepsilon'} f(x).$$

Furthermore, any $s \in \mathbb{R}^n$ is an ε-subgradient of f at x, providing that ε is large enough.

PROOF. Written in the form (1.2.1), an approximate subdifferential of f appears as a sublevel-set of the convex function $f^* - \langle \cdot, x\rangle$; but the 1-coercivity of f means that this function is finite everywhere (Proposition X.1.3.8). For $\varepsilon' > \varepsilon$, the conditions of Proposition VI.1.3.3 are then satisfied to compute the interior of our sublevel-set:

$$\operatorname{int} \partial_{\varepsilon'} f(x) = \{s \in \mathbb{R}^n : f^*(s) + f(x) - \langle s, x\rangle < \varepsilon'\} \supset \partial_\varepsilon f(x).$$

As for the second claim, it comes directly from (1.2.3). □

Proposition 1.2.4 *Let $f \in \overline{\operatorname{Conv}}\, \mathbb{R}^n$ and suppose that $\partial_{\varepsilon_0} f(x_0)$ is a singleton for some $x_0 \in \operatorname{dom} f$ and $\varepsilon_0 > 0$. Then f is affine on $\mathbb{R}^n$.*

PROOF. Denote by s_0 the unique ε_0-subgradient of f at x_0. Let $\varepsilon \in {]0, \varepsilon_0[}$; in view of the monotonicity property (1.1.2), the nonempty set $\partial_\varepsilon f(x_0)$ can only be the same singleton $\{s_0\}$. Then let $\varepsilon' > \varepsilon_0$; the graph-convexity property (1.1.4) easily shows that $\partial_{\varepsilon'} f(x_0)$ is again $\{s_0\}$.

Thus, using the characterization (1.2.1) of an approximate subgradient, we have proved:

$$s \neq s_0 \implies f^*(s) > \varepsilon + \langle s_0, x_0\rangle - f(x_0) \text{ for all } \varepsilon > 0,$$

and not much room is left for f^*: indeed
$$f^* = r + I_{\{s_0\}} \quad \text{for some real number } r.$$
The affine character of f follows; and naturally $r = f^*(s_0) = \langle s_0, x_0 \rangle - f(x_0)$. □

Our next example somehow generalizes this last situation.

Example 1.2.5 (Support and Indicator Functions) Let $\sigma = \sigma_C$ be a closed sublinear function, i.e. the support of the nonempty closed convex set $C = \partial \sigma(0)$. Its conjugate σ^* is just the indicator of C, hence
$$\partial_\varepsilon \sigma_C(d) = \{s \in C : \sigma(d) \leqslant \langle s, d \rangle + \varepsilon\}. \tag{1.2.6}$$
Remembering that $\sigma(0) = 0$, we see that
$$\partial_\varepsilon \sigma_C(d) \subset \partial \sigma_C(0) = \partial_\varepsilon \sigma_C(0) = C \quad \text{for all } d \in \mathbb{R}^n.$$

Keeping Proposition 1.2.4 in mind, the approximate subdifferential of a linear – or even affine – function is a singleton. The set $\partial_\varepsilon \sigma_C(d)$ could be called the "ε-face" of C exposed by d, i.e. the set of ε-maximizers of $\langle \cdot, d \rangle$ over C. As shown in Fig. 1.2.1, it is the intersection of the convex set C with a certain half-space depending on ε. For $\varepsilon = 0$, we do obtain the exposed face itself: $\partial \sigma_C(d) = F_C(d)$.

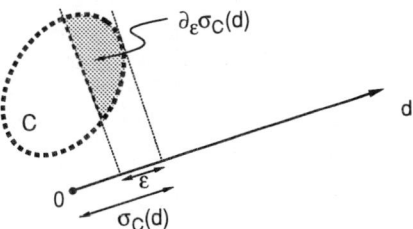

Fig. 1.2.1. An ε-face

Along the same lines, the conjugate of I_C in (1.1.6) is σ_C, so the ε-normal set to C at x is equivalently defined as
$$N_{C,\varepsilon}(x) = \{s \in \mathbb{R}^n : \sigma_C(s) \leqslant \langle s, x \rangle + \varepsilon\}.$$
Beware of the difference with (1.2.6); one set looks like a face, the other like a cone; the relation linking them connotes the polarity relation, see Remark V.3.2.6.

Consider for example a half-space: with $s \neq 0$,
$$H^- := \{x \in \mathbb{R}^n : \langle s, x \rangle \leqslant r\}.$$
Its support function is computed by straightforward calculations, already seen in Example V.2.3.1; we obtain for $x \in H^-$:
$$N_{H^-,\varepsilon}(x) = \{\lambda s : 0 \leqslant \lambda \text{ and } (r - \langle s, x \rangle)\lambda \leqslant \varepsilon\}.$$
In other words, the approximate normal set to a half-space is a segment, which stretches to the whole normal cone when x approaches the boundary. Figure 2.1.1 will illustrate an ε-normal set in a nonlinear situation. □

1.3 Some Useful Properties

(a) Elementary Calculus. First, we list some properties coming directly from the definitions.

Proposition 1.3.1

(i) *For the function* $g(x) := f(x) + r$, $\partial_\varepsilon g(x) = \partial_\varepsilon f(x)$.
(ii) *For the function* $g(x) := \alpha f(x)$ *and* $\alpha > 0$, $\partial_\varepsilon g(x) = \alpha \partial_{\varepsilon/\alpha} f(x)$.
(iii) *For the function* $g(x) := f(\alpha x)$ *and* $\alpha \neq 0$, $\partial_\varepsilon g(x) = \alpha \partial_\varepsilon f(\alpha x)$.
(iv) *More generally, if A is an invertible linear operator,* $\partial_\varepsilon (f \circ A)(x) = A^* \partial_\varepsilon f(Ax)$.
(v) *For the function* $g(x) := f(x - x_0)$, $\partial_\varepsilon g(x + x_0) = \partial f(x)$.
(vi) *For the function* $g(x) := f(x) + \langle s_0, x \rangle$, $\partial_\varepsilon g(x) = \partial_\varepsilon f(x) + \{s_0\}$.
(vii) *If* $f_1 \leqslant f_2$ *and* $f_1(x) = f_2(x)$, *then* $\partial_\varepsilon f_1(x) \subset \partial_\varepsilon f_2(x)$.

PROOF. Apply (1.1.1), or combine (1.2.1) with the elementary calculus rules X.1.3.1, whichever seems easier. □

Our next result expresses how the approximate subdifferential is transformed, when the starting function is restricted to a subspace.

Proposition 1.3.2 *Let H be a subspace containing a point of* dom f *and call* p_H *the operator of orthogonal projection onto H. For all $x \in$ dom $f \cap H$,*

$$s \in \partial_\varepsilon (f + I_H)(x) \iff p_H s \in \partial_\varepsilon (f \circ p_H)(x),$$

i.e.

$$\partial_\varepsilon (f + I_H)(x) = \partial_\varepsilon (f \circ p_H)(x) + H^\perp.$$

PROOF. From the characterization (1.2.1), $s \in \partial_\varepsilon (f + I_H)(x)$ means

$$\begin{aligned}
\varepsilon &\geqslant (f + I_H)(x) + (f + I_H)^*(s) - \langle s, x \rangle = \\
&= f(x) + [(f \circ p_H)^* \circ p_H](s) - \langle s, x \rangle && \text{[from Prop. X.1.3.2]} \\
&= f(p_H(x)) + (f \circ p_H)^*(p_H(s)) - \langle s, p_H(x) \rangle && [x = p_H(x)] \\
&= (f \circ p_H)(x) + (f \circ p_H)^*(p_H(s)) - \langle p_H(s), x \rangle. && [p_H \text{ is symmetric}]
\end{aligned}$$

This just means that $p_H(s)$ is in the ε-subdifferential of $f \circ p_H$ at x. □

A particular case is when dom $f \subset H$: then $f + I_H$ coincides with f and we have $\partial_\varepsilon f = \partial_\varepsilon (f \circ p_H) + H^\perp$.

(b) The Tilted Conjugate Function. From (1.2.1), $\partial_\varepsilon f(x)$ appears as the sublevel-set at level ε of the "tilted conjugate function"

$$\mathbb{R}^n \ni s \mapsto f^*(s) - \langle s, x \rangle + f(x) =: g_x^*(s), \tag{1.3.1}$$

which is clearly in $\overline{\text{Conv}}\, \mathbb{R}^n$ (remember $x \in$ dom f!) and plays an important role. Its infimum on $\mathbb{R}^n$ is 0 (Theorem 1.1.2), attained at the subgradients of f at x, if there are any. Its conjugate $g_x^{**} = g_x$ is easily computed, say via Proposition X.1.3.1, and is simply the "shifted" function

$$\mathbb{R}^n \ni h \mapsto g_x(h) := f(x+h) - f(x). \tag{1.3.2}$$

Indeed, when connecting g_x^* to the approximate subdifferential, we just perform the conjugacy operation on f, but with a special role attributed to the particular point x under consideration.

Proposition 1.3.3 *For $x \in \text{dom } f$, the epigraph of g_x^* is the graph of the multifunction $\varepsilon \mapsto \partial_\varepsilon f(x)$:*

$$\{(s, \varepsilon) \in \mathbb{R}^n \times \mathbb{R} : g_x^*(s) \leqslant \varepsilon\} = \{(s, \varepsilon) \in \mathbb{R}^n \times \mathbb{R} : s \in \partial_\varepsilon f(x)\}. \tag{1.3.3}$$

The support function of this set has, at $(d, -u) \in \mathbb{R}^n \times \mathbb{R}$, the value

$$\sigma_{\text{epi } g_x^*}(d, -u) = \sup_{s,\varepsilon} \{\langle s, d \rangle - \varepsilon u : s \in \partial_\varepsilon f(x)\} = \tag{1.3.4}$$

$$= \begin{cases} u[f(x+d/u) - f(x)] & \text{if } u > 0, \\ \sigma_{\text{dom } f^*}(d) = f'_\infty(d) & \text{if } u = 0, \\ +\infty & \text{if } u < 0. \end{cases}$$

PROOF. The equality of the two sets in (1.3.3) is trivial, in view of (1.2.1) and (1.3.1). Then (1.3.4) comes either from direct calculations, or via Proposition X.1.2.1, with f replaced by g_x^*, whose domain is just dom f^* (Proposition X.1.3.1 may also be used). □

Remember §IV.2.2 and Example I.3.2.2 to realize that, up to the closure operation at $u = 0$, the function (1.3.4) is just the perspective of the shifted function $h \mapsto f(x+h) - f(x)$.

Figure 1.3.1 displays the graph of $\varepsilon \mapsto \partial_\varepsilon f(x)$, with the variable ε plotted along the horizontal axis, as usual (see also the right part of Fig. 1.1.2). Rotating the picture so that this axis becomes vertical, we obtain epi g_x^*.

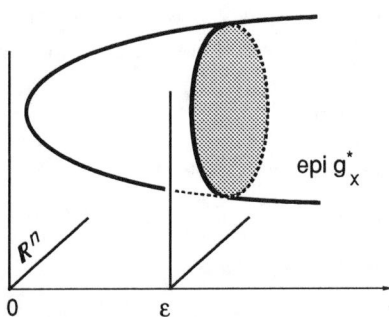

Fig. 1.3.1. Approximate subdifferential as a sublevel set in the dual space

(c) A Critical ε-Value.

Proposition 1.3.4 *Assume that the infimal value $\bar{f}$ of f over $\mathbb{R}^n$ is finite. For all $x \in \mathrm{dom}\, f$, there holds*

$$\inf \{\varepsilon > 0 : 0 \in \partial_\varepsilon f(x)\} = f(x) - \bar{f}. \tag{1.3.5}$$

PROOF. Because g_x^* is a nonnegative function, we always have

$$g_x^*(0) = \inf \{\varepsilon > 0 : \varepsilon \geq g_x^*(0)\}$$

which, in view of (1.3.3) and of the definition of g_x^*, can also be written

$$\inf \{\varepsilon > 0 : 0 \in \partial_\varepsilon f(x)\} = f^*(0) + f(x) = -\bar{f} + f(x). \qquad \square$$

The relation expressed in (1.3.5) is rather natural, and can be compared to Theorem 1.1.5. It defines a sort of "critical" value for ε, satisfying the following property:

Proposition 1.3.5 *Assume that f is 1-coercive. For a non-optimal $x \in \mathrm{dom}\, f$, define*

$$\bar{\varepsilon} := f(x) - \bar{f} > 0.$$

Then the normal cone to $\partial_{\bar{\varepsilon}} f(x)$ at 0 is the set of directions pointing from x to the minimizers of f:

$$N_{\partial_{\bar{\varepsilon}} f(x)}(0) = \mathbb{R}^+(\mathrm{Argmin}\, f - \{x\}).$$

PROOF. We know that $\partial_\varepsilon f(x)$ is the sublevel-set of g_x^* at level ε and we use Theorem VI.1.3.5 to compute its normal cone. From the coercivity of f, g_x^* is finite everywhere (Proposition X.1.3.8); if, in addition, $\varepsilon > 0$, the required Slater assumption is satisfied; hence, for an s satisfying $g_x^*(s) = \varepsilon$ [i.e. $s \in \mathrm{bd}\, \partial_\varepsilon f(x)$],

$$N_{\partial_\varepsilon f(x)}(s) = \mathbb{R}^+ \partial g_x^*(s) = \mathbb{R}^+ (\partial f^*(s) - \{x\}).$$

In this formula, we can set $\varepsilon = \bar{\varepsilon} > 0$ and $s = 0$ because

$$g_x^*(0) = f^*(0) + f(x) - \langle 0, x \rangle = \bar{\varepsilon}.$$

The result follows from the expression (X.1.4.6) of Argmin f as $\partial f^*(0)$. $\qquad \square$

(d) Representations of a Closed Convex Function.

An observation inspired by the function g_x^* is that the multifunctions $\varepsilon \longmapsto F(\varepsilon)$ that are ε-subdifferentials (of a function f at a point x) are the closed convex epigraphs; see §IV.1.3(g). Also, if $\partial_\varepsilon f(x)$ is known for all $\varepsilon > 0$ at a given $x \in \mathrm{dom}\, f$, then f^* is also known, so f itself is completely determined:

Theorem 1.3.6 *For $x \in \mathrm{dom}\, f$, there holds*

$$f(x + h) = f(x) + \sup \{\langle s, h \rangle - \varepsilon : \varepsilon > 0, s \in \partial_\varepsilon f(x)\} \tag{1.3.6}$$

for all $h \in \mathbb{R}^n$; or, using support functions:

$$f(x + h) = f(x) + \sup_{\varepsilon > 0} [\sigma_{\partial_\varepsilon f(x)}(h) - \varepsilon]. \tag{1.3.7}$$

PROOF. Fix $x \in \text{dom } f$. Then set $u = 1$ in (1.3.4) to obtain $\sigma_{\text{epi } g_x^*}(h, -1) = g_x(h)$, i.e. (1.3.7), which is just a closed form for (1.3.6). □

Using the positive homogeneity of a support function, another way of writing (1.3.7) is: if $x \in \text{dom } f$,

$$f(x + td) = f(x) + t \sup_{\alpha > 0}[\sigma_{\partial_\alpha f(x)}(d) - \alpha] \quad \text{for all } t > 0.$$

Theorem 1.3.6 gives a converse to Proposition 1.3.1(vii): if, for a particular $x \in \text{dom } f_1 \cap \text{dom } f_2$,

$$\partial_\varepsilon f_1(x) \subset \partial_\varepsilon f_2(x) \quad \text{for all } \varepsilon > 0,$$

then

$$f_1(y) - f_1(x) \leq f_2(y) - f_2(x) \quad \text{for all } y \in \mathbb{R}^n.$$

Remark 1.3.7 The function g_x^* differs from f^* by the affine function $f(x) - \langle \cdot, x \rangle$ only. In the primal graph-space, g_x is obtained from f by a mere translation of the origin from $(0, 0)$ to $(x, f(x))$. Basically, considering the graph of $\partial_\varepsilon f$, instead of the epigraph of f^*, just amounts to this change of variables. In the framework of numerical optimization, the distinction is significant: it is near the current iterate x of an algorithm that the behaviour of f matters. Here is one more practical motivation for introducing the approximate subdifferential.

Theorem 1.3.6 confirms this observation: numerical algorithms need a model of f near the current x – remember §II.2. A possibility is to consider a model for f^*: it induces via (1.2.1) a model for the multifunction $\varepsilon \mapsto \partial_\varepsilon f(x)$, which in turn gives via (1.3.7) a model for f, hence a possible strategy for finding the next iterate x_+. □

This remark suggests that, in the framework of numerical optimization, (1.3.7) will be especially useful with h close to 0. More globally, we have

$$[f^{**}(x) =] \quad f(x) = \sup\{\langle s, x \rangle - f^*(s) : s \in \mathbb{R}^n\},$$

which also expresses a closed convex function as a supremum, but of *affine* functions. The index s can be restricted to dom f^*, or even to ri dom f^*; another better idea: restrict s to dom ∂f^* ($\supset$ ri dom f^*), in which case $f^*(s) = \langle s, y \rangle - f(y)$ for some $y \in \text{dom } \partial f$. In other words, we can write

$$f(x) = \sup\{f(y) + \langle s, x - y \rangle : y \in \text{dom } \partial f, \ s \in \partial f(y)\}.$$

This formula has a refined form, better suited to numerical optimization where only one subgradient at a particular point is usually known (see again the concept of black box (U1) in §VIII.3.5).

Theorem 1.3.8 *Associated with $f \in \overline{\text{Conv}}\, \mathbb{R}^n$, let $s : \text{ri dom } f \to \mathbb{R}^n$ be a mapping satisfying $s(y) \in \partial f(y)$ for all $y \in \text{ri dom } f$. Then there holds*

$$f(x) = \sup_{y \in \text{ri dom } f} [f(y) + \langle s(y), x - y \rangle] \quad \text{for all } x \in \text{dom } f. \tag{1.3.8}$$

PROOF. The existence of the mapping s is guaranteed by Theorem X.1.4.2; besides, only the inequality $\leq$ in (1.3.8) must be proved. From Lemma III.2.1.6, we can take $d \in \mathbb{R}^n$ such that

$$y_t := x + td \in \operatorname{ri} \operatorname{dom} f \quad \text{for } t \in]0, 1].$$

Then
$$f(x+d) \geq f(y_t) + \langle s(y_t), x+d - y_t \rangle$$

so that
$$\langle s(y_t), d \rangle \leq \frac{f(x+d) - f(y_t)}{1-t}.$$

Then write
$$f(y_t) + \langle s(y_t), x - y_t \rangle = f(y_t) - t\langle s(y_t), d \rangle \geq \frac{f(y_t) - tf(x+d)}{1-t}$$

and let $t \downarrow 0$; use for example Proposition IV.1.2.5 to see that the right-hand side tends to $f(x)$. □

2 The Approximate Directional Derivative

Throughout this section, x is fixed in dom f. As a (nonempty) closed convex set, the approximate subdifferential $\partial_\varepsilon f(x)$ then has a support function, for any $\varepsilon > 0$. We denote it by $f'_\varepsilon(x, \cdot)$:

$$\mathbb{R}^n \ni d \mapsto f'_\varepsilon(x, d) := \sigma_{\partial_\varepsilon f(x)}(d) = \sup_{s \in \partial_\varepsilon f(x)} \langle s, d \rangle, \tag{2.0.1}$$

a closed sublinear function. The notation f'_ε is motivated by §VI.1.1: $f'(x, \cdot)$ supports $\partial f(x)$, so it is natural to denote by $f'_\varepsilon(x, \cdot)$ the function supporting $\partial_\varepsilon f(x)$. The present section is devoted to a study of this support function, which is obtained via an "approximate difference quotient".

2.1 The Support Function of the Approximate Subdifferential

Theorem 2.1.1 *For $x \in$ dom f and $\varepsilon > 0$, the support function of $\partial_\varepsilon f(x)$ is*

$$\mathbb{R}^n \ni d \mapsto f'_\varepsilon(x, d) = \inf_{t>0} \frac{f(x+td) - f(x) + \varepsilon}{t}, \tag{2.1.1}$$

which will be called the ε-directional derivative of f at x.

PROOF. We use Proposition 1.3.3: embedding the set $\partial_\varepsilon f(x)$ in the larger space $\mathbb{R}^n \times \mathbb{R}$, we view it as the intersection of epi g_x^* with the horizontal hyperplane

$$H_\varepsilon := \{(s, \varepsilon) : s \in \mathbb{R}^n\}$$

2 The Approximate Directional Derivative

(rotate and contemplate thoroughly Fig. 1.3.1). Correspondingly, $f'_\varepsilon(x,d)$ is the value at $(d, 0)$ of the support function of our embedded set $\text{epi}\, g_x^* \cap H_\varepsilon$:

$$f'_\varepsilon(x, d) = (I_{\text{epi}\, g_x^*} + I_{H_\varepsilon})^*(d, 0).$$

Our aim is then to apply the calculus rule X.2.3.2, so we need to check the qualification assumption (X.2.3.1). The relative interior of the hyperplane H_ε being H_ε itself, what we have to do is to find in $\text{ri}\,\text{epi}\, g_x^*$ a point of the form (s, ε). Denote by A the linear operator which, to $(s, \alpha) \in \mathbb{R}^n \times \mathbb{R}$, associates $A(s, \alpha) = \alpha$: we want

$$\varepsilon \in A(\text{ri}\,\text{epi}\, g_x^*) = \text{ri}[A(\text{epi}\, g_x^*)],$$

where the last equality comes from Proposition III.2.1.12. But we know from Theorem 1.1.2 and Proposition 1.3.3 that $A(\text{epi}\, g_x^*)$ is $\mathbb{R}^+$ or $\mathbb{R}_*^+$; in both cases, its relative interior is $\mathbb{R}_*^+$, which contains the given $\varepsilon > 0$. Our assumption is checked.

As a result, the following problem

$$\min\{\sigma_{H_\varepsilon}(p, u) + \sigma_{\text{epi}\, g_x^*}(q, v) : p + q = d, u + v = 0\} = f'_\varepsilon(x, d) \quad (2.1.2)$$

has an optimal solution. Now look at its minimand:

- unless $p = 0$, $\sigma_{H_\varepsilon}(p, u) = +\infty$; and $\sigma_{H_\varepsilon}(0, u) = u\varepsilon$;
- unless $v \leqslant 0$, $\sigma_{\text{epi}\, g_x^*}(q, v) = +\infty$ – see (1.3.4).

In a word,

$$f'_\varepsilon(x, d) = \min\{\sigma_{\text{epi}\, g_x^*}(d, -u) + u\varepsilon : u \geqslant 0\}.$$

Remembering that, as a support function, $u \mapsto \sigma_{\text{epi}\, g_x^*}(d, u)$ is in particular continuous for $u \downarrow 0$, and using its value (1.3.4), we finally obtain (2.1.1). □

The above proof strongly suggests to consider the change of variable $t = 1/u$ in (2.1.1), and to set

$$r_\varepsilon(u) := \begin{cases} u[f(x + d/u) - f(x) + \varepsilon] & \text{if } u > 0, \\ f'_\infty(d) = \sigma_{\text{dom} f^*}(d) & \text{if } u = 0, \\ +\infty & \text{if } u < 0. \end{cases} \quad (2.1.3)$$

When $\varepsilon > 0$, this function does achieve its minimum on $\mathbb{R}^+$; in the t-language, this means that the case $t \downarrow 0$ never occurs in the minimization problem (2.1.1). On the other hand, 0 may be the unique minimum point of r_ε, i.e. (2.1.1) may have no solution "at finite distance".

The importance of the assumption $\varepsilon > 0$ cannot be overemphasized; Theorem X.2.3.2 cannot be invoked without it, and the proof breaks down. Consider Fig. 2.1.1 for a counterexample: $B(0, 1)$ is the unit Euclidean ball of $\mathbb{R}^2$, f its indicator function and $x = (0, -1)$. Clearly enough, the directional derivative $f'(x, d)$ for $d = (\gamma, \delta) \neq 0$ is

$$f'(x, d) = \begin{cases} 0 & \text{if } \delta > 0, \\ +\infty & \text{if } \delta \leqslant 0, \end{cases}$$

104   XI. Approximate Subdifferentials of Convex Functions

which cannot be a support function: it is not even closed. On the other hand, the infimum in (2.1.1), which is obtained for $x + td$ on the unit sphere, is easy to compute: for $d \neq 0$,

$$f'_\varepsilon(x, d) = \begin{cases} \dfrac{\gamma^2 + \delta^2}{2\delta}\varepsilon & \text{if } \delta > 0, \\ +\infty & \text{if } \delta \leq 0. \end{cases}$$

It is the support function of the set

$$N_\varepsilon := \{(\xi, \eta) : 2\eta \leq \varepsilon - \xi^2/\varepsilon\} \tag{2.1.4}$$

which, in view of Theorem 2.1.1, is the ε-normal set to $B(0, 1)$ at x.

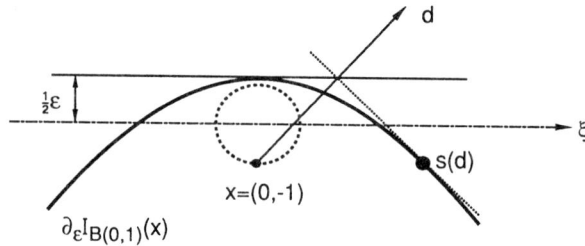

Fig. 2.1.1. The approximate subdifferential of an indicator function

Note: to obtain (2.1.4), it is fun to use elementary geometry (carry out in Fig. 2.1.1 the inversion operation from the pole x), but the argument of Remark VI.6.3.7 is more systematic, and is also interesting.

Remark 2.1.2 We have here an illustration of Example X.2.4.3: closing $f'(x, \cdot)$ is just what is needed to obtain the support function of $\partial f(x)$. For $x \in \text{dom}\, \partial f$, the function

$$\mathbb{R}^n \ni d \mapsto \sigma_{\partial f(x)}(d) = \sup\{\langle s, d\rangle : s \in \partial f(x)\}$$

is the closure of the function

$$\mathbb{R}^n \ni d \mapsto f'(x, d) = \lim_{t \downarrow 0} \frac{f(x + td) - f(x)}{t}.$$

This property appears also in the proof of Theorem 2.1.1: for $\varepsilon = 0$, the closure operation must still be performed after (2.1.2) is solved. □

Figure 2.1.2 illustrates (2.1.1) in a less dramatic situation: for $\varepsilon > 0$, the line representing $t \mapsto f(x) - \varepsilon + tf'_\varepsilon(x, d)$ supports the graph of $t \mapsto f(x + td)$, not at $t = 0$ but at some point $t_\varepsilon > 0$; among all the slopes joining $(0, f(x) - \varepsilon)$ to an arbitrary point on the graph of $f(x + \cdot d)$, the right-hand side of (2.1.1) is the smallest possible one.

On the other hand, Fig. 2.1.2 is the trace in $\mathbb{R} \times \mathbb{R}$ of a picture in $\mathbb{R}^n \times \mathbb{R}$: among all the possible hyperplanes passing through $(x, f(x) - \varepsilon)$ and supporting epi f, there is one touching gr f somewhere along the given d; this hyperplane therefore gives the maximal slope along d, which is the value (2.0.1). The contact $x + t_\varepsilon d$ plays an important role in minimization algorithms and we will return to it later.

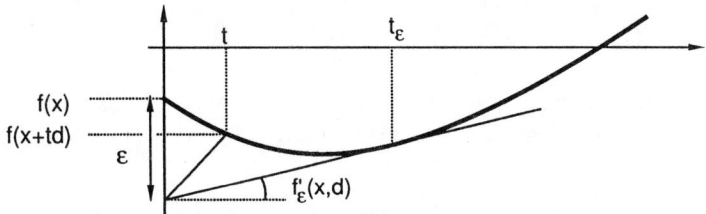

Fig. 2.1.2. The ε-directional derivative and the minimizing t_ε

The same picture illustrates Theorem 1.3.6: consider the point $x + t_\varepsilon d$ as fixed and call it y. Now, for arbitrary $\delta \geqslant 0$, draw a hyperplane supporting epi f and passing through $(0, f(x) - \delta)$. Its altitude at y is $f(x) - \delta + f'_\delta(x, y - x)$ which, by definition of a support, is not larger than $f(y)$, but equal to it when $\delta = \varepsilon$.

In summary: fix $(x, d) \in \text{dom } f \times \mathbb{R}^n$ and consider a pair $(\varepsilon, y) \in \mathbb{R}_*^+ \times \mathbb{R}^n$ linked by the relation $y = x + t_\varepsilon d$.

- To obtain y from ε, use (2.1.1): as a function of the "horizontal" variable $t > 0$, draw the line passing through $(x, f(x) - \varepsilon)$ and $(x + td, f(x + td))$; the resulting slope must be minimized.
- To obtain ε from y, use (1.3.7): as a function of the "vertical" variable $\delta \geqslant 0$, draw a support passing through $(x, f(x) - \delta)$; its altitude at y must be maximized.

Example 2.1.3 Take again the convex quadratic function of Example 1.2.2: $f'_\varepsilon(x, d)$ is the optimal value of the one-dimensional minimization problem

$$\inf_{t>0} \frac{\frac{1}{2}\langle Qd, d\rangle t^2 + \langle \nabla f(x), d\rangle t + \varepsilon}{t}.$$

If $\langle Qd, d\rangle = 0$, this infimum is $\langle \nabla f(x), d\rangle$. Otherwise, it is attained at

$$t_\varepsilon = \sqrt{\frac{2\varepsilon}{\langle Qd, d\rangle}}$$

and

$$f'_\varepsilon(x, d) = \langle \nabla f(x), d\rangle + \sqrt{2\varepsilon \langle Qd, d\rangle}.$$

This is a general formula: it also holds for $\langle Qd, d\rangle = 0$.

It is interesting to observe that $[f'_\varepsilon(x, d) - f'(x, d)]/\varepsilon \to +\infty$ when $\varepsilon \downarrow 0$, but

$$\frac{\frac{1}{2}[f'_\varepsilon(x, d) - f'(x, d)]^2}{\varepsilon} \equiv \langle Qd, d\rangle.$$

This suggests that, for C^2 convex functions,

$$\lim_{\varepsilon \downarrow 0} \frac{\frac{1}{2}[f'_\varepsilon(x, d) - f'(x, d)]^2}{\varepsilon} = f''(x, d)$$

where

$$f''(x,d) := \lim_{t \to 0} \frac{f(x+td) - f(x) - tf'(x,d)}{\frac{1}{2}t^2} = \lim_{t \to 0} \frac{f'(x+td,d) - f'(x,d)}{t}$$

is the second derivative of f at x in the direction d. Here is one more motivation for the approximate subdifferential: it accounts for second order behaviour of f. □

2.2 Properties of the Approximate Difference Quotient

For $x \in \text{dom } f$, the function q_ε encountered in (2.1.1) and defined by

$$]0, +\infty[\ni t \mapsto q_\varepsilon(t) := \frac{f(x+td) - f(x) + \varepsilon}{t} \tag{2.2.1}$$

is called the *approximate difference quotient*. In what follows, we will set $q := q_0$; to avoid the trivial case $q_\varepsilon \equiv +\infty$, we assume

$$[x, x+td] \subset \text{dom } f \quad \text{for some } t > 0.$$

Our aim is now to study the minimization of q_ε, a relevant problem in view of Theorem 2.1.1.

(a) Behaviour of q_ε. In order to characterize the set of minimizers

$$T_\varepsilon := \{t > 0 : q_\varepsilon(t) = f'_\varepsilon(x,d)\}, \tag{2.2.2}$$

two numbers are important. One is

$$f'_\infty(d) := \sup\{q(t) : t > 0\}$$

which describes the behaviour of $f(x+td)$ for $t \to +\infty$ (see §IV.3.2 and Example X.2.4.6 for the asymptotic function f'_∞). The other,

$$t^\ell := \sup\{t \geq 0 : f(x+td) - f(x) = tf'(x,d)\},$$

concerns the case $t \downarrow 0$: $t^\ell = 0$ if $f(x+td)$ has a "positive curvature for $t \downarrow 0$". When $t^\ell = +\infty$, i.e. f is affine on the half-line $x + \mathbb{R}^+ d$, we have $q_\varepsilon(t) = f'(x,d) + \varepsilon/t$; then it is clear that $f'_\varepsilon(x,d) = f'(x,d) = f'_\infty(d)$ and that T_ε is empty for $\varepsilon > 0$.

Example 2.2.1 Before making precise statements, let us see in Fig. 2.2.1 what can be expected, with the help of the example

$$0 \leq t \mapsto f(x+td) := \max\{t, 3t-1\}.$$

The lower-part of the picture gives the correspondence $\varepsilon \leftrightarrows T_\varepsilon$, with the same abscissa-axis as the upper-part (namely t). Some important properties are thus introduced:

– As already known, q increases from $f'(x,d) = 1$ to $f'_\infty(d) = 3$;
– indeed, $q(t)$ is constantly equal to its minimal value $f'(x,d)$ for $0 < t \leq 1/2$, and $t^\ell = 1/2$;

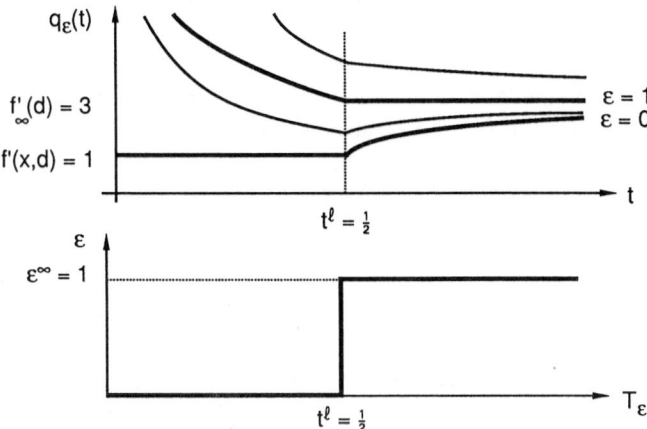

Fig. 2.2.1. A possible q_ε and T_ε

- $f'_\varepsilon(x, d)$ stays between $f'(x, d)$ and $f'_\infty(d)$, and reaches this last value for $\varepsilon \geq 1$;
- T_0 is the segment $]0, 1/2]$; T_1 is the half-line $[1/2, +\infty[$, and T_ε is empty for $\varepsilon > 1$. □

This example reveals another important number associated with large t, which we will call ε^∞ (equal to 1 in the example). The statements (i) and (ii) in the next result are already known, but their proof is more natural in the t-language.

Proposition 2.2.2 *The notations and assumptions are as above.*

(i) *If $\varepsilon > 0$, then $q_\varepsilon(t) \to +\infty$ when $t \downarrow 0$.*
(ii) *When $t \to +\infty$, $q_\varepsilon(t) \to f'_\infty(d)$ independently of $\varepsilon \geq 0$.*
(iii) *The set*

$$E := \{\varepsilon \geq 0 : f'_\varepsilon(x, d) < f'_\infty(d)\} \qquad (2.2.3)$$

is empty if and only if $t^\ell = +\infty$. Its upper bound

$$\sup E =: \varepsilon^\infty \in [0, +\infty]$$

satisfies

$$T_\varepsilon \neq \emptyset \text{ if } \varepsilon < \varepsilon^\infty \quad \text{and} \quad T_\varepsilon = \emptyset \text{ if } \varepsilon > \varepsilon^\infty.$$

PROOF. For $\varepsilon > 0$, let $s_0 \in \partial_{\varepsilon/2} f(x)$: for all $t > 0$,

$$q_\varepsilon(t) \geq \frac{f(x) + t\langle s_0, d\rangle - \varepsilon/2 - f(x) + \varepsilon}{t} = \langle s_0, d\rangle + \frac{\varepsilon}{2t}$$

and this takes care of (i). For $t \to +\infty$, $q_\varepsilon(t) = q(t) + \varepsilon/t$ has the same limit as $q(t)$, so (ii) is straightforward.

Thus $f'_\varepsilon(x, d) \leq f'_\infty(d)$, and non-vacuousness of T_ε depends only on the behaviour of q_ε at infinity. To say that E is empty means from its definition (2.2.3):

$$f'_\infty(d) \leq \inf_{\varepsilon \geq 0} \inf_{t > 0} q_\varepsilon(t) = \inf_{t > 0} \inf_{\varepsilon \geq 0} q_\varepsilon(t) = f'(x, d).$$

Because $f'(x, \cdot) \leq f'_\infty$, this in turn means that $q(t)$ is constant for $t > 0$, or that $t^\ell = +\infty$.

Now observe that $\varepsilon \mapsto f'_\varepsilon(x, d)$ is increasing (just as $\partial_\varepsilon f$); so E, when nonempty, is an interval containing 0: $\varepsilon < \varepsilon^\infty$ implies $\varepsilon \in E$, hence $T_\varepsilon \neq \emptyset$. Conversely, if $\varepsilon > \varepsilon^\infty$, take ε' in $]\varepsilon^\infty, \varepsilon[$ (so $\varepsilon' \notin E$) and $t > 0$ arbitrary:

$$q_\varepsilon(t) > q_{\varepsilon'}(t) \geq f'_{\varepsilon'}(x, d) \geq f'_\infty(d) \geq f'_\varepsilon(x, d);$$

t cannot be in T_ε, which is therefore empty. $\square$

The example $t \mapsto f(x + td) = \sqrt{1 + t^2}$ illustrates the case $\varepsilon^\infty < +\infty$ (here $\varepsilon^\infty = 1$); it corresponds to $f(x+\cdot d)$ having the asymptote $t \mapsto f(x) - \varepsilon^\infty + f'_\infty(d)t$. Also, one directly sees in (2.2.3) that $\varepsilon^\infty = +\infty$ whenever $f'_\infty(d)$ is infinite. With the example f of (1.1.5) (and taking $d = 1$), we have $t^\ell = 0$, $f'_\infty(1) = 0$ and $\varepsilon^\infty = +\infty$.

(b) The Closed Convex Function r_ε. For a more accurate study of T_ε, we use now the function r_ε of (2.1.3), obtained via the change of variable $u = 1/t$: $r_\varepsilon(u) = q_\varepsilon(1/u)$ for $u > 0$. It is the trace on $\mathbb{R}$ of the function (1.3.4), which is known to be in $\overline{\text{Conv}}(\mathbb{R}^n \times \mathbb{R})$; therefore

$$r_\varepsilon \in \overline{\text{Conv}}\,\mathbb{R}.$$

We know from Theorem 2.1.1 that r_ε is minimized on a closed interval

$$U_\varepsilon := \{u \geq 0 : r_\varepsilon(u) = f'_\varepsilon(x, d)\} \tag{2.2.4}$$

which, in view of Proposition 2.2.2(i), is nonempty and bounded if $\varepsilon > 0$. Likewise,

$$T_\varepsilon = \{t = 1/u : u \in U_\varepsilon \text{ and } u > 0\}$$

is a closed interval, empty if $U_\varepsilon = \{0\}$, a half-line if $\{0\} \subsetneq U_\varepsilon$. Knowing that

$$T_0 = \{t \in \mathbb{R} : 0 \leq t \leq t^\ell\},$$

it will be convenient to set, with the conventions $1/0 = +\infty$, $1/\infty = 0$:

$$U_0 = \{u \in \mathbb{R} : u \geq 1/t^\ell\}. \tag{2.2.5}$$

There remains to characterize T_ε, simply by expressing the minimality condition $0 \in \partial r_\varepsilon(u)$.

Lemma 2.2.3 *Let φ be the convex function $0 \leq t \mapsto \varphi(t) := f(x + td)$. For $\varepsilon \geq 0$ and $0 < u \in \text{dom}\, r_\varepsilon$, the subdifferential of r_ε at u is given by*

$$\partial r_\varepsilon(u) = \frac{r_\varepsilon(u) - \partial\varphi(1/u)}{u}. \tag{2.2.6}$$

2 The Approximate Directional Derivative   109

PROOF. The whole point is to compute the subdifferential of the convex function $u \mapsto \psi(u) := u\varphi(1/u)$, and this amounts to computing its one-sided derivatives. Take positive $u' = 1/t'$ and $u = 1/t$, with $u \in \mathrm{dom}\, r_\varepsilon$ (hence $\varphi(1/u) < +\infty$), and compute the difference quotient of ψ (cf. the proof of Theorem I.1.1.6)

$$\frac{u'\varphi(1/u') - u\varphi(1/u)}{u' - u} = \varphi(t) - t\frac{\varphi(t') - \varphi(t)}{t' - t}.$$

Letting $u' \downarrow u$, i.e. $t' \uparrow t$, we obtain the right-derivative

$$D_+\psi(u) = \varphi(t) - tD_-\varphi(t);$$

the left-derivative is obtained likewise.

Thus, the subdifferential of ψ at u is the closed interval

$$\partial \psi(u) = \varphi(t) - t\partial\varphi(t).$$

Knowing that $r_\varepsilon(u') = \psi(u') - u'[f(x) - \varepsilon]$ for all $u' > 0$, we readily obtain

$$\partial r_\varepsilon(u) = \varphi(t) - t\partial\varphi(t) - f(x) + \varepsilon = r_\varepsilon(u)/u - t\partial\varphi(t).$$ □

The minimality condition $0 \in \partial r_\varepsilon(u)$ is therefore $r_\varepsilon(u) \in \partial\varphi(1/u)$. In t-language, we say: $t \in T_\varepsilon$ if and only if there is some $\alpha \in \partial\varphi(t)$ such that

$$f(x + td) - t\alpha = f(x) - \varepsilon. \qquad (2.2.7)$$

Remark 2.2.4 Using the calculus rule from Theorem 3.2.1 below, we will see that $\partial\varphi$ can be expressed with the help of ∂f. More precisely, if $x + \mathbb{R}^+d$ meets $\mathrm{ri}\, \mathrm{dom}\, f$, we have

$$\partial r_\varepsilon(u) = \frac{r_\varepsilon(u) - \langle \partial f(x + d/u), d\rangle}{u}.$$

Then Fig. 2.1.2 gives a nice interpretation of (2.2.7): associated with $t \geqslant 0$ and $s \in \partial f(x + td)$, consider the affine function

$$0 \leqslant \tau \mapsto \bar{f}(\tau) := f(x + td) + \langle s, d\rangle(\tau - t),$$

whose graph supports $\mathrm{epi}\, \varphi$ at $\tau = t$. Its value $\bar{f}(0)$ at the vertical axis $\tau = 0$ is a subderivative of $u \mapsto u\varphi(1/u)$ at $u = 1/t$, and t is optimal in (2.1.1) when $\bar{f}(0)$ reaches the given $f(x) - \varepsilon$. In this case, T_ε is the contact-set between $\mathrm{gr}\, \bar{f}$ and $\mathrm{gr}\, \varphi$. Note: convexity and Proposition 2.2.2(iii) tell us that $f(x) - \varepsilon^\infty \leqslant \bar{f}(0) \leqslant f(x)$. □

Let us summarize the results of this section concerning the optimal set T_ε, or U_ε of (2.2.4).

– First, we have the somewhat degenerate case $t^\ell = +\infty$, meaning that f is affine on the half-line $x + \mathbb{R}^+d$. This can be described by one of the following equivalent statements:

$$f'(x, d) = f'_\infty(d);$$
$$f'_\varepsilon(x, d) = f'(x, d) \quad \text{for all } \varepsilon > 0;$$
$$\forall t > 0,\ q_\varepsilon(t) > f'_\infty(d) \quad \text{for all } \varepsilon > 0;$$
$$T_\varepsilon = \emptyset \quad \text{for all } \varepsilon > 0;$$
$$U_\varepsilon = \{0\} \quad \text{for all } \varepsilon > 0.$$

- The second situation, more interesting, is when $t^\ell < +\infty$; then three essentially different cases may occur, according to the value of ε.
  - When $0 < \varepsilon < \varepsilon^\infty$, one has equivalently
  $$f'(x,d) < f'_\varepsilon(x,d) < f'_\infty(d);$$
  $$\exists t > 0 \quad \text{such that} \quad q_\varepsilon(t) < f'_\infty(d);$$
  $$T_\varepsilon \text{ is a nonempty compact interval};$$
  $$0 \notin U_\varepsilon.$$
  - When $\varepsilon^\infty < \varepsilon$:
  $$f'_\varepsilon(x,d) = f'_\infty(d);$$
  $$\forall t > 0, q_\varepsilon(t) > f'_\infty(d);$$
  $$T_\varepsilon = \emptyset;$$
  $$U_\varepsilon = \{0\}.$$
  - When $\varepsilon = \varepsilon^\infty$:
  $$f'_\varepsilon(x,d) = f'_\infty(d);$$
  $$T_\varepsilon \text{ is empty or unbounded};$$
  $$0 \in U_\varepsilon.$$

Note: in the last case, T_ε nonempty but unbounded means that $f(x+td)$ touches its asymptote $f(x) - \varepsilon + t f'_\infty(d)$ for t large enough.

2.3 Behaviour of f'_ε and T_ε as Functions of ε

In this section, we assume again $[x, x+td] \subset \text{dom } f$ for some $t > 0$. From the fundamental relation (2.1.1), the function $\varepsilon \mapsto -f'_\varepsilon(x,d)$ appears as a supremum of affine functions, and is therefore closed and convex: it is the conjugate of some other function. Following our general policy, we extend it to the whole real line, setting

$$v(\varepsilon) := \begin{cases} -f'_\varepsilon(x,d) & \text{if } \varepsilon \geq 0, \\ +\infty & \text{otherwise}. \end{cases}$$

Then $v \in \overline{\text{Conv}}\, \mathbb{R}$: in fact, dom v is either $[0, +\infty[$ or $]0, +\infty[$. When $\varepsilon \downarrow 0$, $v(\varepsilon)$ tends to $-f'(x,d) \in \mathbb{R} \cup \{+\infty\}$.

Lemma 2.3.1 *With $r := r_0$ of (2.1.3) and for all real numbers ε and u:*

$$v(\varepsilon) = r^*(-\varepsilon), \quad \text{i.e.} \quad r(u) = v^*(-u).$$

PROOF. For $\varepsilon \geq 0$, just apply the definitions:

$$-v(\varepsilon) = f'_\varepsilon(x,d) = \inf_{u \in \text{dom } r}[r(u) + \varepsilon u] = -r^*(-\varepsilon).$$

For $-\varepsilon > 0$, we have trivially (remember that $f'(x,d) < +\infty$)

$$r^*(-\varepsilon) \geq \lim_{u \to +\infty}[-\varepsilon u - r(u)] = +\infty - f'(x,d) = +\infty = v(\varepsilon). \quad \square$$

Observe that the case $u = 1$ is just Theorem 1.3.6; $u = 0$ gives

$$[f'_\infty(d) =] \; r(0) = \sup_{\varepsilon > 0} f'_\varepsilon(x, d) \; [= \lim_{\varepsilon \to +\infty} f'_\varepsilon(x, d)],$$

and this confirms the relevance of the notation f'_∞ for an asymptotic function.

Theorem 2.3.2 *With the notation of Proposition 2.3.1 and (2.2.4),*

$$\partial v(\varepsilon) = -U_\varepsilon \quad \text{for all } \varepsilon \geq 0. \tag{2.3.1}$$

Furthermore, $\varepsilon \mapsto f'_\varepsilon(x, d)$ is strictly increasing on $[0, \varepsilon^\infty[$ and is constantly equal to $f'_\infty(d)$ for $\varepsilon \geq \varepsilon^\infty$.

PROOF. By definition and using Lemma 2.3.1, $-u \in \partial v(\varepsilon)$ if and only if

$$-u\varepsilon = v(\varepsilon) + v^*(-u) = -f'_\varepsilon(x, d) + r(u) \quad \Longleftrightarrow \quad f'_\varepsilon(x, d) = r_\varepsilon(u),$$

which exactly means that $u \in U_\varepsilon$. The rest follows from the conclusions of §2.2. □

Figure 2.3.1 illustrates the graph of $\varepsilon \mapsto f'_\varepsilon(x, d)$ in the example of Fig. 2.2.1. Since the derivative of $\varepsilon \mapsto q_\varepsilon(t)$ is $1/t$, a formal application of Theorem VI.4.4.2 would give directly (2.3.2); but the assumptions of that theorem are hardly satisfied – and certainly not for $\varepsilon \geq \varepsilon^\infty$.

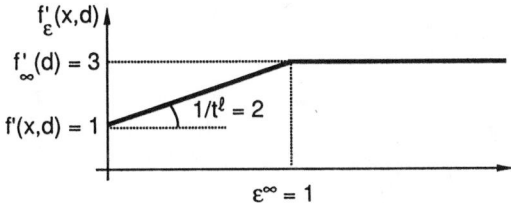

Fig. 2.3.1. The ε-directional derivative as a function of ε

Remark 2.3.3 Some useful formulae follow from (2.3.2): whenever $T_\varepsilon \neq \emptyset$, we have

$$f'_\eta(x, d) \leq f'_\varepsilon(x, d) + \frac{\eta - \varepsilon}{t} \quad \text{for all } t \in T_\varepsilon,$$

$$f'_\eta(x, d) = f'_\varepsilon(x, d) + \frac{\eta - \varepsilon}{\bar{t}_\varepsilon} - o(\eta - \varepsilon) \quad \text{if } \eta \geq \varepsilon,$$

$$f'_\eta(x, d) = f'_\varepsilon(x, d) + \frac{\eta - \varepsilon}{\underline{t}_\varepsilon} - o(\eta - \varepsilon) \quad \text{if } \eta \leq \varepsilon,$$

where $\underline{t}_\varepsilon$ and $\bar{t}_\varepsilon$ are respectively the smallest and largest elements of T_ε, and the remainder terms $o(\cdot)$ are nonnegative (of course, $\underline{t}_\varepsilon = \bar{t}_\varepsilon$ except possibly for countably many values of ε). We also have the integral representation (I.4.2.6)

$$f'_\varepsilon(x, d) = f'(x, d) + \int_0^\varepsilon U_\alpha d\alpha \quad \text{for all } 0 \leq \varepsilon < \varepsilon^\infty. \qquad \square$$

A natural question is now: what happens when $\varepsilon \downarrow 0$? We already know that $f'_\varepsilon(x, d) \to f'(x, d)$ but at what speed? A qualitative answer is as follows (see also Example 2.1.3).

Proposition 2.3.4 *Assuming* $-\infty < f'(x, d)$, *there holds*

$$\lim_{\varepsilon \downarrow 0} \frac{f'_\varepsilon(x, d) - f'(x, d)}{\varepsilon} = \frac{1}{t^\ell} \in [0, +\infty],$$

$$\lim_{\varepsilon \downarrow 0} T_\varepsilon = \{t^\ell\} \quad \text{when} \quad t^\ell < +\infty.$$

PROOF. We know from (2.2.5) and (2.3.2) that $\partial v(0) = -U_0 =]-\infty, -1/t^\ell]$, so $D_+ v(0) = -1/t^\ell$ and everything comes from the elementary results of §I.4.2. □

For fixed x and d, use the notation of Remark 2.2.4 and look again at Fig. 2.1.2, with the results of the present section in mind: it is important to meditate on the correspondence between the horizontal set of stepsizes and the vertical set of f-decrements.

To any stepsize $t \geq 0$ and slope $\langle s, d \rangle$ of a line supporting epi φ at $(t, \varphi(t))$, is associated a value

$$\varepsilon_{t,s} := f(x) - \bar{f}(0) = f(x) - f(x + td) + t\langle s, d \rangle \geq 0. \tag{2.3.2}$$

Likewise, to any decrement $\varepsilon \geq 0$ from $f(x)$ is associated a stepsize $t_\varepsilon \geq 0$ via the slope $f'_\varepsilon(x, d)$. This defines a pair of multifunctions $t \mapsto -\partial r(1/t)$ and $\varepsilon \mapsto T_\varepsilon$, inverse to each other, and monotone in the sense that

$$\text{for } t \in T_\varepsilon \text{ and } t' \in T_{\varepsilon'}, \quad \varepsilon > \varepsilon' \iff t > t'.$$

To go analytically from t to ε, one goes first to the somewhat abstract set of inverse stepsizes u, from which ε is obtained by the duality correspondence of Lemma 2.3.1. See also the lower part of Fig. 2.2.1, for an instance of a mapping T (or rather its inverse).

To finish this section, we give an additional first-order development: there always holds the estimate

$$f(x + td) = f(x) + tf'(x, d) + o(t),$$

and various ways exist to eliminate the remainder term $o(t)$: one is (1.3.7); the mean-value Theorem VI.2.3.3 is more classical; here is a third one.

Proposition 2.3.5 *With the hypotheses and notations given so far, assume* $t^\ell < +\infty$ *(i.e. f is not affine on the whole half-line $x + \mathbb{R}^+ d$). For $t > 0$, there is a unique $\varepsilon(t) \in [0, \varepsilon^\infty[$ such that*

$$f(x + td) = f(x) + tf'_{\varepsilon(t)}(x, d). \tag{2.3.3}$$

PROOF. We have to solve the equation in ε: $v(\varepsilon) = -q(t)$. In view of the second part of Theorem 2.3.2, this equation has a unique solution provided that the right-hand side $-q(t)$ is in the interval

$$\{v(\alpha) : 0 \leqslant \alpha < \varepsilon^\infty\} = -[f'(x,d), f'_\infty(d)[\,.$$

As a result, the unique $\varepsilon(t)$ of (2.3.3) is

$$\varepsilon(t) = \begin{cases} 0 & \text{for } t \in \,]0, t^\ell], \\ (-v)^{-1}(q(t)) & \text{for } t > t^\ell\,. \end{cases}$$

□

Figure 2.3.2 emphasizes the difference between this $\varepsilon(t)$ and the $\varepsilon_{t,s}$ defined in (2.3.2).

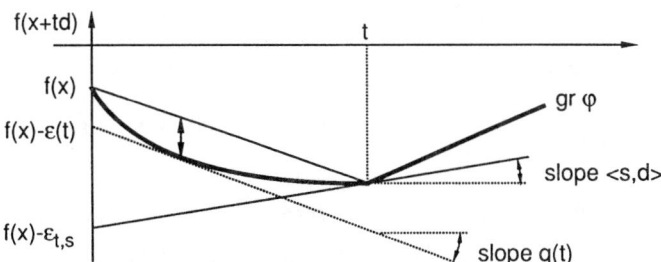

Fig. 2.3.2. The numbers $\varepsilon(t)$ and $\varepsilon_{t,s}$

3 Calculus Rules on the Approximate Subdifferential

In §VI.4, we developed a calculus to compute the subdifferential of a finite-valued convex function f constructed from other functions f_i. Here, we extend the results of §VI.4 to the case of $\partial_\varepsilon f$, with $\varepsilon \geqslant 0$ and $f \in \overline{\text{Conv}}\,\mathbb{R}^n$. Some rudimentary such calculus has already been given in Proposition 1.3.1.

The reader may have observed that the approximate subdifferential is a *global concept* (in contrast to the exact subdifferential, which is purely local): for ε arbitrarily large, $\partial_\varepsilon f(x)$ depends on the behaviour of f arbitrarily far from x. It can therefore be guessed that $\partial_\varepsilon f$ has to depend on the global behaviour of the f_i's, while $\partial f(x)$ depends exclusively on the $\partial f_i(x)$'s. It turns out that the conjugacy operation gives a convenient tool for taking this global behaviour into account. Indeed, knowing that f results from some operation on the f_i's, the characterization (1.2.1) in terms of f^* shows that the whole issue is to determine the effect of the conjugacy on this operation: it is the set of calculus rules of §X.2 that is at stake.

Recall that all the functions involved are in $\overline{\text{Conv}}\,\mathbb{R}^n$.

3.1 Sum of Functions

From Theorem X.2.3.1, the conjugate of a sum of two functions $f_1 + f_2$ is the *closure* of the infimal convolution $f_1^* \mathbin{\dot{\triangledown}} f_2^*$. Expressing $\partial_\varepsilon (f_1 + f_2)(s)$ will require an expression for this infimal convolution, which in turn requires the following basic assumption:

When $s \in \operatorname{dom}(f_1 + f_2)^*$,
$$\begin{array}{l} (f_1 + f_2)^*(s) = f_1^*(\bar{p}_1) + f_2^*(\bar{p}_2) \\ \text{for some } \bar{p}_i \text{ satisfying } \bar{p}_1 + \bar{p}_2 = s. \end{array} \quad (3.1.1)$$

This just expresses that the inf-convolution of f_1^* and f_2^* is *exact* at $s = \bar{p}_1 + \bar{p}_2$: the couple $(\bar{p}_1, \bar{p}_2)$ actually minimizes the function $(p_1, p_2) \mapsto f_1^*(p_1) + f_2^*(p_2)$. Furthermore, we know (Theorem X.2.3.2) that this property holds under various conditions on f_1 and f_2; one of them is

$$\operatorname{ri} \operatorname{dom} f_1 \cap \operatorname{ri} \operatorname{dom} f_2 \neq \emptyset, \quad (3.1.2)$$

which is slightly more stringent than the minimal assumption

$$\operatorname{dom} f_1 \cap \operatorname{dom} f_2 \neq \emptyset \quad [\iff f_1 + f_2 \not\equiv +\infty].$$

Theorem 3.1.1 *For any $\varepsilon \geq 0$ and $x \in \operatorname{dom}(f_1 + f_2) = \operatorname{dom} f_1 \cap \operatorname{dom} f_2$, there holds*

$$\partial_\varepsilon (f_1 + f_2)(x) \supset \cup \left\{ \partial_{\varepsilon_1} f_1(x) + \partial_{\varepsilon_2} f_2(x) : \varepsilon_i \geq 0, \varepsilon_1 + \varepsilon_2 \leq \varepsilon \right\} \quad (3.1.3)$$

with equality under assumption (3.1.1), for example if (3.1.2) holds.

PROOF. Let $\varepsilon \geq 0$ be given. For arbitrary nonnegative $\varepsilon_1, \varepsilon_2$ with $\varepsilon_1 + \varepsilon_2 \leq \varepsilon$, Definition 1.1.1 clearly implies

$$\partial_{\varepsilon_1} f_1(x) + \partial_{\varepsilon_2} f_2(x) \subset \partial_\varepsilon (f_1 + f_2)(x).$$

Conversely, take $s \in \partial_\varepsilon (f_1 + f_2)(x)$, i.e.

$$(f_1 + f_2)^*(s) + (f_1 + f_2)(x) - \langle s, x \rangle \leq \varepsilon. \quad (3.1.4)$$

This s is therefore in $\operatorname{dom}(f_1 + f_2)^*$ and we can apply (3.1.1): with the help of some $\bar{p}_1$ and $\bar{p}_2$, we write (3.1.4) as $\varepsilon_1 + \varepsilon_2 \leq \varepsilon$, where we have set

$$\varepsilon_i := f_i^*(\bar{p}_i) + f_i(x) - \langle \bar{p}_i, x \rangle \quad \text{for } i = 1, 2.$$

We see that $\bar{p}_i \in \partial_{\varepsilon_i} f_i(x)$ for $i = 1, 2$, and the required converse inclusion is proved. □

Naturally, if $f = \sum_{i=1}^m f_i$, the right-hand side in (3.1.3) becomes

$$\cup \left\{ \sum_{i=1}^m \partial_{\varepsilon_i} f_i(x) : \varepsilon_i \geq 0, \sum_{i=1}^m \varepsilon_i \leq \varepsilon \right\}. \quad (3.1.5)$$

Since $\partial_\varepsilon f(x)$ increases with ε, the constraints $\sum_i \varepsilon_i \leq \varepsilon$ *can be replaced* by $\sum_i \varepsilon_i = \varepsilon$ in (3.1.3) and (3.1.5). Also, when $\varepsilon = 0$, there is only one possibility for ε_i in (3.1.3):

Corollary 3.1.2 *For $x \in \operatorname{dom} f_1 \cap \operatorname{dom} f_2$, there holds*

$$\partial (f_1 + f_2)(x) \supset \partial f_1(x) + \partial f_2(x)$$

with equality under, for example, assumption (3.1.2). □

To emphasize the need of an assumption such as (3.1.2), consider in $\mathbb{R}^2$ the Euclidean balls C_1 and C_2, of radius 1 and centered respectively at $(-1, 0)$ and $(1, 0)$: they meet at the unique point $x = (0, 0)$. Then take for f_i the indicator function of C_i, so f is the indicator function $I_{\{0\}}$. We have $\partial f(0) = \mathbb{R}^2$, while $\partial f_1(0) + \partial f_2(0) = \mathbb{R} \times \{0\}$.

Example 3.1.3 (ε-Normal Sets to Closed Convex Polyhedra) Let C be a closed convex polyhedron described by its supporting hyperplanes:

$$H_j^- := \{x \in \mathbb{R}^n : \langle s_j, x \rangle \leqslant r_j\} \quad \text{for } j = 1, \ldots, m,$$
$$C := \cap H_j^- = \{x \in \mathbb{R}^n : \langle s_j, x \rangle \leqslant r_j \text{ for } j = 1, \ldots, m\}. \tag{3.1.6}$$

Its indicator function is clearly

$$I_C = \sum_{j=1}^{m} I_{H_j^-}.$$

Let us compute the ε-subdifferential of this function, i.e. the ε-normal set of Definition 1.1.3. The approximate normal set to H_j^- has been given in Example 1.2.5, we obtain for $x \in C$

$$N_{C,\varepsilon}(x) = \left\{ \sum_{j=1}^{m} \lambda_j s_j : \lambda_j \geqslant 0, \ \sum_{j=1}^{m} c_j(x)\lambda_j \leqslant \varepsilon \right\}, \tag{3.1.7}$$

where we have set $c_j(x) := r_j - \langle s_j, x \rangle \geqslant 0$. In fact, I_C is a sum of polyhedral functions, and Theorem X.2.3.2 – with its qualification assumption (X.2.3.Qj) – does imply (3.1.1) in this case.

Still for $x \in C$, set $K(x) := \max_j c_j(x)$. Clearly enough

$$\tfrac{\varepsilon}{K(x)} \operatorname{co}\{s_1, \ldots, s_m\} \subset N_{C,\varepsilon}(x)$$

(and the normal cone $N_C(x)$ can even be added to the left-hand set). Likewise, set $k(x) := \min_j c_j(x)$. If $k(x) > 0$,

$$N_{C,\varepsilon}(x) \subset \tfrac{\varepsilon}{k(x)} \operatorname{co}\{s_1, \ldots, s_m\}. \qquad \square$$

Example 3.1.4 (Approximate Minimality Conditions) Let be given a convex function $f : \mathbb{R}^n \to \mathbb{R}$ and a nonempty closed convex set C; assume that

$$\bar{f}_C := \inf\{f(x) : x \in C\} \tag{3.1.8}$$

is not $-\infty$. The ε-minimizers of f on C are those $x \in C$ such that $f(x) \leqslant \bar{f}_C + \varepsilon$; clearly enough, an ε-minimizer is an x such that (remember Theorem 1.1.5)

$$(f + I_C)(x) \leqslant \inf_{\mathbb{R}^n}(f + I_C) + \varepsilon \quad \text{or equivalently} \quad 0 \in \partial_\varepsilon(f + I_C)(x).$$

Here $\operatorname{dom} f = \mathbb{R}^n$: we conclude from Theorem 3.1.1 that an ε-minimizer is an x such that

$$0 \in \partial_\alpha f(x) + N_{C,\varepsilon-\alpha}(x) \quad \text{for some } \alpha \in [0, \varepsilon],$$

i.e. f has at x an α-subgradient whose opposite lies in the $(\varepsilon - \alpha)$-normal set to C. The situation simplifies in some cases.

– Set $\varepsilon = 0$ to obtain the standard minimality condition of (VII.1.1.3):

$$x \text{ solves } (3.1.8) \iff -\partial f(x) \cap N_C(x) \neq \emptyset.$$

– Another case is when $C = \{x_0\} + H$ is an affine manifold. Then $N_{C,\beta}(x) = H^\perp$ for all $\beta \geq 0$ and $\partial_\alpha f(x)$ increases with α: our ε-minimality condition becomes $-\partial_\varepsilon f(x) \cap H^\perp \neq \emptyset$.

– Also, when $f = \langle s_0, \cdot \rangle$ is linear, $\partial_\alpha f(x) = \partial f(x) = \{s_0\}$ for all $\alpha \geq 0$, while $N_{C,\beta}(x)$ increases with β; so our ε-minimality condition is: $-s_0 \in N_{C,\varepsilon}(x)$, a triviality if we replace $N_{C,\varepsilon}(x)$ by its definition (1.1.6).

– If C is a closed convex polyhedron as in (3.1.6), its ε-normal set can be specified as in (3.1.7). Take for (3.1.8) a linear programming problem

$$\min\{\langle s_0, x \rangle : \langle s_j, x \rangle \leq r_j \text{ for } j = 1, \ldots, m\}.$$

An ε-minimizer is then an $x \in C$ for which there exists $\mu = (\mu_1, \ldots, \mu_m) \in \mathbb{R}^m$ such that

$$\left.\begin{array}{l} \sum_{j=1}^m \mu_j s_j + s_0 = 0, \\ \sum_{j=1}^m \mu_j r_j + \langle s_0, x \rangle \leq \varepsilon, \\ \mu_j \geq 0 \quad \text{for } j = 1, \ldots, m. \end{array}\right\} \quad \Box$$

We leave it as an exercise to redo the above examples when

$$C := \{x \in K : \langle s_j, x \rangle = r_j \text{ for } j = 1, \ldots, m\}$$

is described in standard form (K being a closed convex polyhedral cone, say the nonnegative orthant).

Remark 3.1.5 In the space $\mathbb{R}^{n_1} \times \mathbb{R}^{n_2}$ equipped with the scalar product of a product-space, take a decomposable function:

$$\mathbb{R}^{n_1} \times \mathbb{R}^{n_2} \ni (x_1, x_2) = x \mapsto f(x) = f_1(x_1) + f_2(x_2).$$

Because of the calculus rule X.1.3.1(ix), the basic assumption (3.1.1) holds automatically, so we always have

$$\partial_\varepsilon f(x) = \cup \{\partial_{\varepsilon_1} f_1(x_1) \times \partial_{\varepsilon_2} f_2(x_2) : \varepsilon_i \geq 0, \varepsilon_1 + \varepsilon_2 \leq \varepsilon\} ;$$

but beware that this set is not a product-set in $\mathbb{R}^{n_1} \times \mathbb{R}^{n_2}$, except for $\varepsilon = 0$. $\quad\Box$

3.2 Pre-Composition with an Affine Mapping

Given $g \in \overline{\text{Conv}}\, \mathbb{R}^m$ and an affine mapping $A : \mathbb{R}^n \to \mathbb{R}^m$ ($Ax = A_0 x + y_0 \in \mathbb{R}^m$ with A_0 linear), take $f := g \circ A \in \overline{\text{Conv}}\, \mathbb{R}^n$. As in §3.1, we need an assumption, which is in this context:

$$\text{When } s \in \text{dom}(g \circ A)^*, \left|\begin{array}{l} (g \circ A)^*(s) = g^*(\bar{p}) - \langle \bar{p}, y_0 \rangle \\ \text{for some } \bar{p} \text{ such that } A_0^* \bar{p} = s \end{array}\right. \quad (3.2.1)$$

($\langle \cdot, \cdot \rangle$ will denote indifferently the scalar product in $\mathbb{R}^n$ or $\mathbb{R}^m$). As was the case with (3.1.1), Theorem X.2.2.1 tells us that the above $\bar{p}$ actually minimizes the function

3 Calculus Rules on the Approximate Subdifferential   117

$p \mapsto g^*(p) - \langle p, y_0 \rangle$ on the affine manifold of equation $A_0^* p = s$. Furthermore we know (Theorem X.2.2.3) that (3.2.1) holds under various conditions on f and A_0; one of them is

$$A(\mathbb{R}^n) \cap \mathrm{ri}\, \mathrm{dom}\, g \neq \emptyset. \tag{3.2.2}$$

Once again, note that removing the words "relative interior" from (3.2.2) amounts to assuming $g \circ A \not\equiv +\infty$.

Theorem 3.2.1 *Let g and A be defined as above. For all $\varepsilon \geqslant 0$ and x such that $Ax \in \mathrm{dom}\, g$, there holds*

$$\partial_\varepsilon (g \circ A)(x) \supset A_0^* \partial_\varepsilon g(Ax), \tag{3.2.3}$$

with equality under assumption (3.2.1), for example if (3.2.2) holds.

PROOF. Fix x such that $Ax \in \mathrm{dom}\, g$ and let $p \in \partial_\varepsilon g(Ax) \subset \mathbb{R}^m$:

$$g(z) \geqslant g(Ax) + \langle p, z - Ax \rangle - \varepsilon \quad \text{for all } z \in \mathbb{R}^m.$$

Taking in particular $z = Ay$ with y describing $\mathbb{R}^n$:

$$g(Ay) \geqslant g(Ax) + \langle A_0^* p, y - x \rangle - \varepsilon \quad \text{for all } y \in \mathbb{R}^n,$$

where we have used the property $A(y - x) = A_0(y - x)$. Thus we have proved that $A_0^* p \in \partial_\varepsilon (g \circ A)(x)$.

Conversely, let $s \in \partial_\varepsilon (g \circ A)(x)$, i.e.

$$(g \circ A)^*(s) + (g \circ A)(x) - \langle s, x \rangle \leqslant \varepsilon. \tag{3.2.4}$$

Apply (3.2.1): with the help of some $\bar{p}$ such that $A_0^* \bar{p} = s$, (3.2.4) can be written

$$\varepsilon \geqslant g^*(\bar{p}) - \langle \bar{p}, y_0 \rangle + g(Ax) - \langle \bar{p}, A_0 x \rangle = g^*(\bar{p}) + g(Ax) - \langle \bar{p}, Ax \rangle.$$

This shows that $\bar{p} \in \partial_\varepsilon g(Ax)$. Altogether, we have proved that our s is in $A_0^* \partial_\varepsilon g(Ax)$. □

Naturally, only the linear part of A counts in the right-hand side of (3.2.3): the translation is taken care of by Proposition X.1.3.1(v).

As an illustration of this calculus rule, take $x_0 \in \mathrm{dom}\, g$, a direction $d \neq 0$, and compute the approximate subdifferential of the function

$$\mathbb{R} \ni t \mapsto \varphi(t) := g(x_0 + td).$$

If $x_0 + \mathbb{R}d$ meets $\mathrm{ri}\, \mathrm{dom}\, g$, we can write

$$\partial_\varepsilon \varphi(t) = \langle \partial_\varepsilon g(x_0 + td), d \rangle \quad \text{for all } t \in \mathrm{dom}\, \varphi.$$

3.3 Image and Marginal Functions

We recall that, for $g \in \overline{\text{Conv}}\,\mathbb{R}^m$ and A linear from $\mathbb{R}^m$ to $\mathbb{R}^n$, the image of g under A is the function Ag defined by

$$\mathbb{R}^n \ni x \mapsto (Ag)(x) := \inf\{g(y) : Ay = x\}. \tag{3.3.1}$$

Once again, we need an assumption for characterizing the ε-subdifferential of Ag, namely that the infimum in (3.3.1) is attained "at finite distance". A sufficient assumption for this is

$$\text{Im}\, A^* \cap \text{ri}\,\text{dom}\, g^* \neq \emptyset, \tag{3.3.2}$$

which implies at the same time that $Ag \in \overline{\text{Conv}}\,\mathbb{R}^n$ (see Theorem X.2.2.3). As already seen for condition (X.2.2.Q.iii), this assumption is implied by

$$g'_\infty(d) > 0 \quad \text{for all nonzero } d \in \text{Ker}\, A.$$

Theorem 3.3.1 *Let $\varepsilon \geqslant 0$ and $x \in \text{dom}\, Ag = A(\text{dom}\, g)$. Suppose that there is some $\bar{y} \in \mathbb{R}^m$ with $A\bar{y} = x$ and $g(\bar{y}) = Ag(x)$; for example assume (3.3.2). Then*

$$\partial_\varepsilon(Ag)(x) = \{s \in \mathbb{R}^n : A^*s \in \partial_\varepsilon g(\bar{y})\}. \tag{3.3.3}$$

PROOF. To say that $s \in \partial_\varepsilon(Ag)(x)$ is to say that

$$(Ag)^*(s) + g(\bar{y}) - \langle s, A\bar{y}\rangle \leqslant \varepsilon,$$

where we have made use of the existence and properties of $\bar{y}$. Then apply Theorem X.2.1.1: $(Ag)^* = g^* \circ A^*$, so

$$s \in \partial_\varepsilon(Ag)(x) \iff g^*(A^*s) + g(\bar{y}) - \langle A^*s, \bar{y}\rangle \leqslant \varepsilon. \qquad \square$$

This result can of course be compared to Theorem VI.4.5.1: the hypotheses are just the same – except for the extended-valuedness possibility. Thus, we see that the inverse image under A^* of $\partial_\varepsilon g(y_x)$ does not depend on the particular y_x optimal in (3.3.1).

We know that a particular case is the marginal function:

$$\mathbb{R}^n \ni x \mapsto f(x) := \inf\{g(x, z) : z \in \mathbb{R}^p\}, \tag{3.3.4}$$

where $g \in \overline{\text{Conv}}(\mathbb{R}^n \times \mathbb{R}^p)$. Indeed, f is the image of g under the projection mapping from $\mathbb{R}^{n+p}$ to $\mathbb{R}^n$ defined by $A(x, z) = x$. The above result can be particularized to this case:

Corollary 3.3.2 *With $g \in \overline{\text{Conv}}(\mathbb{R}^n \times \mathbb{R}^p)$, let g^* be associated with a scalar product preserving the structure of $\mathbb{R}^m = \mathbb{R}^n \times \mathbb{R}^p$ as a product space, namely:*

$$\langle \cdot, \cdot \rangle_m = \langle \cdot, \cdot \rangle_n + \langle \cdot, \cdot \rangle_p \tag{3.3.5}$$

and consider the marginal function f of (3.3.4). Let $\varepsilon \geqslant 0$, $x \in \text{dom}\, f$; suppose that there is some $\bar{z} \in \mathbb{R}^p$ such that $g(x, \bar{z}) = f(x)$; $\bar{z}$ exists for example when

$$\exists s_0 \in \mathbb{R}^n \quad \text{such that} \quad (s_0, 0) \in \text{ri}\,\text{dom}\, g^*. \tag{3.3.6}$$

Then

$$\partial_\varepsilon f(x) = \{s \in \mathbb{R}^n : (s, 0) \in \partial_\varepsilon g(x, \bar{z})\}. \tag{3.3.7}$$

3 Calculus Rules on the Approximate Subdifferential

PROOF. Set $A : (x, z) \mapsto x$. With the scalar product (3.3.5), $A^* : \mathbb{R}^n \to \mathbb{R}^n \times \mathbb{R}^p$ is defined by $A^*s = (s, 0)$, $\operatorname{Im} A^* = \mathbb{R}^n \times \{0\}$, so (3.3.6) and (3.3.7) are just (3.3.2) and (3.3.3) respectively. □

3.4 A Study of the Infimal Convolution

As another example of image functions, consider the infimal convolution:

$$\mathbb{R}^n \ni x \mapsto (f_1 \,\square\, f_2)(x) := \inf\{f_1(y_1) + f_2(y_2) \,:\, y_1 + y_2 = x\},$$

where f_1 and f_2 are both in $\overline{\operatorname{Conv}}\, \mathbb{R}^n$. With $m = 2n$, and $\mathbb{R}^{2n}$ being equipped with the Euclidean structure of a product-space, this is indeed an image function:

$$\left.\begin{aligned}
\langle(s_1, s_2), (y_1, y_2)\rangle_{2n} &:= \langle s_1, y_1\rangle + \langle s_2, y_2\rangle, \\
g(y_1, y_2) &:= f_1(y_1) + f_2(y_2); \quad A(y_1, y_2) := y_1 + y_2, \\
g^*(s_1, s_2) &= f_1^*(s_1) + f_2^*(s_2); \quad A^*s = (s, s).
\end{aligned}\right\} \quad (3.4.1)$$

Theorem 3.4.1 *Let $\varepsilon \geq 0$ and $x \in \operatorname{dom}(f_1 \,\square\, f_2) = \operatorname{dom} f_1 + \operatorname{dom} f_2$. Suppose that there are y_1 and y_2 such that the inf-convolution is exact at $x = y_1 + y_2$; this is the case for example when*

$$\operatorname{ri} \operatorname{dom} f_1^* \cap \operatorname{ri} \operatorname{dom} f_2^* \neq \emptyset. \quad (3.4.2)$$

Then

$$\partial_\varepsilon (f_1 \,\square\, f_2)(x) = \cup \left\{ \partial_{\varepsilon_1} f_1(y_1) \cap \partial_{\varepsilon_2} f_2(y_2) \,:\, \varepsilon_i \geq 0, \varepsilon_1 + \varepsilon_2 \leq \varepsilon \right\}. \quad (3.4.3)$$

PROOF. We apply Theorem 3.3.1 to g and A of (3.4.1). First of all,

$$\operatorname{dom} g^* = \operatorname{dom} f_1^* \times \operatorname{dom} f_2^*$$

so (Proposition III.2.1.11)

$$\operatorname{ri} \operatorname{dom} g^* = \operatorname{ri} \operatorname{dom} f_1^* \times \operatorname{ri} \operatorname{dom} f_2^*$$

and (3.3.2) means:

$$\exists s \in \mathbb{R}^n \quad \text{such that} \quad (s, s) \in \operatorname{ri} \operatorname{dom} f_1^* \times \operatorname{ri} \operatorname{dom} f_2^*$$

which is nothing other than (3.4.2). Now, with (y_1, y_2) as stated and $s \in \mathbb{R}^n$, set

$$\varepsilon_i := f_i^*(s) + f_i(y_i) - \langle s, y_i\rangle \geq 0 \quad \text{for } i = 1, 2, \quad (3.4.4)$$

so that $s \in \partial_{\varepsilon_i} f_i(y_i)$ for $i = 1, 2$. Particularizing (3.3.3) to our present situation:

$$s \in \partial_\varepsilon (f_1 \,\square\, f_2)(x) \quad \Longleftrightarrow \quad (s, s) \in \partial_\varepsilon g(y_1, y_2)$$

which, in view of the definitions (3.4.1), means

$$f_1(y_1) + f_2(y_2) + f_1^*(s) + f_2^*(s) \leqslant \langle s, y_1 \rangle + \langle s, y_2 \rangle + \varepsilon .$$

With (3.4.4), this is just (3.4.3). □

As with Theorem 3.1.1, nothing is changed if we impose $\varepsilon_1 + \varepsilon_2 = \varepsilon$ in (3.4.3). The particular value $\varepsilon = 0$ in the above result brings

$$\partial (f_1 \dotplus f_2)(x) = \partial f_1(y_1) \cap \partial f_2(y_2) \quad (3.4.5)$$
when the inf-convolution is exact at $x = y_1 + y_2$.

Some more information, not directly contained in (3.4.3), can also be derived: a sort of converse result ensuring that the inf-convolution is exact.

Proposition 3.4.2 *With* $f := f_1 \dotplus f_2$, *consider* $y_i \in \operatorname{dom} f_i$ *for* $i = 1, 2$ *and* $x := y_1 + y_2 \ (\in \operatorname{dom} f)$. *Then*

$$\partial f_1(y_1) \cap \partial f_2(y_2) \subset \partial f(x) . \quad (3.4.6)$$

If $\partial f_1(y_1) \cap \partial f_2(y_2) \neq \emptyset$, *the inf-convolution is exact at* $x = y_1 + y_2$ *and equality holds in* (3.4.6).

PROOF. Let $s \in \partial f_1(y_1) \cap \partial f_2(y_2)$ (hence $s \in \operatorname{dom} f_1^* \cap \operatorname{dom} f_2^*$):

$$f_1^*(s) + f_1(y_1) + f_2^*(s) + f_2(y_2) = \langle s, y_1 + y_2 \rangle = \langle s, x \rangle . \quad (3.4.7)$$

Then, using $f_1^* + f_2^* = (f_1 \dotplus f_2)^*$ (Corollary X.2.1.3) and the definition of an inf-convolution,

$$(f_1 \dotplus f_2)^*(s) + (f_1 \dotplus f_2)(x) \leqslant \langle s, x \rangle ;$$

in view of the Fenchel inequality (X.1.1.3), this is actually an equality, i.e. $s \in \partial f(x)$. Now use this last equality as a value for $\langle s, x \rangle$ in (3.4.7) to obtain

$$f_1(y_1) + f_2(y_2) = (f_1 \dotplus f_2)(x) ,$$

i.e. the inf-convolution is exact at $x = y_1 + y_2$; equality in (3.4.6) follows from (3.4.5). □

In summary, fix $x \in \operatorname{dom}(f_1 \dotplus f_2)$ and denote by $H \subset \mathbb{R}^n \times \mathbb{R}^n$ the hyperplane of equation $y_1 + y_2 = x$. When (y_1, y_2) describes H, the set $D(y_1, y_2) := \partial f_1(y_1) \cap \partial f_2(y_2)$ assumes at most two values: the empty set and $\partial (f_1 \dotplus f_2)(x)$. Actually, there are two possibilities:

- either $D(y_1, y_2) = \emptyset$ for all $(y_1, y_2) \in H$;
- or $D(\bar{y}_1, \bar{y}_2) \neq \emptyset$ for some $(\bar{y}_1, \bar{y}_2) \in H$. This implies $\partial (f_1 \dotplus f_2)(x) \neq \emptyset$ and then, we have for all $(y_1, y_2) \in H$:

$$D(y_1, y_2) \neq \emptyset \iff D(y_1, y_2) = \partial (f_1 \dotplus f_2)(x) \iff$$
$$\iff \text{the inf-convolution is exact at } x = y_1 + y_2 .$$

Remark 3.4.3 Beware that D may be empty on the whole of H while $\partial(f_1 \triangledown f_2)(x) \neq \emptyset$. Take for example
$$f_1(y) = \exp y \quad \text{and} \quad f_2 = \exp(-y);$$
then $f_1 \triangledown f_2 \equiv 0$, hence $\partial(f_1 \triangledown f_2) \equiv \{0\}$. Yet $D(y_1, y_2) = \emptyset$ for all $(y_1, y_2) \in H$: the inf-convolution is nowhere exact.

Note also that (3.4.5) may express the equality between empty sets: take for example $f_1 = I_{[0,+\infty[}$ and
$$\mathbb{R} \ni y_2 \mapsto f_2(y_2) = \begin{cases} -\sqrt{y_2} & \text{if } y_2 \geq 0, \\ +\infty & \text{otherwise}. \end{cases}$$

It is easy to see that
$$(f_1 \triangledown f_2)(x) = \inf\{-\sqrt{y_2} : 0 \leq y_2 \leq x\} = -\sqrt{x} \quad \text{for all } x \geq 0.$$

This inf-convolution is exact in particular at $0 = 0 + 0$, yet $\partial(f_1 \triangledown f_2)(0) = \emptyset$. □

Example 3.4.4 (Moreau-Yosida Regularizations) For $c > 0$ and $f \in \overline{\text{Conv}}\,\mathbb{R}^n$, let $f_{(c)} := f \triangledown (1/2 c \|\cdot\|^2)$, i.e.
$$f_{(c)}(x) = \min\{f(y) + \tfrac{1}{2}c\|x - y\|^2 : y \in \mathbb{R}^n\};$$
this $f_{(c)}$ is called the Moreau-Yosida regularization of f. Denote by x_c the unique minimal y above, characterized by
$$0 \in \partial f(x_c) + c(x_c - x).$$

Using the approximate subdifferential of $1/2 \|\cdot\|^2$ (see Example 1.2.2 if necessary), Theorem 3.4.1 gives
$$\partial_\varepsilon f_{(c)}(x) = \bigcup_{0 \leq \alpha \leq \varepsilon} \left[\partial_{\varepsilon-\alpha} f(x_c) \cap B(c(x - x_c), \sqrt{2c\alpha})\right].$$

It is interesting to note that, when $\varepsilon = 0$, this formula reduces to
$$\nabla f_{(c)}(x) = c(x - x_c) \quad [\in \partial f(x_c)]. \tag{3.4.8}$$

Thus $f_{(c)}$ is a differentiable convex function. It can even be said that $\nabla f_{(c)}$ is Lipschitzian with constant c on $\mathbb{R}^n$. To see this, recall that the conjugate
$$f^*_{(c)} = f^* + \tfrac{1}{2c}\|\cdot\|^2$$
is strongly convex with modulus $1/c$; then apply Theorem X.4.2.1.

In the particular case where $c = 1$ and f is the indicator function of a closed convex set C, $f_{(c)} = 1/2\, d_C^2$ (the squared distance to C) and $x_c = p_C(x)$ (the projection of x onto C). Using the notation of (1.1.6):
$$\partial_\varepsilon(\tfrac{1}{2} d_C^2)(x) = \bigcup_{0 \leq \alpha \leq \varepsilon} \left[N_{C,\varepsilon-\alpha}(p_C(x)) \cap B(x - p_C(x), \sqrt{2\alpha})\right]. \quad □$$

Another regularization is $f_{[c]} := f \mathbin{\underset{\vee}{+}} c\|\cdot\|$, i.e.

$$f_{[c]}(x) := \inf \{f(y) + c\|x - y\| : y \in \mathbb{R}^n\}, \qquad (3.4.9)$$

which will be of importance in §4.1.

Proposition 3.4.5 *For $f \in \overline{\mathrm{Conv}}\,\mathbb{R}^n$, define $f_{[c]}$ as above; assume*

$$c > \underline{c} := \inf\{\|s\| : s \in \partial f(x),\ x \in \mathrm{dom}\,\partial f\}$$

and consider

$$C_c[f] := \{x \in \mathbb{R}^n : f_{[c]}(x) = f(x)\},$$

the coincidence set of f and $f_{[c]}$. Then

(i) *$f_{[c]}$ is convex, finite-valued, and Lipschitzian with Lipschitz constant c on $\mathbb{R}^n$;*
(ii) *the coincidence set is nonempty and characterized by*

$$C_c[f] = \{x \in \mathbb{R}^n : \partial f(x) \cap B(0,c) \neq \emptyset\};$$

(iii) *for all $x \in C_c[f]$, there holds*

$$\partial_\varepsilon f_{[c]}(x) = \partial_\varepsilon f(x) \cap B(0,c).$$

PROOF. [(i)] There exist by assumption x_0 and $s_0 \in \partial f(x_0)$ with $\|s_0\| \leqslant c$, hence

$$f \geqslant f(x_0) + \langle s_0, \cdot - x_0 \rangle \quad \text{and} \quad c\|\cdot\| \geqslant \langle s_0, \cdot \rangle;$$

f and $c\|\cdot\|$ are minorized by a common affine function: from Proposition IV.2.3.2, $f_{[c]} \in \mathrm{Conv}\,\mathbb{R}^n$. Furthermore, $f_{[c]}$ is finite everywhere by construction.

Now take x and x' in $\mathbb{R}^n$. For any $\eta > 0$, there is y_η such that

$$f(y_\eta) + c\|x - y_\eta\| \leqslant f_{[c]}(x) + \eta$$

and, by definition of $f_{[c]}$,

$$\begin{aligned} f_{[c]}(x') &\leqslant f(y_\eta) + c\|x' - y_\eta\| \leqslant f(y_\eta) + c\|x' - x\| + c\|x - y_\eta\| \\ &\leqslant f_{[c]}(x) + c\|x' - x\| + \eta. \end{aligned}$$

This finishes the proof of (i), since η was arbitrary.

[(ii) and (iii)] To say that $x \in C_c[f]$ is to say that the inf-convolution of f and $c\|\cdot\|$ is exact at $x = x + 0$. From Theorem 3.4.2, this means that $\partial f(x) \cap B(0,c)$ is nonempty: indeed

$$\partial(c\|\cdot\|)(0) = cB(0,1) = B(0,c) = \partial_\varepsilon(c\|\cdot\|)(0).$$

The last equality comes from the sublinearity of $c\|\cdot\|$ (cf. Example 1.2.5) and implies with the aid of Theorem 3.4.1:

$$\partial_\varepsilon f_{[c]}(x) = \cup\{\partial_\alpha f(x) \cap B(0,c) : 0 \leqslant \alpha \leqslant \varepsilon\} = \partial_\varepsilon f(x) \cap B(0,c). \qquad \Box$$

Just as in Example 3.4.4, we can particularize $f_{[c]}$ to the case where f is the indicator function of a closed convex set C: we get $f_{[c]} = cd_C$ and therefore

$$\partial_\varepsilon(cd_C)(x) = N_{C,\varepsilon}(x) \cap B(0,c) \quad \text{for all } x \in C.$$

Thus, when ε increases, the set K' in Fig. V.2.3.1 increases but stays confined within $B(0,c)$.

Remark 3.4.6 For $x \notin C_c[f]$, there is some $y_c \neq x$ yielding the infimum in (3.4.9). At such a y_c, the Euclidean norm is differentiable, so we obtain that $f_{[c]}$ is differentiable as well and
$$\nabla f_{[c]}(x) = c \frac{x - y_c}{\|x - y_c\|} \in \partial f(y_c).$$
Compare this with Example 3.4.4: here y_c need not be unique, but the *direction* $y_c - x$ depends only on x. □

3.5 Maximum of Functions

In view of Theorem X.2.4.4, computing the approximate subdifferential of a supremum involves the closed convex hull of an infimum. Here, we limit ourselves to the case of *finitely many* functions, say $f := \max_{j=1,\ldots,p} f_j$, a situation complicated enough. Thus, to compute $f^*(s)$, we have to solve

$$\left| \begin{array}{l} \displaystyle\inf_{\alpha_j, s_j} \sum_{j=1}^p \alpha_j f_j^*(s_j), \\ \displaystyle\sum_{j=1}^p \alpha_j s_j = s, \\ \alpha_j \geq 0, \ \displaystyle\sum_{j=1}^p \alpha_j = 1 \quad [\text{i.e. } \alpha \in \Delta_p]. \end{array} \right. \tag{3.5.1}$$

An important issue is whether this minimization problem has a solution, and whether the infimal value is a closed function of s. Here again, the situation is not simple when the functions are extended-valued.

Theorem 3.5.1 *Let $f_1, \ldots, f_p$ be a finite number of convex functions from $\mathbb{R}^n$ to $\mathbb{R}$ and let $f := \max_j f_j$; set $m := \min\{p, n+1\}$. Then, $s \in \partial_\varepsilon f(x)$ if and only if there exist m vectors $s_j \in \text{dom } f_j^*$, convex multipliers α_j, and nonnegative ε_j such that*

$$\left.\begin{array}{l} s_j \in \partial_{\varepsilon_j/\alpha_j} f_j(x) \quad \text{for all } j \text{ such that } \alpha_j > 0, \\ s = \sum_j \alpha_j s_j, \\ \sum_j [\varepsilon_j - \alpha_j f_j(x)] + f(x) \leq \varepsilon. \end{array}\right\} \tag{3.5.2}$$

PROOF. Theorem X.2.4.7 tells us that
$$f^*(s) = \sum_{j=1}^m \alpha_j f_j^*(s_j),$$
where $(\alpha_j, s_j) \in \mathbb{R}^+ \times \text{dom } f_j^*$ form a solution of (3.5.1) (after possible renumbering of the j's). By the characterization (1.2.1), the property $s \in \partial_\varepsilon f(x)$ is therefore equivalent to the existence of $(\alpha_j, s_j) \in \mathbb{R}^+ \times \text{dom } f_j^*$ – more precisely a solution of (3.5.1) – such that

$$\sum_{j=1}^{m} \alpha_j f_j^*(s_j) + f(x) \leqslant \varepsilon + \sum_{j=1}^{m} \alpha_j \langle s_j, x \rangle,$$

which we write

$$\sum_{j=1}^{m} \alpha_j [f_j^*(s_j) + f_j(x) - \langle s_j, x \rangle] \leqslant \varepsilon + \sum_{j=1}^{m} \alpha_j f_j(x) - f(x). \tag{3.5.3}$$

Thus, if $s \in \partial_\varepsilon f(x)$, i.e. if (3.5.3) holds, we can set

$$\varepsilon_j := \alpha_j [f_j^*(s_j) + f_j(x) - \langle s_j, x \rangle]$$

so that $s_j \in \partial_{\varepsilon_j/\alpha_j} f_j(x)$ if $\alpha_j > 0$: $(s_j, \alpha_j, \varepsilon_j)$ are exhibited for (3.5.2). Conversely, if (3.5.2) holds, multiply by α_j each of the inequalities

$$f_j^*(s_j) + f_j(x) - \langle s_j, x \rangle \leqslant \frac{\varepsilon_j}{\alpha_j};$$

add to the list thus obtained those inequations having $\alpha_j = 0$ (they hold trivially!); then sum up to obtain (3.5.3). □

It is important to realize what (3.5.2) means. Its third relation (which, incidentally, could be replaced by an equality) can be written

$$\sum_{j=1}^{m} (\varepsilon_j + \alpha_j e_j) \leqslant \varepsilon, \tag{3.5.4}$$

where, for each j, the number

$$e_j := f(x) - f_j(x)$$

is nonnegative and measures how close f_j comes to the maximal f at x. Using elementary calculus rules for approximate subdifferentials, a set-formulation of (3.5.2) is

$$\partial_\varepsilon f(x) = \cup \left\{ \sum_{j=1}^{m} \partial_{\varepsilon_j}(\alpha_j f_j)(x) : \alpha \in \Delta_m, \; \sum_{j=1}^{m} (\varepsilon_j + \alpha_j e_j) \leqslant \varepsilon \right\} \tag{3.5.5}$$

(remember that the ε-subdifferential of the zero function is identically $\{0\}$). Another observation is as follows: those elements (α_j, s_j) with $\alpha_j = 0$ do not matter and can be dropped from the combination making up s in (3.5.2). Then set $\eta_j := \varepsilon_j/\alpha_j$ to realize that $s \in \partial_\varepsilon f(x)$ if and only if there are positive α_j summing up to 1, $\eta_j \geqslant 0$ and $s_j \in \partial_{\eta_j} f(x)$ such that

$$\sum_j \alpha_j (\eta_j + e_j) \leqslant \varepsilon \quad \text{and} \quad \sum_j \alpha_j s_j = s. \tag{3.5.6}$$

The above formulae are rather complicated, even more so than (3.1.5) corresponding to a sum – which was itself not simple. In Theorem 3.5.1, denote by

$$J_\varepsilon(x) := \{j : f_j(x) \geqslant f(x) - \varepsilon\}$$

the set of ε-active indices; then we have the inclusion

$$\cup_{j \in J_\varepsilon(x)} \partial f_j(x) \subset \partial_\varepsilon f(x),$$

which is already simpler. Unfortunately, equality need not hold: with $n = 1$, take $f_1(\xi) = \xi$, $f_2(\xi) = -\xi$. At $\xi = 1$, the left-hand side is $\{1\}$ for all $\varepsilon \in [0, 2[$, while the right-hand side has been computed in §1.1: $\partial_\varepsilon f(1) = [1 - \varepsilon, 1]$.

In some cases, the exact formulae simplify:

Example 3.5.2 Consider the function $f^+ := \max\{0, f\}$, where $f : \mathbb{R}^n \to \mathbb{R}$ is convex. We get
$$\partial_\varepsilon(f^+)(x) = \cup\{\partial_\delta(\alpha f)(x) : 0 \leq \alpha \leq 1, \ \delta + f^+(x) - \alpha f(x) \leq \varepsilon\}.$$
Setting $\varepsilon(\alpha) := \varepsilon - f^+(x) + \alpha f(x)$, this can be written as
$$\partial_\varepsilon(f^+)(x) = \cup\{\partial_{\varepsilon(\alpha)}(\alpha f)(x) : \alpha \in [0, 1]\}$$
with the convention $\partial_\eta f(x) = \emptyset$ if $\eta < 0$. □

Another important simplification is obtained when each f_j is affine:

Example 3.5.3 (Piecewise Affine Functions) Take
$$f(x) := \max\{\langle s_j, x\rangle + b_j : j = 1, \ldots, p\}. \tag{3.5.7}$$
Each ε_j/α_j-subdifferential in (3.5.2) is constantly $\{s_j\}$: the ε_j's play no role and can be eliminated from (3.5.4), which can be used as a mere definition
$$0 \ [\leq \sum_{j=1}^p \varepsilon_j] \ \leq \varepsilon - \sum_{j=1}^p \alpha_j e_j.$$
Thus, the approximate subdifferential of the function (3.5.7) is the compact convex polyhedron
$$\partial_\varepsilon f(x) = \left\{\sum_{j=1}^p \alpha_j s_j : \alpha \in \Delta_p, \ \sum_{j=1}^p \alpha_j e_j \leq \varepsilon\right\}.$$

In this formulation, $e_j = f(x) - \langle s_j, x\rangle - b_j$. The role of e_j appears more explicitly if the origin of the graph-space is carried over to $(x, f(x))$ (remember Example VI.3.4): f of (3.5.7) can be alternatively defined by
$$\mathbb{R}^n \ni y \mapsto f(y) = f(x) + \max\{-e_j + \langle s_j, y - x\rangle : j = 1, \ldots, p\}.$$
The constant term $f(x)$ is of little importance, as far as subdifferentials are concerned. Neglecting it, e_j thus appears as the value at x (the point where $\partial_\varepsilon f$ is computed) of the jth affine function making up f.

Geometrically, $\partial_\varepsilon f(x)$ dilates when ε increases, and describes a sort of spider web with $\partial f(x)$ as "kernel". When ε reaches the value $\max_j e_j$, $\partial_\varepsilon f(x)$ stops at $\text{co}\{s_1, \ldots, s_p\}$. □

Finally, if $\varepsilon = 0$, (3.5.4) can be satisfied only by $\varepsilon_j = 0$ and $\alpha_j e_j = 0$ for all j. We thus recover the important Corollary VI.4.3.2. Comparing it with (3.5.6), we see that, for larger e_j or η_j, the η_j-subgradient s_j is "more remote from $\partial f(x)$", and as a result, its weight α_j must be smaller.

3.6 Post-Composition with an Increasing Convex Function

Theorem 3.6.1 *Let $f : \mathbb{R}^n \to \mathbb{R}$ be convex, $g \in \overline{\text{Conv}}\,\mathbb{R}$ be increasing, and assume the qualification condition $f(\mathbb{R}^n) \cap \text{int dom}\, g \neq \emptyset$. Then, for all x such that $f(x) \in \text{dom}\, g$,*
$$\left.\begin{array}{l} s \in \partial_\varepsilon(g \circ f)(x) \iff \\ \exists \varepsilon_1, \varepsilon_2 \geq 0 \ \text{and}\ \alpha \geq 0 \ \text{such that} \\ \varepsilon_1 + \varepsilon_2 = \varepsilon, \ s \in \partial_{\varepsilon_1}(\alpha f)(x), \ \alpha \in \partial_{\varepsilon_2} g(f(x)). \end{array}\right\} \tag{3.6.1}$$

PROOF. [⇒] Let $\alpha \geqslant 0$ be a minimum of the function ψ_s in Theorem X.2.5.1; there are two cases:

(a) $\alpha = 0$. Because dom $f = \mathbb{R}^n$, this implies $s = 0$ and $(g \circ f)^*(0) = g^*(0)$. The characterization of $s = 0 \in \partial_\varepsilon (g \circ f)(x)$ is

$$g^*(0) + g(f(x)) \leqslant \varepsilon, \quad \text{i.e.} \quad 0 \in \partial_\varepsilon g(f(x)).$$

Thus, (3.6.1) holds with $\varepsilon_2 = \varepsilon$, $\varepsilon_1 = 0$ – and $\alpha = 0$ (note: $\partial(0f) \equiv \{0\}$ because f is finite everywhere).

(b) $\alpha > 0$. Then $(g \circ f)^*(s) = \alpha f^*(s/\alpha) + g^*(\alpha)$ and the characterization of $s \in \partial_\varepsilon(g \circ f)(x)$ is

$$\alpha f^*(s/\alpha) + g^*(\alpha) + g(f(x)) - \langle s, x \rangle \leqslant \varepsilon,$$

i.e.

$$(\alpha f)^*(s) + \alpha f(x) - \alpha f(x) + g^*(\alpha) + g(f(x)) - \langle s, x \rangle \leqslant \varepsilon.$$

Split the above left-hand side into

$$(\alpha f)^*(s) + \alpha f(x) - \langle s, x \rangle =: \varepsilon_1 \geqslant 0,$$
$$g^*(\alpha) + g(f(x)) - \alpha f(x) =: \varepsilon_2 \geqslant 0,$$

which can be enunciated as: $s \in \partial_{\varepsilon_1}(\alpha f)(x)$, $\alpha \in \partial_{\varepsilon_2} g(f(x))$, $\varepsilon_1 + \varepsilon_2 \leqslant \varepsilon$.

Because, once again, η-subdifferentials increase with η, (3.6.1) is established.

[⇐] (3.6.1) means:

$$\alpha[f(y) - f(x)] \geqslant \langle s, y - x \rangle - \varepsilon_1 \quad \text{for all } y \in \mathbb{R}^n,$$
$$g(r) \geqslant g(f(x)) + \alpha[r - f(x)] - \varepsilon_2 \quad \text{for all } r \in \mathbb{R},$$

hence, with $r = f(y)$:

$$g(f(y)) \geqslant g(f(x)) + \langle s, y - x \rangle - \varepsilon \quad \text{for all } y \in \mathbb{R}^n. \qquad \square$$

As an application, consider the set

$$C := \{x \in \mathbb{R}^n : c(x) \leqslant 0\} \quad \text{with} \quad c : \mathbb{R}^n \to \mathbb{R} \text{ convex}. \tag{3.6.2}$$

Writing I_C as the composition $I_{]-\infty,0]} \circ c$, the ε-normal set $N_{C,\varepsilon}(x)$ of Definition 1.1.3 can be characterized in terms of approximate subdifferentials of c at x. Theorem 3.6.1 is valid under our qualification condition $c(\mathbb{R}^n) \cap \,]-\infty, 0[\neq \emptyset$, in which we recognize Slater's assumption. Furthermore, the approximate subdifferential of $I_{]-\infty,0]} = g$ has been computed in Example 1.2.5:

$$\alpha \in \partial_{\varepsilon_2} g(r) \quad \Longleftrightarrow \quad \alpha \geqslant 0 \text{ and } \alpha r + \varepsilon_2 \geqslant 0.$$

Then we obtain: $s \in N_{C,\varepsilon}(x)$ if and only if there are nonnegative α, ε_1, ε_2, with $\varepsilon_1 + \varepsilon_2 = \varepsilon$, such that

$$s \in \partial_{\varepsilon_1}(\alpha c)(x) \quad \text{and} \quad \alpha c(x) + \varepsilon_2 \geqslant 0. \tag{3.6.3}$$

Corollary 3.6.2 *Let C be described by (3.6.2) and assume that $c(x_0) < 0$ for some x_0. Then*

$$N_{C,\varepsilon}(x) = \cup \{\partial_\delta(\alpha c)(x) \,:\, \alpha \geqslant 0,\ \delta \geqslant 0,\ \delta - \alpha c(x) \leqslant \varepsilon\}. \tag{3.6.4}$$

In particular, if $x \in \operatorname{bd} C$, i.e. if $c(x) = 0$,

$$N_{C,\varepsilon}(x) = \cup\{\partial_\varepsilon(\alpha c)(x) \,:\, \alpha \geqslant 0\}. \tag{3.6.5}$$

PROOF. Eliminate ε_2 and let $\delta := \varepsilon_1$ in (3.6.3) to obtain the set-formulation (3.6.4). Then remember (VI.1.3.6) and use again the monotonicity of the multifunction $\delta \mapsto \partial_\delta(\alpha c)(x)$. □

Naturally, (3.6.5) with $\varepsilon = 0$ reduces to Theorem VI.1.3.5.

All the necessary material to reproduce Chap. VII is now at hand. For example, Corollary 3.6.2 and various calculus rules from the present Section 3 allow the derivation of necessary and sufficient conditions for *approximate* minimality in a constrained convex minimization problem. With the help of Example 3.5.2, the theory of exact penalty can be reproduced, etc. It is interesting to note that the various (Q)-type assumptions, giving the calculus rules in §X.2.2 and §X.2.3, are intimately related to the constraint qualification conditions of §VII.2.

Let us conclude with a general remark: in §III.5.3, we have alluded to some calculus rules for normal and tangent cones. They can be developed in a rigorous manner, by an application of the present calculus to functions of the form $f_j = I_{C_j}$. All these exercises are left to the reader.

4 The Approximate Subdifferential as a Multifunction

4.1 Continuity Properties of the Approximate Subdifferential

We will see in this section that the multifunction $\partial_\varepsilon f$ is much more regular when $\varepsilon > 0$ than the exact subdifferential, studied in §VI.6.2. We start with two useful properties, stating that the approximate subdifferential $(\varepsilon, x) \mapsto \partial_\varepsilon f(x)$ has a closed graph, and is locally bounded on the interior of $\operatorname{dom} f$; see §A.5 for the terminology.

Proposition 4.1.1 *Let $\{(\varepsilon_k, x_k, s_k)\}$ be a sequence converging to (ε, x, s), with $s_k \in \partial_{\varepsilon_k} f(x_k)$ for all k. Then $s \in \partial_\varepsilon f(x)$.*

PROOF. By definition,

$$f(y) \geqslant f(x_k) + \langle s_k, y - x_k \rangle - \varepsilon_k \quad \text{for all } y \in \mathbb{R}^n.$$

Pass to the limit on k and use the lower semi-continuity of f. □

Proposition 4.1.2 *Assume $\operatorname{int} \operatorname{dom} f \neq \emptyset$; let $\delta > 0$ and L be such that f is Lipschitzian with constant L on some ball $B(x, \delta)$, where $x \in \operatorname{int} \operatorname{dom} f$. Then, for all $\delta' < \delta$,*

$$\|s\| \leqslant L + \frac{\varepsilon}{\delta - \delta'} \tag{4.1.1}$$

whenever $s \in \partial_\varepsilon f(y)$, with $y \in B(x, \delta')$.

PROOF. We know (Theorem IV.3.1.2) that f is locally Lipschitzian on int dom f, so we can take x, δ, L as stated. To prove (4.1.1), take δ', y and s as stated, assume $s \neq 0$ and set $z := y + (\delta - \delta')s/\|s\|$ in the definition

$$f(z) \geq f(y) + \langle s, z - y \rangle - \varepsilon.$$

Observe that z and y are in $B(x, \delta)$ and conclude

$$L(\delta - \delta') \geq (\delta - \delta')\|s\| - \varepsilon. \qquad \square$$

As a result, the multifunction $\partial_\varepsilon f$ is outer semi-continuous, just as is the exact subdifferential. But for $\varepsilon > 0$, it also enjoys fairly strong continuity properties. We first assume that f is finite everywhere, and then we consider the general case via the regularized version $f \mathbin{\triangledown} c\|\cdot\|$ of f. In the result below, recall from Theorem V.3.3.8 that the Hausdorff distance between two (nonempty) compact convex sets A and B has the expression

$$\Delta_H(A, B) = \max\{|\sigma_A(d) - \sigma_B(d)| : \|d\| = 1\}.$$

Theorem 4.1.3 *Let $f: \mathbb{R}^n \to \mathbb{R}$ be a convex Lipschitzian function on $\mathbb{R}^n$. Then there exists $K > 0$ such that, for all x, x' in $\mathbb{R}^n$ and $\varepsilon, \varepsilon'$ positive:*

$$\Delta_H(\partial_\varepsilon f(x), \partial_{\varepsilon'} f(x')) \leq \frac{K}{\min\{\varepsilon, \varepsilon'\}}(\|x - x'\| + |\varepsilon - \varepsilon'|). \qquad (4.1.2)$$

PROOF. With d of norm 1, use (2.1.1): for any $\eta > 0$, there is $t_\eta > 0$ such that

$$q_\varepsilon(x, t_\eta) \leq f'_\varepsilon(x, d) + \eta \qquad (4.1.3)$$

where we have used the notation $q_\varepsilon(x, t)$ for the approximate difference quotient. By assumption, we can let $\delta \to +\infty$ in Proposition 4.1.2: there is a global Lipschitz constant, say L, such that $f'_\varepsilon(x, d) \leq L$. From

$$-L + \frac{\varepsilon}{t_\eta} \leq q_\varepsilon(x, t_\eta) \leq L + \eta,$$

we therefore obtain

$$\frac{1}{t_\eta} \leq \frac{2L + \eta}{\varepsilon}.$$

Then we can write, using (2.1.1) and (4.1.3) again:

$$f'_{\varepsilon'}(x', d) - f'_\varepsilon(x, d) - \eta \leq q_{\varepsilon'}(x', t_\eta) - q_\varepsilon(x, t_\eta) =$$
$$= \frac{f(x' + t_\eta d) - f(x + t_\eta d) + f(x) - f(x') + \varepsilon' - \varepsilon}{t_\eta} \leq$$
$$\leq \frac{2L\|x' - x\| + |\varepsilon' - \varepsilon|}{t_\eta} \leq \frac{2L+\eta}{\varepsilon}(2L\|x' - x\| + |\varepsilon' - \varepsilon|).$$

Remembering that $\eta > 0$ is arbitrary and inverting (x, ε) with (x', ε'), we do obtain

$$|f'_{\varepsilon'}(x',d) - f'_\varepsilon(x,d)| \leq \tfrac{2L}{\min(\varepsilon,\varepsilon')}(2L\|x'-x\| + |\varepsilon'-\varepsilon|)$$

so the theorem is proved, for example with $K = \max\{2L, 4L^2\}$. □

This result definitely implies the inner semi-continuity of $(x,\varepsilon) \mapsto \partial_\varepsilon f(x)$ for a Lipschitz-continuous f. In particular, for fixed $\varepsilon > 0$,

$$\partial_\varepsilon f(y) \subset \partial_\varepsilon f(x) + \|y-x\| B(0, \tfrac{K}{\varepsilon}) \quad \text{for all } x \text{ and } y,$$

a property already illustrated by Fig. 1.1.2. Remember from §VI.6.2 that the multifunction ∂f need not be inner semi-continuous; when $\varepsilon = 0$, no inclusion resembling the above can hold, unless x is *fixed*.

A local version of Theorem 4.1.3 can similarly be proved: (4.1.2) holds on the compact sets included in int dom f. Here, we consider an extension of the result to unbounded subdifferentials. Recall that the Hausdorff distance is not convenient for unbounded sets. A better distance is obtained by comparing the *bounded parts* of closed convex sets: for $c \geq 0$, we take

$$\Delta_{H,c}(C_1, C_2) := \Delta_H(C_1 \cap B(0,c), C_2 \cap B(0,c)).$$

Corollary 4.1.4 *Let* $f \in \overline{\text{Conv}}\,\mathbb{R}^n$. *Suppose that* $S \subset \text{dom}\, f$ *and* $\underline{c} > 0$ *are such that* $\partial f(x) \cap B(0, \underline{c}) \neq \emptyset$ *for all* $x \in S$. *Then, for all* $c > \underline{c}$, *there exists* K_c *such that, for all* x, x' *in* S *and* $\varepsilon, \varepsilon'$ *positive,*

$$\Delta_{H,c}(\partial_\varepsilon f(x), \partial_{\varepsilon'} f(x')) \leq \tfrac{K_c}{\min\{\varepsilon,\varepsilon'\}}(\|x'-x\| + |\varepsilon'-\varepsilon|).$$

PROOF. Consider $f_{[c]} := f \mathbin{\dot\nabla} c\|\cdot\|$ and observe that we are in the conditions of Proposition 3.4.5: $f_{[c]}$ is Lipschitzian on $\mathbb{R}^n$ and Theorem 4.1.3 applies. The rest follows because the coincidence set of $f_{[c]}$ and f contains S. □

Applying this result to an f finite everywhere, we obtain for example the following local Lipschitz continuity:

Corollary 4.1.5 *Let* $f : \mathbb{R}^n \to \mathbb{R}$ *be convex. For any* $\delta \geq 0$, *there is* $K_\delta > 0$ *such that*

$$\Delta_H(\partial_\varepsilon f(x), \partial_\varepsilon f(x')) \leq \tfrac{K_\delta}{\varepsilon}\|x - x'\| \quad \text{for all } x \text{ and } x' \text{ in } B(0, \delta).$$

PROOF. We know from Proposition 4.1.2 that $\partial_\varepsilon f$ is bounded on $B(0, \delta)$, so the result is a straightforward application of Corollary 4.1.4. □

4.2 Transportation of Approximate Subgradients

In §VI.6.3, we have seen that $\partial f(x)$ can be constructed by piecing together limits of subgradients along directional sequences. From a practical point of view, this is an important property: for example, it is the basis for descent schemes in nonsmooth optimization, see Chap. IX. This kind of property is even more important for approximate subdifferentials: remember from §II.1.2 that the only information obtainable from f is a black box (U1), which computes an *exact* subgradient at designated points. The concept of approximate subgradient is therefore of no use, as long as there is no "black box" to compute one. Starting from this observation, we study here the problem of constructing $\partial_\varepsilon f(x)$ with the sole help of the same (U1).

Theorem 4.2.1 (A. Brøndsted and R.T. Rockafellar) *Let be given $f \in \overline{\text{Conv}}\, \mathbb{R}^n$, $x \in \text{dom}\, f$ and $\varepsilon \geq 0$. For any $\eta > 0$ and $s \in \partial_\varepsilon f(x)$, there exist $x_\eta \in B(x, \eta)$ and $s_\eta \in \partial f(x_\eta)$ such that $\|s_\eta - s\| \leq \varepsilon/\eta$.*

PROOF. The data are $x \in \text{dom}\, f$, $\varepsilon > 0$ (if $\varepsilon = 0$, just take $x_\eta = x$ and $s_\eta = s$!), $\eta > 0$ and $s \in \partial_\varepsilon f(x)$. Consider the closed convex function

$$\mathbb{R}^n \ni y \mapsto \varphi(y) := f(y) + f^*(s) - \langle s, y \rangle.$$

It is nonnegative (Fenchel's inequality), satisfies $\varphi(x) \leq \varepsilon$ (cf. (1.2.1)), and its subdifferential is

$$\partial \varphi(y) = \partial f(y) - \{s\} \quad \text{for all } y \in \text{dom}\, \varphi = \text{dom}\, f.$$

Perturb φ to the closed convex function

$$\mathbb{R}^n \ni y \mapsto \psi(y) := \varphi(y) + \frac{\varepsilon}{\eta}\|y - x\|,$$

whose subdifferential at $y \in \text{dom}\, f$ is (apply Corollary 3.1.2: (3.1.2) obviously holds)

$$\partial \psi(y) = \partial \varphi(y) + \tfrac{\varepsilon}{\eta}\partial(\|\cdot - x\|)(y) \subset \partial \varphi(y) + B(0, \tfrac{\varepsilon}{\eta}).$$

Because φ is bounded from below, the 0-coercivity of the norm implies the 0-coercivity of ψ; there exists a point, say x_η, minimizing ψ on $\mathbb{R}^n$; then $0 \in \partial \psi(x_\eta)$ is written

$$0 \in \partial f(x_\eta) - \{s\} + B(0, \tfrac{\varepsilon}{\eta}).$$

It remains to prove that $x_\eta \in B(x, \eta)$. Using the nonnegativity of φ and optimality of x_η:

$$\tfrac{\varepsilon}{\eta}\|x_\eta - x\| \leq \psi(x_\eta) \leq \psi(x) = \varphi(x) \leq \varepsilon. \qquad \square$$

This result can be written in a set formulation:

$$\partial_\varepsilon f(x) \subset \bigcap_{\eta > 0} \bigcup_{\|y - x\| \leq \eta} \left\{ \partial f(y) + B(0, \tfrac{\varepsilon}{\eta}) \right\}. \tag{4.2.1}$$

It says that any ε-subgradient at x can be approximated by some exact subgradient, computed at some y, possibly different from x. For y close to x (η small) the approximation may be coarse; an accurate approximation (η large), may require seeking y far from x. The value $\eta = \sqrt{\varepsilon}$ is a compromise, which equates the deviation from x and the degree of approximation: (4.2.1) implies

$$\partial_\varepsilon f(x) \subset \cup \{\partial f(y) + B(0, \sqrt{\varepsilon}) : y \in B(0, \sqrt{\varepsilon})\}.$$

An illustrative example is the one-dimensional function $f: x \mapsto x + 1/x$, with $x = 1$ and $\varepsilon = 1$ (which is the value ε^∞ of §2.2). Then $1 \in \partial_\varepsilon f(x)$ but 1 is nowhere the derivative of f: in the above proof, x_η is unbounded for $\eta \to +\infty$.

We also see from (4.2.1) that $\partial_\varepsilon f(x)$ is contained in the closure of $\partial f(\mathbb{R}^n)$ – and the example above shows that the closure operation is necessary. Our next question is then: given an exact subgradient s computed somewhere, how can we recognize whether this s is in the set $\partial_\varepsilon f(x)$ that we are interested in? The answer turns out to be very simple:

Proposition 4.2.2 (Transportation Formula) *With x and x' in dom f, let $s' \in \partial f(x')$. Then $s' \in \partial_\varepsilon f(x)$ if and only if*

$$f(x') \geqslant f(x) + \langle s', x' - x \rangle - \varepsilon. \tag{4.2.2}$$

PROOF. The condition is obviously necessary, since the relation of definition (1.1.1) must in particular hold at $y = x'$. Conversely, for $s' \in \partial f(x')$, we have for all y

$$f(y) \geqslant f(x') + \langle s', y - x' \rangle =$$
$$= f(x) + \langle s', y - x \rangle + [f(x') - f(x) + \langle s', x - x' \rangle].$$

If (4.2.2) holds, $s' \in \partial_\varepsilon f(x)$. □

The next chapters, dealing with numerical algorithms, will use (4.2.2) intensively. We call it the *transportation formula*, since it "transports" at x a given subgradient at x'. Geometrically, Fig. 4.2.1 shows that it is very natural, and it reveals an important concept:

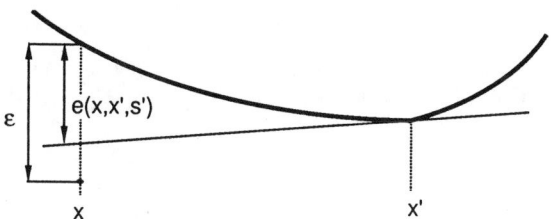

Fig. 4.2.1. Linearization errors and the transportation formula

Definition 4.2.3 (Linearization Error) *For $(x, x', s') \in \text{dom } f \times \text{dom } f \times \mathbb{R}^n$, the linearization error made at x, when f is linearized at x' with slope s', is the number*

$$e(x, x', s') := f(x) - f(x') - \langle s', x - x' \rangle.$$

This linearization error is of particular interest when $s' \in \partial f(x')$: then, calling

$$\bar{f}_{x',s'}(y) := f(x') + \langle s', y - x' \rangle$$

the corresponding affine approximation of f, there holds the relations
$$\bar{f}_{x',s'} \leq f, \quad \bar{f}_{x',s'}(x') = f(x'), \quad \bar{f}_{x',s'}(x) = f(x) - e(x, x', s').$$
The definition of $s' \in \partial f(x')$ via the conjugate function can also be used:
$$e(x, x', s') = f(x) + f^*(s') - \langle s', x \rangle \geq 0 \quad \text{if } s' \in \partial f(x'). \qquad \square$$

The content of the transportation formula (4.2.2), illustrated by Fig. 4.2.1, is that any $s' \in \partial f(x')$ is an $e(x, x', s')$-subgradient of f at x; and also that it is not in a tighter approximate subdifferential, i.e.
$$\varepsilon < e(x, x', s') \quad \Longrightarrow \quad s' \notin \partial_\varepsilon f(x).$$
This latter property relies on the contact (at x') between gr $\bar{f}_{x',s'}$ and gr f. In fact, a result slightly different from Proposition 4.2.2 is:

Proposition 4.2.4 *Let* $s' \in \partial_\eta f(x')$. *Then* $s' \in \partial_\varepsilon f(x)$ *if*
$$f(x') \geq f(x) + \langle s', x' - x \rangle - \varepsilon + \eta$$
or equivalently
$$e(x, x', s') + \eta \leq \varepsilon.$$

PROOF. Proceed as for the "if"-part of Proposition 4.2.2. $\qquad \square$

The transportation formula gives an easy answer to the pragmatic question: "Let a subgradient be computed at some point by the black box (U1), is it an approximate subgradient at some other point?" Answer: just compare f- and $\bar{f}$-values. Returning to a more theoretical framework, we now ask: given x and ε, what are those x' such that (4.2.2) holds? This question is ambiguous when $\partial f(x')$ is not a singleton, so we define two sets:
$$V_\varepsilon(x) := \{x' \in \text{dom}\, \partial f \;:\; \partial f(x') \subset \partial_\varepsilon f(x)\}$$
$$\overline{V}_\varepsilon(x) := \{x' \in \mathbb{R}^n \;:\; \partial f(x') \cap \partial_\varepsilon f(x) \neq \emptyset\}.$$
Equivalent definitions are
$$V_\varepsilon(x) := \{x' \in \text{dom}\, \partial f \;:\; e(x, x', s') \leq \varepsilon \text{ for all } s' \in \partial f(x')\}$$
$$\overline{V}_\varepsilon(x) := \{x' \in \mathbb{R}^n \;:\; e(x, x', s') \leq \varepsilon \text{ for some } s' \in \partial f(x')\}, \qquad (4.2.3)$$
and it is clear that $V_\varepsilon(x) \subset \overline{V}_\varepsilon(x)$. If f is differentiable on $\mathbb{R}^n$, the two sets obviously coincide:
$$V_\varepsilon(x) = \overline{V}_\varepsilon(x) = \{x' \in \mathbb{R}^n \;:\; e(x, x', \nabla f(x')) \leq \varepsilon\}.$$
The next result motivates our notation.

Proposition 4.2.5 *Suppose that* $x \in \text{int dom}\, f$. *Then*
(i) $V_\varepsilon(x)$ *is a neighborhood of x if $\varepsilon > 0$.*
(ii) $\overline{V}_\varepsilon(x)$ *is the closure of $V_\varepsilon(x)$.*

4 The Approximate Subdifferential as a Multifunction 133

PROOF. [*(i)*] Apply Proposition 4.1.2: there exist $\delta > 0$ and a constant K such that, for all $x' \in B(x, \delta)$ and $s' \in \partial f(x')$,

$$e(x, x', s') \leqslant |f(x') - f(x)| + \|s'\| \|x' - x\| \leqslant 2K \|x' - x\|;$$

so $V_\varepsilon(x)$ contains the ball of center x and radius $\min\{\delta, \varepsilon/(2K)\}$.

[*(ii)*] Take a point $x' \in \overline{V}_\varepsilon(x)$. We claim that

$$\partial f(x + t(x' - x)) \subset \partial_\varepsilon f(x) \quad \text{for all } t \in]0, 1[\,; \qquad (4.2.4)$$

see Fig. 4.2.1: we insert a point between x and x'. A consequence of (4.2.4) will be $\overline{V}_\varepsilon(x) \subset \text{cl } V_\varepsilon(x)$ (let $t \uparrow 1$).

To prove our claim, set $d := x' - x$ and use the function $r = r_0$ of (2.1.3) to realize with Remark 2.2.4 that

$$\{e(x, x + td, s) : s \in \partial f(x + td)\} = \partial r(1/t) \quad \text{for all } t > 0. \qquad (4.2.5)$$

By definition of $\overline{V}_\varepsilon(x)$, we can pick $s' \in \partial f(x')$ such that

$$-\partial r(1) \ni e(x, x', s') \leqslant \varepsilon.$$

Then we take arbitrary $1/t = u > 1$ and $s'' \in \partial f(x + td)$, so that $-e(x, x + td, s'') \in \partial r(u)$. The monotonicity property (VI.6.1.1) or (I.4.2.1) of the subdifferential ∂r gives

$$[-e(x, x + td, s'') + e(x, x', s')](u - 1) \geqslant 0$$

hence $e(x, x + td, s'') \leqslant e(x, x', s') \leqslant \varepsilon$, which proves (4.2.4).

There remains to prove that $\overline{V}_\varepsilon(x)$ is closed, so let $\{x'_k\}$ be a sequence of $\overline{V}_\varepsilon(x)$ converging to some x'. To each x'_k, we associate $s_k \in \partial f(x'_k) \cap \partial_\varepsilon f(x)$; $\{s_k\}$ is bounded by virtue of Theorem 1.1.4: extracting a subsequence if necessary, we may assume that $\{s_k\}$ has a limit s'. Then, Proposition 4.1.1 and Theorem 1.1.4 show that $s' \in \partial f(x') \cap \partial_\varepsilon f(x)$, which means that $x' \in \overline{V}_\varepsilon(x)$. □

From a practical point of view, the set $V_\varepsilon(x)$ and the property (i) above are both important: provided that y is close enough to x, the $s(y) \in \partial f(y)$ computed by the black box (U1) is *guaranteed* to be an ε-subgradient at x. Now, $V_\varepsilon(x)$ and $\overline{V}_\varepsilon(x)$ differ very little (they are numerically indistinguishable), and the latter has a strong intrinsic value: by definition, $x' \in \overline{V}_\varepsilon(x)$ if and only if

$$\exists s' \in \partial_\varepsilon f(x) \quad \text{such that} \quad s' \in \partial f(x'), \quad \text{i.e. such that} \quad x' \in \partial f^*(s').$$

In the language of multifunctions, this can be written

$$\overline{V}_\varepsilon(x) = \partial f^*(\partial_\varepsilon f(x)). \qquad (4.2.6)$$

Furthermore the above proof, especially (4.2.5), establishes a connection between $\overline{V}_\varepsilon(x)$ and §2.2. First $\overline{V}_\varepsilon(x) - \{x\}$ is star-shaped. Also, consider the intersection of $\overline{V}_\varepsilon(x)$ with a direction d issuing from x. Keeping (4.2.3) in mind, we see that this set is the closed interval where the perturbed difference quotient q_ε is decreasing. In a word, let $\bar{t}_\varepsilon(d)$ be the largest element of $T_\varepsilon = T_\varepsilon(d)$ defined by (2.2.2), with the convention $\bar{t}_\varepsilon(d) = +\infty$ if T_ε is empty or unbounded (meaning that the approximate difference quotient is decreasing on the whole of $\mathbb{R}_*^+$). Then

$$\bar{V}_\varepsilon(x) = \{x + td \,:\, t \in [0, \bar{t}_\varepsilon(d)], \, d \in B(0, 1)\}. \tag{4.2.7}$$

See again the geometric interpretations of §2, mainly the end of §2.3.

Our neighborhoods $V_\varepsilon(x)$ and $\bar{V}_\varepsilon(x)$ enjoy some interesting properties if additional assumptions are made on f:

Proposition 4.2.6

(i) If f is 1-coercive, then $\bar{V}_\varepsilon(x)$ is bounded.
(ii) If f is differentiable on $\mathbb{R}^n$, then $V_\varepsilon(x) = \bar{V}_\varepsilon(x)$ and $\nabla f(V_\varepsilon(x)) \subset \partial_\varepsilon f(x)$.
(iii) If f is 1-coercive and differentiable, then $\nabla f(V_\varepsilon(x)) = \partial_\varepsilon f(x)$.

PROOF. In case (i), f^* is finite everywhere, so the result follows from (4.2.6) and the local boundedness of the subdifferential mapping (Proposition VI.6.2.2).

When f is differentiable, the equality between the two neighborhoods has already been observed, and the stated inclusion comes from the very definition of $V_\varepsilon(x)$.

Finally, let us establish the converse inclusion in case (iii): for $s \in \partial_\varepsilon f(x)$, pick $y \in \partial f^*(s)$, i.e. $s = \nabla f(y)$. The result follows from (4.2.6): $y \in \bar{V}_\varepsilon(x) = V_\varepsilon(x)$. □

Example 4.2.7 Let f be convex quadratic as in Example 1.2.2. Substitute in (1.2.5)

$$\nabla f(x) + Qy = Qx + b + Qy = Q(x + y) + b = \nabla f(x + y)$$

so as to obtain the form

$$\partial_\varepsilon f(x) = \{\nabla f(x') \,:\, \tfrac{1}{2}\langle Q(x' - x), x' - x\rangle \leqslant \varepsilon\}, \tag{4.2.8}$$

which discloses the neighborhood

$$\bar{V}_\varepsilon(x) = \{x' \in \mathbb{R}^n \,:\, \tfrac{1}{2}\langle Q(x' - x), x' - x\rangle \leqslant \varepsilon\}. \tag{4.2.9}$$

When Q is invertible, we have a perfect duality correspondence: $\bar{V}_\varepsilon(x)$ [resp. $\partial_\varepsilon f(x)$] is the ball of radius $\sqrt{2\varepsilon}$ associated with the metric of Q [resp. Q^{-1}], and centered at x [resp. at $\nabla f(x)$].

Finally note that we can use a pseudo-inverse in (4.2.9):

$$x' = Q^-(Qx' + b - b) = Q^-[\nabla f(x') - b]$$

and obtain the form

$$\bar{V}_\varepsilon(x) = \{Q^-[\nabla f(x') - b] \,:\, \tfrac{1}{2}\langle Q(x' - x), x' - x\rangle \leqslant \varepsilon\}.$$

This illustrates (4.2.6), knowing that $\partial f^*(s) = Q^-(s - b) + \text{Ker } Q$ for $s - b \in \text{Im } Q$ and that $\text{Ker } Q \subset \bar{V}_\varepsilon(x)$. □

A final remark: we have seen in §II.2 the importance of defining a neighborhood of a given iterate $x = x_k$, to construct suitable directions of search. For the sake of efficiency, this neighborhood had to reflect the behaviour of f near x. Here, $\overline{V}_\varepsilon(x)$ is a possible candidate: interpret the constraint in (II.2.3.1) as defining a neighborhood to be compared with (4.2.9). However, it is only because ∇f and ∇f^* are affine mappings that $\overline{V}_\varepsilon(x)$ has such a nice expression in the quadratic case. For a general f, $\overline{V}_\varepsilon(x)$ is hardly acceptable; (4.2.7) shows that it need not be bounded, and (4.2.6) suggests that it need not be convex, being the image of a convex set by a mapping which is rather complicated.

Example 4.2.8 Let f be defined by

$$f(x) = \max\{0, \xi^2 + \eta^2 - 1\} \quad \text{for all } x = (\xi, \eta) \in \mathbb{R}^2 \,.$$

Consider $x = (1, 1)$, at which $f(x) = 1$ and $\nabla f(x) = (2, 2)$, and take $\varepsilon = 1/2$.

Look at Fig. 4.2.2 to see how $\overline{V}_\varepsilon(x)$ is constructed: if f were the quadratic function $\xi^2 + \eta^2 - 1$, we would obtain the ball

$$D = \{(\xi, \eta) : (\xi - 1)^2 + (\eta - 1)^2 \leq 1/2\}$$

(see Example 4.2.7). However, f is minimal on the unit ball $B(0, 1)$, i.e. $0 \in \partial f(x')$ for all $x' \in B(0, 1)$; furthermore, $0 \notin \partial_{1/2} f(x)$. In view of its definition (4.2.3), $V_\varepsilon(x)$ therefore does not meet $B(0, 1)$:

$$V_{1/2}(x) \subset D \backslash B(0, 1) \,.$$

On the other hand, it *suffices* to remove $B(0, 1)$ from D, and this can be seen from the definition (4.2.3) of $\overline{V}_\varepsilon(x)$: the linearization error $e(x, \cdot, \cdot)$ is left unperturbed by the max-operation defining f. In summary:

$$\overline{V}_\varepsilon(x) = \{(\xi, \eta) : (\xi - 1)^2 + (\eta - 1)^2 \leq 1/2, \ \xi^2 + \eta^2 \geq 1\} \,.$$

We thus obtain a nonconvex neighborhood; observe its star-shaped character. □

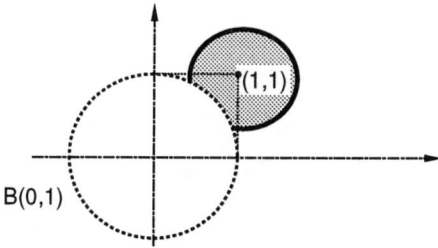

Fig. 4.2.2. A nonconvex neighborhood $V_\varepsilon(x)$

By contrast, another interesting neighborhood is indeed convex: it is obtained by inverting the role of x and x' in (4.2.3). For $x \in \text{int dom } f$, define the set illustrated in Fig. 4.2.3:

$$V_\varepsilon^*(x) := \{x' \in \mathbb{R}^n : e(x', x, s) \leq \varepsilon \text{ for all } s \in \partial f(x)\}.$$

Another expression is

$$V_\varepsilon^*(x) = \cap \{\partial_\varepsilon f^*(s) : s \in \partial f(x)\},$$

which shows that $V_\varepsilon^*(x)$ is closed and convex. Reproduce the proof of Proposition 4.2.5(i) to see that $V_\varepsilon^*(x)$ is a neighborhood of x. For the quadratic function of Example 4.2.7, $V_\varepsilon^*(x)$ and $V_\varepsilon(x)$ coincide.

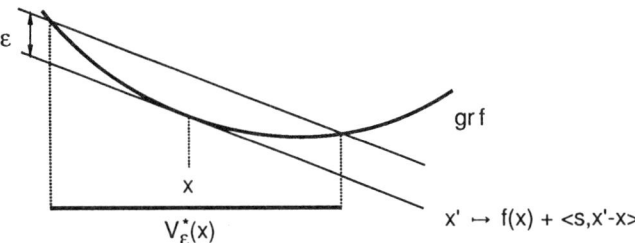

Fig. 4.2.3. A second-order neighborhood

XII. Abstract Duality for Practitioners

Prerequisites. Subdifferentials of finite convex functions (Chap. VI); minimality conditions and elementary duality theory (Chap. VII, but much less than one might think); and for the last part of the chapter, definition and elementary properties of conjugate convex functions (Chap. X).

Introduction. The subject of this chapter is by far the most important application of convex minimization, namely *general decomposition* in mathematical programming (more exactly: *price decomposition*), and *dual algorithms*. It can be safely ascertained that this subject nearly coincides with convex minimization, studied in Chaps. VIII and IX:

– on the one hand, algorithms for convex minimization have their best-suited field of application in decomposition, more precisely the problem of price-adjustment (another important field concerns eigenvalue optimization of a varying matrix but there, nonconvexity comes quickly into play);
– on the other hand, when decomposing a problem – more exactly: when adjusting prices via a decentralization algorithm in a (usually large-scale) optimization problem –, one is primarily minimizing a certain convex function, namely the dual function associated with the problem; and there is no way around this.

1 The Problem and the General Approach

1.1 The Rules of the Game

We consider in this chapter a constrained optimization problem characterized by a nonempty set U of admissible control variables u, an objective function $\varphi : U \to \mathbb{R}$, and constraint-functions $c_1, \ldots, c_m : U \to \mathbb{R}$; then we want to solve

$$\left|\begin{array}{ll} \sup \varphi(u) & u \in U, \\ c_j(u) = 0 & \text{for } j = 1, \ldots, m, \end{array}\right. \tag{1.1.1}$$

which we will call the *primal problem*. As usual, a $u \in U$ satisfying the constraints $c_j(u) = 0$ will be called feasible. Throughout this chapter, without further mention, the following assumption will be in force:

$$\boxed{U \neq \emptyset; \quad \varphi, c_1, \ldots c_m \text{ are finite everywhere on } U.}$$

So far, we are not assuming any structure in U whatsoever. For example, U is not at all supposed to be (a subset of) a vector space like $\mathbb{R}^n$; we will see in

§1.2 how abstract, or how concrete U can be. This implies in particular that the objective- and constraint-functions have no structure either, such as convexity, or a fortiori differentiability: in U, these words are meaningless for the moment. They will of course appear (to be solvable, an optimization problem such as (1.1.1) must enjoy some structure), but much later and it is useful to see how far the theory can be developed in abstracto. Observe also that assuming φ and each c_j to be finite everywhere is not really a restriction: if these functions were extended-valued, U could be replaced by its intersection with their domains.

Actually, the only structure assumed concerns the sets of objective and constraint values, which are $\mathbb{R}$ and $\mathbb{R}^m$ respectively. Then we equip $\mathbb{R}^m$ with the ordinary dot-product: for $(\lambda, c) \in \mathbb{R}^m \times \mathbb{R}^m$,

$$\lambda^\top c := \sum_{j=1}^m \lambda_j c_j . \tag{1.1.2}$$

The associated norm will be $\|\cdot\|$. As far as the space of constraint-values is concerned, and *this space only*, we are therefore in the general framework of this book, with scalar products, duality, metric, and so on.

Then, denoting by $c(u) := (c_1(u), \ldots, c_m(u)) \in \mathbb{R}^m$ the vector of constraint-values at $u \in U$, we can consider the Lagrange function or *Lagrangian*, defined by

$$L(u, \lambda) := \varphi(u) - \lambda^\top c(u) \quad \text{for all } \lambda \in \mathbb{R}^m \text{ and } u \in U . \tag{1.1.3}$$

For a given $\lambda \in \mathbb{R}^m$, this appears as a perturbation of the objective function of (1.1.1), which simply says the following: a violation of the constraints is accepted but then, a price $\lambda^\top c(u)$ must be paid when the control value is $u \in U$.

Remark 1.1.1 Naturally, (1.1.3) is the same Lagrange function that played a central role in Chap. VII. Once again, however, it is important for a better understanding to realize that L is defined here on $U \times \mathbb{R}^m$. By contrast, Chap. VII heavily relied on the situation $U = \mathbb{R}^n$ – so as to define subdifferentiability of φ and c – and little on the concept of a varying λ.

With respect to Chap. VII, beware of the minus-sign in (1.1.3); it comes from the fact that we start from a maximization problem, and will be motivated later. □

No theoretical property is assumed on the data (U, φ, c), other than the duality structure induced by (1.1.2). For *practical* purposes, however, we do make a heavy assumption:

Assumption 1.1.2 (Practical) We assume that the optimization problem in u

$$\sup \{L(u, \lambda) : u \in U\}, \tag{1.1.4}_\lambda$$

where λ is fixed in $\mathbb{R}^m$, is *considerably simpler* than the primal problem (1.1.1).

We will call $(1.1.4)_\lambda$ the *Lagrange problem* associated with λ. □

To quantify a little bit the wording "considerably simpler", we can say for example:
– the simpler $(1.1.4)_\lambda$ is, the more efficient the approach of this chapter will be;

- Assumption 1.1.2 holds when an efficient methodology exists for $(1.1.4)_\lambda$, but not for (1.1.1);
- or when it costs more to solve (1.1.1) once than to solve $(1.1.4)_\lambda$ 10^2–10^4 times, say;
- this still loose quantification can be slightly sharpened by saying that we will rather be in the 10^2-range if $(1.1.4)_\lambda$ enjoys some more theoretical properties (convexity), and rather in the 10^4-range if not.

However vague it is, Assumption 1.1.2 is fundamental; it has to be supplemented by one more practical assumption, hardly more meaningful (but certainly fairly usual in mathematical programming):

Assumption 1.1.3 One will be content with an approximate solution of (1.1.1), i.e. with $\bar{u} \in U$ such that
- $\varphi(\bar{u})$ is possibly slightly less than the supremal value in (1.1.1),
- and/or $c(\bar{u})$ is possibly not exactly $0 \in \mathbb{R}^m$. □

When faced with a practical optimization problem, there are several ways of formulating the data (U, φ, c) of (1.1.1): constraints may be incorporated in φ via some penalty term (§VII.3.2); or its extreme version, which is the indicator function of a feasible set; they can also be considered as making up U; or as making up c, and so on. To choose a formulation adapted to our present framework, Assumptions 1.1.2 and 1.1.3 must be the central concern. The case of Assumption 1.1.2 is linked to the *decomposability* of (1.1.1) and will be seen more closely in §1.2; as for Assumption 1.1.3, it distinguishes two categories of constraints:

- the *hard* constraints, which must absolutely be satisfied by $\bar{u}$; they will most conveniently appear in the definition of U;
- the *soft* constraints, for which some tolerance is accepted; they can go in the c-set.

The methods of this chapter are aimed at solving (1.1.1) with the help of (1.1.3). A *black box* (the black box (U1) of Fig. II.2.1 that keeps appearing in this book) is available which, given $\lambda \in \mathbb{R}^m$, solves $(1.1.4)_\lambda$ and returns appropriate information. Our problem in this chapter is therefore as follows:

Problem 1.1.4 Find some suitable value of λ such that the associated Lagrange problem, as solved by the black box, provides a solution of the primal problem.
The unknown λ will be called the *dual variable*. □

This is the so-called *coordination*, or *decentralization* problem, because $(1.1.4)_\lambda$ is usually a decomposed form of (1.1.1), see §1.2 below. From its very definition (1.1.2), λ varies in the dual of the space of constraint-values, hence 1.1.4 can also be called the "dual problem"; but we will see in §2.1 that this terminology is ambiguous.

Then comes an important point: what information is available from the black box, to be used in our search for λ? The answer depends first on the basic question: does the Lagrange problem have a solution at all? It may happen that $\sup L(\cdot, \lambda) = +\infty$, or that $(1.1.4)_\lambda$ has no optimal solution "at finite distance"; in either case, the black box is unlikely to return any valuable information. The exposition is much easier if this never happens, so we consider the easy situation, at least for the moment:

Assumption 1.1.5 (Temporary)

(i) For all $\lambda \in \mathbb{R}^m$, the Lagrange problem $(1.1.4)_\lambda$ has some optimal solution.
(ii) The black box computes one such optimal solution, say $u_\lambda \in U$, and returns the maximal value $L(u_\lambda, \lambda)$, together with its corresponding constraint-value $c(u_\lambda) \in \mathbb{R}^m$. □

Seen from the point of view of 1.1.4, the situation is therefore as illustrated by Fig. 1.1.1: the coordinator (sometimes called the *master program*) chooses a price-vector λ, sends it to the black box (sometimes called the *local problem*), and receives in return information consisting of a vector $(c, L) \in \mathbb{R}^m \times \mathbb{R}$. The duty of the coordinator is then to decide about optimality in terms of (1.1.1), and if not satisfied, to modify λ accordingly.

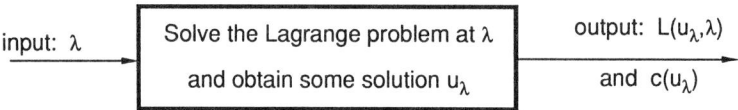

Fig. 1.1.1. Useful local black box for a dual-solver

Remark 1.1.6 The temporary Assumption 1.1.5(i) is of a theoretical nature; it holds essentially if U is a compact set, on which $L(\cdot, \lambda)$ is upper semi-continuous. A particular and important case is when U is simply a finite set.

Part (ii), concerning the nature and amount of information computed by the black box, is typical and useful. We have selected it because it serves best our purpose, mainly illustrative. Other situations are possible, though:

– A very powerful black box, able to return the maximal L together with all $c(u)$ for all optimal u's, could be conceived of; but we will see in the next sections that it usually makes little sense.
– By contrast, more meaningful would be a weaker black box, only able to compute some suboptimal solution, i.e. an $u_{\lambda,\varepsilon} \in U$ satisfying, for some $\varepsilon > 0$:

$$L(u_{\lambda,\varepsilon}, \lambda) \geqslant L(u, \lambda) - \varepsilon \quad \text{for all } u \in U.$$

This situation is of theoretical interest: an ε-optimal solution exists whenever the supremum in $(1.1.4)_\lambda$ is finite; also, from a practical point of view, it demands less from the black box. This last advantage should not be over-rated, though: for efficiency, ε must be decided *by the coordinator*; and just for this reason, life may not be easier for the black box.
– We mention that the values of L and c are readily available after $(1.1.4)_\lambda$ is solved: a situation in which only L is returned is hardly conceivable; this poor information would result in a poor coordinator anyway. □

Let us mention one last point: $L(u_\lambda, \lambda)$ is a well-defined number, namely the optimal value in $(1.1.4)_\lambda$; but $c(u_\lambda)$ is not a well-defined vector, since it depends on the particular solution selected by the black box. We will see in the next sections that $(1.1.4)_\lambda$ has a unique solution for almost all $\lambda \in \mathbb{R}^m$ – and this property holds independently of the data (U, φ, c). Nevertheless, there may exist "critical" values of λ for which this uniqueness does not hold.

1 The Problem and the General Approach

This question of uniqueness is *crucial*, and comes on the top of the practical Assumption 1.1.2 to condition the success of the approach. If $(1.1.4)_\lambda$ has a unique maximizer for each $\lambda \in \mathbb{R}^m$, then the dual approach is extremely powerful; difficulties begin in case of non-uniqueness, and then some structure in (U, φ, c) becomes necessary.

1.2 Examples

Practically all problems suited to our framework are *decomposable*, i.e.:

- $U = U^1 \times U^2 \times \cdots \times U^n$, where each U^i is "simpler" (a word that matches the practical Assumption 1.1.2); the control variables will be denoted by $u = (u^1, \ldots, u^n)$, where $u^i \in U^i$ for $i = 1, \ldots, n$;
- φ is a *sum* of "individual objective functions": $u = (u^1, \ldots, u^n) \mapsto \varphi(u) = \sum_{i=1}^n \varphi^i(u^i)$;
- likewise, each constraint is a sum: $c_j(u) = \sum_{i=1}^n c_j^i(u^i)$.

Then (1.1.1) is

$$\left| \begin{array}{l} \sup \sum_{i=1}^n \varphi^i(u^i) \quad u^i \in U^i \text{ for } i = 1, \ldots, n \\ \sum_{i=1}^n c_j^i(u^i) = 0 \quad \text{for } j = 1, \ldots, m \quad \left[\text{in short } \sum_{i=1}^n c^i(u^i) = 0 \in \mathbb{R}^m \right]. \end{array} \right. \tag{1.2.1}$$

In these problems, $c(u) = 0$ appears as a (vector-valued) *coupling* constraint, which links the individual control variables u^i. If this constraint were not present, (1.2.1) would reduce to n problems, each posed in the simpler set U^i, and might thus become "considerably simpler". This is precisely what happens to the Lagrange problem, which splits into

$$\sup \left\{ \varphi^i(u^i) - \lambda^\top c^i(u^i) : u^i \in U^i \right\} \quad \text{for } i = 1, \ldots, n. \tag{1.2.2}$$

Such decomposable problems form a wide variety in the world of optimization; usually, each u^i is called a *local control variable*, hence the word "local problems" to designate the black box of Fig. 1.1.1. We choose three examples for illustration.

(a) The Knapsack Problem. Our first example is the simplest instance of *combinatorial* problems, in which U is a finite or countable set. One has a knapsack and one considers putting in it n objects (toothbrush, saucepan, TV set,...) each of which has a price p^i – expressing how much one would like to take it – and a volume v^i. Then one wants to make the knapsack of maximal price, knowing that it has a limited volume v.

For each object, two decisions are possible: take it or not; U is the set of all possible such decisions for the n objects, i.e. U can be identified with the set $\{1, \ldots, 2^n\}$. To each $u \in U$, are associated the objective- and constraint-values, namely the sum of respectively the prices and volumes of all objects taken. The problem is solvable in a finite time, but quite a large time if n is large; actually, problems of this kind are extremely difficult.

Now consider the Lagrange problem: $m = 1$ and the number λ is a penalty coefficient, or the "price of space" in the knapsack: when the ith object is taken, the payoff is no longer p^i but $p^i - \lambda v^i$. The Lagrange problem is straightforward: for each object i,

$$\left| \begin{array}{l} \text{if } p^i - \lambda v^i > 0, \text{ take the object}; \\ \text{if } p^i - \lambda v^i < 0, \text{ leave it}; \\ \text{if } p^i - \lambda v^i = 0, \text{ do what you want}. \end{array} \right.$$

While making each of these n decisions, the corresponding terms are added to the Lagrange- and constraint-values and the job to be done in Fig. 1.1.1 is clear enough.

Some preliminary observations are worth mentioning.

– U is totally unstructured: for example, what is the sum of two decisions? On the other hand, the Lagrange function defines, via the coefficients $p^i - \lambda v^i$, an *order* in U, which depends on λ;
– from its interpretation, λ should be nonnegative: there is no point in giving a bonus to bulky objects like TV sets;
– there always exists an optimal decision in the Lagrange problem $(1.1.4)_\lambda$; and it is well-defined (unique) except when λ is one of the n values p^i/v^i.

Here is a case where the practical Assumption 1.1.2 is "very true". It should not be hastily concluded, however, that the dual approach is going to solve the problem easily: in fact, the rule is that Problem 1.1.4 is itself hard when (1.1.1) is combinatorial (so the "law of conservation of difficulty" applies). The reason comes from the conjunction of two bad things: uniqueness in the Lagrange problem does not hold – even though it holds "most of the time" – and U has no nice structure. On the other hand, the knapsack problem is useful to us, as it illustrates some important points of the approach.

The problem can be given an analytical flavour: assign to the control variable u^i the value 1 if the ith object is taken, 0 if not. Then we have to solve

$$\max \sum_{i=1}^n p^i u^i \quad \text{subject to} \quad u^i \in \{0, 1\}, \quad \sum_{i=1}^n v^i u^i \leqslant v.$$

In order to fit with the equality-constrained form (1.1.1), a nonnegative *slack* variable u^0 is appended and we obtain the 0–1 programming problem

$$\left| \begin{array}{l} \max \varphi(u) := \sum_{i=1}^n p^i u^i \quad [+0 \cdot u^0] \\ c(u) := \sum_{i=1}^n v^i u^i + u^0 - v = 0 \\ u^0 \geqslant 0, \quad u^i \in \{0, 1\} \text{ for } i = 1, \ldots, n \quad [\Leftrightarrow u \in U], \end{array} \right. \quad (1.2.3)$$

which, incidentally, certainly has a solution.

1 The Problem and the General Approach

Associated with $\lambda \in \mathbb{R}$, the Lagrange problem is then

$$\sup\left\{\lambda v - \lambda u^0 + \sum_{i=1}^{n}(p^i - \lambda v^i)u^i \,:\, u^0 \geq 0,\ u^i \in \{0, 1\} \text{ for } i = 1, \ldots, n\right\}.$$

Maximization with respect to u^0 is easy, and $L(u_\lambda, \lambda) = +\infty$ for $\lambda < 0$; Assumption 1.1.5(i) is satisfied via a simple trick: impose the (natural) constraint $\lambda \geq 0$, somehow eliminating the absurd negative values of λ. Thus, for any $\lambda \geq 0$, we will have

$$L(u_\lambda, \lambda) = \sum_{i \in I(\lambda)}(p^i - \lambda v^i) + \lambda v, \quad c(u_\lambda) = \sum_{i \in I(\lambda)} v^i - v,$$

where, for example,

$$I(\lambda) = \{i \,:\, p^i - \lambda v^i \geq 0\} \tag{1.2.4}$$

but strict inequality would also define a correct $I(\lambda)$: here lies the ambiguity of u_λ.

Needless to say, the formulation (1.2.3) should not hide the fact that $U = \mathbb{R}^+ \times \{0, 1\}^n$ is still unstructured: one can now define, say $1/2\,(u_1 + u_2)$ for $u_1 \in U$ and $u_2 \in U$; but the half of a toothbrush has little to do with a toothbrush. Another way of saying the same thing is as follows: the constraints $u^i \in \{0, 1\}$ can be formulated as

$$0 \leq u^i \leq 1 \quad \text{and} \quad u^i(1 - u^i) = 0.$$

Then these constraints are hard; by contrast, the constraint $c(u) = 0$ can be considered as soft: if some tolerance on the volume of the knapsack is accepted, the constraint does not have to be strictly satisfied.

(b) The Constrained Lotsizing Problem. Suppose a machine produces objects, for example bottles, "in lots", i.e. at a high rate. Producing one bottle costs p (French Francs, say); but first, the machine must be installed and there is a *setup cost* S to produce one lot. Thus, the cost for producing u bottles in v lots is

$$Sv + pu.$$

When produced, the bottles are sold at a slow rate, say r bottles per unit time. A stock is therefore formed, which costs s per bottle and per unit time. Denoting by $u(t)$ the number of bottles in the stock at time t, the total inventory cost over a period $[t_1, t_2]$ is

$$s \int_{t_1}^{t_2} u(t)\,dt.$$

Suppose that a total amount u must be produced, to be sold at constant rate of r bottles per day. The idea of lot-production is to accept an increase in the setup cost, while reducing the inventory cost: if a total of u bottles must be produced, the production being split in v lots of u/v bottles each, the stock evolves as in Fig. 1.2.1. The total cost, including setup, production and inventory, is then

$$pu + Sv + \tfrac{1}{2}\frac{s}{r}\frac{u^2}{v}. \tag{1.2.5}$$

Between the two extremes (one lot = high inventory vs. many lots = high setup), there is an optimum, the *economic lotsize*, obtained when v minimizes (1.2.5), i.e. when

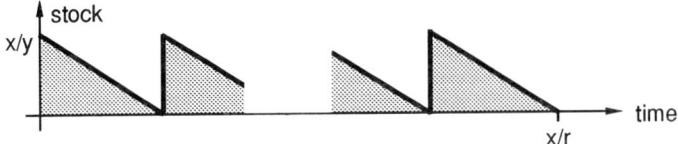

Fig. 1.2.1. Evolution of the stock when producing in lots

$$v = \sqrt{\frac{s}{2rS}} u \, . \tag{1.2.6}$$

Furthermore, production and setup take time, and the total availability of the machine may be limited, say

$$t_p u + t_s v \leqslant T \, .$$

N.B. These times are not taken into account in Fig. 1.2.1 (which assumes $t_p = t_s = 0$), but they do not change the essence of (1.2.5); also, v is allowed non-integer values, which simply means that sales will continue to deplete the stock left at time t_2.

Now suppose that several machines, and possibly several kinds of bottles, are involved, each with its own cost and time characteristics. The problem is schematically formulated as

$$\left|\begin{array}{ll} \min \sum_{i=1}^{n} \sum_{j=1}^{m} \varphi_{ij}(u_{ij}, v_{ij}) & \text{(i)} \\[4pt] \sum_{j=1}^{m} u_{ij} = D_i & \text{for } i = 1, \ldots, n \quad \text{(ii)} \\[4pt] \sum_{i=1}^{n} (a_{ij} u_{ij} + b_{ij} v_{ij}) \leqslant T_j & \text{for } j = 1, \ldots, m \quad \text{(iii)} \\[4pt] u_{ij} \geqslant 0, \; v_{ij} \geqslant 0 & \text{for all } i \text{ and } j \, , \quad \text{(iv)} \end{array}\right. \tag{1.2.7}$$

where the cost is given as in (1.2.5), say

$$\varphi_{ij}(u, v) := \begin{cases} 0 & \text{if } u = v = 0 \, , \\ \alpha_{ij} u + \beta_{ij} v + \gamma_{ij} \dfrac{u^2}{v} & \text{if } u, \; v > 0, \\ +\infty \text{ otherwise} \, . \end{cases}$$

The notation is clear enough; D_i is the total (known) demand in bottles of type i, T_j is the total availability of the machine j; all data are positive. To explain the $+\infty$-value for φ_{ij}, observe that a positive number of units cannot be processed in 0 lots.

Here, the practical Assumption 1.1.2 comes into play because, barring the time-constraints (iii) in (1.2.7), the problem would be very easy. First of all, the variables v_{ij} would become unconstrained; assuming each u_{ij} fixed, an economic lotsize formula like (1.2.6) would give the optimal v_{ij}. The resulting problem in u would then split into independent problems, one for each product i.

More specifically, define the Lagrange function by "dualizing" the constraints (1.2.7)(iii), and optimize the result with respect to v to obtain

1 The Problem and the General Approach   145

$$v_{ij} = v_{ij}(u, \lambda) = \sqrt{\frac{\gamma_{ij}}{\beta_{ij} + \lambda_j b_{ij}}}\, u_{ij}\,. \tag{1.2.8}$$

The ith local control domain is then the simplex

$$U^i := \left\{ u \in \mathbb{R}^m\ :\ u_j \geqslant 0 \text{ for all } j,\ \sum_{j=1}^m u_j = D_i \right\}, \tag{1.2.9}$$

and the ith local problem is a linear program whose constraint-set is U^i:

$$\left| \begin{array}{l} \displaystyle\min \sum_{j=1}^m \left[\alpha_{ij} + \lambda_j a_{ij} + 2\sqrt{\gamma_{ij}(\beta_{ij} + \lambda_j b_{ij})}\right] u_j \\ \displaystyle\sum_{j=1}^m u_j = D_i\,,\quad u_j \geqslant 0 \text{ for } j = 1,\ldots, m\,, \end{array}\right. \tag{1.2.10}_\lambda$$

admitting that u_j stands for u_{ij}, but with i fixed. As was the case with the knapsack problem, the dualized constraints are intrinsic inequalities and the infimum is $-\infty$ if $\lambda_j < 0$ for some j. Otherwise, we have to minimize a linear objective function on the convex hull of the m extreme points of U^i; it suffices to take one best such extreme point, associated with some $j(i)$ giving the smallest of the following m numbers:

$$\alpha_{ij} + \lambda_j a_{ij} + \sqrt{\gamma_{ij}(\beta_{ij} + \lambda_j b_{ij})}\quad\text{for } j = 1,\ldots, m\,. \tag{1.2.11}$$

In summary, the Lagrange problem $(1.2.10)_\lambda$ can be solved as follows: for $i = 1,\ldots, n$, set each u_{ij} to 0, except $u_{i,j(i)} = D_i$. Economically, this amounts to producing each product i on one single machine, namely the cheapest one depending on the given λ. Then the lotsizes $v_{i,j(i)}$ are given by (1.2.8) for each i. The answer from the black box in Fig. 1.1.1 is thus computed in a straightforward way (do not forget to change signs, since one is supposed to maximize, and to add $\sum_j \lambda_j T_j$).

Here, Assumption 1.1.5(i) is again "mildly violated" and dual constraints $\lambda_j \geqslant 0$ will be in order. Uniqueness is again not assured, but again it holds almost everywhere. Despite this absence of uniqueness, the situation is now extremely favourable, Problem 1.1.4 is reasonably easy. The reason is that U has a structure: it is a *convex set*; and besides, the objective function is convex.

(c) Entropy Maximization. The set of control variables is now an infinite-dimensional space, say $U = L_1(\Omega, \mathbb{R})$ where Ω is an interval of $\mathbb{R}$. We want to solve

$$\left|\begin{array}{l} \displaystyle\sup \int_\Omega \psi(x, u(x))\, dx \quad u \in L_1(\Omega, \mathbb{R})\,, \\ \displaystyle\int_\Omega \gamma_j(x, u(x))\, dx = 0 \quad \text{for } j = 1,\ldots, m\,. \end{array}\right. \tag{1.2.12}$$

In addition to Ω, the data of the problem is the $m+1$ functions $\psi, \gamma_1, \ldots, \gamma_m$, which send $(x, t) \in \Omega \times \mathbb{R}$ to $\mathbb{R}$. The Lagrangian is the integral over Ω of the function

$$\ell(x, u, \lambda) := \psi(x, u(x)) - \sum_{j=1}^{m} \lambda_j \gamma_j(x, u(x)).$$

This type of problem appears all the time when some measurements (characterized by the functions γ_j) are made on u; one then seeks the "most probable" u (in the sense of the *entropy* $-\psi$) compatible with these measurements.

In a sense, we are in the decomposable situation (1.2.1), but with "sums of infinitely many terms". Indeed, without going into technical details from functional analysis, suppose that the maximization of ℓ with respect to $u \in L_1(\Omega, \mathbb{R})$ makes sense, and that the standard optimality conditions hold: given $\lambda \in \mathbb{R}^m$, we must solve for each $x \in \Omega$ the equation in $t \in \mathbb{R}$:

$$\frac{\partial \ell}{\partial u}(x, t, \lambda) = \frac{\partial \psi}{\partial u}(x, t) - \sum_{j=1}^{m} \lambda_j \frac{\partial \gamma_j}{\partial u}(x, t) = 0. \tag{1.2.13}$$

Take a solution giving the largest ℓ (if there are several), call $u_\lambda(x)$ the result, and this gives the output from the black box.

The favourable cases are those where the function $t \mapsto \psi(x, t)$ is strictly concave and each function $t \mapsto \gamma_j(x, t)$ is affine. A typical example is one in which ψ is an entropy (usually not depending on x), for example

$$\psi(x, t) = \begin{cases} -t \log t & \text{for } t > 0, \\ 0 & \text{for } t = 0, \\ -\infty & \text{for } t < 0, \end{cases} \tag{1.2.14}$$

or

$$\psi(x, t) = \begin{cases} \log t & \text{if } t > 0 \\ -\infty & \text{otherwise}. \end{cases} \tag{1.2.15}$$

Being affine, the constraints have the general form:

$$\int_\Omega a_j(x) u(x) \, dx - b_j = 0$$

where each $a_j \in L_\infty(\Omega, \mathbb{R})$ is a given function, for example $a_j(x) = \cos(2\pi j x)$ (Fourier constraints).

Here, the practical Assumption 1.1.2 is satisfied, at least in the case (1.2.14): the Lagrange problem (1.2.13) has the explicit solution

$$u_\lambda(x) = \exp\left(-1 - \sum_{j=1}^{m} \lambda_j a_j(x)\right). \tag{1.2.16}$$

By contrast, (1.2.12) has "infinitely many" variables; even with a suitable discretization, it will likely remain a cumbersome problem.

Remark 1.2.1 On the other hand, the technical details already alluded to must not be forgotten. Take for example (1.2.15): the optimality conditions (1.2.13) for u_λ give

$$u_\lambda(x) = \frac{1}{\sum_{j=1}^{m} \lambda_j a_j(x)}. \tag{1.2.17}$$

The set of λ for which this function is nice and log-integrable is now much more complicated than the nonnegative orthant of the two previous examples (a) and (b). Actually, several delicate questions have popped up in this problem: when does the Lagrange problem have a solution? and when do the optimality conditions (1.2.13) hold at such a solution? These questions have little to do with our present concern, and are left unanswered here. □

As long as the entropy ψ is strictly concave and the constraints γ_j are affine (and barring the possible technical difficulties from functional analysis), the dual approach is here very efficient. The reason, now, is that the u_λ of the black box is unambiguous: the Lagrange problem has a well-defined unique solution, which behaves itself when λ varies.

2 The Necessary Theory

Throughout this section, which requires some attention from the reader, our leading motivation is as follows: we want to solve the u-problem (1.1.1); but we have replaced it by the λ-problem 1.1.4, which is vaguely stated. Our first aim is to formulate a more explicit λ-problem (hereafter called the *dual* problem), which will turn out to be very well-posed. Then, we will examine questions regarding its *solvability*, and its *relevance*: to what extent does the dual problem really solve the primal u-problem that we started from?

The data (U, φ, c) will still be viewed as totally unstructured, up to the point where we need to require some specific properties from them; in particular, Assumption 1.1.5 will not be in force, unless otherwise specified.

2.1 Preliminary Results: The Dual Problem

To solve our problem, we must at least find a feasible u in (1.1.1). This turns out to be also sufficient, and the reason lies in what is probably the most important practical link between (1.1.1) and (1.1.4)$_\lambda$:

Theorem 2.1.1 (H. Everett) *Fix $\lambda \in \mathbb{R}^m$; suppose that* (1.1.4)$_\lambda$ *has an optimal solution $u_\lambda \in U$ and set $c_\lambda := c(u_\lambda) \in \mathbb{R}^m$. Then u_λ is also an optimal solution of*

$$\left| \begin{array}{l} \max \varphi(u) \quad u \in U, \\ c(u) = c_\lambda \in \mathbb{R}^m. \end{array} \right. \tag{2.1.1}$$

PROOF. Take an arbitrary $u \in U$. By definition of u_λ,

$$\varphi(u) - \lambda^\top c(u) = L(u, \lambda) \leqslant L(u_\lambda, \lambda) = \varphi(u_\lambda) - \lambda^\top c_\lambda.$$

If, in addition, $c(u) = c_\lambda$, then u is feasible in (2.1.1) and the above relations become $\varphi(u) \leqslant \varphi(u_\lambda)$. □

Thus, once we have solved the Lagrange problem (1.1.4)$_\lambda$ for some particular λ, we have at the same time solved a perturbation of (1.1.1), in which 0 is replaced by an a posteriori right-hand side. Immediate consequence:

148   XII. Abstract Duality for Practitioners

Corollary 2.1.2 *If, for some $\lambda \in \mathbb{R}^m$, (1.1.4)$_\lambda$ happens to have a solution u_λ which is feasible in (1.1.1), then this u_λ is a solution of (1.1.1).* □

To solve Problem 1.1.4, we of course *must* solve the system of equations

$$c_j(u_\lambda) = 0 \quad \text{for } j = 1, \ldots, m, \tag{2.1.2}$$

where u_λ is given by the black box; Corollary 2.1.2 then says that it *suffices* to solve this system. As stated, the problem is not simple: the mapping $\lambda \mapsto c_\lambda$ is not even well-defined since u_λ, as returned by the black box, is ambiguous.

Example 2.1.3 Take the knapsack problem (1.2.4). In view of the great simplicity of the black box, c_λ is easy to compute. For $\lambda \geqslant 0$ (otherwise there is no u_λ and no c_λ), $u^0 = 0$ does not count; with the index-set $I(\lambda)$ defined in (1.2.4), the left-hand side of our (univariate) equation (2.1.2) is obviously

$$c_\lambda = v - \sum_{i \in I(\lambda)} v_i\,.$$

When λ varies, c_λ stays constant except at the branch-points where $I(\lambda)$ changes; then c_λ jumps to another constant value. Said otherwise (see also the text following Remark 1.1.6), the mapping $\lambda \mapsto c_\lambda$, which is ill-defined as it depends on the black box, is *discontinuous*. Such wild behaviour makes it hard to imagine an efficient numerical method. □

Fortunately, another result, just as simple as Theorem 2.1.1, helps a great deal. First of all, we consider a problem more tolerant than (2.1.2):

$$\text{Find } \lambda \text{ such that there is } u \in U \;\left|\; \begin{array}{l} \text{maximizing } L(\cdot, \lambda) \text{ and} \\ \text{satisfying } c(u) = 0\,. \end{array} \right. \tag{2.1.3}$$

This formulation is mathematically more satisfactory: we have now to solve a "set-valued equation" with respect to λ. Note the difference, though: even if λ solves (2.1.3), we still have to find the correct u, while this $u = u_\lambda$ mentioned in (2.1.2) was automatically given by the black box. Equivalence between (2.1.2) and (2.1.3) holds in two cases: if the Lagrange problems (1.1.4)$_\lambda$ have unique solutions, or if the black box can solve "intelligently" (1.1.4)$_\lambda$.

Now we introduce a fundamental notation.

Definition 2.1.4 (Dual Function) In the Lagrange problem (1.1.4)$_\lambda$, the optimal value is called the *dual function*, denoted by Θ:

$$\mathbb{R}^m \ni \lambda \mapsto \Theta(\lambda) := \sup\{L(u, \lambda) \,:\, u \in U\}\,. \tag{2.1.4}$$

□

Note that Θ is just the opposite (cf. Remark 1.1.1) of the function ψ of §VII.4.5; it is a direct and unambiguous output of the black box. If Assumption 1.1.5 holds, then $\Theta(\lambda) = L(u_\lambda, \lambda)$ but existence of an optimal solution is by no means necessary for the definition of Θ. Indeed, the content of the practical Assumption 1.1.2 is just that $\Theta(\lambda)$ is easy to compute for given λ, even if an optimal solution u_λ is not easy to obtain, or even if none exists.

Theorem 2.1.5 (Weak Duality) *For all $\lambda \in \mathbb{R}^m$ and all u feasible in* (1.1.1), *there holds*
$$\Theta(\lambda) \geqslant \varphi(u). \tag{2.1.5}$$

PROOF. For any u feasible in (1.1.1), we have
$$\varphi(u) = \varphi(u) - \lambda^\top c(u) = L(u, \lambda) \leqslant \Theta(\lambda). \qquad \square$$

Thus, each value of the dual function gives an upper bound on the primal optimal value; a sensible idea is then to find the best such upper bound:

Definition 2.1.6 (Dual Problem) The optimization problem in λ
$$\inf \{\Theta(\lambda) : \lambda \in \mathbb{R}^m\} \tag{2.1.6}$$
is called the *dual problem* associated with (U, φ, c) of (1.1.1). $\qquad \square$

The following crucial result assesses the importance of this dual problem.

Theorem 2.1.7 *If $\bar{\lambda}$ is such that the associated Lagrange problem* $(1.1.4)_{\bar{\lambda}}$ *is maximized at a feasible $\bar{u}$, i.e. if $\bar{\lambda}$ solves* (2.1.3), *then $\bar{\lambda}$ is also a solution of the dual problem* (2.1.6).

Conversely, if $\bar{\lambda}$ solves (2.1.6), *and if* (2.1.3) *has a solution at all, then $\bar{\lambda}$ is such a solution.*

PROOF. The property $c(\bar{u}) = 0$ means
$$\Theta(\bar{\lambda}) = L(\bar{u}, \bar{\lambda}) = \varphi(\bar{u}),$$
which shows that Θ assumes the least of its values allowed by inequality (2.1.5).

Conversely, let (2.1.3) have a solution λ^*: some $u^* \in U$ satisfies $c(u^*) = 0$ and $L(u^*, \lambda^*) = \Theta(\lambda^*) = \varphi(u^*)$. Again from (2.1.5), $\Theta(\lambda^*)$ has to be the minimal value $\Theta(\bar{\lambda})$ of the dual function and we write
$$L(u^*, \bar{\lambda}) = \varphi(u^*) - \bar{\lambda}^\top c(u^*) = \varphi(u^*) = \Theta(\bar{\lambda}).$$
In other words, our feasible u^* maximizes $L(\cdot, \bar{\lambda})$. $\qquad \square$

Existence of a solution to (2.1.3) means full success of the dual approach, in the following sense:

Corollary 2.1.8 *Suppose* (2.1.3) *has a solution. Then, for any solution λ of the dual problem* (2.1.6), *the primal solutions are those u maximizing $L(\cdot, \lambda)$ that are feasible in* (1.1.1).

PROOF. When (2.1.3) has a solution, we already know that λ minimizes Θ if and only if: λ solves (2.1.3), and there is some u such that $\varphi(u) = L(u, \lambda) = \Theta(\lambda)$.

Then, for any $\bar{u}$ solving (1.1.1), we have $c(\bar{u}) = 0$ and $\varphi(\bar{u}) = \varphi(u)$, hence $L(\bar{u}, \lambda) = \Theta(\lambda)$. $\qquad \square$

In the above two statements, never forget the property "(2.1.3) has a solution", which need not hold in general. We will see in §2.3 what it implies, in terms of the data (U, φ, c).

Remark 2.1.9 When (2.1.3) has no solution, at least a *heuristic resolution* of (1.1.1) can be considered, inspired by Everett's Theorem 2.1.1 and the weak duality Theorem 2.1.5.

Apply an iterative algorithm to minimize Θ and, during the course of this algorithm, store the primal points computed by the black box. After the dual problem is solved, select among these primal points one giving a best compromise between its φ-value and constraint-violation.

This heuristic heavily depends on Assumption 1.1.3. Its quality can be judged a posteriori from the infimal value of Θ; but nothing can be guaranteed in advance: φ may remain far from Θ, or the constraints may doggedly refuse to approach 0. □

2.2 First Properties of the Dual Problem

Let us summarize our development so far: starting from the vaguely stated problem 1.1.4, we have introduced a number of formulations, interconnected as follows:

$$\boxed{\begin{array}{l} \lambda \text{ solves } 1.1.4 \iff \lambda \text{ solves } (2.1.2) \implies \\ \implies \lambda \text{ solves } (2.1.3) \implies \lambda \text{ solves } (2.1.6). \end{array}} \qquad (2.2.1)$$

Thus, to solve the ugly equation (2.1.2), we *must* solve the perfectly well-stated problem (2.1.6). Furthermore, we have *nothing to lose*:

– If no primal solution is thus produced, the task was hopeless from the very beginning: no other value of λ could give a primal solution; this is the converse part in Theorem 2.1.7.
– If the technique works, no primal solution can be missed: they all solve the Lagrange problem associated with any dual optimum; this is Corollary 2.1.8.

Remark 2.2.1 From now on, we will use the word "dual" for everything concerning (2.1.6): we have to minimize the dual function Θ, with respect to the dual variable λ (u being the primal variable), possibly via a dual algorithm, yielding a dual solution, and so on.

The symmetry between (1.1.1) and (2.1.6) becomes more suggestive if the primal constraints are incorporated into the objective function: (1.1.1) can be formulated as

$$\sup \{\varphi(u) - I_C(u) : u \in U\},$$

where C is the domain described by $c(u) = 0$, I_C being its indicator function. □

The applications described in §1.2 give good illustrations of the logical chain (2.2.1).

(a) The knapsack problem is a typical example where (2.1.6) has little to do with (2.1.2), or even (2.1.3). Take the (particularly simple!) case $n = 1$; for example

$$\max u \quad \text{subject to} \quad 2u \leqslant 1, \ u \in \{0, 1\}. \qquad (2.2.2)$$

The dual function is a nice polyhedral convex function (cf. Example 2.1.3 if necessary):

2 The Necessary Theory

$$\mathbb{R} \ni \lambda \mapsto \Theta(\lambda) = \begin{cases} +\infty & \text{if } \lambda < 0, \\ 1 - \lambda & \text{if } 0 \leqslant \lambda \leqslant 1/2, \\ \lambda & \text{if } \lambda \geqslant 1/2; \end{cases}$$

by contrast, the left-hand side in (2.1.2) is the discontinuous mapping

$$c_\lambda = \begin{cases} 1 & \text{if } 0 \leqslant \lambda < 1/2, \\ \text{ambiguously } \pm 1 & \text{if } \lambda = 1/2, \\ -1 & \text{if } \lambda > 1/2. \end{cases}$$

For (2.1.3), $\lambda = 1/2$ gives the doubleton $\{-1, +1\}$.

Here, none of the converse inclusions hold in 2.2.1; indeed (2.1.3) – and a fortiori (2.1.2) – has no solution.

(b) The case of economic lotsize is more involved but similar: c_λ jumps when λ is such that the optimal j switches in (1.2.11) for some i. A discontinuity in c_λ has an interesting economic meaning, with λ_j viewed as the *marginal price* to be paid when the jth machine is occupied. For most values of λ, it is economically cheaper to produce the ith product entirely on one single machine $j(i)$. At some "critical" values of λ, however, two machines at least, say $j_1(i)$ and $j_2(i)$, are equally cheap for some product i; then a splitting of the production between these machines can be considered, without spoiling optimality. The u_i-part of a solution $(u, v)_\lambda$ becomes a subsimplex of (1.2.9); and the corresponding v_i-part is also a simplex, as described by (1.2.8).

Here we will see that (2.1.2) $\neq$ 2.1.3 $\Rightarrow$ (2.1.6).

(c) As for the entropy-problem, (2.1.2) and (2.1.6) have similar complexity. Take for example the entropy (1.2.14): from (1.2.16), straightforward calculations give Θ:

$$\mathbb{R}^m \ni \lambda \mapsto \Theta(\lambda) = b^\top \lambda + \int_\Omega \exp\left(-1 - \sum_{j=1}^m \lambda_j a_j(x)\right) dx,$$

$$c_\ell(u_\lambda) = \int_\Omega a_\ell(x) \exp\left(-1 - \sum_{j=1}^m \lambda_j a_j(x)\right) dx - b_\ell \quad \text{for } \ell = 1, \ldots, m.$$

A good exercise is to do the same calculations with the entropy (1.2.15) (without bothering too much with Remark 1.2.1); observe in passing that, in both cases, Θ is convex and $-c = \nabla \Theta$.

Here full equivalence holds in (2.2.1).

In all three examples including (c), the dual problem will be more advantageously viewed as minimizing Θ than solving (2.1.2) or (2.1.3). See the considerations at the end of §II.1: having a "potential" Θ to minimize makes it possible to stabilize the resolution algorithm, whatever it is. Indeed, to some extent, the dual problem (2.1.6) is well posed:

Proposition 2.2.2 *If not identically $+\infty$, the dual function Θ is in $\overline{\text{Conv}}\,\mathbb{R}^m$. Furthermore, for any u_λ solution of the Lagrange problem $(1.1.4)_\lambda$, the corresponding $-c_\lambda = -c(u_\lambda)$ is a subgradient of Θ at λ.*

152   XII. Abstract Duality for Practitioners

PROOF. Direct proofs are straightforward; for example, write for each λ and μ:

$$\begin{aligned}
\Theta(\mu) &\geqslant \varphi(u_\lambda) - \mu^\top c_\lambda & &\text{[by definition of } \Theta(\mu)\text{]} \\
&= \varphi(u_\lambda) - \lambda^\top c_\lambda + (\lambda - \mu)^\top c_\lambda \\
&= \Theta(\lambda) + (\mu - \lambda)^\top(-c_\lambda), & &\text{[by definition of } u_\lambda\text{]}
\end{aligned}$$

so $-c_\lambda \in \partial\Theta(\lambda)$. The other claims can be proved similarly. Proposition IV.2.1.2 and Lemma VI.4.4.1 can also be invoked: the dual function

$$\mathbb{R}^m \ni \lambda \mapsto \Theta(\lambda) = \sup_{u \in U} \left[\varphi(u) - \lambda^\top c(u)\right]$$

is the pointwise supremum of affine functions indexed by u.   □

Note that the case $\Theta \equiv +\infty$ is quite possible: take with $U = \mathbb{R}$, $\varphi(u) = u^2$, $c(u) = u$. Then, no matter how λ is chosen, the Lagrange function $L(u, \lambda) = u^2 - \lambda u$ goes to $+\infty$ with $|u|$. Note also that Θ has no chance to be differentiable, except if the Lagrange problem has a unique solution.

Thus the dual problem enjoys several important properties:

– It makes sense to minimize on $\mathbb{R}^m$ a *lower semi-continuous* function such as Θ.
– The set of minimizers of a *convex* function such as Θ is well delineated: there is no ambiguity between local and global minima, and any suitably designed minimization algorithm will converge to such a minimum (if any). Furthermore, the weak duality Theorem 2.1.5 tells us that Θ is *bounded from below*, unless there is no feasible u (in which case the primal problem does not make sense anyway).
– Another consequence, of practical value, is that our dual problem (2.1.6) fits within the framework of Chap. IX, at least in the situation described by Fig. 1.1.1 (which in particular implies Assumption 1.1.5): the only available information about our objective function Θ is a black box which, for given λ, computes the value $\Theta(\lambda)$ and *some subgradient* which we are entitled to call $s(\lambda) := -c_\lambda = -c(u_\lambda)$.

The property $-c_\lambda \in \partial\Theta(\lambda)$ is fundamental for applications. Note that $-c_\lambda$ is just the coefficient of λ in the Lagrange function, and remember Remark VI.4.4.6 explaining why this partial derivative of L becomes a total derivative of Θ.

Remark 2.2.3 Once again, it is important to understand what is lost by the dual problem. Altogether, a proper resolution of our Problem 1.1.4 entails two steps:

(i) First solve (2.1.6), i.e. use the one-way implications in (2.2.1); this must be done anyway, whether (2.1.6) is equivalent to (2.1.2) or not.
(ii) Then take care of the lost converse implications in (2.2.1) to recover a primal solution from a dual one.

In view of Proposition 2.2.2, (i) is an easy task; a first glance at how it might be solved was given in Chap. IX, and we will see in §4 below some other traditional algorithms. The status of (ii) is quite different and we will see in §2.3 that there are three cases:

(ii$_1$) (2.1.2) has no solution – and hence, (i) was actually of little use; example: knapsack.

(ii$_2$) A primal solution can be found after (i) is done, but this requires some extra work because (2.1.3) $\Leftrightarrow$ (2.1.6) but (2.1.3) $\not\Rightarrow$ (2.1.2); example: economic lotsize.
(ii$_3$) A primal solution is readily obtained from the black box after (i) is done, (2.1.6) $\Leftrightarrow$ (2.1.2); example: entropy (barring difficulties from functional analysis). □

The dual approach amounts to exploring a certain primal set *parameterized* by $\lambda \in \mathbb{R}^m$, namely

$$\widetilde{U} := \{u \in U : u \text{ solves the Lagrange problem } (1.1.4)_\lambda \text{ for some } \lambda\}. \quad (2.2.3)$$

Usually, $\widetilde{U}$ is properly contained in U: for example, the functions (1.2.16) and (1.2.17) certainly describe a very small part of $L_1(\Omega, \mathbb{R})$ when λ describes $\mathbb{R}^m$ – and the interest of duality precisely lies there. If (and only if) $\widetilde{U}$ contains a feasible point, the dual approach will produce a primal solution; otherwise, it will break down. How much it breaks down depends on the problem. Consider the knapsack example (2.2.2): it has a unique dual solution $\bar{\lambda} = 1/2$, at which the dual optimal value provides the upper bound $\Theta(1/2) = 1/2$; but the primal optimal value is 0. At $\bar{\lambda} = 1/2$, the Lagrange problem has two (slackened) solutions:

$$(u, u^0) = (0, 0) \quad \text{and} \quad (u, u^0) = (1, 0),$$

with constraint-values -1 and 1: none of them is feasible, of course. On the other hand, the non-slackened solution $u = 0$ is feasible with respect to the *inequality* constraint.

This last example illustrates the following important concept:

Definition 2.2.4 (Duality Gap) The difference between the optimal primal and dual values

$$\inf_{\lambda \in \mathbb{R}^m} \Theta(\lambda) - \sup_{u \in U} \{\varphi(u) : c(u) = 0\}, \quad (2.2.4)$$

when it is not zero, is called the *duality gap*. □

By now, the following facts should be clear:

– The number (2.2.4) is always nonnegative (weak duality Theorem 2.1.5).
– The presence of a duality gap definitely implies that $\widetilde{U}$ contains no feasible point: failure of the dual approach.
– The absence of a duality gap, i.e. having 0 in (2.2.4), is not quite sufficient; to be on the safe side, each of the two extremization problems in (2.2.4) must have a solution.
– In this latter case, we are done: Corollary 2.1.8 applies.

To close this section, we emphasize once more the fact that all the results so far are valid without any specific assumption on the primal data (U, φ, c). Such assumptions become relevant only for questions like: does the primal problem have a solution? and how can it be recovered from a dual solution? Does the dual problem have a solution? These questions are going to be addressed now.

2.3 Primal-Dual Optimality Characterizations

Suppose the dual problem is solved: some $\lambda \in \mathbb{R}^m$ has been found, minimizing the dual function Θ on the whole space; a solution of the primal problem (1.1.1) is still to be found. It is now that some assumptions must be made – if only to make sure that (1.1.1) has a solution at all!

A dual solution λ is characterized by $0 \in \partial\Theta(\lambda)$; now the subdifferentials of Θ lie in the dual of the λ-space, i.e. in the space of constraint-values. Indeed, consider the (possibly empty) optimal set in the Lagrange problem $(1.1.4)_\lambda$:

$$U(\lambda) := \{u \in U : L(u, \lambda) = \Theta(\lambda)\}; \tag{2.3.1}$$

remembering Proposition 2.2.2,

$$\partial\Theta(\lambda) \supset \operatorname{co}\{-c(u) : u \in U(\lambda)\}.$$

The converse inclusion is therefore crucial, and motivates the following definition.

Definition 2.3.1 (Filling Property) With $U(\lambda)$ defined in (2.3.1), we say that the *filling property* holds at $\lambda \in \mathbb{R}^m$ when

$$\partial\Theta(\lambda) = -\operatorname{co}\{c(u) : u \in U(\lambda)\}. \tag{2.3.2}$$

Observe that the right-hand side in (2.3.2) is then closed, since the left-hand side is.
□

Lemma 2.3.2 *Suppose that U in (1.1.1) is a compact set, on which φ is upper semi-continuous, and each c_j is continuous. Then the filling property (2.3.2) holds at each $\lambda \in \mathbb{R}^m$.*

PROOF. Under the stated assumptions, the Lagrange function $L(\cdot, \lambda)$ is upper semi-continuous and has a maximum for each λ: $\operatorname{dom}\Theta = \mathbb{R}^m$. For the supremum of affine functions $\lambda \mapsto \varphi(u) - \lambda^\top c(u)$, indexed by $u \in U$, the calculus rule VI.4.4.4 applies.
□

We have therefore a sufficient condition for an easy description of $\partial\Theta$ in terms of primal points. It is rather "normal", in that it goes in harmony with two other important properties: existence of an optimal solution in the primal problem (1.1.1), and in the Lagrange problems (1.1.4).

Remark 2.3.3 In case of inequality constraints, nonnegative slack variables can be appended to the primal problem, as in §1.2(a). Then, the presence of the (unbounded) nonnegative orthant in the control space kills the compactness property required by the above result. However, we will see in §3.2 that this is a "mild" unboundedness. Just remember here that the calculus rule VI.4.4.4 is still valid when λ has all its coordinates positive. Lemma 2.3.2 applies in this case, to the extent that $\lambda \in \operatorname{int} \operatorname{dom}\Theta$. The slacks play no role: for $\mu \in (\mathbb{R}^+)^m$ (close enough to λ), maximizing the slackened Lagrangian

$$\varphi(u) - \mu^\top c(u) + \mu^\top v$$

with respect to $(u, v) \in U \times (\mathbb{R}^+)^m$ just amounts to maximizing the ordinary Lagrangian $L(\cdot, \mu)$ over U.
□

The essential result concerning primal-dual relationships can now be stated.

Theorem 2.3.4 *Let the filling property* (2.3.2) *hold (for example, make the assumptions of Lemma 2.3.2) and denote by*

$$C(\lambda) := \{c(u) \in \mathbb{R}^m : u \in U(\lambda)\} \quad (2.3.3)$$

the image by c of the set (2.3.1). *A dual optimum $\bar{\lambda}$ is characterized by the existence of $k \leq m+1$ points $u_1, \ldots, u_k$ in $U(\bar{\lambda})$ and convex multipliers $\alpha_1, \ldots, \alpha_k$ such that*

$$\sum_{i=1}^{k} \alpha_i c(u_i) = 0.$$

In particular, if $C(\bar{\lambda})$ is convex for some optimal $\bar{\lambda}$ then, for any optimal λ^*, the feasible points in $U(\lambda^*)$ make up all the solutions of the primal problem (1.1.1).

PROOF. When the filling property holds, the minimality condition $0 \in \partial \Theta(\bar{\lambda})$ is exactly the existence of $\{u_i, \alpha_i\}$ as stated, i.e. $0 \in \text{co}\, C(\bar{\lambda})$. If $C(\bar{\lambda})$ is convex, the "co"-operation is useless, 0 is already the constraint-value of some u maximizing $L(\cdot, \bar{\lambda})$. The rest follows from Corollary 2.1.8. □

This is the first instance where *convexity* comes into play, to guarantee that the set $\widetilde{U}$ of (2.2.3) contains a feasible point. Convexity is therefore crucial to rule out a duality gap.

Remark 2.3.5 Once again, a good illustration is obtained from the simple knapsack problem (2.2.2). At the unique dual optimum $\bar{\lambda} = 1/2$, $C(\bar{\lambda}) = \{-1, +1\}$ is not convex and there is a duality gap equal to $1/2$. Nevertheless, the corresponding (slackened) set of solutions to the Lagrange problem is $U(1/2) = \{(0, 0), (1, 0)\}$ and the convex multipliers $\alpha_1 = \alpha_2 = 1/2$ do make up the point $(1/2, 0)$, which satisfies the constraint. Unfortunately, this convex combination is not in the (slackened) control space $U = \{0, 1\} \times \mathbb{R}^+$.

Remember the observation made in Remark VIII.2.1.3: even though the kinks of a convex function (such as Θ) form a set of measure zero, a minimum point is usually in this set. Here $\partial\Theta(1/2) = [-1, 1]$ and the observation is confirmed: the minimum $\bar{\lambda} = 1/2$ is the only kink of Θ. □

In practice, the convexity property needed for Theorem 2.3.4 implies that the original problem itself has the required convex structure: roughly speaking, the only cases in which there is no duality gap are those described by the following result.

Corollary 2.3.6 *Suppose that the filling property* (2.3.2) *holds. In either of the following situations* (i), (ii) *below, there is no duality gap; for every dual solution λ^* (if any), the feasible points in $U(\lambda^*)$ make up all the solutions of the primal problem* (1.1.1).

  (i) *For some dual solution $\bar{\lambda}$, the associated Lagrange function $L(\cdot, \bar{\lambda})$ is maximized at a unique $\bar{u}$; then, $\bar{u}$ is the unique solution of* (1.1.1).
  (ii) *In* (1.1.1), *U is convex, φ is concave and $c : U \to \mathbb{R}^m$ is affine.*

PROOF. Straightforward. □

Case (i) means that Θ is actually *differentiable* at $\bar{\lambda}$; we are in the situation (ii$_3$) of Remark 2.2.3. Differentiability of the dual function thus appears as an important property, not only for convenience when minimizing it, but also for a harmonious primal-dual relationship. The entropy problem of §1.2(c) fully enters this framework, at least with the entropy (1.2.14). The calculations in §2.2(c) confirm that the dual function is then differentiable and its minimization amounts to solving $\nabla\Theta(\lambda) = -c(u_\lambda) = 0$. For this example, if Lemma 2.3.2 applies (cf. difficulties from functional analysis), the black box automatically produces the unique primal solution at any optimal λ (if any).

The knapsack problem gives an illustration *a contrario*: U is not convex and Θ is, as a rule, not differentiable at a minimum (see §2.2(a) for example).

These two situations are rather clear, the intermediate case 2.2.3(ii$_2$) is more delicate. At a dual solution $\bar{\lambda}$, the Lagrange problem has several solutions; some of them solve (1.1.1), but they may not be produced by the black box. A *constructive* character must still be given to Corollary 2.3.6(ii).

Example 2.3.7 Take the lotsizing problem of §1.2(b), and remember that we have only considered a simple black box: for any λ, each product i is assigned entirely to one single machine $j(i)$, even in the ambiguous cases where some product has the same cost for two different machines. For example, the black box is totally unable to yield an allocation in which some product is split between several machines; and it may well be that only such an allocation is optimal in (1.2.7); actually, since Θ is piecewise affine, there is almost certainly such an ambiguity at any optimal $\bar{\lambda}$; remember Remark 2.3.5.

On the other hand, imagine a sophisticated black box, computing the full optimal set in the Lagrange problem (1.2.10)$_\lambda$. This amounts to determining, for each i, all the optimal indices j in (1.2.11); let there be m_i such indices. Each of them defines an extreme point in the simplex (1.2.9); piecing them together, they make up the full solution-set of (1.2.10)$_\lambda$, which is the convex hull of $p = m_1 \times \cdots \times m_n$ points in $U^1 \times \cdots \times U^n$; let us call them $u^{(1)}, \ldots, u^{(p)}$. Now comes the fundamental property: to each $u^{(k)}$ corresponds via (1.2.8) a point $v^{(k)} := v(u^{(k)}, \lambda) \in \mathbb{R}^{n\times m}$; and because $u \mapsto v(u, \lambda)$ in (1.2.8) is *linear*, the solution-set of (1.2.10)$_\lambda$ is the convex hull of the points $(u^{(1)}, v^{(1)}), \ldots, (u^{(p)}, v^{(p)})$ thus obtained.

Assume for simplicity that λ is optimal and has all its coordinates positive. From Theorem 2.3.4 or Corollary 2.3.6, there is a convex combination of the $(u^{(k)}, v^{(k)})$, $k = 1, \ldots, p$ (hence a solution of the Lagrange problem) which satisfies the constraints (1.2.7)(iii) as equalities: this is a primal optimum. □

Remark 2.3.8 It is interesting to interpret the null-step mechanism of Chap. IX, in our present duality framework. Use this mechanism to find a descent direction of the present function Θ, but start it from a dual optimum, say $\bar{\lambda}$. To make things simple, just consider the knapsack problem (2.2.2).

– At the optimal $\bar{\lambda} = 1/2$, the index-set (1.2.4) is $I(1/2) = \{1\}$, thus $s_1 = -1$; the black box suggests that the object is worth taking and that, to decrease Θ, λ must be increased: $1/2$ is an insufficient "price of space".

– For all $\lambda > 1/2$, $I(\lambda) = \emptyset$; the black box now suggests to leave the object and to decrease λ. During the first line-search in Algorithm IX.1.6, no improvement is obtained from Θ and $s_2 = 1 \in \partial\Theta(1/2)$ is eventually produced.

– At this stage, the direction-finding problem (IX.1.8) detects optimality of $\bar{\lambda}$; but more importantly, it produces $\alpha = 1/2$ for the convex combination

$$\alpha s_1 + (1-\alpha)s_2 = 0.$$

Calling $u^{(1)}$ ($= 1$) and $u^{(2)}$ ($= 0$) the decisions corresponding to s_1 and s_2, the convex combination $1/2\,[u^{(1)} + u^{(2)}]$ is the one revealed by Theorem 2.3.4.

Some important points are raised by this demonstration.

(i) In the favourable cases (no duality gap), constructing the zero vector in $\partial\Theta(\lambda)$ corresponds to constructing an optimal solution in the primal problem.

(ii) Seen with "dual glasses", a problem with a duality gap behaves just as if Corollary 2.3.6(ii) applied. In the above knapsack problem, $u = 1/2$ is computed by the dual algorithm, even though a half of a TV set is useless. This $u = 1/2$ would become a primal optimum if $U = \{0, 1\}$ in (2.2.2) were replaced by its *convex relaxation* $[0, 1]$; but the dual algorithm would see no difference.

(iii) The bundling mechanism is able to play the role of the sophisticated black box wanted in Example 2.3.7. For this, an access is needed to the primal points computed by a simple black box, in addition to their constraint-values.

(iv) Remember Example IX.3.1.2: it can be hoped that the bundling mechanism generates a primal optimum with only a few calls to the simple black box, when it is started on an optimal $\bar{\lambda}$; and it should do this independently of the complexity of the set $U(\bar{\lambda})$. Suppose we had illustrated (iii) with the lotsizing problem of Example 2.3.7, rather than a trivial knapsack problem. We would have obtained $0 \in \partial\Theta(\lambda)$ with probably an order of m iterations; by contrast, the $m_1 \times \cdots \times m_n$ primal candidates furnished by a sophisticated black box should all be collected for a direct primal construction. □

We conclude with a comment concerning the two assumptions needed for Theorem 2.3.4:

– The behaviour of the dual function must be fully described by the optimal solutions of the Lagrange problem. This is the filling property (2.3.2), of topological nature, and fairly hard to establish (see §VI.4.4). Let us add that situations in which it does not hold can be considered as pathological.

– Some algebraic structure must exist in the solution-set of the Lagrange problem, at least at an optimal λ (Corollary 2.3.6). It can be considered as a "gift of God": it has no reason to exist a priori but when it does, everything becomes straightforward.

2.4 Existence of Dual Solutions

The question is now whether the dual problem (2.1.6) has a solution $\bar{\lambda} \in \mathbb{R}^m$. This implies first that the dual function Θ is bounded from below; furthermore, it must attain its infimum. For our study, the relevant object is the image

$$C(U) := \{\gamma \in \mathbb{R}^m \,:\, \gamma = c(u) \text{ for some } u \in U\} \qquad (2.4.1)$$

of the control set U under the constraint mapping $c : U \to \mathbb{R}^m$. Needless to say, an essential prerequisite for the primal problem (1.1.1) to make sense is $0 \in C(U)$; as far as the dual problem is concerned, however, cases in which $0 \notin C(U)$ are also of interest.

First of all, denote by Γ the affine hull of $C(U)$:

$$\Gamma := \left\{ \gamma \in \mathbb{R}^m : \gamma = \sum_{i=1}^k \alpha_i c(u_i),\ u_i \in U,\ \sum_{i=1}^k \alpha_i = 1 \text{ for } k = 1, 2, \ldots \right\},$$

and by Γ_0 the subspace parallel to Γ. Fixing $u_0 \in U$, we have $c(u) - c(u_0) \in \Gamma_0$ for all $u \in U$, hence

$$L(u, \lambda + \mu) = L(u, \lambda) - \mu^\top c(u) = L(u, \lambda) - \mu^\top c(u_0) \quad \text{for all } (\lambda, \mu) \in \mathbb{R}^m \times \Gamma_0^\perp$$

and this equality is inherited by the supremal values:

$$\Theta(\lambda + \mu) = \Theta(\lambda) - \mu^\top c(u_0) \quad \text{for all } \lambda \in \mathbb{R}^m \text{ and } \mu \in \Gamma_0^\perp. \tag{2.4.2}$$

In other words, Θ is affine in the subspace $\Gamma_0^\perp$. This observation clarifies two cases:

– If $0 \notin \Gamma$, let $\mu_0 \neq 0$ be the projection of the origin onto Γ; in (2.4.2), fix λ and take $\mu = t\mu_0$ with $t \to +\infty$. Because $c(u_0) \in \Gamma$, we have with (2.4.2)

$$\Theta(\lambda + t\mu_0) = \Theta(\lambda) - t\mu_0^\top c(u_0) = \Theta(\lambda) - t\|\mu_0\|^2 \to -\infty.$$

In this case, the primal problem cannot have a feasible point (weak duality Theorem 2.1.5); to become meaningful, (1.1.1) should have its constraints perturbed, say to $c(u) = \mu_0$.

– The only interesting case is therefore $0 \in \Gamma$. The dual optimal set is then $\Lambda + \Gamma_0^\perp$, where $\Lambda \subset \Gamma_0$ is the (possibly empty) optimal set of

$$\inf \{\Theta(\lambda) : \lambda \in \Gamma_0 = \operatorname{aff} C(U) = \operatorname{lin} C(U)\}. \tag{2.4.3}$$

In a way, (2.4.3) is the relevant dual problem to consider – admitting that Γ_0 is known. Alternatively, the essence of the dual problem would remain unchanged if we assumed $\Gamma = \Gamma_0 = \mathbb{R}^m$.

In §2.3, some convexity structure came into play for solving the primal problem by duality; the same phenomenon occurs here. The next result contains the essential conditions for existence of a dual solution.

Proposition 2.4.1 *Assume $\Theta \not\equiv +\infty$. With the definition (2.4.1) of $C(U)$,*

(i) *if $0 \notin \overline{\operatorname{co}}\, C(U)$, then $\inf_\lambda \Theta(\lambda) = -\infty$;*
(ii) *if $0 \in \operatorname{co} C(U)$, then $\inf_\lambda \Theta(\lambda) > -\infty$;*
(iii) *if $0 \in \operatorname{ri} \operatorname{co} C(U)$, then the dual problem has a solution.*

PROOF. [(i)] The closed convex set $\overline{\operatorname{co}}\, C(U)$ is separated from $\{0\}$ (Theorem III.4.1.3): for some $\mu_0 \neq 0$ and $\delta > 0$,

$$-\mu_0^\top c(u) \leqslant -\delta < 0 \quad \text{for all } u \in U.$$

Now let λ_0 be such that $\Theta(\lambda_0) < +\infty$, and write for all $u \in U$:

$$\begin{aligned}
L(u, \lambda_0 + t\mu_0) &= \varphi(u) - \lambda_0^\mathsf{T} c(u) - t\mu_0^\mathsf{T} c(u) &&\text{[definition of } L\text{]}\\
&\leqslant \Theta(\lambda_0) - t\mu_0^\mathsf{T} c(u) &&\text{[definition of } \Theta\text{]}\\
&\leqslant \Theta(\lambda_0) - t\delta. &&\text{[definition of } \mu_0\text{]}
\end{aligned}$$

Since u was arbitrary, this implies $\Theta(\lambda_0 + t\mu_0) \leqslant \Theta(\lambda_0) - t\delta$, and the latter term tends to $-\infty$ when $t \to +\infty$.

[(ii)] There are finitely many points in U, say $u_1, \ldots, u_p$, and some α in the unit simplex of $\mathbb{R}^p$, such that $0 = \sum_{i=1}^p \alpha_i c(u_i)$. Then write for all $\lambda \in \mathbb{R}^m$:

$$\Theta(\lambda) \geqslant \varphi(u_i) - \lambda^\mathsf{T} c(u_i) \quad \text{for } i = 1, \ldots, p,$$

and obtain by convex combination

$$\Theta(\lambda) \geqslant \sum_{i=1}^p \alpha_i \varphi(u_i) - \lambda^\mathsf{T} \sum_{i=1}^p \alpha_i c(u_i) \geqslant \min_{1 \leqslant i \leqslant p} \varphi(u_i),$$

where the second inequality holds by construction of $\{\alpha_i, u_i\}$.

[(iii)] We know from Theorem III.2.1.3 (see Fig. III.2.1.1) that, if $0 \in \operatorname{ri co} C(U)$, then 0 is also in the relative interior of some simplex contained in $\operatorname{co} C(U)$. Using the notation of (2.4.3), this means that there are finitely many points in U, say $u_1, \ldots, u_p$, and $\delta > 0$ such that

$$B(0, \delta) \cap \Gamma_0 \subset \operatorname{co} \{c(u_1), \ldots, c(u_p)\}.$$

For all $\lambda \in \mathbb{R}^m$, still by definition of $\Theta(\lambda)$,

$$\Theta(\lambda) \geqslant \varphi(u_i) - \lambda^\mathsf{T} c(u_i) \quad \text{for } i = 1, \ldots, p. \tag{2.4.4}$$

If $\Gamma_0 = \{0\}$, i.e. $C(U) = \{0\}$, $L(u, \lambda) = \varphi(u)$ for all $u \in U$, so Θ is a constant function; and this constant is not $+\infty$ by assumption. If $\Gamma_0 \neq \{0\}$, take $0 \neq \lambda \in \Gamma_0$ in (2.4.4) and

$$\gamma := -\delta \frac{\lambda}{\|\lambda\|} \in B(0, \delta) \cap \Gamma_0;$$

this γ is therefore a convex combination of the $c(u_i)$'s. The same convex combination in (2.4.4) gives

$$\Theta(\lambda) \geqslant \sum_{i=1}^p \alpha_i \varphi(u_i) + \delta \|\lambda\| \geqslant \min_{1 \leqslant i \leqslant p} \varphi(u_i) + \delta \|\lambda\|.$$

We conclude that $\Theta(\lambda) \to +\infty$ if λ grows unboundedly in Γ_0: the closed function Θ (Proposition 2.2.2) does have a minimum on Γ_0, and on $\mathbb{R}^m$ as well, in view of (2.4.3). □

Figure 2.4.1 summarizes the different situations revealed by our above analysis. Thus, it is the *convex hull* of the image-set $C(U)$ which is relevant; and, just as in §2.3, this convexification is crucial:

160   XII. Abstract Duality for Practitioners

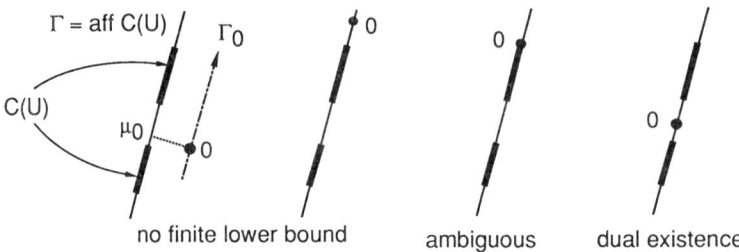

Fig. 2.4.1. Various situations for dual existence

Example 2.4.2 In view of the weak duality Theorem 2.1.5, existence of a feasible u in (1.1.1) implies boundedness of Θ from below. To confirm the prediction of (ii), that this is not necessary, take a variant of the knapsack problem (2.2.2), in which the knapsack should be completely filled: we want to solve

$$\sup u, \quad 2u = 1, \quad u \in \{0, 1\}.$$

This problem clearly has no feasible point: its supremal value is $-\infty$. Nevertheless, the dual problem is essentially the same, and it still has the optimal value $\Theta(1/2) = 1/2$. Here, the duality gap is infinite. □

The entropy problem of §1.2(c) provides a few examples to show that the topological operations also play their role in Theorem 2.4.1.

Example 2.4.3 Take $\Omega = \{0\}$ (so that $L_1(\Omega, \mathbb{R}) = \mathbb{R}$) and one equality constraint: our entropy problem is

$$\sup \varphi(u), \quad u \in U, \quad u - u_0 = 0.$$

where u_0 is fixed. We consider two particular cases:

$$U_1 = [0, +\infty[\quad \varphi_1(u) = -u \log u, \qquad U_2 =]0, +\infty[\quad \varphi_2(u) = \log u.$$

The corresponding sets $C(U)$ are both convex: in fact $C(U_1) = [-u_0, +\infty[$ and $C(U_2) =]-u_0, +\infty[$. The dual functions are easy to compute:

$$\mathbb{R} \ni \lambda \mapsto \begin{cases} \Theta_1(\lambda) = e^{-1-\lambda} + \lambda u_0 \\ \Theta_2(\lambda) = -\log \lambda + \lambda u_0 \quad \text{(for } \lambda > 0\text{)}. \end{cases}$$

As predicted by Theorem 2.4.1, the cases $u_0 > 0$ and $u_0 < 0$ are clear; but for $u_0 = 0$, we have $0 \in \text{bd}\, C(U_i)$ for $i = 1, 2$. Two situations are then illustrated: $\inf \Theta_1 = 0$ and $\inf \Theta_2 = -\infty$; there is no dual solution. If we took

$$U_3 = [0, +\infty[\quad \varphi_3(u) = -\tfrac{1}{2} u^2,$$

we would have

$$\Theta_3(\lambda) = \begin{cases} \tfrac{1}{2}\lambda^2 + u_0 \lambda & \text{for } \lambda \leq 0, \\ u_0 \lambda & \text{for } \lambda \geq 0 \end{cases}$$

and a dual solution would exist for all $u_0 \geq 0$. □

3 Illustrations

3.1 The Minimax Point of View

The Lagrange function (1.1.3) is not the only possibility to replace the primal problem (1.1.1) by something supposedly simpler. In the previous sections, we have followed a path

$$\text{primal problem (1.1.1)} \mapsto \text{Lagrange problem (1.1.4)}_\lambda \mapsto$$
$$\text{feasibility problem (2.1.2) or (2.1.3)} \mapsto \text{dual problem (2.1.6)}.$$

This was just one instance of the following formal construction:

- define some set V (playing the role of $\mathbb{R}^m$, the dual of the space of constraint-values), whose elements $v \in V$ play the role of $\lambda \in \mathbb{R}^m$;
- define some bivariate function $\ell : U \times V \to \mathbb{R} \cup \{+\infty\}$, playing the role of the Lagrangian L;
- V and ℓ must satisfy the following property, for all $u \in U$,

$$\inf_{v \in V} \ell(u, v) = \begin{cases} \varphi(u) & \text{if } c(u) = 0, \\ -\infty & \text{otherwise}; \end{cases} \quad (3.1.1)$$

then it is clear that (1.1.1) is equivalent to

$$\sup_{u \in U} [\inf_{v \in V} \ell(u, v)]. \quad (3.1.2)$$

- In addition, the function

$$\theta(v) := \sup_{u \in U} \ell(u, v)$$

must be easy to compute, as in the practical Assumption 1.1.2;
- the whole business in this general pattern is therefore to invert the inf- and sup-operations, replacing (1.1.1) = (3.1.2) by (2.1.6), which reads here

$$\inf_{v \in V} [\sup_{u \in U} \ell(u, v)], \quad \text{i.e.} \quad \inf_{v \in V} \theta(v). \quad (3.1.3)$$

- In order to make this last problem really easy, an extra requirement is therefore:

$$\text{the function } \theta \text{ and the set } V \text{ are closed convex.} \quad (3.1.4)$$

The above construction may sound artificial but the ordinary Lagrange technique, precisely, discloses an instance where it is not. Its starting ingredient is the pairing (1.1.2) (a natural "pricing" in the space of constraint-values); then (3.1.1) is obviously satisfied by the Lagrange function $\ell = L$ of (1.1.3); furthermore, the latter is affine with respect to the dual variable $v = \lambda$, and this takes care of the requirement (3.1.4) for the dual function $\theta = \Theta$, provided that $\theta \not\equiv +\infty$ (Proposition IV.2.1.2).

Remark 3.1.1 In the dual approach, the basic idea is thus to *replace* the sup-inf problem (1.1.1) = (3.1.2) by its inf-sup form (2.1.6) = (3.1.3) (after having defined an appropriate scalar product in the space of constraint-values). The theory in §VII.4 tells us when this is going to work, and explains the role of the assumptions appearing in Sections 2.3 and 2.4:

- The function ℓ (i.e. L) is already closed and convex in λ; what is missing is its concavity with respect to u.
- Some topological properties (of L, U and/or V) are also necessary, so that the relevant extrema are attained.

Then we can apply Theorem VII.4.3.1: the Lagrange function has a saddle-point, the values in (3.1.2) and (3.1.3) are equal, there is no duality gap; the feasibility problem (2.1.3) is equivalent to the dual problem (2.1.6). This explains also the "invariance" property resulting from Corollary 2.1.8: the set of saddle-points was seen to be a Cartesian product. □

This point of view is very useful to "dualize" appropriately an optimization problem. The next subsections demonstrate it with some specific examples.

3.2 Inequality Constraints

Suppose that (1.1.1) has rather the form

$$\sup_{u \in U} \{\varphi(u) \, : \, c_j(u) \leq 0 \text{ for } j = 1, \ldots, p\}, \qquad (3.2.1)$$

in which the constraint can be condensely written $c(u) \in -(\mathbb{R}^+)^p$. Two examples were seen in §1.2; in both of them, nonnegative slack variables were included, so as to recover the equality-constrained framework; then the dual function was $+\infty$ outside the nonnegative orthant.

The ideas of §3.1 can also be used: form the same Lagrange function as before,

$$U \times V \ni (u, \mu) \mapsto L(u, \mu) := \varphi(u) - \mu^\top c(u),$$

but take

$$V := (\mathbb{R}^+)^p = \{\mu \in \mathbb{R}^p \, : \, \mu_j \geq 0 \text{ for } j = 1, \ldots, p\}.$$

This construction satisfies (3.1.1): if u violates some constraint, say $c_i(u) > 0$, fix all μ_j at 0 except μ_i, which is sent to $+\infty$; then the Lagrange function tends to $-\infty$. The dual problem associated with (3.2.1) is therefore as before:

$$\inf \left\{ \sup_{u \in U} L(u, \mu) \, : \, \mu \in (\mathbb{R}^+)^p \right\}$$

(note that the nonnegative orthant $(\mathbb{R}^+)^p$ is closed convex: (3.1.4) is preserved if $\Theta \not\equiv +\infty$).

Instead of introducing slack variables, we thus have a shortcut allowing the primal set U to remain unchanged; the price is modest: just add simple constraints in the dual problem.

Remark 3.2.1 More generally, a problem such as

$$\sup_{u \in U} \{\varphi(u) \, : \, a_i(u) = 0 \text{ for } i = 1, \ldots, m, \quad c_j(u) \leq 0 \text{ for } j = 1, \ldots, p\}$$

can be dualized as follows: supremize over $u \in U$ the Lagrange function

$$L(u, \lambda, \mu) := \varphi(u) - \sum_{i=1}^{m} \lambda_i a_i(u) - \sum_{j=1}^{p} \mu_j c_j(u)$$

to obtain the dual function $\Theta : \mathbb{R}^m \times \mathbb{R}^p \to \mathbb{R} \cup \{+\infty\}$ as before; then solve the dual problem

$$\inf \{\Theta(\lambda, \mu) : (\lambda, \mu) \in \mathbb{R}^m \times (\mathbb{R}^+)^p\}. \tag{3.2.2}$$

Naturally, other combinations are possible: the primal problem may be a minimization one, the linking constraints may be introduced with a "+" sign in the Lagrange function (1.1.3), and so on. Some mental exercise is required to make the link with §VII.4.5.

It is wise to adopt a definite strategy, so as to develop *unconscious automatisms*.

(i) We advise proceeding as follows: formulate the primal problem as

$$\inf \{\varphi(u) : u \in U, \; c_j(u) \leqslant 0 \text{ for } j = 1, \ldots, p\}$$

(and also with some equality constraints if applicable). Then take the dual function

$$\mu \mapsto \Theta(\mu) = \inf_{u \in U} \left[\varphi(u) + \sum_{j=1}^{p} \mu_j c_j(u) \right],$$

which must be maximized over $(\mathbb{R}^+)^p$. It is concave and $-c(u_\mu) \in \partial(-\Theta)(\mu)$. This was the strategy adopted in Chap. VII.

(ii) It is only for local needs that we adopt a different strategy in the present chapter: we prefer to *minimize* a *convex* dual function. The reason for the minus-sign in the Lagrange function (1.1.3) is to keep nonnegative dual variables.

In either technique (i), (ii), equality [resp. inequality] constraints give birth to unconstrained [resp. nonnegative] dual variables. □

Depending on the context, it may be simpler to impose directly dual nonnegativity constraints, or to use slack variables. Let us review the main results of §2 in an inequality-constrained context.

(a) Modified Everett's Theorem. For $\mu \in (\mathbb{R}^+)^p$, let u_μ maximize the Lagrangian $\varphi(u) - \mu^\top c(u)$ associated with (3.2.1). Then u_μ solves

$$\begin{vmatrix} \max \varphi(u) & u \in U, \\ c_j(u) \leqslant c_j(u_\mu) & \text{if } \mu_j > 0, \\ c_j(u) \text{ unconstrained} & \text{otherwise}. \end{vmatrix} \tag{3.2.3}$$

A way to prove this is to use a slackened formulation of (3.2.1):

$$\sup_{u,v} \{\varphi(u) : u \in U, \; v \in (\mathbb{R}^+)^p, \; c(u) + v = 0\}. \tag{3.2.4}$$

Then the original Everett theorem directly applies; observe that any v_j maximizes the associated Lagrangian if $\mu_j = 0$.

Naturally, u_μ is a primal point all the more interesting when (3.2.3) is less constrained; but constraints may be appended ad libitum, under the sole condition that they do not eliminate u_μ. For example, if u_μ is feasible, the last line of (3.2.3) can be replaced by

$$c_j(u) \leqslant 0 \quad [\geqslant c_j(u_\mu)].$$

We see that, if u_μ is feasible in (3.2.1) and satisfies

$$c_j(u_\mu) = 0 \quad \text{if } \mu_j > 0,$$

then u_μ solves (3.2.1). The wording "and satisfies the complementarity slackness" must be added to "feasible" wherever applicable in Sections 2.1 and 2.3 (for example in Corollary 2.1.8).

(b) Modified Filling Property. For convenient primal-dual relationship in problems with inequality constraints, the filling assumption (2.3.2) is not sufficient. Let again $\Theta(\mu) = \sup_u L(u, \mu)$ be the dual function associated with (3.2.1); the basic result reproducing §2.3 is then as follows:

Proposition 3.2.2 *With the above notation, assume that*

$$\partial \Theta(\mu) = -\operatorname{co}\{c(u) : L(u, \mu) = \Theta(\mu)\} \quad \text{for all } \mu \in (\mathbb{R}^+)^p$$

and

$$\Theta(\mu) < +\infty \quad \text{for some } \mu \text{ with } \mu_j > 0 \text{ for } j = 1, \ldots, p. \tag{3.2.5}$$

Then $\bar{\mu}$ solves the dual problem if and only if: there are $k \leqslant p + 1$ primal points $u^1, \ldots, u^k$ maximizing $L(\cdot, \bar{\mu})$, and $\alpha = (\alpha_1, \ldots, \alpha_k) \in \Delta_k$ such that

$$\sum_{i=1}^k \alpha_i c_j(u^i) \leqslant 0 \quad \text{and} \quad \bar{\mu}_j \sum_{i=1}^k \alpha_i c_j(u^i) = 0 \text{ for } j = 1, \ldots, k.$$

PROOF. The dual optimality condition is $0 \in \partial(\Theta + I_{(\mathbb{R}^+)^p})(\bar{\mu})$, so we need to compute the subdifferential of this sum of functions. With μ as stated in (3.2.5), there is a ball $B(\mu, \delta)$ contained in $(\mathbb{R}_*^+)^p$; and any such ball intersects ri dom Θ (μ is adherent to ri dom Θ). Then Corollary XI.3.1.2 gives

$$\partial\left(\Theta + I_{(\mathbb{R}^+)^p}\right) = \partial \Theta + N_{(\mathbb{R}^+)^p} \quad \text{on } \operatorname{dom} \Theta \cap (\mathbb{R}^+)^p.$$

By assumption, a subgradient of Θ is opposite to a convex combination $\sum_{i=1}^k \alpha_i c(u^i)$ of constraint-values at points u^i maximizing the Lagrangian. The rest follows without difficulty; see in particular Example VII.1.1.6 to compute the normal cone $N_{(\mathbb{R}^+)^p}$. □

(c) Existence of Dual Solutions. Finally, let us examine how Proposition 2.4.1 can handle inequality constraints. Using slack variables, the image-set associated with (3.2.4) is

$$C'(U) = C(U) + (\mathbb{R}^+)^p,$$

where $C(U)$ is defined in (2.4.1). Clearly, aff $C'(U) = \mathbb{R}^p$ and we obtain:

(i) If $0 \notin \overline{\text{co}}[C(U) + (\mathbb{R}^+)^p]$, then inf $\Theta = -\infty$; unfortunately this closed convex hull is not easy to describe in simpler terms.

(ii) If $0 \in \text{co}\, C(U) + (\mathbb{R}^+)^p$, then inf $\Theta > -\infty$. This comes from the easy relation

$$\text{co}\, C'(U) \ [= \text{co}[C(U) + (\mathbb{R}^+)^p]] \ = \text{co}\, C(U) + (\mathbb{R}^+)^p.$$

(iii) If $0 \in \text{ri}[\text{co}\, C(U) + (\mathbb{R}^+)^p]$, then a dual solution exists. Interestingly enough, this is a Slater-type assumption:

Proposition 3.2.3 *If there are $k \leq p+1$ points $u^1, \ldots, u^k$ in U and convex multipliers $\alpha_1, \ldots, \alpha_k$ such that*

$$\sum_{i=1}^{k} \alpha_i c_j(u^i) < 0 \quad \text{for } j = 1, \ldots, p,$$

then the dual function associated with the optimization problem (3.2.1) has a minimum over the nonnegative orthant $(\mathbb{R}^+)^p$.

PROOF. Our assumption means that 0 lies in co $C(U) + (\mathbb{R}_*^+)^p$, an open convex set:

$$0 \in \text{co}\, C(U) + (\mathbb{R}_*^+)^p = \text{ri}[\text{co}\, C(U) + (\mathbb{R}_*^+)^p].$$

The result follows from Proposition 2.4.1(iii). □

3.3 Dualization of Linear Programs

As a particular case of §3.2, suppose that U is $\mathbb{R}^n$, with some scalar product $\langle \cdot, \cdot \rangle$, and that (3.2.1) is

$$\begin{vmatrix} \sup \langle q, u \rangle \\ \langle a_j, u \rangle \leq b_j \quad \text{for } j = 1, \ldots, p, \end{vmatrix} \qquad (3.3.1)$$

where q and each a_j are in $\mathbb{R}^n$, $b = (b_1, \ldots, b_p) \in \mathbb{R}^p$. The Lagrange function

$$\mathbb{R}^n \times \mathbb{R}^p \ni (u, \mu) \mapsto L(u, \mu) := \left\langle q - \sum_{j=1}^{p} \mu_j a_j, u \right\rangle + \sum_{j=1}^{p} \mu_j b_j$$

is "easy to maximize": $\Theta(\mu) = +\infty$ if $[\mu \notin (\mathbb{R}^+)^p$ or] $q - \sum_j \mu_j a_j \neq 0$. The dual problem is therefore

$$\inf \left\{ \sum_{j=1}^{p} \mu_j b_j : \mu_j \geq 0 \text{ for } j = 1, \ldots, p, \ \sum_{j=1}^{p} \mu_j a_j = q \right\}, \qquad (3.3.2)$$

another linear program, posed in $\mathbb{R}^p$.

Here, neither Assumption 1.1.5, nor the concept of a black box giving u_μ has much relevance: if μ satisfies the dual constraints (i.e. if $\mu \in \text{dom}\,\Theta$, the only interesting case), any $u \in \mathbb{R}^n$ is a u_μ.

Remark 3.3.1 The dual constraints that appear in (3.2.2) and (3.3.2) simply express that the study can be restricted to dom Θ, since Θ must be minimized. Instead of (3.3.2), for example, the dual of (3.3.1) can obviously be formulated as the unconstrained minimization of the function

$$\mathbb{R}^p \ni \mu \mapsto \begin{cases} \sum_{j=1}^p \mu_j b_j & \text{if } \sum_{j=1}^p \mu_j a_j = q \text{ and } \mu \in (\mathbb{R}^+)^p, \\ +\infty & \text{otherwise}. \end{cases}$$

□

Continuing with this example, consider the *standard form* of a linear program: assume that $\langle \cdot, \cdot \rangle$ is the usual dot-product and that our primal problem is now

$$\sup \{q^\top u : Au = b, \ u_i \geqslant 0 \text{ for } i = 1, \ldots, n\}. \tag{3.3.3}$$

Here the matrix A has m rows, $b \in \mathbb{R}^m$. It is natural to take as primal set $U := (\mathbb{R}^+)^n$, so that the dual space is a priori $\mathbb{R}^m$, m being the number of equalities. Then the Lagrange function

$$\mathbb{R}^n \times \mathbb{R}^m \ni (u, \lambda) \mapsto L(x, \lambda) := \left(q^\top - \lambda^\top A\right) u - \lambda^\top b$$

is again "easy to maximize" over U: the maximal value is finite if and only if the vector $q - A^\top \lambda$ has all its coordinates nonpositive. In summary, the dual of (3.3.3) is

$$\inf \{b^\top \lambda : \lambda \in \mathbb{R}^m, \ A^\top \lambda - q \in (\mathbb{R}^+)^n\}. \tag{3.3.4}$$

The nonnegativity constraints in (3.3.3) can be dualized as well, even though the example in (1.2.1) does not suggest doing so because they have already a decomposed form. In fact, nothing much interesting comes out of this extra dualization: the Lagrange function becomes

$$\mathbb{R}^n \times \mathbb{R}^m \times \mathbb{R}^n \ni (u, \lambda, \mu) \mapsto (q^\top + \mu^\top - \lambda^\top A)u - \lambda^\top b,$$

and we obtain the new dual problem

$$\inf \{b^\top \lambda : q^\top + \mu^\top - \lambda^\top A = 0, \ \mu \in (\mathbb{R}^+)^n\},$$

in which μ is redundant and can be eliminated: the result is exactly (3.3.4).

3.4 Dualization of Quadratic Programs

Let again U be $\mathbb{R}^n$, equipped with the dot-product for simplicity, and consider

$$\sup \{q^\top u - \tfrac{1}{2} u^\top Q u : Au \leqslant b\}. \tag{3.4.1}$$

Here $q \in \mathbb{R}^n$, $b \in \mathbb{R}^p$ and the $n \times n$ matrix Q is symmetric positive definite. This problem has a unique solution if the feasible domain is nonempty. Choose the Lagrange function

$$\mathbb{R}^n \times \mathbb{R}^p \ni (u, \mu) \mapsto L(u, \mu) := \left(q - A^\top \mu\right)^\top u - \tfrac{1}{2} u^\top Q u + b^\top \mu.$$

As a consequence of our assumptions, its maximum is attained at the unique

$$u_\mu = Q^{-1}(q - A^\top \mu), \tag{3.4.2}$$

for which the corresponding constraint-value is

$$c(u_\mu) = AQ^{-1}(q - A^\top \mu) - b.$$

Plugging the value (3.4.2) into L, we obtain the dual function

$$\mathbb{R}^p \ni \mu \mapsto \Theta(\mu) = \tfrac{1}{2}(q - A^\top \mu)^\top Q^{-1}(q - A^\top \mu) + b^\top \mu,$$

to be minimized over $(\mathbb{R}^+)^p$. Note that it is differentiable and its gradient illustrates Proposition 2.2.2:

$$\nabla \Theta(\mu) = AQ^{-1}(A^\top \mu - q) + b = -c(u_\mu) = \nabla_\mu L(u_\mu, \mu).$$

On the other hand, suppose that the constraints in (3.4.1) are equalities: $Au = b$; then the dual minimization is performed on the whole of $\mathbb{R}^p$. If they exist, the dual solutions are the solutions of $\nabla \Theta(\mu) = 0$, which is the linear system

$$\nabla \Theta(\mu) = AQ^{-1}(A^\top \mu - q) + b = 0. \tag{3.4.3}$$

Existence of such a solution implies first that $b \in \operatorname{Im} A$, i.e. that there exists a primal feasible u. In this case we claim that the dual problem has one solution at least, and that all such solutions make up via (3.4.2) a unique point (the unique primal solution). This will illustrate Corollary 2.1.8.

The key lies in the following facts:

(i) The two subspaces $\operatorname{Im} A^\top$ and $\operatorname{Ker} A$ are two orthogonal generators of $\mathbb{R}^n$. This implies:

(ii) when applying the positive definite operator Q or Q^{-1} to one of them, we never obtain a vector in the other:

$$Q^{-1}(\operatorname{Im} A^\top) \cap \operatorname{Ker} A = \{0\} = \operatorname{Im} A^\top \cap Q(\operatorname{Ker} A). \tag{3.4.4}$$

Hence:

(iii) we have further decompositions of $\mathbb{R}^n$ into two subspaces:

$$Q^{-1}(\operatorname{Im} A^\top) \oplus \operatorname{Ker} A = \mathbb{R}^n = \operatorname{Im} A^\top \oplus Q(\operatorname{Ker} A). \tag{3.4.5}$$

Then the proof of our claim goes as follows:

[*Existence*] If $b \in \operatorname{Im} A$, take $u_0 \in \mathbb{R}^n$ such that $b = -AQ^{-1}u_0$; then use (3.4.5) to write $q + u_0 = A^\top \mu_0 + Qv_0$, with $v_0 \in \operatorname{Ker} A$. In summary (3.4.3) becomes

$$0 = AQ^{-1}(A^\top \mu - q - u_0) = AQ^{-1}A^\top(\mu - \mu_0) - Av_0,$$

which has the obvious solution $\mu = \mu_0$.

[*Uniqueness*] If μ_1 and μ_2 solve (3.4.3), with corresponding u_1 and u_2 maximizing L over $\mathbb{R}^n$,
$$0 = AQ^{-1}A^\top(\mu_1 - \mu_2) = A(u_2 - u_1)$$
and we deduce
$$\operatorname{Ker} A \ni u_2 - u_1 = Q^{-1}A^\top(\mu_1 - \mu_2) \in Q^{-1}\left(\operatorname{Im} A^\top\right);$$
in view of (3.4.4), this implies $u_1 - u_2 = 0$.

The essence of the above "proof" is the fact that $\langle \mu, \nu \rangle := \mu^\top AQ^{-1}A^\top \nu$ is a scalar product in $\operatorname{Ker} A^\top = \operatorname{Im} A$. When (3.4.3) has no solution, which means that the primal feasible set is empty, Θ is not bounded from below: apply Proposition 2.4.1(i), knowing that $C(U)$ is here a (closed convex) affine manifold.

3.5 Steepest-Descent Directions

Consider the direction-finding problem of Chapters VIII and IX, assuming the Euclidean norming on $\mathbb{R}^n$: $\|\cdot\|_* = \|\cdot\|$. We had a number of (sub)gradients $s_1, \ldots, s_k$ and we wanted to solve, for a given normalization parameter $\kappa > 0$:

$$\left|\begin{array}{ll} \min r & r \in \mathbb{R}, \; d \in \mathbb{R}^n, \\ r \geq \langle s_i, d \rangle & \text{for } i = 1, \ldots, k, \\ \|d\| = \kappa. & \end{array}\right. \quad (3.5.1)$$

Adapting the notations (u is the couple (d, r), the dual variable is $\alpha \in \mathbb{R}^k$, we have now a minimization problem), define the Lagrange function

$$\mathbb{R}^n \times \mathbb{R} \times \mathbb{R}^k \ni (d, r, \alpha) \mapsto L(d, r, \alpha) := \left(1 - \sum_{i=1}^k \alpha_i\right) r + \left\langle \sum_{i=1}^k \alpha_i s_i, d \right\rangle, \quad (3.5.2)$$

to be minimized, with fixed α, over

$$U := \left\{(d, r) \in \mathbb{R}^{n+1} : \|d\| = \kappa\right\}. \quad (3.5.3)$$

The dual function (to be maximized) is "easy to compute": the dual constraint $1 - \sum_{i=1}^k \alpha_i = 0$ takes care of the unconstrained linear variable r; altogether, the dual variable has to vary over the unit simplex Δ_k of $\mathbb{R}^k$.

Setting $\hat{s}(\alpha) := \sum_{i=1}^k \alpha_i s_i$, the Lagrange problem associated with $\alpha \in \Delta_k$ then has solutions (d_α, r_α) defined as follows:

$$d_\alpha = \begin{cases} -\frac{\kappa}{\|\hat{s}(\alpha)\|} \hat{s}(\alpha) & \text{if } \hat{s}(\alpha) \neq 0, \\ \text{arbitrary of norm } \kappa & \text{otherwise}, \end{cases} \quad (3.5.4)$$

and r_α is arbitrary, its coefficient in (3.5.2) being 0. Finally the dual problem is

$$\max\left\{-\kappa \|\hat{s}(\alpha)\| : \alpha \in \Delta_k\right\}.$$

Now, once this dual problem is solved, say at some $\bar{\alpha}$, there are two cases. If the corresponding optimal $\hat{s}(\bar{\alpha})$ is nonzero, the primal solution $\bar{d}$ is recovered from (3.5.4) (the corresponding $\bar{r}$ is easy to recover). If the optimal $\hat{s}(\bar{\alpha})$ is zero, nothing useful is obtained from the dual approach, except the information that 0 is in the convex hull of the s_i's – and this was good enough in the case of interest.

In this example, Corollary 2.3.6(i) applies if $\hat{s}(\bar{\alpha}) \neq 0$, which means that $\hat{s}(\alpha) \neq 0$ for all $\alpha \in \Delta_k$: even though the original problem (3.5.1) is not convex, the dual function is differentiable everywhere (at least relative to its domain).

By contrast, suppose $\hat{s}(\bar{\alpha}) = 0$. Then the dual function is differentiable at no dual optimum, simply because $\|\cdot\|$ is not differentiable at 0; its subdifferential can be computed with the help of the calculus developed in §VI.4: to within the normalization coefficient κ, we obtain the ellipsoid

$$\left\{ \sum_{i=1}^{k} \sigma_i s_i : \sum_{i=1}^{k} \sigma_i^2 \leq 1 \right\}.$$

Having a dual optimal solution $\bar{\alpha}$ is absolutely of no help to find a primal solution: the Lagrange function $L(\cdot, \cdot, \bar{\alpha})$ of (3.5.2) is optimal at all feasible points of (3.5.1), duality has killed the role of the objective function.

Remark 3.5.1 Formulating (3.5.1) with slack variables, say ρ_i, we obtain the new Lagrange function (with obvious notation)

$$L(d, r, \rho, \alpha) := \left(1 - \sum_{i=1}^{k} \alpha_i\right) r + \sum_{i=1}^{k} \alpha_i \rho_i + \left\langle \sum_{i=1}^{k} \alpha_i s_i, d \right\rangle,$$

to be minimized over the new primal set

$$U := \left\{ (d, r, \rho) \in \mathbb{R}^{n+k+1} : \|d\| = \kappa, \ \rho_i \geq 0 \text{ for } i = 1, \ldots, k \right\}.$$

Now, the minimizers of $L(\cdot, \cdot, \cdot, \bar{\alpha})$ must have $\rho_i = 0$ if $\bar{\alpha}_i > 0$ and the normal situation is that none of these minimizers satisfies the (slackened) constraints

$$r = \rho_i + \langle s_i, d \rangle \quad \text{for } i = 1, \ldots, k.$$

Take for example, with $n = 1$, $k = 2$ and $\kappa = 1$: $s_1 = -1$ and $s_2 = p \in [0, 1[$, so that the steepest-descent direction is $d = 1$, the unique dual solution (α_1, α_2) is $\alpha_1 = p/(1 + p)$, $\alpha_2 = 1/(1 + p)$, and the difference between the primal and dual optimal values is $p \geq 0$.

– If $p > 0$ (there is a duality gap), the slacks must be $\rho_1 = \rho_2 = 0$; no matter how we choose d in $\{-1, +1\}$, r cannot be adjusted to satisfy the two dualized equalities.
– If $p = 0$ (no duality gap), ρ_1 is arbitrary nonnegative and we can for example choose $d = 1$, $r = 0 = pd$ and $\rho_1 = 1$; this minimizes the Lagrange function and solves the primal problem as well. □

Needless to say, all gap-problems would disappear if the normalization constraint in (3.5.1) were $\|d\| \leq \kappa$; it is this idea that was exploited in §VIII.1.2.

Finally, there is an interesting variant: dualize in (3.5.1) the normalization constraint only, to obtain the Lagrange function

$$\mathbb{R}^n \times \mathbb{R} \times \mathbb{R} \ni (d, r, \alpha_0) \mapsto L(d, r, \alpha_0) := r + \alpha_0(\|d\|^2 - \kappa^2)$$

which has the single dual variable α_0. Put

$$\Theta(\alpha_0) := - \inf_{(d,r) \in U} L(d, r, \alpha_0), \tag{3.5.5}$$

where U is the primal domain

$$U := \{(d, r) \in \mathbb{R}^{n+1} : r \geq \langle s_i, d \rangle \text{ for } i = 1, \ldots, k\}. \tag{3.5.6}$$

In contrast to (3.5.4), the Lagrange minimization problem of (3.5.5) cannot be solved explicitly, so this approach does not do much good for solving (3.5.1) numerically. It provides some interesting illustrations, though: σ_S denoting the support function of $S := \text{co}\{s_1, \ldots, s_k\}$, the following properties hold for the Lagrange problem

$$\Theta(\alpha_0) = - \inf_{d \in \mathbb{R}^n} \left[\alpha_0(\|d\|^2 - \kappa^2) + \sigma_S(d) \right]. \tag{3.5.7}$$

– dom $\Theta \subset \mathbb{R}^+$;
– $\Theta(0) = 0$ if $0 \in S$, $+\infty$ if not;
– Θ is finite on $\mathbb{R}_*^+$;
– since $\Theta \in \overline{\text{Conv}}\,\mathbb{R}$, $\lim_{\alpha_0 \downarrow 0} \Theta(\alpha_0) = \Theta(0)$;
– at any $\alpha_0 > 0$, Θ has the derivative $\Theta'(\alpha_0) = \kappa^2 - \|d(\alpha_0)\|^2$, where $d(\alpha_0)$ is the unique optimal solution in (3.5.7);
– if Θ is minimal at some $\alpha_0 > 0$, then $d(\alpha_0)$ has norm κ and solves the original problem (Corollary 2.3.6);
– if Θ is minimal at 0, there are two cases, illustrated by Example 3.5.1: either $\|d(\alpha_0)\| \uparrow \kappa$ when $\alpha_0 \downarrow 0$; there is no duality gap, this was the case $p > 0$. If $\|d(\alpha_0)\|$ has a limit strictly smaller than κ, the original problem (3.5.1) is not solvable via duality.

4 Classical Dual Algorithms

In this section, we review the most classical algorithms aimed at solving the dual problem (2.1.6). We neglect for the moment the primal aspect of the question, namely the task (ii) mentioned in Remark 2.2.3. According to the various results of Sections 2.1, 2.2, our situation is as follows:

– we must minimize a convex function (the dual function Θ);
– the only information at hand is the black box that solves the Lagrange problem $(1.1.4)_\lambda$, for each given $\lambda \in \mathbb{R}^m$;
– Assumption 1.1.5 is in force; thus, for each $\lambda \in \mathbb{R}^m$, the black box computes the number $\Theta(\lambda)$ and some $s(\lambda) \in \partial \Theta(\lambda)$.

The present problem is therefore just the one considered in Chap. IX; formally, it is also that of Chap. II, with the technical difference that the function $\lambda \mapsto s(\lambda)$ is not continuous. The general approach will also be the same: at the kth iteration, the black box is called (i.e. the Lagrange problem is solved at λ_k) to obtain $\Theta(\lambda_k)$ and $s(\lambda_k)$; then the $(k+1)^{\text{st}}$ iterate λ_{k+1} is computed. A dual algorithm is thus characterized by the *set of rules* giving λ_{k+1}.

Note here that our present task is after all nothing but developing general algorithms which minimize a convex function finite everywhere, and which work with the help of the above-mentioned information (function- and subgradient-values). It is fair to say that most of the known algorithms of this type have been actually motivated precisely by the need to solve dual problems.

Note also that all this is valid again with no assumption on the primal problem (1.1.1), except Assumption 1.1.5.

4.1 Subgradient Optimization

The simplest algorithm for convex minimization is directly inspired by the gradient method of Definition II.2.2.2: the next iterate is sought along the direction $-s(\lambda_k)$ issuing from λ_k. There is a serious difficulty, however, which has been the central issue in Chap. IX: no line-search is possible based on decreasing Θ, simply because $-s(\lambda_k)$ may not be a descent direction – or so weak that the resulting sequence $\{\lambda_k\}$ would not minimize Θ. The relevant algorithm is then as follows:

Algorithm 4.1.1 (Basic Subgradient Algorithm) A sequence $\{t_k\}$ is given, with $t_k > 0$ for $k = 1, 2, \ldots$

STEP 0 (initialization). Choose $\lambda_1 \in \mathbb{R}^m$ and obtain $s_1 \in \partial\Theta(\lambda_1)$. Set $k = 1$.
STEP 1. If $s_k = 0$ stop. Otherwise set

$$\lambda_{k+1} = \lambda_k - t_k \frac{s_k}{\|s_k\|}. \qquad (4.1.1)$$

STEP 2. Obtain $s_{k+1} \in \partial\Theta(\lambda_{k+1})$. Replace k by $k+1$ and loop to Step 1. □

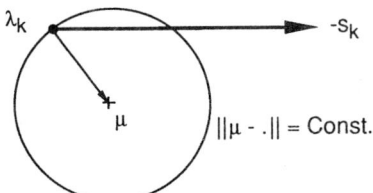

Fig. 4.1.1. An anti-subgradient gets closer to any better point

Needless to say, each subgradient is obtained via some u_λ solving the Lagrange problem at the corresponding λ – hence the need for Assumption 1.1.5. Note that the function-values $\Theta(\lambda_k)$ are never used by this algorithm, whose motivation is as follows (see Fig. 4.1.1): if μ is better than λ_k, we obtain from the subgradient inequality

$$s_k^\top(\mu - \lambda_k) \leqslant \Theta(\mu) - \Theta(\lambda_k) < 0,$$

which means that the angle between the direction of move $-s_k$ and the "nice" direction $\mu - \lambda_k$ is acute. If our move along that direction is small enough, we get closer to μ. From this interpretation, the stepsize t_k should be small, and we will require

$$t_k \to 0 \quad \text{for } k \to +\infty. \qquad (4.1.2)$$

On the other hand, observe that $\|\lambda_{k+1} - \lambda_k\| = t_k$ and the triangle inequality implies that all iterates are confined in some ball: for $k = 1, 2, \ldots$

$$\lambda_k \in B(\lambda_1, T), \quad \text{where} \quad T := \sum_{k=1}^{\infty} t_k.$$

Suppose Θ has a nonempty set of minima; then Algorithm 4.1.1 has no chance of producing a minimizing sequence if $B(\lambda_1, T)$ does not intersect this set: in spite of (4.1.2), the stepsizes should not be too small, and we will require

$$\sum_{k=1}^{\infty} t_k = +\infty. \tag{4.1.3}$$

Our proof of convergence is not the simplest, nor the oldest; but it motivates Lemma 4.1.3 below, which is of interest in its own right.

Lemma 4.1.2 *Let $\{t_k\}$ be a sequence of positive numbers satisfying (4.1.2), (4.1.3) and set*

$$\tau_n := \sum_{k=1}^{n} t_k, \quad \rho_n := \sum_{k=1}^{n} t_k^2. \tag{4.1.4}$$

Then ($\tau_n \to +\infty$ and) $\dfrac{\rho_n}{\tau_n} \to 0$ when $n \to +\infty$.

PROOF. Fix $\delta > 0$; there is some $n(\delta)$ such that $t_k \leqslant \delta$ for all $n > n(\delta)$; then

$$\rho_n = \rho_{n(\delta)-1} + \sum_{k=n(\delta)}^{n} t_k^2 \leqslant \rho_{n(\delta)-1} + \delta \sum_{k=n(\delta)}^{n} t_k \leqslant \rho_{n(\delta)-1} + \delta \tau_n,$$

so that

$$\frac{\rho_n}{\tau_n} \leqslant \frac{\rho_{n(\delta)-1}}{\tau_n} + \delta \quad \text{for all } n > n(\delta);$$

thus, $\limsup \dfrac{\rho_n}{\tau_n} \leqslant \delta$. The result follows since $\delta > 0$ was arbitrary. □

Lemma 4.1.3 *Let $\Theta : \mathbb{R}^m \to \mathbb{R}$ be convex and fix $\bar{\lambda} \in \mathbb{R}^m$. For all $\lambda \in \mathbb{R}^m$ such that $\Theta(\lambda) > \Theta(\bar{\lambda})$ and for all $s \in \partial\Theta(\lambda)$, set*

$$\bar{d}(\lambda) := \frac{\langle s, \lambda - \bar{\lambda} \rangle}{\|s\|} > 0. \tag{4.1.5}$$

Given $M > 0$, there exists $L > 0$ such that

$$\bar{d}(\lambda) \leqslant M \implies [0 <] \ \Theta(\lambda) - \Theta(\bar{\lambda}) \leqslant L\bar{d}(\lambda).$$

PROOF. Positivity of $\bar{d}(\lambda)$ is clear from the subgradient inequality written at λ. Let $\mu(\lambda)$ be the projection of $\bar{\lambda}$ onto the hyperplane

$$\{\mu \in \mathbb{R}^m : \langle s, \mu - \lambda \rangle = 0\}.$$

An easy calculation gives $\|\mu(\lambda) - \bar{\lambda}\| = \bar{d}(\lambda)$ and Fig. 4.1.2 shows that $\Theta(\mu(\lambda)) \geqslant \Theta(\lambda)$. If $\mu(\lambda) \in B(\bar{\lambda}, M)$ as assumed, the local Lipschitz property of Θ (Theorem IV.3.1.2) implies the existence of L such that

$$[\Theta(\lambda) \leqslant] \ \Theta(\mu(\lambda)) \leqslant \Theta(\bar{\lambda}) + L\|\mu(\lambda) - \bar{\lambda}\|,$$

which is the required inequality. □

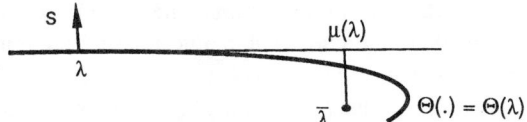

Fig. 4.1.2. A technical majoration

This last result says that $\bar{d}(\lambda)$ is a sort of "distance" between λ and $\bar{\lambda}$, for which Θ has a locally Lipschitzian behaviour.

Finally, we introduce the sequence of best values generated by Algorithm 4.1.1:

$$\overline{\Theta}_k := \min\{\Theta(\lambda_i) : i = 1, \ldots, k\},$$

which is needed because $\{\Theta(\lambda_k)\}$ is not monotonic. The whole issue is whether these best function-values tend to the infimum of Θ over $\mathbb{R}^m$ (a number in $\mathbb{R} \cup \{-\infty\}$).

Theorem 4.1.4 *Apply Algorithm 4.1.1 to the convex function $\Theta : \mathbb{R}^m \to \mathbb{R}$, and let the stepsizes satisfy (4.1.2), (4.1.3). Then*

$$\overline{\Theta}_k \to \inf_{\lambda \in \mathbb{R}^m} \Theta(\lambda) \quad \text{when } k \to +\infty.$$

PROOF. Assume for contradiction the existence of $\bar{\lambda} \in \mathbb{R}^m$ and $\eta > 0$ such that

$$[\Theta(\lambda_k) \geqslant] \ \overline{\Theta}_k \geqslant \Theta(\bar{\lambda}) + \eta \quad \text{for all } k; \qquad (4.1.6)$$

then develop

$$\begin{aligned}
\|\bar{\lambda} - \lambda_{k+1}\|^2 &= \|\bar{\lambda} - \lambda_k\|^2 + 2(\bar{\lambda} - \lambda_k)^\top (\lambda_k - \lambda_{k+1}) + \|\lambda_k - \lambda_{k+1}\|^2 \\
&= \|\bar{\lambda} - \lambda_k\|^2 - 2t_k \bar{d}(\lambda_k) + t_k^2,
\end{aligned}$$

where the notation (4.1.5) is used: the triple $(\bar{\lambda}, \lambda_k, s_k)$ enters the framework of Lemma 4.1.3. For $n \geqslant k$, set $\delta_n := \min_{k=1,\ldots,n} \bar{d}(\lambda_k)$ so that

$$\|\bar{\lambda} - \lambda_{k+1}\|^2 \leqslant \|\bar{\lambda} - \lambda_k\|^2 - 2t_k \delta_n + t_k^2 \quad \text{for } k = 1, \ldots, n;$$

summing from 1 to n:

$$[\|\bar{\lambda} - \lambda_{n+1}\|^2 +] \ 2\delta_n \sum_{k=1}^n t_k \leqslant \|\bar{\lambda} - \lambda_1\|^2 + \sum_{k=1}^n t_k^2 \quad \text{for all } n.$$

From Lemma 4.1.2, it follows that $\delta_n \to 0$ when $n \to +\infty$.

Thus, we have an infinite subset K of integers such that $\lim_{k \in K} \bar{d}(\lambda_k) = 0$. Now apply Lemma 4.1.3: $\{\bar{d}(\lambda_k)\}_{k \in K}$ is bounded and

$$\lim_{k \in K} [\Theta(\lambda_k) - \Theta(\bar{\lambda})] = 0,$$

which is a suitable contradiction to (4.1.6). □

Algorithm 4.1.1 is often called the *subgradient* method, or sometimes relaxation; both terminologies are rather ambiguous. Note a peculiarity: there is no convenient stopping criterion (in particular, s_k has no reason to tend to 0!). The algorithm must in fact be stopped "manually", when t_k is small compared to the scale of the problem.

Remark 4.1.5 We mention here that the convergence assumptions concerning (4.1.4) are somewhat bizarre. A computer cannot represent positive numbers under a certain threshold, say $m_0 > 0$. As a result,

– *Charybdis*: to satisfy (4.1.3), the computer must set $t_k \geqslant m_0$ for all k: (4.1.2) becomes impossible.
– *Scylla*: alternatively, setting $t_k = 0$ for k large enough is even worse. No update is performed by (4.1.1) and the algorithm has to stop; but at this point, the series $\sum t_k$ has not diverged yet.

On the other hand, this kind of argument is not totally convincing: the mere concept "$\varepsilon_k \to 0$" is important in numerical analysis; yet, who can be patient enough to check whether a sequence is convergent, and to obtain its limit? We shall only say that requiring a property resembling (4.1.2), (4.1.3) is a bit sloppy with respect to finite arithmetic. □

Naturally, if the dual function appears to be differentiable everywhere, any of the "normal" methodologies of Chap. II can be used. In this framework, the steepest-descent method (in the dual space) is usually called Uzawa's method, and is valid in the "ideal" situation: U convex, φ strictly concave and c affine.

4.2 The Cutting-Plane Algorithm

To ease our exposition, we assume now that a compact convex set $C \subset \mathbb{R}^m$ is known to contain a dual solution. In view of the sup-form of Θ, our dual problem (2.1.6) can then be written

$$\left| \begin{array}{ll} \min \theta & \theta \in \mathbb{R}, \ \lambda \in C, \\ \theta \geqslant L(u, \lambda) & \text{for all } u \in U. \end{array} \right. \quad (4.2.1)$$

This is a semi-infinite programming problem, which would become easy if it had only finitely many constraints, i.e. if U were a finite set; taking for example a compact convex polyhedron for C, (4.2.1) would become an ordinary linear program. Then the basic idea of the cutting-plane algorithm is quite natural: accumulate the constraints one after the other in (4.2.1). Furthermore take advantage of the fact that the constraint-index u can be restricted to the smaller set $\widetilde{U}$ of (2.2.3).

Algorithm 4.2.1 (Basic Cutting-Plane Algorithm) The compact convex set C and the stopping tolerance $\varepsilon \geqslant 0$ are given.

STEP 0 (initialization). Choose $\lambda_0 \in C$ and solve the Lagrange problem at λ_0 to obtain $u_0 := u_{\lambda_0}$. Set $k = 0$.

STEP 1 (master problem). Solve the following relaxation of (4.2.1):

$$\left|\begin{array}{ll} \min \theta & \theta \in \mathbb{R},\ \lambda \in C, \\ \theta \geqslant L(u_i, \lambda) & \text{for } i = 0, \ldots, k-1 \end{array}\right. \quad (4.2.2)$$

to obtain a solution (θ_k, λ_k).

STEP 2 (local problem). Solve the Lagrange problem $(1.1.4)_{\lambda_k}$ to obtain a next primal point $u_k := u_{\lambda_k}$.

STEP 3 (stopping criterion and loop). If

$$\Theta(\lambda_k) \leqslant \theta_k + \varepsilon, \quad (4.2.3)$$

then stop. Otherwise replace k by $k+1$ and loop to Step 1. □

Typically, C is a parallelotope characterized by known bounds on the components of λ; then (4.2.2) is a linear program with $2m + k$ constraints. It is important to realize that the above description can be made without any reference to primal points u: the Lagrange function being affine in λ, we have

$$L(u_i, \lambda) = L(u_i, \lambda_i) - [c(u_i)]^\top(\lambda - \lambda_i) = \Theta(\lambda_i) + [s(\lambda_i)]^\top(\lambda - \lambda_i), \quad (4.2.4)$$

valid for all λ and $i = 0, 1, \ldots$ Thus, (4.2.2) can be written

$$\left|\begin{array}{ll} \min \theta & \theta \in \mathbb{R},\ \lambda \in C, \\ \theta \geqslant \Theta(\lambda_i) + [s(\lambda_i)]^\top(\lambda - \lambda_i) & \text{for } i = 0, \ldots, k-1, \end{array}\right. \quad (4.2.5)$$

which involves only dual objects, namely Θ- and s-values.

Remark 4.2.2 When we write (4.2.2) in the form (4.2.5), we do nothing but prove again Proposition 2.2.2: $c(u_\lambda) \in -\partial\Theta(\lambda)$. When we hope that (4.2.5) does approximate (4.2.1), we rely upon the property proved in Theorem XI.1.3.8: the convex function Θ can be expressed as a supremum of affine functions, namely

$$\Theta(\lambda) = \sup\left\{\Theta(\mu) + [s(\mu)]^\top(\lambda - \mu) : \mu \in \mathbb{R}^m\right\} \quad \text{for all } \lambda \in \mathbb{R}^m,$$

$s(\mu)$ being arbitrary in $\partial\Theta(\mu)$. □

The convergence of Algorithm 4.2.1 is easy to establish:

Theorem 4.2.3 *With $C \subset \mathbb{R}^m$ convex compact and Θ convex from $\mathbb{R}^m$ to $\mathbb{R}$, consider the optimal value*

$$\overline{\Theta}_C := \min\{\Theta(\lambda) : \lambda \in C\}.$$

In Algorithm 4.2.1, $\theta_k \leqslant \overline{\Theta}_C$ for all k and the following convergence properties hold:

– *If $\varepsilon > 0$, the stop occurs at some iteration k_ε with λ_{k_ε} satisfying*

$$\Theta(\lambda_{k_\varepsilon}) \leqslant \overline{\Theta}_C + \varepsilon. \quad (4.2.6)$$

– *If $\varepsilon = 0$, the sequences $\{\theta_k\}$ and $\{\Theta(\lambda_k)\}$ tend to $\overline{\Theta}_C$ when $k \to +\infty$.*

PROOF. Because (4.2.1) is less constrained than (4.2.2) = (4.2.5), the inequality $\theta_k \leqslant \overline{\Theta}_C$ is clear and the optimality condition (4.2.6) does hold when the stopping criterion (4.2.3) is satisfied.

Now suppose for contradiction that, for $k = 1, 2, \ldots$

$$\Theta(\lambda_k) - \varepsilon > \theta_k \geqslant \Theta(\lambda_i) + [s(\lambda_i)]^\top (\lambda_k - \lambda_i) \quad \text{for all } i < k,$$

hence

$$-\varepsilon > \Theta(\lambda_i) - \Theta(\lambda_k) - \|s(\lambda_i)\| \, \|\lambda_k - \lambda_i\|.$$

Because Θ is Lipschitzian on the bounded set C (Theorem IV.3.1.2 and Proposition VI.6.2.2), there is L such that

$$2L\|\lambda_k - \lambda_i\| > \varepsilon \quad \text{for all } i < k = 1, 2, \ldots$$

This is incompatible with the boundedness of $\{\lambda_k\} \subset C$.

Because $\varepsilon > 0$ was arbitrary and because $\theta_k \leqslant \overline{\Theta}_C \leqslant \Theta(\lambda_k)$ for all k, we have actually just proved that $\Theta(\lambda_k) - \theta_k \to 0$ if $k \to +\infty$ (which can happen only if $\varepsilon = 0$); then $\overline{\Theta}_C$ is the common limit of $\{\theta_k\}$ and $\{\Theta(\lambda_k)\}$. □

If $\varepsilon = 0$, Algorithm 4.2.1 normally loops forever. An interesting case, however, is when U is a finite set: then the algorithm stops anyway, even if $\varepsilon = 0$. This is implied by the following result.

Proposition 4.2.4 *Here no assumption is made on C in Algorithm 4.2.1. The u_k generated at Step 2 is different from all points $u_0, \ldots, u_{k-1}$, unless the stop in Step 3 is going to operate.*

PROOF. If $u_k = u_i$ for some $i \leqslant k - 1$, then

$$\Theta(\lambda_k) = L(u_i, \lambda_k) \leqslant \theta_k,$$

where the second relation relies on (4.2.2). □

The presence of a compact set C in the basic cutting-plane Algorithm 4.2.1 is not only motivated by Theorem 4.2.3. More importantly, (4.2.5) would normally have no solution if C were unbounded: with $C = \mathbb{R}^m$, think for example of the first iteration, having only one constraint on θ and $s(\lambda_0) \neq 0$. This is definitely a drawback of the algorithm, which corresponds to an inherent *instability*: the sequence $\{\lambda_k\}$ must be artificially stabilized.

In some favourable circumstances, C can eventually be eliminated. This usually happens when: (i) the dual function Θ has a bounded set of minima, and (ii) enough dual iterations have been performed, so that the "linearized sublevel-set"

$$\{\lambda \in \mathbb{R}^m : \Theta(\lambda_i) + [s(\lambda_i)]^\top (\lambda - \lambda_i) \leqslant \theta \text{ for } i = 0, \ldots, k-1\}$$

is bounded (for some θ: remember Proposition IV.3.2.5); it may even be included in C. Under these circumstances, it becomes clear that (4.2.5) can be replaced by

$$\min \left\{ \theta \; : \; \Theta(\lambda_i) + [s(\lambda_i)]^T (\lambda - \lambda_i) \leq \theta \text{ for } i = 0, \ldots, k-1 \right\}. \quad (4.2.7)$$

The dualization of (4.2.7) is interesting. Call $\alpha \in \mathbb{R}^k$ the dual variable, form the Lagrange function $L(\theta, \lambda, \alpha)$, and apply the technique of §3.3 to obtain the dual linear program

$$\left| \begin{array}{l} \max \sum_{i=0}^{k-1} \alpha_i \left[\Theta(\lambda_i) - [s(\lambda_i)]^T \lambda_i \right] \quad \alpha \in \mathbb{R}^k, \\ \sum_{i=0}^{k-1} \alpha_i s(\lambda_i) = 0, \\ \sum_{i=0}^{k-1} \alpha_i = 1 \quad \text{and} \quad \alpha_i \geq 0 \text{ for } i = 0, \ldots, k-1. \end{array} \right.$$

Not unexpectedly, this problem can be written in primal notation: since $s(\lambda_i) = -c(u_i)$ and knowing that Δ_k is the unit simplex of $\mathbb{R}^k$, we actually have to solve

$$\max_{\alpha \in \Delta_k} \left\{ \sum_{i=0}^{k-1} \alpha_i \varphi(u_i) \; : \; \sum_{i=0}^{k-1} \alpha_i c(u_i) = 0 \right\}. \quad (4.2.8)$$

When stated with (4.2.8) instead of (4.2.7), the cutting-plane Algorithm 4.2.1 (a *row-generation* mechanism) is called *Dantzig-Wolfe*'s algorithm (a *column-generation* mechanism). In terms of the original problem (1.1.1), it is much more suggestive: basically, $\varphi(\cdot)$ and $c(\cdot)$ are replaced in (4.2.8) by appropriate convex combinations. Accordingly, the point

$$\Delta_k \ni \alpha \mapsto u(\alpha) := \sum_{i=0}^{k-1} \alpha_i u_i$$

can reasonably be viewed as an approximate primal solution:

Theorem 4.2.5 *Let the convexity conditions of Corollary 2.3.6(ii) hold (U convex, φ concave, c affine) and suppose that Algorithm 4.2.1, applied to the convex function $\Theta : \mathbb{R}^m \to \mathbb{R}$, can be used with (4.2.7) instead of (4.2.5). When the stop occurs, denote by $\bar{\alpha} \in \Delta_k$ an optimal solution of (4.2.8). Then $u(\bar{\alpha})$ is an ε-solution of the primal problem (1.1.1).*

PROOF. Observe first that $u(\bar{\alpha})$ is feasible by construction. Also, there is no duality gap between the two linear programs (4.2.7) and (4.2.8), so we have (use the concavity of φ)

$$\theta_k = \sum_{i=0}^{k-1} \bar{\alpha}_i \varphi(u_i) \leq \varphi(u(\bar{\alpha})).$$

The stopping criterion and the weak duality Theorem 2.1.5 terminate the proof. □

A final comment: the convergence results above heavily rely on the fact that *all* the iterates u_i are stored and taken into account for (4.2.2). No trick similar to the compression mechanism of §IX.2.1 seems possible in general. We will return to the cutting-plane algorithm in Chap. XV.

178 XII. Abstract Duality for Practitioners

5 Putting the Method in Perspective

On several occasions, we have observed connections between the present dual approach and various concepts from the previous chapters. These connections were by no means fortuitous, and exploiting them allows fruitful interpretations and extensions. In this section, Assumption 1.1.5 is not in force; in particular, Θ may assume the value $+\infty$.

5.1 The Primal Function

From its definition itself, the dual function strongly connotes the conjugacy operation of Chap. X. This becomes even more suggestive if we introduce the artificial variable $\gamma = c(u)$ (the coefficient of λ); such a trick is always very fruitful when dealing with duality. Here we consider the following function (remember §IV.2.4):

$$\mathbb{R}^m \ni \gamma \mapsto P(\gamma) := -\sup\{\varphi(u) : u \in U, \ c(u) = \gamma\}. \tag{5.1.1}$$

In other words, the right-hand side of the constraints in the primal problem (1.1.1) is considered as a varying parameter γ, the supremal value being a function of that parameter; and the original problem is recovered when γ is fixed to 0. It is this *primal function* that yields the expected conjugacy correspondence. The reason for the change of sign is that convexity comes in in a more handy way.

Theorem 5.1.1 *Suppose* $\operatorname{dom}\Theta \neq \emptyset$. *Then*

$$P^*(\lambda) = \Theta(-\lambda) \quad \text{for all } \lambda \in \mathbb{R}^m.$$

PROOF. By definition,

$$\begin{aligned} P^*(\lambda) &= \sup_\gamma \left[\lambda^\top \gamma - P(\gamma)\right] \\ &= \sup_\gamma \left\{\sup_{u \in U}[\lambda^\top \gamma + \varphi(u) : c(u) = \gamma]\right\} \\ &= \sup_\gamma \left\{\sup_{u \in U}[\lambda^\top c(u) + \varphi(u)]\right\} \\ &= \sup_{u \in U} \left[\lambda^\top c(u) + \varphi(u)\right] = \Theta(-\lambda). \end{aligned}$$
□

Remark 5.1.2 The property $\operatorname{dom}\Theta \neq \emptyset$ should be expressed by a condition on the primal data (U, φ, c). As seen in the beginning of Chap. X, $\operatorname{dom}\Theta = -\operatorname{dom}P^*$ is (up to a symmetry) the set of slopes minorizing P. The condition in question is therefore: for some $(\theta, \mu) \in \mathbb{R} \times \mathbb{R}^m$,

$$P(\gamma) \geqslant \theta + \mu^\top \gamma \quad \text{for all } \gamma \in \mathbb{R}^m.$$

By definition of P, this exactly means

$$\varphi(u) \leqslant -\theta - \mu^\top c(u) \quad \text{for all } u \in U,$$

i.e. $\Theta(-\mu) \leqslant -\theta < +\infty$.

This property holds for example when φ is bounded from above on U (take $\mu = 0$). In view of Proposition IV.1.2.1, it also holds if $P \in \operatorname{Conv} \mathbb{R}^m$, which in turn is true if (use the notation (2.4.1), Theorem IV.2.4.2, and remember that φ is assumed finite-valued on $U \neq \emptyset$):

$$\left.\begin{array}{l} U \subset \mathbb{R}^n \text{ is a convex set, } -\varphi \in \operatorname{Conv} \mathbb{R}^n, \ c : \mathbb{R}^n \to \mathbb{R}^m \text{ is linear} \\ \text{and, for all } \gamma \in C(U), \ \varphi \text{ is bounded from above on } c^{-1}(\gamma). \end{array}\right\} \quad (5.1.2)$$

$\square$

Theorem 5.1.1 explains why convexity popped up in Sections 2.3 and 2.4: being a conjugate function, Θ *does not distinguish* the starting function P from its closed convex hull. As a first consequence, the closed convex hull of P is readily obtained:

$$P^{**}(\gamma) = \overline{\operatorname{co}} P(\gamma) = \Theta^*(-\gamma) \quad \text{for all } \gamma \in \mathbb{R}^m.$$

This sheds some light on the connection between the primal-dual pair of problems (1.1.1) and (2.1.6):

(i) When it exists, the duality gap is the number

$$\inf_{\mathbb{R}^m} \Theta - (-P)(0) = -\Theta^*(0) + P(0) = P(0) - \overline{\operatorname{co}} P(0),$$

i.e. the discrepancy at 0 between P and its closed convex hull. There is no duality gap if $P \in \overline{\operatorname{Conv}} \mathbb{R}^m$, or at least if P has a "closed convex behaviour near 0"; see (iii) below.

In the situation (5.1.2), $P \in \operatorname{Conv} \mathbb{R}^m$, so $\overline{\operatorname{co}} P = \operatorname{cl} P$. In this case, there is no duality gap if $P(0) = \operatorname{cl} P(0)$.

(ii) The minima of a closed convex function have been characterized in (X.1.4.6), which becomes here

$$\operatorname{Argmin}\{\Theta(\lambda) : \lambda \in \mathbb{R}^m\} = \partial \Theta^*(0) = -\partial(\overline{\operatorname{co}} P)(0).$$

Using Theorem X.1.4.2, a sufficient condition for existence of a dual solution is therefore $0 \in \operatorname{ri dom}(\overline{\operatorname{co}} P)$, and this explains Theorem 2.4.1(iii).

(iii) Absence of a duality gap means $P(0) = \overline{\operatorname{co}} P(0) \ [< +\infty]$, and we know from Proposition X.1.4.3(ii) that $\partial P(0) = \partial(\overline{\operatorname{co}} P)(0)$ in this case. Thus, the primal problems that are amenable to duality are those where P has a sublinearization at 0:

$$\left.\begin{array}{l} \text{a dual solution exists and} \\ \text{there is no duality gap} \end{array}\right\} \iff \partial P(0) \neq \emptyset.$$

Do not believe, however, that the λ-problem (2.1.3) has in this case a solution $\bar{\lambda} \ [\in -\partial P(0)]$: existence of a solution to the Lagrange problem $(1.1.4)_{\bar\lambda}$ and the filling property (2.3.2) at $\bar\lambda$ are still needed for this.

(iv) The optimization problem (5.1.1) has a dualization in its own right; it gives

$$L^\gamma(u, \lambda) := L(u, \lambda) + \lambda^\top \gamma, \quad \Theta^\gamma(\lambda) := \Theta(\lambda) + \lambda^\top \gamma,$$

where $L = L^0$ and $\Theta = \Theta^0$ are the Lagrangian and dual function associated with the original problem (U, φ, c). Assume that there is no duality gap, not only at $\gamma = 0$, but in a whole neighborhood of 0; in other words,

$$-P(\gamma) = \inf_\lambda \Theta^\gamma(\lambda) = \inf_\lambda \left[\Theta(\lambda) + \lambda^\top \gamma\right] \quad \text{for } \|\gamma\| \text{ small enough}.$$

This proves once more Theorem 5.1.1, but also explains Theorem VII.3.3.2, giving an expression for the subdifferential of a closed convex primal function: even though (5.1.1) does not suggest it, P is in this case a sup-function and its subdifferential can be obtained from the calculus rules of §VI.4.4.

(v) The most interesting observation is perhaps that the dual problem actually solves the "closed convex version" of the primal problem.

This latter object is rather complicated, but it simplifies in some cases, for example when the data (U, φ, c) are as follows:

$$\left| \begin{array}{l} U \text{ is a bounded subset of } \mathbb{R}^n, \\ \varphi(u) = \langle q, u \rangle \text{ is linear}, \\ c(u) = Au - b \text{ is affine}. \end{array} \right. \tag{5.1.3}$$

Proposition 5.1.3 *Consider the dual problem associated with (U, φ, c) in (5.1.3). Its infimal value is the supremal value in*

$$\sup\{\langle q, u\rangle \,:\, u \in \overline{\mathrm{co}}\, U,\ Au - b = 0\}. \tag{5.1.4}$$

Furthermore, assume that this dual problem has some optimal solution $\bar\lambda$; then the solutions of (5.1.4) are those

$$u \in (\overline{\mathrm{co}}\, U)(\bar\lambda) = \{u \in \overline{\mathrm{co}}\, U \,:\, L(u, \bar\lambda) = \Theta(\bar\lambda)\}$$

that satisfy $Au = b$.

PROOF. In the case of (5.1.3), we recognize in (5.1.1) the definition of an image-function:

$$P(\gamma - b) = \inf\{I_U(u) - \langle q, u\rangle \,:\, Au = \gamma\} \quad \text{for all } \gamma \in \mathbb{R}^m;$$

let us compute its biconjugate. Thanks to the boundedness of U, Theorem X.2.1.1 applies and, using various calculus rules in X.1.3.1, we obtain

$$P^*(\lambda) = (I_U - \langle q, \cdot\rangle)^*(A^*\lambda) + \lambda^\top b = \sigma_U(A^*\lambda + q) + \lambda^\top b \quad \text{for all } \lambda \in \mathbb{R}^m.$$

The support function $\sigma_U = \sigma_{\overline{\mathrm{co}}\, U}$ is finite everywhere; Theorem X.2.2.1 applies and, using again X.1.3.1, we conclude

$$\begin{aligned}(\overline{\mathrm{co}}\, P)(\gamma - b) &= \inf\{I_{\overline{\mathrm{co}}\, U}(u) - \langle q, u\rangle \,:\, Au = \gamma\} \\ &= -\sup\{\langle q, u\rangle \,:\, u \in \overline{\mathrm{co}}\, U,\ Au = \gamma\} \quad \text{for all } \gamma \in \mathbb{R}^m.\end{aligned}$$

In other words, closing the duality gap amounts to replacing U by $\overline{\mathrm{co}}\, U$.

To finish the proof, observe that the primal problem (5.1.4) satisfies the assumptions of Lemma 2.3.2 and Corollary 2.3.6(ii). □

Note: it is only for simplicity that we have assumed U bounded; the result would still hold with finer hypotheses relating Ker A with the asymptotic cone of $\overline{\mathrm{co}}\, U$. An interesting instance of (5.1.3) is *integer linear programming*. Consider the primal problem in $\mathbb{N}^n$ (a sort of generalized knapsack problem)

$$\left| \begin{array}{l} \inf c^\top x \\ Ax = a \in \mathbb{R}^m, \quad Bx - b \in -(\mathbb{R}^+)^p \quad \text{[i.e. } Bx \leqslant b \in \mathbb{R}^p\text{]} \\ x^i \in \mathbb{N}, \; x^i \leqslant \bar{x} \quad \text{for } i = 1, \ldots, n. \end{array} \right. \quad (5.1.5)$$

Here $\bar{x}$ is some positive integer, introduced again to avoid technicalities; then the next result uses the duality scheme

$$\mathbb{N}^n \times \mathbb{R}^m \times (\mathbb{R}^+)^p \ni (x, \lambda, \mu) \mapsto L(x, \lambda, \mu) = (c + A^\top \lambda + B^\top \mu)^\top x - \lambda^\top a - \mu^\top b.$$

Corollary 5.1.4 *The dual optimal value associated with (5.1.5) is the optimal value of*

$$\left| \begin{array}{l} \inf c^\top x \\ Ax = a, \quad Bx - b \in -(\mathbb{R}^+)^p, \\ 0 \leqslant x^i \leqslant \bar{x} \quad \text{for } i = 1, \ldots, n. \end{array} \right. \quad (5.1.6)$$

PROOF. Introduce slack variables in (5.1.5) to describe the inequality constraints as $Bx + z = b$, $z \geqslant 0$. Because x is bounded, z is bounded as well, say $z^j \leqslant \bar{z}$ for $j = 1, \ldots, p$. Then we are in the situation (5.1.3) with

$$U := \{(x, z) \in \mathbb{N}^n \times [0, \bar{z}]^p \; : \; x^i \leqslant \bar{x} \text{ for } i = 1, \ldots, n\},$$

whose closed convex hull is clearly $[0, \bar{x}]^n \times [0, \bar{z}]^p$. Thus Proposition 5.1.3 applies, and it suffices to eliminate the slacks to obtain (5.1.6). □

Naturally, (5.1.6) is a standard linear program, to which the primal-dual relations of §3.3 can be applied. The duality technique studied in the present chapter is usually called *Lagrangian relaxation*: the constraint $c(u) = 0$ in (1.1.1) is somehow relaxed, when we maximize the Lagrangian (1.1.3). The linear program (5.1.6) is likewise called the *convex relaxation* of (5.1.5): the integrality constraints are relaxed, to form a convex minimization problem. The message of Corollary 5.1.4 is that both techniques are equivalent for integer linear programming.

5.2 Augmented Lagrangians

We saw in §2 that an important issue was *uniqueness* of a solution to the Lagrange problem $(1.1.4)_\lambda$. More precisely, the important property was single-valuedness of the multifunction $\lambda \mapsto C(\lambda)$ of (2.3.3). Such a property would imply, under the filling property (2.3.2):

– differentiability of the dual function,
– direct availability of a primal solution, from the black box called at a dual optimum.

To get a singleton for $C(\lambda)$, a wise idea is to make the Lagrangian *strictly concave* with respect to the "variable" $c(u)$; and this in turn might become possible if we subtract for example $\|c(u)\|^2$ from the Lagrangian. Indeed, for given $t \geqslant 0$, the problem

$$\left| \begin{array}{l} \sup \left[\varphi(u) - \tfrac{1}{2} t \|c(u)\|^2 \right] \quad u \in U, \\ c(u) = 0 \in \mathbb{R}^m \end{array} \right. \quad (5.2.1)$$

is evidently equivalent to (1.1.1): it has the same feasible set, and the same objective function there. In the dual space, however, this equivalence no longer holds: the Lagrangian associated with (5.2.1) is

$$L_t(u, \lambda) := \varphi(u) - \tfrac{1}{2}t\|c(u)\|^2 - \lambda^\top c(u) = L(u, \lambda) - \tfrac{1}{2}t\|c(u)\|^2,$$

called the *augmented Lagrangian* associated with (1.1.1) (knowing that $L = L_0$ is the "ordinary" Lagrangian). Correspondingly, we have the "augmented dual function"

$$\Theta_t(\lambda) := \sup_{u \in U} L_t(u, \lambda). \tag{5.2.2}$$

Remark 5.2.1 (Inequality Constraints) When the problem has inequality constraints, as in (3.2.1), the augmentation goes as follows: with slack variables, we have

$$U \times (\mathbb{R}^+)^p \times \mathbb{R}^p \ni (u, v, \lambda) \mapsto L_t(u, v, \lambda) = \varphi(u) - \lambda^\top[c(u) + v] - \tfrac{1}{2}t\|c(u) + v\|^2,$$

which can be maximized with respect to $v \in (\mathbb{R}^+)^p$: we obtain

$$v_j = v_j(u, \lambda_j) = \max\left\{0, -c_j(u) - \tfrac{1}{t}\lambda_j\right\} \quad \text{for } j = 1, \ldots, p.$$

Working out the calculations, the "non-slackened augmented Lagrangian" boils down to

$$U \times \mathbb{R}^p \ni (u, \lambda) \mapsto \ell_t(u, \lambda) = \varphi(u) - \sum_{j=1}^p \pi_t(c_j(u), \lambda_j),$$

where the function $\pi_t : \mathbb{R}^2 \to \mathbb{R}$ is defined by

$$\pi_t(\gamma, \mu) = \begin{cases} \mu\gamma + \tfrac{1}{2}t\gamma^2 & \text{if } \gamma \geqslant \tfrac{1}{t}\mu, \\ -\tfrac{1}{2t}\|\mu\|^2 & \text{if } \gamma \leqslant \tfrac{1}{t}\mu. \end{cases} \tag{5.2.3}$$

In words: the jth constraint is "augmentedly dualized" if it is frankly violated; otherwise, it is neglected, but a correcting term is added to obtain continuity in u. Note: the dual variables μ_j are no longer constrained, and they appear in a "more strictly convex" way. □

A mere application of the weak duality theorem to the primal-dual pair (5.2.1), (5.2.2) gives

$$\Theta_t(\lambda) \geqslant \varphi(u) - \tfrac{1}{2}t\|c(u)\|^2 = \varphi(u) \tag{5.2.4}$$

for all t, all $\lambda \in \mathbb{R}^m$ and all u feasible in (5.2.1) or (1.1.1). On the other hand, the obvious inequality $L_t \leqslant L$ extends to dual functions: $\Theta_t \leqslant \Theta$, i.e. the augmented Lagrangian approach cannot worsen the duality gap. Indeed, this approach turns out to be efficient when the "lack of convexity" of the primal function can be corrected by a quadratic term:

Theorem 5.2.2 *With P defined by* (5.1.1), *suppose that there are $t_0 \geqslant 0$ and $\bar{\lambda} \in \mathbb{R}^m$ such that*

$$P(\gamma) \geqslant P(0) - \bar{\lambda}^\top \gamma - \tfrac{1}{2}t_0\|\gamma\|^2 \quad \text{for all } \gamma \in \mathbb{R}^m. \tag{5.2.5}$$

Then, for all $t \geqslant t_0$, there is no duality gap associated with the augmented Lagrangian L_t, and actually

$$-P(0) = \Theta_t(\bar{\lambda}) \leqslant \Theta_t(\lambda) \quad \text{for all } \lambda \in \mathbb{R}^m.$$

PROOF. When (5.2.5) holds, it holds also with t_0 replaced by any $t \geq t_0$, and then we have for all γ:

$$\begin{aligned}-P(0) &\geq -P(\gamma) - \bar{\lambda}^T \gamma - \tfrac{1}{2}t\|\gamma\|^2 = \\ &= \sup_{u \in U}\left\{\varphi(u) - \bar{\lambda}^T\gamma - \tfrac{1}{2}t\|\gamma\|^2 \,:\, c(u) = \gamma\right\} \\ &= \sup_{u \in U}\{L_t(u, \bar{\lambda}) \,:\, c(u) = \gamma\}.\end{aligned}$$

Since γ was arbitrary, we conclude that

$$-P(0) \geq \sup_{u,\gamma}\{L_t(u,\bar{\lambda}) \,:\, c(u) = \gamma\} = \Theta_t(\bar{\lambda}).$$

Remember the definition (5.1.1) of P: this means precisely that there is no duality gap, and that $\bar{\lambda}$ minimizes Θ_t. □

This result is to be compared with our comment in §5.1(iii): $\bar{\lambda}$ satisfies (5.2.5) if and only if it minimizes Θ_{t_0}. Needless to say, (5.2.5) is more tolerant than the property $\partial P(0) \neq \emptyset$. Actually, the primal function associated with (5.2.1) is

$$-P_t(\gamma) := \sup_{u \in U}\left\{\varphi(u) - \tfrac{1}{2}t\|c(u)\|^2 \,:\, c(u) = \gamma\right\} = -P(\gamma) - \tfrac{1}{2}t\|\gamma\|^2$$

and (5.2.5) just means $\partial P_{t_0}(0) \neq \emptyset$ (attention: the calculus rule on a sum of subdifferentials is absurd for $P_t = P + 1/2\,t\|\cdot\|^2$, simply because P is not convex!).

The perturbed primal function establishes an interesting connection between the augmented Lagrangian and the Moreau-Yosida regularization of Example XI.3.4.4:

Proposition 5.2.3 *Suppose $P \in \overline{\mathrm{Conv}}\,\mathbb{R}^m$; then, for all $t > 0$,*

$$\Theta_t(\lambda) = \left(\Theta \,\underset{\vee}{+}\, \tfrac{1}{2t}\|\cdot\|^2\right)(\lambda) = \min\left\{\Theta(\mu) + \tfrac{1}{2t}\|\mu - \lambda\|^2 \,:\, \mu \in \mathbb{R}^m\right\}. \quad (5.2.6)$$

PROOF. Apply Theorem 5.1.1 to the perturbed primal function P_t:

$$\Theta_t(-\lambda) = P_t^*(\lambda) = \left(P + \tfrac{1}{2}t\|\cdot\|^2\right)^*(\lambda).$$

The squared-norm function is finite-valued everywhere, so the inf-convolution of this sum of closed convex functions is given by Theorem X.2.3.2:

$$\Theta_t(\lambda) = \left(P^* \,\underset{\vee}{+}\, \tfrac{1}{2t}\|\cdot\|^2\right)(-\lambda) = \min\left\{\Theta(-\mu) + \tfrac{1}{2t}\|\nu\|^2 \,:\, \mu + \nu = -\lambda\right\},$$

and this is exactly (5.2.6). □

Thus, suppose that our initial problem (1.1.1) is such that $P + 1/2\,t_0\|\cdot\|^2 \in \overline{\mathrm{Conv}}\,\mathbb{R}^m$ for some $t_0 \geq 0$. Then the augmented Lagrangian approach becomes extremely efficient:

– It suppresses the duality gap (Theorem 5.2.2).
– It smoothes out the dual function: use (5.2.6) to realize that, for $t > t_0$,

$$\Theta_t = \Theta_{t_0} \,\underset{\vee}{+}\, \tfrac{1}{2(t-t_0)}\|\cdot\|^2$$

is a Moreau-Yosida regularization; as such, it has a Lipschitzian gradient mapping with constant $1/(t - t_0)$ on $\mathbb{R}^m$, see Example XI.3.4.4.

184   XII. Abstract Duality for Practitioners

– We will see in Chap. XV that there are sound numerical algorithms computing a Moreau-Yosida regularization, at least approximately; then the gradient $\nabla \Theta_t$ (or at least approximations of it) will be available "for free", see (XI.3.4.8).

However, the augmented Lagrangian technique suffers a serious drawback: it usually kills the practical Assumption 1.1.2. The reader can convince himself that, in all examples of §1.2, the augmented Lagrangian no longer has a decomposed structure; roughly speaking, if a method is conceived to maximize it, the same method will solve the original problem (1.1.1) as well, or at least its penalized form

$$\sup_{u \in U} \left[\varphi(u) - \tfrac{1}{2} t \|c(u)\|^2 \right]$$

(see §VII.3.2 for example). We conclude that, *in practice*, a crude use of the augmented Lagrangian is rarely possible. On the other hand, it can be quite useful in theory, particularly for various interpretational exercises: remember our comments at the end of §VII.3.2.

Example 5.2.4 Consider the simple knapsack problem (2.2.2), whose optimal value is 0, but for which the minimal value of Θ is 1/2 (see Remark 2.3.5 again). The primal function is plotted on Fig. 5.2.1, observe that the value at 0 of its convex hull is $-1/2$ (the opposite of the optimal dual value). The graph of P_t is obtained by bending upwards the graph of P, i.e. adding $1/2\,t\|\cdot\|^2$; for $t \geq t_0 = 2$, the discontinuity at $\gamma = 1$ is lifted high enough to yield (5.2.5) with $\bar{\lambda} = 0$. In view of Theorem 5.2.2, the duality gap is suppressed.

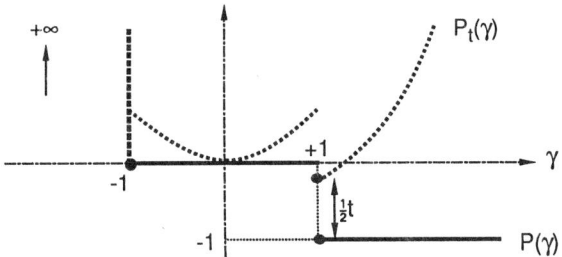

Fig. 5.2.1. The primal function in a knapsack problem

Calculations confirm this happy event: with the help of (5.2.3), the augmented dual function is (we neglect large values of λ)

$$\Theta_t(\lambda) = \max\left\{ \tfrac{1}{2t}\lambda^2, 1 - \tfrac{1}{2}t - \lambda \right\} \quad \text{for } |\lambda| \leq t;$$

this shows that, if $t \geq 2$ and λ is close to 0, $\Theta_t(\lambda) = 1/(2t)\,\lambda^2$, whose minimal value is 0.

Examining Fig. 5.2.1, the reader can convince himself that the same phenomenon occurs for an arbitrary knapsack problem; and even more generally, for any integer program such as (5.1.5). In other words, any integer linear programming problem is equivalent to minimizing the (convex) augmented dual function: an easy problem. The price for this "miracle" is the (expensive) maximization of the augmented Lagrangian.   □

To conclude, we mention that (5.2.1) is not the only possible augmentation. Closed convexity of P_t implies $\partial P_t(\gamma) \neq \emptyset$ for "many" γ (namely on the whole relative interior of dom P_t = dom P), a somewhat luxurious property: as far as closing the duality gap is concerned, it suffices to have $\partial P_t(0) \neq \emptyset$ (Theorem 5.2.2). In terms of (1.1.1), this latter property means that it is possible to add some stiff enough quadratic term to P, thus obtaining a perturbed primal function which "looks convex near 0". Other devices may be more appropriate in some other situations:

Example 5.2.5 Suppose that there are $\bar\lambda \in \mathbb{R}^m$ and $t_0 > 0$ such that

$$P(\gamma) \geqslant P(0) - \bar\lambda^T \gamma - t_0 \|\gamma\| \quad \text{for all } \gamma \in \mathbb{R}^m;$$

in other words, the same apparent convexity near $\gamma = 0$ is yielded by some steep enough sublinear term added to P. Then consider for $t > 0$ the class of problems

$$-\hat{P}_t(\gamma) := \sup_{u \in U} \{\varphi(u) - t\|c(u)\| : c(u) = \gamma\} = -P(\gamma) - t\|\gamma\|, \quad (5.2.7)$$

going together with the augmented dualization

$$\hat{L}_t(u, \lambda) := L(u, \lambda) - t\|c(u)\|, \quad \hat{\Theta}_t(\lambda) := \sup_{u \in U} \hat{L}_t(u, \lambda).$$

This latter augmentation does suppress the duality gap at $\gamma = 0$: reproduce the proof of Theorem 5.2.2 to obtain

$$-\hat{P}_t(0) = \inf_{\mathbb{R}^m} \hat{\Theta} = \hat{\Theta}(\bar\lambda) \quad \text{for } t \geqslant t_0.$$

Concerning the regularization ability proved by Proposition 5.2.3, we have the following: $t\|\cdot\|$ is the support function of the ball $B(0, t)$ (§V.2.3); its conjugate is the indicator function of $B(0, t)$ (Example X.1.1.5); computing the infimal convolution of the latter with a function such as Θ gives at λ the value $\inf_{\mu - \lambda \in B(0,t)} \Theta(\mu)$. In summary, Proposition 5.2.3 can be reproduced to establish the correspondence

$$\hat{\Theta}_t(\lambda) = \min\{\Theta(\mu) : \|\mu - \lambda\| \leqslant t\} \quad \text{if } P \in \overline{\text{Conv}}\, \mathbb{R}^m.$$

Finally note the connection with the exact penalty technique of §VII.1.2. We know from Corollary VII.3.2.3 that, under appropriate assumptions and for t large enough, $\hat{P}_t(0)$ is not changed if the constraint $c(u) = 0$ is removed from (5.2.7). □

5.3 The Dualization Scheme in Various Situations

In §1.2, we have applied duality for the actual solution of some practical optimization problems; more examples have been seen in §3. We now review a few situations where duality can be extremely useful also for theoretical purposes.

(a) Constraints with Values in a Cone. Consider abstractly a primal problem posed under the form

$$\left| \begin{array}{l} \sup \varphi(u) \quad u \in U, \\ c(u) \in K. \end{array} \right. \quad (5.3.1)$$

In (1.1.1) for example, we had $K = \{0\} \subset \mathbb{R}^m$; in (3.2.1), K was the nonpositive orthant of $\mathbb{R}^p$. More generally, we take here for K a *closed convex cone* in some finite-dimensional vector space, call it $\mathbb{R}^m$, equipped with a scalar product $\langle \cdot, \cdot \rangle$; and K° will be the polar cone of K.

The Lagrange function is then conveniently defined as

$$U \times K^\circ \ni (u, \lambda) \mapsto L(u, \lambda) = \varphi(u) - \langle \lambda, c(u) \rangle, \qquad (5.3.2)$$

and the dual function is $\Theta = \sup_{u \in U} L(u, \cdot)$ as before. This dualization enters the framework of §3.1, as is shown by an easy check of (3.1.1):

$$\inf_{\lambda \in K^\circ} L(u, \lambda) = \varphi(u) - \sup_{\lambda \in K^\circ} \langle \lambda, c(u) \rangle = \varphi(u) - I_K(c(u)),$$

where the last equality simply relates the support and indicator functions of a closed convex cone; see Example V.2.3.1. The weak duality theorem

$$\Theta(\lambda) \geqslant \varphi(u) \quad \text{for all } (u, \lambda) \in U \times K^\circ \text{ with } u \text{ feasible}$$

follows (or could be checked in the first place).

Slack variables can be used in (5.3.1):

$$c(u) \in K \quad \Longleftrightarrow \quad [c(u) - v = 0, \ v \in K],$$

a formulation useful to adapt Everett's Theorem 2.1.1. Indeed the slackened Lagrangian is

$$U \times K \times K^\circ \ni (u, v, \lambda) \mapsto \varphi(u) - \langle \lambda, c(u) \rangle + \langle \lambda, v \rangle,$$

and a key is to observe that $\langle \lambda, v \rangle$ stays constant when v describes the face $F_K(\lambda)$ of K exposed by $\lambda \in K^\circ$. In other words, if u_λ maximizes the Lagrangian (5.3.2), the whole set $\{u_\lambda\} \times F_K(\lambda) \subset U \times K$ maximizes its slackened version. Everett's theorem becomes: a maximizer u_λ of the Lagrangian (5.3.2) solves the perturbed primal problem

$$\sup_{u \in U} \{\varphi(u) : c(u) - c(u_\lambda) \in K - F_K(\lambda)\}.$$

If $c(u_\lambda) \in F_K(\lambda)$, the feasible set in this problem contains that of (5.3.1). Thus, Corollary 2.1.2 becomes: if some u_λ maximizing the Lagrangian is feasible and satisfies the complementarity slackness $\langle \lambda, c(u_\lambda) \rangle = 0$, then this u_λ is optimal in (5.3.1).

The results from Sections 2.3 and 2.4 can also be adapted via a slight generalization of §3.2(b) and (c), which is left to the reader. We just mention that in §3.2, the cones K and K° were both full-dimensional, yielding some simplification in the formulae.

(b) Penalization of the Constraints. We mentioned in §1 that the constraints in (1.1.1) should be "soft", i.e. some tolerance should be accepted on the feasibility of primal solutions; this allows an approximate minimization of Θ. A sensible idea is then to penalize these soft constraints, for example replacing (1.1.1) by

$$\sup_{u \in U} \left[\varphi(u) - \tfrac{1}{2} t \| c(u) \|^2 \right], \tag{5.3.3}$$

for a given penalty parameter $t > 0$ (supposed to be "large"); any concept of feasibility has disappeared. In terms of the practical problem to be solved, constraint violations are replaced by a price to pay; a quadratic price may not be particularly realistic, but other penalizations could be chosen as well (remember Chap. VII, more specifically its Sections 1.2 and 3.2).

Even under this form, our primal problem is still amenable to duality, thus preserving the possible advantage of the approach, for example in decomposable cases. Introduce the additional variable $v \in \mathbb{R}^m$ and formulate (5.3.3) as

$$\left| \begin{array}{l} \sup \left[\varphi(u) - \tfrac{1}{2} t \| v \|^2 \right] \quad u \in U, \ v \in \mathbb{R}^m, \\ c(u) = v; \end{array} \right. \tag{5.3.4}$$

note the difference with (5.2.1): the extra variable v is free, while it was frozen to 0 in the case of an augmented Lagrangian. We find ourselves in a constrained optimization framework, and we can define the Lagrangian

$$\widetilde{L}_t(u, v, \lambda) := \varphi(u) - \tfrac{1}{2} t \| v \|^2 - \lambda^\top [c(u) - v] = L(u, \lambda) + \lambda^\top v - \tfrac{1}{2} t \| v \|^2$$

for all $(u, v, \lambda) \in U \times \mathbb{R}^m \times \mathbb{R}^m$. The corresponding dual function is easy to compute: the *same* Lagrangian black box as before is used for u (another difference with §5.2), and the maximization in v is explicit:

$$\widetilde{\Theta}_t(\lambda) := \sup_{u, v} \widetilde{L}_t(u, v, \lambda) = \Theta(\lambda) + \tfrac{1}{2t} \| \lambda \|^2 .$$

A first interesting observation is that $\widetilde{\Theta}_t$ always has a minimum (provided that Θ itself was not identically $+\infty$). This was indeed predicted by Proposition 2.4.1(iii): for our problem (5.3.4), the image-set defined by (2.4.1) is the whole space $\mathbb{R}^m$.

Second, consider the primal function associated with (5.3.4):

$$\mathbb{R}^m \ni \gamma \mapsto -\widetilde{P}_t(\gamma) := \sup_{u \in U} \left[\varphi(u) - \tfrac{1}{2} t \| c(u) - \gamma \|^2 \right] ; \tag{5.3.5}$$

once again, the Moreau-Yosida regularization comes into play (cf. Proposition 5.2.3):

Proposition 5.3.1 *Let $t > 0$; with the notation (5.1.1), (5.3.5), there holds*

$$\widetilde{P}_t(\gamma) = \inf_{z \in \mathbb{R}^m} \left[P(z) + \tfrac{1}{2} t \| z - \gamma \|^2 \right] .$$

PROOF. Associativity of infima is much used. By definition of P,

$$\begin{aligned} \inf_z \left[P(z) + \tfrac{1}{2} t \| z - \gamma \|^2 \right] &= \inf_{z,u} \left\{ -\varphi(u) + \tfrac{1}{2} t \| z - \gamma \|^2 \ : \ c(u) = z \right\} \\ &= \inf_u \left[-\varphi(u) + \tfrac{1}{2} t \inf_{z = c(u)} \| z - \gamma \|^2 \right] \\ &= \inf_u \left[-\varphi(u) + \tfrac{1}{2} t \| c(u) - \gamma \|^2 \right] = \widetilde{P}_t(\gamma) . \quad \square \end{aligned}$$

This result was not totally unexpected: in view of Theorem 5.1.1, the conjugate of $\widetilde{P}_t$ is the sum $\widetilde{\Theta}_t$ of two closed convex functions (barring the change of sign). No wonder, then, that $\widetilde{P}_t$ is an infimal convolution: remember Corollary X.2.1.3. Note, however, that the starting function P is not in $\overline{\mathrm{Conv}}\,\mathbb{R}^m$, so a proof is really needed.

Remark 5.3.2 Compare Proposition 5.3.1 with Proposition 5.2.3. Augmenting the Lagrangian and penalizing the primal constraints are operations somehow conjugate to each other: in one case, the primal function is a sum and the dual function is an infimal convolution; and vice versa. Note an important difference, however: closed convexity of P_t was essential for Proposition 5.2.3, while Proposition 5.3.1 is totally general.

This difference has a profound reason, outlined in Example 5.2.4: the augmented Lagrangian is able to suppress the duality gap, a very powerful property, which therefore requires some assumption; by contrast, penalization brings nothing more than existence of a dual solution. □

(c) Mixed Optimality Conditions for Constrained Minimization Problems. In Chap. VII, we have studied constrained minimization problems from two viewpoints: §VII.1 made an abstract study, the feasible set being just $x \in C$; §VII.2 was analytical, with a feasible set described by equalities and inequalities. Consider now a problem where both forms occur:

$$\left| \begin{array}{l} \inf f(x) \quad x \in C_0, \\ Ax = b, \\ c_j(x) \leqslant 0 \quad \text{for } j = 1, \ldots, p \quad [c(x) \leqslant 0 \text{ for short}]. \end{array} \right. \quad (5.3.6)$$

Here $f \in \overline{\text{Conv}}\, \mathbb{R}^n$, $C_0 \subset \mathbb{R}^n$ is a nonempty closed convex set intersecting dom f, A is linear from $\mathbb{R}^n$ to $\mathbb{R}^m$, and each $c_j \colon \mathbb{R}^n \to \mathbb{R}$ is convex. Thus we now accept an extended-valued objective function; but the constraint-functions are still assumed finite-valued.

This problem enters the general framework of the present chapter if we take the Lagrangian

$$\mathbb{R}^n \times \mathbb{R}^m \times (\mathbb{R}^+)^p \ni (x, \lambda, \mu) \mapsto L(x, \lambda, \mu) = f(x) + \lambda^\top (Ax - b) + \mu^\top c(x)$$

and the corresponding dual function

$$\mathbb{R}^m \times (\mathbb{R}^+)^p \ni (\lambda, \mu) \mapsto \Theta(\lambda, \mu) = -\inf\{L(x, \lambda, \mu) : x \in C_0\}.$$

The control space is now $U = C_0 \cap \text{dom } f$. The whole issue is then whether there is a dual solution, and whether the filling property (2.3.2) holds; altogether, these properties will guarantee the existence of a saddle-point, i.e. of a primal-dual solution-pair.

Theorem 5.3.3 *With the above notation, make the following Slater-type assumption:*

There is $x_0 \in (\text{ri dom } f) \cap \text{ri } C_0$ such that $\left| \begin{array}{l} Ax_0 = b \text{ and} \\ c_j(x_0) < 0 \quad \text{for } j = 1, \ldots, p. \end{array} \right.$

A solution $\bar{x}$ of (5.3.6) is characterized by the existence of $\lambda = (\lambda_1, \ldots, \lambda_m) \in \mathbb{R}^m$ and $\mu = (\mu_1, \ldots, \mu_p) \in \mathbb{R}^p$ such that

$$0 \in \partial f(\bar{x}) + A^\top \lambda + \sum_{j=1}^p \mu_j \partial c_j(\bar{x}) + N_{C_0}(\bar{x}), \quad (5.3.7)$$

$$\mu_j \geqslant 0 \quad \text{and} \quad \mu_j c_j(\bar{x}) = 0 \text{ for } j = 1, \ldots, p.$$

PROOF. Use the notation
$$C_1 := \{x \in \mathbb{R}^n : Ax = b\}, \quad C_2 := \{x \in \mathbb{R}^n : -c(x) \in (\mathbb{R}^+)^p\},$$
so that the solutions of (5.3.6) are those $\bar{x}$ satisfying
$$0 \in \partial(f + I_{C_0} + I_{C_1} + I_{C_2})(\bar{x}).$$
The x_0 postulated in our Slater assumption is in the intersection of the three sets

ri dom $f \cap$ ri C_0	[by assumption]
ri C_1	[$= C_1$]
ri C_2.	[$=$ int C_2: see (VI.1.3.5)]

The calculus rule XI.3.1.2 can therefore be invoked: $\bar{x}$ solves (5.3.6) if and only if
$$0 \in \partial f(\bar{x}) + N_{C_0}(\bar{x}) + N_{C_1}(\bar{x}) + N_{C_2}(\bar{x}).$$
Then it suffices to express the last two normal cones, which was done for example in §VII.2.2. □

Of course, (5.3.7) means that $L(\cdot, \lambda, \mu)$ has a subgradient at $\bar{x}$ whose opposite is normal to C_0 at $\bar{x}$; thus, $\bar{x}$ solves the Lagrangian problem associated with (λ, μ) – and (λ, μ) solves the dual problem.

In view of §2.3, a relevant question is now the following: suppose we have found a dual solution $(\bar{\lambda}, \bar{\mu})$, can we reconstruct a primal solution from it? For this, we need the filling property, which in turn calls for the calculus rule VI.4.4.2. The trick is to realize that the dual function is not changed (and hence, neither are its subdifferentials) if the minimization of the Lagrangian is restricted to some sublevel-set: for r large enough and (λ, μ) in a neighborhood of $(\bar{\lambda}, \bar{\mu})$,
$$-\Theta(\lambda, \mu) = \inf_{x \in C_0} \{L(x, \lambda, \mu) : L(x, \lambda, \mu) \leq r\}.$$
If this new formulation forces x into a compact set, we are done.

Proposition 5.3.4 *With the notation adopted in this subsection, assume that f is 1-coercive on C_0:*
$$\frac{f(x)}{\|x\|} \to +\infty \quad \text{when} \quad \|x\| \to +\infty, \quad x \in C_0.$$
Then, for any bounded set $B \subset \mathbb{R}^m \times (\mathbb{R}^+)^p$, there are a number r and a compact set K such that
$$\{x \in C_0 : L(x, \lambda, \mu) \leq r\} \subset K \quad \text{for all } (\lambda, \mu) \in B.$$

PROOF. Each function c_j is minorized by some affine function: $c_j \geq \langle s_j, \cdot \rangle + \beta_j$ for $j = 1, \ldots, p$. Then write
$$L(x, \lambda, \mu) \geq f(x) - \|A^\top \lambda\| \|x\| - \sum_{j=1}^m \mu_j [\|s_j\| \|x\| + |\beta_j|];$$
the result follows from the 1-coercivity of f on C_0. □

5.4 Fenchel's Duality

On several occasions in this Section 5, connections have appeared between the *Lagrangian duality* of the present chapter and the conjugacy operation of Chap. X. On the other hand, it has been observed in (X.2.3.2) that, for two closed convex functions g_1 and g_2, the optimal value in the "primal problem"

$$m := \inf \{g_1(x) + g_2(x) : x \in \mathbb{R}^n\} \tag{5.4.1}$$

is opposite to that in the "dual problem"

$$\inf \{g_1^*(s) + g_2^*(-s) : s \in \mathbb{R}^n\} \tag{5.4.2}$$

under some appropriate assumption, for example: m is a finite number and

$$\text{ri dom } g_1 \cap \text{ri dom } g_2 \neq \emptyset. \tag{5.4.3}$$

This assumption implies also that (5.4.2) has a solution.

The construction (5.4.1) $\mapsto$ (5.4.2) is called *Fenchel's duality*, whose starting idea is to conjugate the sum $g_1 + g_2$ in (5.4.1). Convexity is therefore required from the very beginning, so that Theorem X.2.3.1 applies. About the relationship between the associated primal-dual optimal sets, the following can be said:

Proposition 5.4.1 *Let g_1 and g_2 be functions of $\overline{\text{Conv}} \, \mathbb{R}^n$ satisfying (5.4.3). If s is an arbitrary solution of (5.4.2), the (possibly empty) solution-set of (5.4.1) is*

$$\partial g_1^*(s) \cap \partial g_2^*(-s). \tag{5.4.4}$$

PROOF. Apply (XI.3.4.5) to the functions $f_i = g_i^*$ for $i = 1, 2$: the infimal convolution $g_1^* \vee\!\!\!\!\wedge\, g_2^*$ is exact at $0 = s - s$ and the subdifferential $\partial(g_1^* \vee\!\!\!\!\wedge\, g_2^*)(0)$ is then (5.4.4). Now apply Theorem X.2.3.2: this last subdifferential is $\partial(g_1 + g_2)^*(0)$, which in turn is just the solution-set of (5.4.1), see (X.1.4.6). □

(a) From Fenchel to Lagrange. As was done in §5.3 in different situations, the approach of this chapter can be applied to Fenchel's duality. It suffices to formulate (5.4.1) as

$$\inf \{g_1(x_1) + g_2(x_2) : x_1 - x_2 = 0\}.$$

This is a minimization problem posed in $\mathbb{R}^n \times \mathbb{R}^n$, with constraint-values in $\mathbb{R}^n$, which lends itself to Lagrangian duality: taking the dual variable $\lambda \in \mathbb{R}^n$, we form the Lagrangian

$$L(x_1, x_2, \lambda) = g_1(x_1) + g_2(x_2) + \lambda^\top (x_1 - x_2).$$

The associated closed convex dual function (to be minimized)

$$\mathbb{R}^n \ni \lambda \mapsto \Theta(\lambda) = -\inf_{x_1, x_2} L(x_1, x_2, \lambda)$$

can be written

$$\Theta(\lambda) = -\inf_{x_1}\left[g_1(x_1) + \lambda^\top x_1\right] - \inf_{x_2}\left[g_2(x_2) - \lambda^\top x_2\right] = g_1^*(-\lambda) + g_2^*(\lambda),$$

a form which blatantly displays Fenchel's dual problem (5.4.2).

(b) From Lagrange to Fenchel. Conversely, suppose we would like to apply Fenchel duality to the problems encountered in this chapter. This can be done at least formally, with the help of appropriate functions in (5.4.1):

(i) $g_2(x) = I_{\{0\}}(Ax - b)$ models affine equality constraints, as in convex instances of (1.1.1);
(ii) $g_2(x) = I_K(c(x))$, where K is a closed convex cone, plays the same role for the problem of §5.3(a); needless to say, inequality constraints correspond to the nonpositive orthant K of $\mathbb{R}^p$;
(iii) $g_2(x) = 1/2\,t\,\|Ax - b\|^2$ is associated to penalized affine constraints, as in §5.3(b);
(iv) in the case of the augmented Lagrangian (5.2.1), we have a sum of three functions:

$$\inf\left[-\varphi(x) + \tfrac{1}{2}t\|Ax - b\|^2 + I_{\{0\}}(Ax - b)\right].$$

Many other situations can be imagined; let us consider the case of affine constraints in some detail.

(c) Qualification Property. With g_1 and f_2 closed convex, A linear from $\mathbb{R}^n$ to $\mathbb{R}^m$, and $\mathbb{R}^m$ equipped with the usual dot-product, consider the following form of (5.4.1):

$$\inf_{x \in \mathbb{R}^n} [g_1(x) + f_2(Ax - b)]. \tag{5.4.5}$$

The role of the control set U is played by dom g_1, and f_2 can be an indicator, a squared norm, etc.; g_2 of (5.4.1) is given by

$$\mathbb{R}^n \ni x \mapsto g_2(x) := f_2(Ax - b).$$

Use Proposition III.2.1.12 to write (5.4.3) in the form

$$\text{there exists } x \in \text{ri dom } g_1 \text{ such that } Ax - b \in \text{ri dom } f_2. \tag{5.4.6}$$

With $f_2 = I_{\{0\}}$ for example, this means $0 \in A(\text{ri dom } g_1) - b = \text{ri}\,[A(\text{dom } g_1) - b]$; compare with condition (iii) in Proposition 2.4.1.

(d) Dual Problem. The conjugate of g_2 can be computed with the help of the calculus rule X.2.2.3: assume (5.4.6), so that in particular $[A(\mathbb{R}^n) - b] \cap \text{ri dom } f_2 \neq \emptyset$,

$$g_2^*(-s) = \min\left\{f_2^*(\lambda) + b^\top \lambda \,:\, A^\top \lambda = -s\right\}, \tag{5.4.7}$$

and the dual problem (5.4.2) associated with (5.4.5) is

$$\min_s \min_\lambda \left\{g_1^*(s) + f_2^*(\lambda) + b^\top \lambda \,:\, A^\top \lambda = -s\right\}.$$

Invert λ and s to obtain $s = -A^\top \lambda$, where λ solves

$$\min_\lambda \left[g_1^*(-A^\top \lambda) + f_2^*(\lambda) + b^\top \lambda\right].$$

By the definition of g_1^*, a further equivalent formulation is
$$\min_\lambda \sup_x \left[-\lambda^\top Ax - g_1(x) + f_2^*(\lambda) + b^\top \lambda\right] \quad \text{or} \quad \min_\lambda \sup_x [f_2^*(\lambda) - L(x,\lambda)].$$

In summary, we have proved the following result:

Proposition 5.4.2 *Assuming (5.4.6), the solutions of Fenchel's dual problem associated with (5.4.5) are $s = -A^\top \lambda$, where λ describes the nonempty solution-set of*
$$\min\{f_2^*(\lambda) + \Theta(\lambda) \,:\, \lambda \in \mathbb{R}^m\}. \tag{5.4.8}$$

Here,
$$\lambda \mapsto \Theta(\lambda) = -\inf\{L(x,\lambda) \,:\, x \in \mathbb{R}^n\}$$

is the dual function associated with the Lagrangian
$$\mathbb{R}^n \times \mathbb{R}^m \ni (x,\lambda) \mapsto L(x,\lambda) = g_1(x) + \lambda^\top (Ax - b). \tag{5.4.9}$$

Furthermore, the optimal values in (5.4.5) and (5.4.8) are equal. □

(e) Primal-Dual Relationship. Proposition 5.4.1 characterizes the solutions of the primal problem (5.4.5) (if any): they are those x satisfying simultaneously

* $x \in \partial g_1^*(-A^\top \lambda)$ with λ solving (5.4.8); equivalently, $0 \in \partial g_1(x) + A^\top \lambda$, i.e. x minimizes the Lagrangian $L(\cdot, \lambda)$ of (5.4.9);
* $x \in \partial g_2^*(A^\top \lambda)$ with λ solving (5.4.8) and g_2^* given by (5.4.7); using the calculus rule XI.3.3.1, this means $Ax - b \in \partial f_2^*(\lambda)$.

To cut a long story short, when we compute the dual function
$$\Theta(\lambda) := \sup_{u \in U} [\varphi(u) - \lambda^\top(Au - b)]$$

associated with the primal problem in the format (1.1.1), namely
$$\sup_{u \in U} \{\varphi(u) \,:\, Au - b = 0\},$$

we obtain precisely
$$\Theta(\lambda) = (-\varphi + I_U)^*(-A^\top \lambda) + \lambda^\top b.$$

(f) Nonlinear Constraints. The situation corresponding to (ii) in §5.4(b) gives also an interesting illustration: with p convex functions $c_1, \ldots, c_p$ from $\mathbb{R}^n$ to $\mathbb{R}$, define
$$\mathbb{R}^n \ni x \mapsto g_j(x) := I_{]-\infty,0]}(c_j(x)) \quad \text{for } j = 1, \ldots, p.$$

Then take an objective function $g_0 \in \overline{\text{Conv}}\,\mathbb{R}^n$, and consider the primal problem inspired by the form (3.2.1) with inequality constraints:
$$\inf_x \left[g_0(x) + \sum_{j=1}^p g_j(x)\right].$$

Its dual in Fenchel's style is
$$\inf\left\{\sum_{j=0}^p g_j^*(s_j) \,:\, \sum_{j=0}^p s_j = 0\right\}; \tag{5.4.10}$$

in "normal" situations, this can also be obtained from Lagrange duality:

Proposition 5.4.3 *With the above notation, assume the existence of $\bar{x}_1, \ldots, \bar{x}_p$ such that $c_j(\bar{x}_j) < 0$ for $j = 1, \ldots, p$. Then the optimal value in (5.4.10) is*

$$\inf\{\Theta(\mu) : \mu \in (\mathbb{R}_*^+)^p\}$$

with

$$\mathbb{R}^p \ni \mu \mapsto \Theta(\mu) := -\inf_{x \in \mathbb{R}^n}\left[g_0(x) + \sum_{j=1}^p \mu_j c_j(x)\right].$$

PROOF. Our assumptions allow the use of the conjugate calculus given in Example X.2.5.3:

$$g_j^*(s) = \min_{\mu > 0} \mu c_j^*(\tfrac{1}{\mu}s) \quad \text{for } j = 1, \ldots, p.$$

Then (5.4.10) has actually two (groups of) minimization variables: $(s_1, \ldots, s_p)$ and $\mu = (\mu_1, \ldots, \mu_p)$. We minimize with respect to $\{s_j\}$ first: the value (5.4.10) is the infimum over $\mu \in (\mathbb{R}_*^+)^p$ of the function

$$\mu \mapsto \inf\left\{g_0^*(s_0) + \sum_{j=1}^p \mu_j c_j^*(\tfrac{1}{\mu_j}s_j) : \sum_{j=0}^p s_j = 0\right\}. \tag{5.4.11}$$

The key is to realize that this is the value at $s = 0$ of the infimal convolution

$$\mathbb{R}^n \ni s \mapsto \left(g_0^* \, \dot{\triangledown} \, \pi_1^* \, \dot{\triangledown} \cdots \dot{\triangledown} \, \pi_p^*\right)(s) =: g(s)$$

where, for $j = 1, \ldots, p$,

$$\mathbb{R}^n \ni s \mapsto \pi_j^*(s) := \mu_j c_j^*(\tfrac{1}{\mu_j}s)$$

is the conjugate of $x \mapsto \pi_j(x) = \mu_j c_j(x)$, a convex function finite everywhere.

Then Theorem X.2.3.2 tells us that g is the conjugate of $g_0 + \pi_1 + \cdots + \pi_p$; its value at zero is opposite to the infimum of this last function. In a word, the function of (5.4.11) reduces to

$$\mu \mapsto -\inf_{x \in \mathbb{R}^n}\left[g_0(x) + \sum_{j=1}^p \mu_j c_j(x)\right]. \qquad \square$$

Note that this is an abstract result, which says nothing about primal-dual relationships, nor existence of primal-dual solutions; in particular, it does not rule out the case dom $\Theta = \emptyset$.

Let us conclude: there is a two-way correspondence between Lagrange and Fenchel duality schemes, even though they start from different primal problems; the difference is mainly a matter of taste. The Lagrangian approach may be deemed more natural and flexible; in particular, it is often efficient when the initial optimization problem contains "intermediate" variables, say $y_j = c_j(x)$, which one wants to single out for some reason. On the other hand, Fenchel's approach is often more direct in theoretical developments.

XIII. Inner Construction of the Approximate Subdifferential: Methods of ε-Descent

Prerequisites. Basic concepts of numerical optimization (Chap. II); bundling mechanism for minimizing convex functions (Chap. IX, essential); definition and elementary properties of approximate subdifferentials and difference quotients (Chap. XI, especially §XI.2.2).

Introduction. In this chapter, we study a first minimization algorithm in detail, including its numerical implementation. It is able to minimize rather general convex functions, without any smoothness assumptions – in contrast to the algorithms of Chap. II. On the other hand, it is fully implementable, which was not the case of the bundling algorithm exposed in Chap. IX.

We do not study this algorithm for its practical value: in our opinion, it is not suitable for "real life" problems. However, it is the one that comes first to mind when starting from the ideas developed in the previous chapters. Furthermore, it provides a good introduction to the (more useful) methods to come in the next chapters: rather than a dead end, it is a point of departure, comparable to the steepest-descent method for minimizing smooth functions, which is bad but basic. Finally, it can be looked at in a context different from optimization proper, namely that of separating closed convex sets.

The situation is the same as in Chap. IX:
- $f : \mathbb{R}^n \to \mathbb{R}$ is a (finite-valued) convex function to be minimized;
- $f(x)$ and $s(x) \in \partial f(x)$ are computed when necessary in a black box (U1), as in §.II.1.2.
- the norm $\|\cdot\|$ used to compute steepest-descent directions and projections (see §VIII.1 and §IX.1) is the Euclidean norm: $\|\cdot\| = \|\cdot\| = \sqrt{\langle\cdot,\cdot\rangle}$.

In other words, we want to minimize f even though the information available is fairly poor: one subgradient at each point, instead of the full subdifferential as in Chap. VIII. Recall that the notation $s(x)$ is misleading but can be used in practice; this was explained in Remark VIII.3.5.1.

1 Introduction. Identifying the Approximate Subdifferential

1.1 The Problem and Its Solution

In Chap. IX, we have studied a mechanism for constructing the subdifferential of f at a given x, or at least to resolve the following alternatives:

– If there exists a hyperplane separating $\partial f(x)$ and $\{0\}$, i.e. a *descent direction*, i.e. a d with $f'(x, d) < 0$, find one.

– Or, if there does not exist any such direction, explain why, i.e. find a subgradient which is (approximately) 0 – the difficulty being that the black box (U1) is not supposed to ever answer $s(x) = 0$.

This mechanism consisted in collecting information about $\partial f(x)$ in a *bundle*, and worked as follows (see Algorithm IX.1.6 for example):

– Given a compact convex polyhedron S under-estimating $\partial f(x)$, i.e. with $S \subset \partial f(x)$, one computed a hyperplane separating S and $\{0\}$. This hyperplane was essentially defined by its normal vector d, interpreted as a direction, and one actually computed *the best* such hyperplane, namely the projection of 0 onto S.
– Then one made a line-search along this d, with two possible exits:

  (a) The hyperplane actually separated $\partial f(x)$ and $\{0\}$, in which case the process was terminated. The line-search was successful and f could be improved in the direction d, to obtain the next iterate, say x_+.

  (b) Or the hyperplane did not separate $\partial f(x)$ and $\{0\}$, in which case the line-search produced a new subgradient, say $s_+ \in \partial f(x)$, to improve the current S – the line-search was unsuccessful and one looped to redo it along a new direction, issuing from the same x, but obtained from the better S.

We explained that this mechanism could be grafted onto each iteration of a descent scheme. It would thus allow the construction of descent directions *without* computing the full subdifferential explicitly, a definite advantage over the steepest-descent method of Chap. VIII. A fairly general algorithm would then be obtained, which could for example minimize dual functions associated with abstract optimization problems; such an algorithm would thus be directly comparable to those of §XII.4.

Difficulty 1.1.1 If we do such a grafting, however, two difficulties appear, which were also mentioned in Remark IX.1.7:

(i) According to the whole idea of the process, the descent direction found in case (a) will be "at best" the steepest-descent direction. As a result, the sequence of iterates will very probably not be a minimizing sequence: this is the message of §VIII.2.2.

(ii) Finding the subgradient s_+ in case (b) requires performing an endless line-search: the stepsize must tend to 0 so as to compute the directional derivative $f'(x, d)$ and to obtain s_+ by virtue of Lemma VI.6.3.4. Unless f has some very specific properties (such as being piecewise affine), this cannot be implemented.

In other words, the grafting will result in a non-convergent and non-implementable algorithm. □

Yet the idea is good if a simple precaution is taken: rather than the *purely local* set $\partial f(x)$, it is $\partial_\varepsilon f(x)$ that we must identify. It gathers differential information from a specific, "finitely small", neighborhood of x – called $\overline{V}_\varepsilon(x)$ in Theorem XI.4.2.5 – instead of a neighborhood shrinking to $\{x\}$, as is the case for $\partial f(x)$ – see Theorem VI.6.3.1. Accordingly, we can guess that this will fix (i) and (ii) above. In fact:

1 Introduction. Identifying the Approximate Subdifferential

(i$_\varepsilon$) A direction d such that $f'_\varepsilon(x,d) < 0$ is downhill not only at x but at any point in $\overline{V}_\varepsilon(x)$, so a line-search along such a direction is able to drive the next iterate out of $\overline{V}_\varepsilon(x)$ and the difficulty 1.1.1(i) will be eliminated.

(ii$_\varepsilon$) A subgradient at a point of the form $x + td$ is in $\partial_\varepsilon f(x)$ for t small enough, more precisely whenever $x + td \in \overline{V}_\varepsilon(x)$; so it is no longer necessary to force the stepsize in (ii) to tend to 0.

Definition 1.1.2 A nonzero $d \in \mathbb{R}^n$ is said to be a direction of ε-*descent* for f at x if $f'_\varepsilon(x,d) < 0$; in other words d defines a hyperplane separating $\partial_\varepsilon f(x)$ and $\{0\}$.

A point $x \in \mathbb{R}^n$ is said to be ε-*minimal* if there is no such separating d, i.e. $f'_\varepsilon(x,d) \geq 0$ for all d, i.e. $0 \in \partial_\varepsilon f(x)$. □

The reason for this terminology is straightforward:

Proposition 1.1.3 *A direction d is of ε-descent if and only if*

$$f(x+td) < f(x) - \varepsilon \quad \text{for some } t > 0.$$

A point x is ε-minimal if and only if it minimizes f within ε, i.e.

$$f(y) \geq f(x) - \varepsilon \quad \text{for all } y \in \mathbb{R}^n.$$

PROOF. Use for example Theorem XI.2.1.1: for given $d \neq 0$, let $\eta := f'_\varepsilon(x,d)$. If $\eta < 0$, we can find $t > 0$ such that

$$f(x+td) - f(x) + \varepsilon \leq \tfrac{1}{2}t\eta < 0.$$

Conversely, $\delta \geq 0$ means that

$$f(x+td) - f(x) + \varepsilon \geq \eta \geq 0 \quad \text{for all } t > 0. \qquad \square$$

Then consider the following algorithmic scheme, of "ε-descent". The descent iterations are denoted here by a superscript p; the subscript k that was used in Chap. VIII will be reserved to the bundling iterations to come later.

Algorithm 1.1.4 (Conceptual Algorithm of ε-Descent) Start from some $x^1 \in \mathbb{R}^n$. Choose $\varepsilon > 0$. Set $p = 1$.

STEP 1. If $0 \in \partial_\varepsilon f(x^p)$ stop. Otherwise compute a direction of ε-descent, say d^p.

STEP 2. Make a line-search along d^p to obtain a stepsize $t^p > 0$ such that

$$f(x^p + t^p d^p) < f(x^p) - \varepsilon.$$

Set $x^{p+1} := x^p + t^p d^p$. Replace p by $p + 1$ and loop to Step 1. □

If this scheme is implementable at all, it will be an interesting alternative to the steepest-descent scheme, clearly eliminating the difficulty 1.1.1(i).

Proposition 1.1.5 *In Algorithm 1.1.4, either $f(x^p) \to -\infty$, or the stop of Step 1 occurs for some finite iteration index p_*, at which x^{p_*} is ε-minimal.*

PROOF. Fairly obvious: one has by construction

$$f(x^p) < f(x^1) - (p-1)\varepsilon \quad \text{for } p = 1, 2, \ldots$$

Therefore, if f is bounded from below, say by $\bar{f} := \inf_x f(x)$, the number p_* of iterations cannot be larger than $[f(x^1) - \bar{f}]/\varepsilon$, after which the stopping criterion of Step 1 plays its role. □

Remark 1.1.6 This is just the basic scheme, for which several variants can easily be imagined. For example, $\varepsilon = \varepsilon^p$ can be chosen at each iteration. A stop at iteration p_* will then mean that x^{p_*} is ε^{p_*}- minimal. Various conditions will ensure that this stop does eventually occur; for example

$$\sum_{p=1}^{p_*} \varepsilon^p \to +\infty \quad \text{when } p_* \to \infty. \tag{1.1.1}$$

It is not the first time that we encounter this idea of taking a divergent series: the subgradient algorithm of §XII.4.1 also used one for its stepsize. It is appropriate to recall Remark XII.4.1.5: here again, a condition like (1.1.1) makes little sense in practice. This is particularly visible with p_* appearing explicitly in the summation: when choosing ε^p, we do not know if the present p^{th} iteration is going to be the last one. In the present context, we just mention one simple "online" rule, more reasonable: take a fixed ε but, when $0 \in \partial_\varepsilon f(x^p)$, diminish ε, say divide it by 2, until it reaches a final acceptable threshold. It is straightforward to extend the proof of Proposition 1.1.5 to such a strategy.

A general convergence *theory* of these schemes with varying ε's is rather trivial and is of little interest; we will not insist on this aspect here. The real issue would be of a *practical* nature, consisting of a study of the values of ε ($= \varepsilon^p$), and/or of the norming $\|\!|\cdot|\!\|$, to reach maximal efficiency – whatever this means! □

Remark 1.1.7 To solve a numerical problem (for example of optimization) one normally constructs a sequence – say $\{x^p\}$, usually infinite – for which one proves some desired asymptotic property: for example that any, or some, cluster point of $\{x^p\}$ solves the initial problem – recall Chap. II, and more particularly (II.1.1.8). In this sense, the statement of Proposition 1.1.5 appears as rather non-classical: we construct a *finite* sequence $\{x^p\}$, and then we establish how close the last iterate comes to optimality. This point of view will be systematically adopted here. It makes convergence results easier to prove in our context, and furthermore we believe that it better reflects what actually happens on the computer.

Observe the double role played by ε at each iteration of Algorithm 1.1.2: an "active" role to compute d^p, and a "passive" role as a stopping criterion. We should really have two different epsilons: one, possibly variable, used to compute the direction; and the other to stop the algorithm. Such considerations will be of importance for the next chapter. □

1 Introduction. Identifying the Approximate Subdifferential

Having eliminated the difficulty 1.1.1(i), we must now check that 1.1.1(ii) is eliminated as well; this is not quite trivial and motivates the present chapter. We therefore study one single iteration of Algorithm 1.1.4, i.e. one execution of Step 1, regardless of any sophisticated choice of ε. In other words, we are interested in the static aspect of the algorithm, in which the current iterate x^p is considered as fixed.

Dropping the index p from our notation, we are given fixed $x \in \mathbb{R}^n$ and $\varepsilon > 0$, and we apply the mechanism of Chap. IX to construct compact convex polyhedra in $\partial_\varepsilon f(x)$.

If we look at Step 1 of Algorithm 1.1.4 in the light of Algorithm IX.1.6, we see that it is going to be an iterative subprocess which works as follows; we use the subscript k as in Chap. IX.

Process 1.1.8 The black box (U1) has computed a first ε-subgradient s_1, and a first polyhedron S_1 is on hand:

$$s_1 := s(x) \in \partial f(x) \subset \partial_\varepsilon f(x) \quad \text{and} \quad S_1 := \{s_1\}.$$

– At stage k, having the current polyhedron $S_k \subset \partial_\varepsilon f(x)$, compute the best hyperplane separating S_k and $\{0\}$, i.e. compute

$$-d_k = \operatorname{Proj} 0/S_k = \operatorname{Argmin}\left\{\tfrac{1}{2}\|\cdot - d\|^2 \,:\, -d \in S_k\right\}$$

(recall that Proj denotes the Euclidean projection onto the closed convex hull of a set).
– Then determine whether

(a$_\varepsilon$) d_k separates not only S_k but even $\partial_\varepsilon f(x)$ from $\{0\}$; then the work is finished, $\partial_\varepsilon f(x)$ is properly identified, d_k is an ε-descent direction and we can pass to the next iteration in Algorithm 1.1.4;

or

(b$_\varepsilon$) d_k does not separate from $\{0\}$; then enlarge S_k with a new $s_{k+1} \in \partial_\varepsilon f(x)$ and loop to the next k. □

Of course, the above alternatives (a$_\varepsilon$) – (b$_\varepsilon$) play the same role as (a) – (b) of §1.1. In Chapter IX, (a) – (b) were resolved by a line-search minimizing f along d_k. In case (b), the optimal stepsize was 0, the event $f(x + td_k) < f(x)$ was impossible to obtain, and stepsizes $t \downarrow 0$ were produced. The new element s_{k+1}, to be appended to S_k, was then a by-product of this line-search, namely a corresponding cluster point of $\{s(x + td_k)\}_{t \downarrow 0}$.

1.2 The Line-Search Function

Here, in order to resolve the alternatives (a$_\varepsilon$) – (b$_\varepsilon$), and detect whether d_k is an ε-descent direction (instead of a mere descent direction), we can again minimize f along d_k and see whether ε can thus be dropped from f. This is no good, however: in case of failure we will not obtain a suitable s_{k+1}.

Instead of simply applying Proposition 1.1.3 and checking whether

$$f(x + td_k) < f(x) - \varepsilon \quad \text{for some } t > 0, \tag{1.2.1}$$

a much better idea is to use Definition 1.1.2 itself: *compute the support function* $f'_\varepsilon(x, d_k)$ and check whether it is negative. From Theorem IX.2.1.1, this means minimizing the perturbed difference quotient, a problem that the present §1.2 is devoted to. Clearly enough, the alternatives $(a_\varepsilon) - (b_\varepsilon)$ will thus be resolved; but in case of failure, when (b_ε) holds, we will see below that a by-product of this minimization is an s_{k+1} suitable for our enlargement problem.

Naturally, the material in this section takes a lot from §XI.2. In the sequel, we drop the index k: given $x \in \mathbb{R}^n$, $d \neq 0$ and $\varepsilon > 0$, the perturbed difference quotient is as in §XI.2.2:

$$q_\varepsilon(t) := \begin{cases} +\infty & \text{if } t \leq 0, \\ \dfrac{f(x+td) - f(x) + \varepsilon}{t} & \text{if } t > 0, \end{cases} \tag{1.2.2}$$

and we are interested in

$$\inf_{t>0} q_\varepsilon(t). \tag{1.2.3}$$

Remark 1.2.1 The above problem (1.2.3) looks like a *line-search*, just as in Chap. II: after all, it amounts to finding a stepsize along the direction d. We should mention, however, that such an interpretation is slightly misleading, as the motivation is substantially different. First, the present "line-search" is aimed at diminishing the perturbed difference quotient, rather than the objective function itself.

More importantly, as we will see in §2.2 below, q_ε must be minimized rather accurately. By contrast, we have insisted long enough in §II.3 to make it clear that a line-search had little to do with one-dimensional minimization: its role was rather to find a "reasonable" stepsize along the given direction, "reasonable" being understood in terms of the objective function.

We just note here that the present "line-search" will certainly not produce a zero-stepsize since $q_\varepsilon(t) \to +\infty$ when $t \downarrow 0$ (cf. Proposition XI.2.2.2). This confirms our observation (ii_ε) in §1.1. □

Minimizing q_ε is a (hidden) convex problem, via the change of variable $u = 1/t$. We recall the following facts from §XI.2.2: the function

$$u \mapsto r_\varepsilon(u) := \begin{cases} q_\varepsilon(\tfrac{1}{u}) = u\left[f\left(x + \tfrac{1}{u}d\right) - f(x) + \varepsilon\right] & \text{if } u > 0, \\ f'_\infty(d) = \lim_{t \to \infty} q_\varepsilon(t) & \text{if } u = 0, \\ +\infty & \text{if } u < 0 \end{cases} \tag{1.2.4}$$

is in $\overline{\text{Conv}}\,\mathbb{R}$. Its subdifferential at $u > 0$ is

$$\partial r_\varepsilon(u) = \left\{\varepsilon - e\left(x, x + \tfrac{1}{u}d, s\right) : s \in \partial f\left(x + \tfrac{1}{u}d\right)\right\};$$

here,

$$e(x, y, s) := f(x) - [f(y) + \langle s, x - y \rangle] \tag{1.2.5}$$

is the linearization error made at x when f is linearized at y with slope s (see the transportation formula of Proposition XI.4.2.2). The function r_ε is minimized over a nonempty compact interval $U_\varepsilon = [\underline{u}_\varepsilon, \bar{u}_\varepsilon]$, with

$$0 \leqslant \underline{u}_\varepsilon \leqslant u \leqslant \bar{u}_\varepsilon < +\infty,$$

and its positive minima are characterized by the property

$$e(x, x + \tfrac{1}{u}d, s) = \varepsilon \quad \text{for some } s \in \partial f(x + \tfrac{1}{u}d).$$

Since we live in the geometrical world of positive stepsizes, we find it convenient to translate these results into the t-language. They allow the following supplement to the classification at the end of §XI.2.2:

Lemma 1.2.2 *The set*

$$T_\varepsilon := \{t = \tfrac{1}{u} : u \in U_\varepsilon \text{ and } u > 0\}$$

of minima of q_ε is a closed interval (possibly empty, possibly not bounded from above), which does not contain 0. Denoting by $\underline{t}_\varepsilon := 1/\bar{u}_\varepsilon$ and $\bar{t}_\varepsilon := 1/\underline{u}_\varepsilon$ the endpoints of T_ε (the convention $1/0 = +\infty$ is used), there holds for any $t > 0$:

(i) $t \in T_\varepsilon \iff e(x, x + td, s) = \varepsilon \ \text{for some } s \in \partial f(x + td)$
(ii) $t < \underline{t}_\varepsilon \iff e(x, x + td, s) < \varepsilon \ \text{for all } s \in \partial f(x + td)$
(iii) $t > \bar{t}_\varepsilon \iff e(x, x + td, s) > \varepsilon \ \text{for all } s \in \partial f(x + td).$

Of course, (iii) is pointless if $\underline{u}_\varepsilon = 0$.

PROOF. All this comes from the change of variable $t = 1/u$ in the convex function r_ε (remember that it is minimized "at finite distance"). The best way to "see" the proof is probably to draw a picture, for example Figs. 1.2.1 and 1.2.2. □

For our present concern of minimizing q_ε, the two cases illustrated respectively by Fig. 1.2.1 ($\bar{u}_\varepsilon > 0$, T_ε nonempty) and Fig. 1.2.2 ($\bar{u}_\varepsilon = \underline{u}_\varepsilon = 0$, T_ε empty) are rather different. For each stepsize $t > 0$, take $s \in \partial f(x + td)$. When t increases,
– the inverse stepsize $u = 1/t$ decreases;
– being the slope of the convex function r_ε, the number $\varepsilon - e(x, x + td, s)$ decreases (not continuously);
– therefore $e(x, x + td, s)$ increases.
– Altogether, $e(x, x + td, s)$ starts from 0 and increases with $t > 0$; a discontinuity occurs at each t such that $\partial f(x + td)$ has a nonzero breadth along d.

The difference between the two cases in Figs. 1.2.1 and 1.2.2 is whether or not $e(x, x + td, s)$ crosses the value ε; this property conditions the non-vacuousness of T_ε.

Lemma 1.2.2 is of course crucial for a one-dimensional search aimed at minimizing q_ε. First, it shows that q_ε is mildly nonconvex (the technical term is *quasi-convex*). Also, the number $\varepsilon - e$, calculated at a given t, contains all the essential information

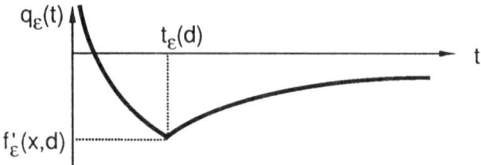

Fig. 1.2.1. A "normal" approximate difference quotient

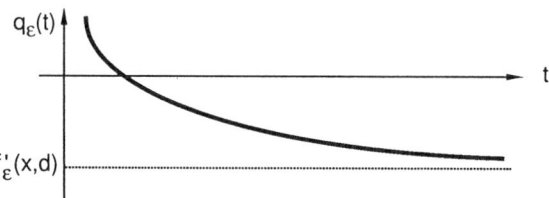

Fig. 1.2.2. An approximate difference quotient with no minimum

of a "derivative". If it happens to be 0, t is optimal; otherwise, it indicates whether q_ε is increasing or decreasing near t.

We must now see why it is more interesting to solve (1.2.3) than (1.2.1). Concerning the test for appropriate descent, minimizing q_ε does some good in terms of diminishing f:

Lemma 1.2.3 *Suppose that d is a direction of ε-descent: $f'_\varepsilon(x, d) < 0$. Then*

$$f(x + td) < f(x) - \varepsilon$$

for $0 < t$ close enough to T_ε (i.e. t large enough when $T_\varepsilon = \emptyset$).

PROOF. Under the stated assumption, $q_\varepsilon(t)$ has the sign of its infimum, namely "$-$".
□

This fixes the (a_ε)-problem in Process 1.1.8. As for the new s_+, necessary in the (b_ε)-case, we remember that it must enjoy two properties:

– It must be an ε-subgradient at x. By virtue of the transportation formula (XI.4.2.2), this requires precisely that q_ε be locally decreasing: we have

$$\partial f(x + td) \ni s \in \partial_\varepsilon f(x) \quad \text{if and only if} \quad e(x, x + td, s) \leqslant \varepsilon. \tag{1.2.6}$$

– It must separate the current S from $\{0\}$ "sufficiently well", which means it must make $\langle s_+, d \rangle$ large enough, see for example Remark IX.2.1.3.

Then the next result will be useful.

Lemma 1.2.4 *Suppose that d is not a direction of ε-descent: $f'_\varepsilon(x, d) \geqslant 0$. Then:*

(j) *If $t \in T_\varepsilon$, there is some $s \in \partial f(x + td)$ such that*

$$\langle s, d \rangle \geqslant 0 \quad \text{and} \quad s \in \partial_\varepsilon f(x).$$

(jj) *Suppose $T_\varepsilon = \emptyset$. Then*

$$\partial f(x+td) \subset \partial_\varepsilon f(x) \text{ for all } t \geq 0, \quad \text{and}$$

for all $\eta > 0$, there exists $M > 0$ such that
$$t \geq M, s \in \partial f(x+td) \implies \langle s, d \rangle \geq -\eta.$$

PROOF. [(j)] Combine Lemma 1.2.2(i) with (1.2.6) to obtain $s \in \partial f(x+td) \cap \partial_\varepsilon f(x)$ such that
$$f(x) = f(x+td) - t\langle s, d\rangle + \varepsilon;$$
since $t > 0$, the property $f(x) \leq f(x+td) + \varepsilon$ implies $\langle s, d \rangle \geq 0$.

[(jj)] In case (jj), every $t > 0$ comes under case (ii) of Lemma 1.2.2, and any $s(t) \in \partial f(x+td)$ is in $\partial_\varepsilon f(x)$ by virtue of (1.2.6). Also, since d is not a direction of ε-descent,

$$\begin{aligned} f(x+td) &\geq f(x) - \varepsilon \\ &\geq f(x+td) - t\langle s(t), d \rangle - \varepsilon. \quad \text{[because } s(t) \in \partial f(x+td)\text{]} \end{aligned}$$

Thus, $\langle s(t), d \rangle + \varepsilon/t \geq 0$, and $\liminf_{t \to \infty} \langle s(t), d \rangle \geq 0$. □

Remark 1.2.5 Compare with §XI.4.2: for all t smaller than the largest element $\bar{t}_\varepsilon(d)$ of T_ε, $\partial f(x+td) \subset \partial_\varepsilon f(x)$ because $x+td \in V_\varepsilon(x)$. For $t = \bar{t}_\varepsilon(d)$, one can only ascertain that some subgradient at $x+td$ is an ε-subgradient at x and can fruitfully enlarge the current polyhedron S. For $t > \bar{t}_\varepsilon(d)$, $\partial f(x+td) \cap \partial_\varepsilon f(x) = \emptyset$ because $x+td \notin \overline{V}_\varepsilon(x)$. □

In summary, when t minimizes q_ε (including the case "t large enough" if $T_\varepsilon = \emptyset$), it is theoretically possible to find in $\partial f(x+td)$ the necessary subgradient s_+ to improve the current approximation of $\partial_\varepsilon f(x)$. We will see in §2 that it is not really possible to find s_+ in practice; rather, it is a convenient approximation of it which will be obtained during the process of minimizing q_ε.

1.3 The Schematic Algorithm

We begin to see how Algorithm 1.1.4 will actually work: Step 1 will be a projection onto a convex polyhedron, followed by a minimization of the convex function r_ε. In anticipation of §2, we restate Algorithm 1.1.4, with a somewhat more detailed Step 1. Also, we incorporate a criterion to stop the algorithm without waiting for the (unlikely) event "$0 \in S_k$".

Algorithm 1.3.1 (Schematic Algorithm of ε-Descent) Start from some $x^1 \in \mathbb{R}^n$. Choose the descent criterion $\varepsilon > 0$ and the convergence parameter $\delta > 0$. Set $p = 1$.

STEP 1.0 (starting to find a direction of ε-descent). Compute $s(x^p) \in \partial f(x^p)$. Set $s_1 = s(x^p)$, $k = 1$.

STEP 1.1 (computing the trial direction). Solve the minimization problem in α

$$\min_{\alpha \in \Delta_k} \tfrac{1}{2} \left\| \sum_{i=1}^{k} \alpha_i s_i \right\|^2, \qquad (1.3.1)$$

where Δ_k is the unit simplex of $\mathbb{R}^k$. Set $d_k = -\sum_{i=1}^{k} \alpha_i s_i$; if $\|d_k\| \leq \delta$ stop.

STEP 1.2 (line-search). Minimize r_ε and conclude:

(a$_\varepsilon$) either $f'_\varepsilon(x^p, d_k) < 0$; then go to Step 2.
(b$_\varepsilon$) or $f'_\varepsilon(x^p, d_k) \geq 0$; then obtain a suitable $s_{k+1} \in \partial_\varepsilon f(x^p)$, replace k by $k+1$ and loop to Step 1.1.

STEP 2 (descent). Obtain $t > 0$ such that $f(x^p + td_k) < f(x^p) - \varepsilon$. Set $x^{p+1} = x^p + td_k$, replace p by $p+1$ and loop to Step 1.0. □

Remark 1.3.2 Within Step 1, k increases by one at each subiteration. The number of such subiterations is not known in advance, so the number of subgradients to be stored can grow arbitrarily large, and the complexity of the projection problem as well. We know, however, that it is not really necessary to keep all the subgradients at each iteration: Theorem IX.2.1.7 tells us that the $(k+1)^{st}$ projection must be made onto a polyhedron which can be as small as the segment $[-d_k, s_{k+1}]$. In other words, the number of subgradients to be stored, i.e. the number of variables in the quadratic problem of Step 1.1, can be as small as 2 (but at the price of a substantial loss in actual performance: remember Fig. IX.2.2.3). □

Remark 1.3.3 If we compare this algorithm to those of Chap. II, we see that it works rather similarly: at each iterate x^p, it constructs a direction d^k, along which it performs a line-search. However there are two new features, both characteristic of all the minimization algorithms to come.

– First, the direction is computed in a rather sophisticated way, as the solution of an auxiliary minimization problem instead of being given explicitly in terms of the "gradient" s_k.
– The second feature brings a rather fundamental difference: it is that the line-search has two possible exits (a$_\varepsilon$) and (b$_\varepsilon$). The first case is the normal one and could be called a *descent step*, in which the current iterate x^p is updated to a better one. In the second case, the iterate is kept where it is and the next line-search will start from the same x^p. As in the bundling Algorithm IX.1.6, this can be called a *null-step*. □

We can guess that there will be two rather different cases in Step 1.2, corresponding to those of Figs. 1.2.1, 1.2.2. This will be seen in more detail in §2, but we give already here an idea of the convergence proof, assuming a simple situation: only the case of Fig. 1.2.1 occurs, q_ε is minimized exactly at each line-search, and correspondingly, an s_{k+1} predicted by Lemma 1.2.4 is found at each null-step. Note that the last two assumptions are rather unrealistic.

Theorem 1.3.4 *Suppose that each execution of Step 1.2 in Algorithm 1.3.1 produces an optimal $u_k > 0$ and a corresponding $s_{k+1} \in \partial f(x^p + 1/u_k d_k)$ such that*

1 Introduction. Identifying the Approximate Subdifferential

$$0 = \varepsilon - e_k \in \partial r_\varepsilon(u_k),$$

where

$$e_k := f(x^p) - f\left(x^p + \tfrac{1}{u_k}d_k\right) + \langle s_{k+1}, \tfrac{1}{u_k}d_k\rangle.$$

Then: either $f(x^p) \to -\infty$, or the stop of Step 1.1 occurs for some finite iteration index p_*, at which there holds

$$f(x^{p_*}) \leq f(y) + \varepsilon + \delta\|y - x^{p_*}\| \quad \text{for all } y \in \mathbb{R}^n. \tag{1.3.2}$$

PROOF. Suppose $f(x^p)$ is bounded from below. As in Proposition 1.1.5, Step 2 cannot be executed infinitely many times. At some finite iteration index p_*, Algorithm 1.3.1 therefore loops between Steps 1.1 and 1.2, i.e. case (b_ε) always occurs at this iteration. Then Lemma 1.2.4(j) applies for each k: first

$$s_{k+1} \in \partial_{e_k} f(x^{p_*}) = \partial_\varepsilon f(x^{p_*}) \quad \text{for all } k;$$

we deduce in particular that the sequence $\{s_k\} \subset \partial_\varepsilon f(x^{p_*})$ is bounded (Theorem XI.1.1.4). Second, $\langle s_{k+1}, d_k\rangle \geq 0$ which, together with the minimality conditions for the projection $-d_k$, yields

$$\langle -d_k, s_j - s_{k+1}\rangle \geq \|d_k\|^2 \quad \text{for } j = 1, \ldots, k.$$

Lemma IX.2.1.1 applies (with $\hat{s}_k = -d_k$ and $m = 1$): $d_k \to 0$ if $k \to \infty$, so the stop must eventually occur. Finally, each $-d_k$ is a convex combination of ε-subgradients $s_1, \ldots, s_k$ of f at x^{p_*} (note that $s_1 \in \partial_\varepsilon f(x^{p_*})$ by construction) and is therefore an ε-subgradient itself:

$$f(y) \geq f(x^{p_*}) - \langle d_k, y - x^{p_*}\rangle - \varepsilon \quad \text{for all } y \in \mathbb{R}^n.$$

This implies (1.3.2) by the Cauchy-Schwarz inequality. □

It is worth mentioning that the division Step 1 – Step 2 is somewhat artificial, especially with Remark 1.3.3 in mind; we will no longer make this division. Actually, it is Step 1.2 which contains the bulk of the work – and Step 1.1 to a lesser extent. Step 1.2 even absorbs the work of Step 2 (the suitable t and its associated s have already been found during the line-search) and of Step 1.0 (the starting s_1 has been found at the end of the previous line-search). We have here one more illustration of the general principle that line-searches are the most important ingredient when implementing optimization algorithms.

Remark 1.3.5 The boundedness of $\{s_k\}$ for each outer iteration p is a key property. Technically, it is interesting to observe that it is automatically satisfied, because of the very fact that $s_k \in \partial_\varepsilon f(x^p)$: no boundedness of $\{x^p\}$ is needed.

Along the same idea, the assumption that f is finite everywhere has little importance: it could be replaced by something more local, for example

$$S_{f(x^1)}(f) := \{x \in \mathbb{R}^n : f(x) \leq f(x^1)\} \subset \operatorname{int} \operatorname{dom} f;$$

Theorem XI.1.1.4 would still apply. This would just imply some sophistication in the line-search of the next section, to cope with the case of (U1) being called at a point $x^p + td_k \notin \operatorname{dom} f$. In this case, the question arises of what (U1) should return for $f(x^p + td_k)$ and $s(x^p + td_k)$. □

2 A Direct Implementation: Algorithm of ε-Descent

This section contains the computational details that are necessary to implement the algorithm introduced in §1, sketched as Algorithm 1.3.1. We mention that these details are by no means fundamental for the next chapters and can therefore be skipped by a casual reader – it has already been mentioned that methods of ε-descent are not advisable for actual use in "real life" problems. On the other hand, these details give a good idea of the kind of questions that arise when methods for nonsmooth optimization are implemented.

To obtain a really implementable form of Algorithm 1.3.1, we need to specify two calculations: in Step 1.1, how the α-problem is solved, and in Step 1.2, how the line-search is performed. The α-problem is a classical convex quadratic minimization problem with linear constraints; as such, it poses no particular difficulty. The line-search, on the contrary, is rather new since it consists of minimizing the nonsmooth function q_ε or r_ε. It forms the subject of the present section, in which we largely use the principles of §II.3.

A totally clear implementation of a line-search implies answering three questions.

(i) *Initialization*: how should the first trial stepsize be chosen? Here, the line-search must be initialized at each new direction d_k. No really satisfactory initialization is known – and this is precisely one of the drawbacks of the present algorithm. We will not study this problem, considering that the choice $t = 1$ is simplest, if not excessively sensible! (instead of 1, one must of course at least choose a number which, when multiplied by $\|d_k\|$, gives a reasonable move from x^p).

(ii) *Iteration*: given the current stepsize, assumed not suitable, how the next one can be chosen? This will be the subject of §2.1.

(iii) *Stopping Criterion*: when can the current stepsize be considered as suitable (§2.2)? As is the case with all line-searches, this is by far the most delicate question in the present context. It is more crucial here than ever: it gives not only the conditions for stopping the line-search, but also those for determining the next subgradient to be added to the current approximation of $\partial_\varepsilon f(x^p)$.

2.1 Iterating the Line-Search

In order to make the notation less cumbersome, when no confusion is possible, we drop the indices p and k, since they are fixed here and in the next subsection. We are therefore given $x \in \mathbb{R}^n$ and $0 \neq d \in \mathbb{R}^n$, the starting point and the direction for the current line-search under study. We are also given $\varepsilon > 0$, which is crucial in all this chapter.

As already mentioned in §1, the aim of the line-search is to minimize the function q_ε of (1.2.2), or equivalently the function r_ε of (1.2.4). We prefer to base our development on the more natural q_ε, despite its nonconvexity. Of course, the reciprocal correspondence (1.2.4) must be kept in mind and used whenever necessary.

At the point $y = x + td$, it is convenient to simplify the notation: $s(t)$ will stand for $s(x + td) \in \partial f(x + td)$ and

2 A Direct Implementation: Algorithm of ε-Descent

$$e(t) := e(x, x + td, s(t)) = f(x) - f(x + td) + t\langle s(t), d\rangle \qquad (2.1.1)$$

for the linearization error (1.2.5). This creates no confusion because x and d are fixed and because, once again, the subgradient s is considered as a single-valued function, depending on the black box (U1). Recall that $\varepsilon - e(t)$ is in $\partial r_\varepsilon(1/t)$ (we remind the reader that the last expression means the subdifferential of the convex function $u \mapsto r_\varepsilon(u)$ at the point $u = 1/t$).

The problem to be solved in this subsection is: during the process of minimizing q_ε, where should we place each trial stepsize? Applying the principles of §II.3.1, we must decide whether a given $t > 0$ is

(0) convenient, so the minimization of q_ε can be stopped;
(L) on the left of the set of convenient stepsizes;
(R) on their right.

The key to designing (L) and (R) lies in the statements 1.2.2 and 1.2.3: in fact, call the black box (U1) to compute $s(t) \in \partial f(x + td)$; then compute $e(t)$ of (2.1.1) and compare it to ε. There are three possibilities:

(0) := " $\{e(t) = \varepsilon\}$ " (an extraordinary event!). Then this t is optimal and the linesearch is finished.

(L) := " $\{e(t) < \varepsilon\}$ ". This t is too small, in the sense that no optimal $\bar{t}$ can lie on its left. It can serve as a lower bound for all subsequent stepsizes, therefore set $t_L = t$ before looping to the next trial.

(R) := " $\{e(t) > \varepsilon\}$ ". Not only T_ε can be but on the left of this t, but also $s(t) \notin \partial_\varepsilon f(x)$, so t is too large (see Proposition XI.4.2.5 and the discussion following it). This makes two reasons to set $t_R = t$ before looping.

Apart from the stopping criterion, we are now in a position to describe the linesearch in some detail. The notation in the following algorithm is exactly that of §II.3.

Algorithm 2.1.1 (Prototype Line-Search for ε-Descent)

STEP 0 (initialization). Set $t_L = 0$, $t_R = +\infty$. Choose an initial $t > 0$.
STEP 1 (work). Obtain $f(x + td)$ and $s(t) \in \partial f(x + td)$; compute $e(t)$ defined by (2.1.1).
STEP 2 (dispatching). If $e(t) < \varepsilon$ set $t_L = t$; if $e(t) > \varepsilon$ set $t_R = t$.
STEP 3 (stopping criterion). Apply the stopping criterion, which must include in particular: stop the line-search if $e(t) = \varepsilon$ or if $f(x + td) < f(x) - \varepsilon$.
STEP 4 (new stepsize). If $t_R = +\infty$ then extrapolate i.e. compute a new $t > t_L$. If $t_R < +\infty$ then interpolate i.e. compute a new $t \in]t_L, t_R[$. □

As explained in §II.3.1, and as can be seen by a long enough contemplation of this algorithm, two sequences of stepsizes $\{t_R\}$ and $\{t_L\}$ are generated. They have the properties that $\{t_L\}$ is increasing, $\{t_R\}$ is decreasing. At each cycle,

$$t_L \leqslant \bar{t} \leqslant t_R$$

for any $\bar{t}$ minimizing q_ε. Once an interpolation is made, i.e. once some real t_R has been found, no extrapolation is ever made again. In this case one can ascertain that q_ε

has a minimum "at finite distance". The new t in Step 4 must be computed so that, if infinitely many extrapolations [resp. interpolations] are made, then $t_L \to \infty$ [resp. $t_R - t_L \to 0$]. This is what was called the safeguard-reduction Property II.3.1.3.

Remark 2.1.2 See §II.3.4 for suitable safeguarded strategies in Step 4. Without entering into technical details, we mention a possibility for the interpolation formula; it is aimed at minimizing a convex function, so we switch to the (u, r)-notation, instead of (t, q).

Suppose that some $u_L > 0$ (i.e. $t_R = 1/u_L < +\infty$) has been generated: we do have an actual bracket $[u_L, u_R]$, with corresponding actual function-and slope-values; let us denote them by r_L and r'_L, r_R and r'_R respectively. We will subscript by G the endpoint that is better than the other. Look for example at Fig. 2.1.1: $u_G = u_L$ because $r_L < r_R$. Finally, call $\bar{u}$ the minimum of r_ε, assumed unique for simplicity.

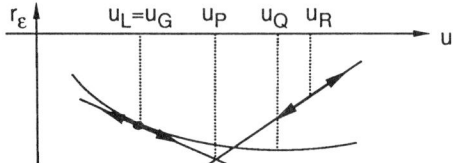

Fig. 2.1.1. Quadratic and polyhedral approximations

To place the next iterate, say u_+, two possible ideas can be exploited.

Idea Q (quadratic). Assume that r_ε is smooth near its minimum. Then it is a good idea to adopt a smooth model for r, for example a convex quadratic function Q:

$$r_\varepsilon(u) \simeq Q(u) := r_G + r'_G(u - u_G) + \tfrac{1}{2}c(u - u_G)^2,$$

where $c > 0$ estimates the local *curvature* of r_ε. This yields a proposal $u_Q := \arg\min Q$ for the next iterate.

Idea P (polyhedral). Assume that r_ε is kinky near its minimum. Then, best results will probably be obtained with a piecewise affine model P:

$$r_\varepsilon(u) \simeq P(u) := \max\{r_L + r'_L(u - u_L), r_R + r'_R(u - u_R)\}$$

yielding a next proposal $u_P := \arg\min P$.

The limited information available from the function r_ε makes it hard to guess a proper value for c, and to choose safely between u_Q and u_P. Nevertheless, the following strategy is efficient:

– Remembering the secant method of Remark II.2.3.5, take for c the difference quotient between slopes computed on the side of $\bar{u}$ where u_G lies: in Fig. 2.1.1, these slopes would have been computed at u_G and at a previous u_L (if any). By virtue of Theorem I.4.2.1(iii), $\{r'_L\}$ has a limit, so normally this c behaves itself.

– Then choose for u_+ either u_Q or u_P, namely the one that is closer to u_G; in Fig. 2.1.1, for example, u_+ would be chosen as u_P.

It can be shown that the resulting interpolation formulae have the following properties: without any additional assumption on r_ε, i.e. on f, u_G *does converge* to $\bar{u}$, even though c may grow without bound. If r_ε enjoys some natural additional assumption concerning difference quotients between slopes, then the convergence is also *fast* in some sense. □

2.2 Stopping the Line-Search

It has been already mentioned again and again that the stopping criterion (Step 3 in Algorithm 2.1.1) is by far the most important ingredient of the line-search. Without it, no actual implementation can even be considered. It is a pivot between Algorithms 2.1.1 and 1.3.1.

As it is clear from Step 1.2 in Algorithm 1.3.1, the stopping test consists in choosing between three possibilities:

- either the present t is not suitable and the line-search must be continued;
- or the present t is suitable because case (a_ε) is detected: $f'_\varepsilon(x, d) < 0$; this simply amounts to testing the descent inequality

$$f(x^p + td_k) < f(x^p) - \varepsilon ; \qquad (2.2.1)$$

if it is true, update x^p to $x^{p+1} := x^p + td_k$ and increase p by 1;
- or the present t is suitable because case (b_ε) is detected: $f'_\varepsilon(x^p, d_k) \geqslant 0$ and also a *suitable* ε-subgradient s_{k+1} has been obtained; then append s_{k+1} to the current polyhedral approximation S_k of $\partial_\varepsilon f(x^p)$ and increase k by 1.

The second possibility is quite trivial to test and requires no special comment. As for the third, it has been studied all along in Chap. IX. Assume that (2.2.1) never happens, i.e. that $k \to \infty$ in Step 1 of Algorithm 1.3.1. Then the whole point is to realize that $0 \in \partial_\varepsilon f(x^p)$, i.e. to have $d_k \to 0$. For this, each iteration k must produce $s_{k+1} \in \mathbb{R}^n$ satisfying *two* properties, already seen before Lemma 1.2.4:

$$s_{k+1} \in \partial_\varepsilon f(x^p), \qquad (2.2.2)$$

$$\langle s_{k+1}, d_k \rangle \geqslant \text{``0''}. \qquad (2.2.3)$$

Here, the symbol "0" can be taken as the mere number 0, as was the case in the framework of Theorem 1.3.4. However, we saw in Remark IX.2.1.3 that "0" could also be a suitable negative tolerance, namely a fraction of $-\|d_k\|^2$. Actually, the following specification of (2.2.3):

$$\langle s_{k+1}, d_k \rangle \geqslant -m'\|d_k\|^2, \quad m' \text{ fixed in }]0, 1[\qquad (2.2.4)$$

preserves the desired property $d_k \to 0$: we are in the framework of Lemma IX.2.1.1.

On the other hand, if s_{k+1} is to be the current subgradient $s(t) = s(x^p + td_k)$ ($t > 0$ being the present trial stepsize, t_L or t_R), we know from (1.2.6) that (2.2.2) is equivalent to $e(x^p, x^p + td_k, s(t)) \leqslant \varepsilon$, i.e.

$$t\langle s(t), d_k \rangle \leqslant \varepsilon - f(x^p + td_k) + f(x^p). \qquad (2.2.5)$$

It can be seen that the two requirements (2.2.4) and (2.2.5) are antagonistic. They may even be incompatible:

Example 2.2.1 With $n = 1$, take the objective function:

$$x \mapsto f(x) = \max\{f_1(x), f_2(x)\} \quad \text{with} \quad \begin{vmatrix} f_1(x) := 2 - x \\ f_2(x) := \exp x \,. \end{vmatrix} \quad (2.2.6)$$

The solution $\bar{x}$ of $\exp x + x = 2$ is the only point where f is not differentiable. It is also the minimum of f, and $f(\bar{x}) = \exp \bar{x}$. Note that $\bar{x} \in]0, 1[$ and $f(\bar{x}) \in]1, 2[$. The subdifferential of f is

$$\partial f(x) = \begin{cases} \{-1\} & \text{if } x < \bar{x}, \\ [-1, \exp \bar{x}] & \text{if } x = \bar{x}, \\ \exp x & \text{if } x > \bar{x}. \end{cases}$$

Now suppose that the ε-descent Algorithm 1.3.1 is initialized at $x^1 = 0$, where the direction of search is $d_1 = 1$. The trace-function $t \mapsto f(0 + t1)$ is differentiable except at a certain $\bar{t} \in [0, 1[$, corresponding to $\bar{x}$. The linearization-error function e of (2.1.1) is

$$e(t) = \begin{cases} 0 & \text{if } 0 \leq t < \bar{t}, \\ 2 + (t-1)\exp t & \text{if } t > \bar{t}. \end{cases}$$

At $t = \bar{t}$, the number $e(\bar{t})$ is somewhere in $[0, e^*]$ (depending on which subgradient $s(\bar{x})$ is actually computed by the black box), where

$$e^* := 2 + (\bar{t} - 1)\exp \bar{t} = 3\bar{t} - \bar{t}^2 > 1.$$

This example is illustrated in Fig. 2.2.1. Finally suppose that $\varepsilon \leq e^*$, for example $\varepsilon = 1$. Clearly enough,

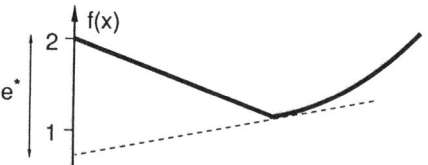

Fig. 2.2.1. Discontinuity of the function e

– if $t \in [0, \bar{t}[$, then $\langle s(t), d_1 \rangle = -1 = -\|d_1\|^2$ so (2.2.4) cannot hold, no matter how m' is chosen in $]0, 1[$;
– if $t > \bar{t}$, then $e(t) > e^* \geq \varepsilon$, so (2.2.5) cannot hold.

In other words, it is impossible to obtain simultaneously (2.2.4) *and* (2.2.5) unless the following two extraordinary events happen:

(i) the line-search must produce the particular stepsize $t = \bar{t}$ (said otherwise, q_ε or r_ε must be *exactly* minimized);
(ii) at this $\bar{x} = 0 + \bar{t} \cdot 1$, the black box must produce a rather particular subgradient s, namely one between the two extreme points -1 and $\exp \bar{x}$, so that $\langle s, d_1 \rangle$ is large enough and (2.2.5) has a chance to hold.

Note also that, if ε is large enough, namely $\varepsilon \geq 2 - f(\bar{x})$, (2.2.1) cannot be obtained, just by definition of $\bar{x}$. □

2 A Direct Implementation: Algorithm of ε-Descent

Remark 2.2.2 A reader not familiar with numerical computations may not consider (i) as so extraordinary. He should however remember how a line-search algorithm has to work (see §II.3, and especially Example II.3.1.5). Furthermore, $\bar{t}$ is given by a non-solvable equation and cannot be computed exactly. This is why we bothered to take a somewhat complicated f in (2.2.6). The example would be just as good, say with $f(x) = |x - 1|$. □

From this example, we draw the following conclusion: it may happen that for the given f, x, d and ε, no line-search algorithm can produce a suitable new element in $\partial_\varepsilon f(x)$; and this no matter how m' is chosen in $]0, 1[$. When this phenomenon happens, Algorithm 2.1.1 is stuck: it loops forever between Step 1 and Step 4. Of course, relaxing condition (2.2.4), say by taking $m' \geq 1$, does not help because then, it is Algorithm 1.3.1 which may be stuck, looping within Step 1: d_k will have no reason to tend to 0 and the stop of Step 1.1 will never occur.

The diagnostic is that (1.2.6), i.e. the transportation formula, does not allow us to reach those ε-subgradients that we need. The remedy is to use a slightly more sophisticated formula, to construct ε-subgradients as *combinations* of subgradients computed at *several* other points:

Lemma 2.2.3 *Let be given: $x \in \mathbb{R}^n$, $y_j \in \mathbb{R}^n$ and $s_j \in \partial f(y_j)$ for $j = 1, \ldots, m$; set*

$$e_j := e(x, y_j, s_j) = f(x) - f(y_j) - \langle s_j, x - y_j \rangle \quad \text{for } j = 1, \ldots, m.$$

With $\alpha = (\alpha_1, \ldots, \alpha_m)$ in the unit simplex Δ_m, set $e := \sum_{j=1}^m \alpha_j e_j$. Then there holds

$$s := \sum_{j=1}^m \alpha_j s_j \in \partial_e f(x).$$

PROOF. Immediate: proceed exactly as for the transportation formula XI.4.2.2. □

This result can be applied to our line-search problem. Suppose that the minima of q_ε have been bracketed by t_L and t_R: in view of Lemma 1.2.2, we have on hand two subgradients $s_L := s(x + t_L d)$ and $s_R := s(x + t_R d)$ satisfying

$$t_L \langle s_L, d \rangle < \varepsilon - f(x) + f(x + t_L d) \qquad [D_- q_\varepsilon(t_L) < 0]$$

$$t_R \langle s_R, d \rangle > \varepsilon - f(x) + f(x + t_R d). \qquad [D_+ q_\varepsilon(t_R) > 0]$$

Then there are two key observations. One is that, since (2.2.1) is assumed not to hold at $t = t_R$ (otherwise we have no problem), the second inequality above implies $0 < \langle s_R, d \rangle$. Second, (1.2.6) implies $s_L \in \partial_\varepsilon f(x)$ and we may just assume $\langle s_L, d \rangle < 0$ (otherwise we have no problem either). In summary,

$$\langle s_L, d \rangle < 0 < \langle s_R, d \rangle. \tag{2.2.7}$$

Therefore, consider a convex combination

$$s^\mu := \mu s_L + (1 - \mu) s_R, \quad \mu \in [0, 1].$$

According to (2.2.7), $\langle s^\mu, d\rangle \geqslant -m'\|d\|^2$ if μ is small enough, namely $\mu \leqslant \bar{\mu}$ with

$$\bar{\mu} := \frac{\alpha\|d\|^2 + \langle s_R, d\rangle}{\langle s_R, d\rangle - \langle s_L, d\rangle} > 0.$$

On the other hand, Lemma 2.2.3 shows that $s^\mu \in \partial_\varepsilon f(x)$ if μ is large enough, namely $\mu \geqslant \underline{\mu}$, with

$$\underline{\mu} := \frac{e_R - \varepsilon}{e_R - e_L} \in \,]0, 1[\,. \tag{2.2.8}$$

It turns out (and this will follow from Lemma 2.3.2 below) that, from the property $f(x+td) \geqslant f(x) - \varepsilon$, we have $\underline{\mu} \leqslant \bar{\mu}$ for $t_R - t_L$ small enough. When this happens, we are done because we can find $s_{k+1} = s^\mu$ satisfying (2.2.4) and (2.2.5). Algorithm 1.3.1 can be stopped, and the current approximation of $\partial_\varepsilon f(x^p)$ can be suitably enlarged.

There remains to treat the case in which no finite t_R can ever be found, which happens if q_ε behaves as in Fig. 1.2.2. Actually, this case is simpler: $t_L \to \infty$ and Lemma 1.2.4 tells us that $s(x + t_L d)$ is eventually convenient.

A simple and compact way to carry out the calculations is to take for μ the maximal value (2.2.8), being understood that $\mu = 1$ if $t_R = \infty$. This amounts to taking systematically s^μ in $\partial_\varepsilon f(x)$. Then the stopping criterion is:

(a_ε) stop the line-search if $f(x+td) < f(x) - \varepsilon$ (and pass to the next x);
(b_ε) stop the line-search if $\langle s^\mu, d\rangle \geqslant \alpha\|d\|^2$ (and pass to the next d).

In all other cases, continue the line-search, i.e. pass to the next trial t.

The consistency of this stopping criterion will be confirmed in the next section.

2.3 The ε-Descent Algorithm and Its Convergence

We are now in a position to realize this Section 2 and to write down the complete organization of the algorithm proposed. The initial iterate x^1 is on hand, together with the black box (U1) which, for each $x \in \mathbb{R}^n$, computes $f(x)$ and $s(x) \in \partial f(x)$. Furthermore, a number $\bar{k} \geqslant 2$ is given, which is the maximal number of n-vectors that can be stored, in view of the memory allocated in the computer. We choose the tolerances $\varepsilon > 0$, $\delta > 0$ and $m' \in \,]0, 1[\,$. Our aim is to furnish a final iterate x^{p*} satisfying

$$f(y) \geqslant f(x^{p*}) - \varepsilon - \delta\|y - x^{p*}\| \quad \text{for all } y \in \mathbb{R}^n. \tag{2.3.1}$$

This gives some hints on how to choose the tolerances: ε is homogeneous to function-values; as in Chap. II, δ is homogeneous to gradient-norms; and the value $m' = 0.1$ is reasonable.

The algorithm below is of course a combination of Algorithms 1.3.1, 2.1.1, and of the stopping criterion introduced at the end of §2.2. Notes such as (^{1}) refer to explanations given afterwards.

Algorithm 2.3.1 (Algorithm of ε-Descent) The initial iterate x^1 is given. Set $p = 1$. Compute $f(x^p)$ and $s(x^p) \in \partial f(x^p)$. Set $f^0 = f(x^p)$, $s_1 = s(x^1)$.(^{1})

2 A Direct Implementation: Algorithm of ε-Descent

STEP 1 (starting to find a direction of ε-descent). Set $k = 1$.

STEP 2 (computing the trial direction). Set $d = -\sum_{i=1}^{k} \alpha_i s_i$, where α solves

$$\min_{\alpha \in \Delta_k} \tfrac{1}{2} \left\| \sum_{i=1}^{k} \alpha_i s_i \right\|^2$$

STEP 3 (final stopping test). If $\|d\| \leq \delta$ stop.[2]

STEP 4 (initializing the line-search). Choose an initial $t > 0$.
  Set $t_L = 0$, $s_L = s_1$, $e_L = 0$; $t_R = 0$, $s_R = 0$.[3]

STEP 5. Compute $f = f(x^p + td)$ and $s = s(x^p + td) \in \partial f(x^p + td)$; set

$$e = f^0 - f + t\langle s, d\rangle.$$

STEP 6 (dispatching). If $e < \varepsilon$, set $t_L = t$, $e_L = e$, $s_L = s$.
  If $e \geq \varepsilon$, set $t_R = t$, $e_R = e$, $s_R = s$.
  If $t_R = 0$, set $\mu = 1$; otherwise set $\mu = (e_R - \varepsilon)/(e_R - e_L)$.[4]

STEP 7 (stopping criterion of the line-search). If $f < f^0 - \varepsilon$ go to Step 11.
  Set $s^\mu = \mu s_L + (1 - \mu)s_R$.[5] If $\langle s^\mu, d\rangle \geq -m'\|d\|^2$ go to Step 9.

STEP 8 (iterating the line-search). If $t_R = 0$, extrapolate i.e. compute a new $t > t_L$.
  If $t_R \neq 0$, interpolate i.e. compute a new t in $]t_L, t_R[$. Loop to Step 5.

STEP 9 (managing the computer-memory). If $k = \bar{k}$, delete at least two (arbitrary) elements from the list $s_1, \ldots, s_k$. Insert in the new list the element $-d$ coming from Step 2 and let $k < \bar{k}$ be the number of elements thus obtained.[6]

STEP 10 (iterating the direction-finding process). Set $s_{k+1} = s^\mu$.[7] Replace k by $k + 1$ and loop to Step 2.

STEP 11 (iterating the descent process). Set $x^{p+1} = x^p + td$, $s_1 = s$.[8] Replace p by $p + 1$ and loop to Step 1. □

Comments

[1] f^0 is the objective-value at the origin-point x^p of each line-search; s^1 is the subgradient computed at this origin-point.

[2] This stop may not be a real stop but a signal to reduce ε and/or δ (cf. Remark 1.3.5). It should be clear that, at this stage of the algorithm, the current iterate satisfies the approximate minimality condition (2.3.1).

[3] The initializations of s_L and e_L are "normal": in view of [1] above, $s_1 \in \partial f(x^p)$ and, as is obvious from (2.1.1), $e(x^p, x^p, s_1) = 0$. On the other hand, the initializations of t_R and s_R are purely artificial (see Algorithm II.3.1.2 and the remark following it): they simply help define s^μ in the forthcoming Step 7. Initializing e_R is not necessary.

[4] In view of the values of e_L and e_R, μ becomes < 1 as soon as some $t_R > 0$ is found.

[5] If $t_R = 0$, then $\mu = 1$ and $s^\mu = s_L$. If $e_R = \varepsilon$, then $\mu = 0$ and $s^\mu = s_R$. Otherwise s^μ is a nontrivial convex combination. In all cases $s^\mu \in \partial_\varepsilon f(x^p)$.

(⁶) We have chosen to give a specific rule for cleaning the bundle, instead of staying abstract as in Algorithm IX.2.1.5. If $k < \bar{k}$, then k can be increased by one and the next subgradient can be appended to the current approximation of $\partial_\varepsilon f(x^p)$. Otherwise, one must make room to store at least the current projection $-d$ and the next subgradient s^μ, thus making it possible to use Theorem IX.2.1.7. Note, however, that if the deletion process turns out to keep every s_i corresponding to $\alpha_i > 0$, then the current projection need not be appended: it belongs to the convex hull of the current subgradients and will not improve the next polyhedron anyway.

(⁷) Here we arrive from Step 7 via Step 10 and s^μ is the convenient convex combination found by the line-search. This s^μ is certainly in $\partial_\varepsilon f(x^p)$: if $\mu = 1$, then $e_L \leqslant \varepsilon$ and the transportation formula (XI.4.2.2) applies; if $\mu \in \,]0, 1[\,$, the corresponding linearization error $\mu e_L + (1-\mu) e_R$ is equal to ε, and it is Lemma 2.2.3 that applies.

(⁸) At this point, s is the last subgradient computed at Step 5 of the line-search. It is a subgradient at the next iterate $x^{p+1} = x^p + td$. □

A loop from Step 11 to Step 1 represents an actual ε-descent, with x^p moved; this was called a *descent-step* in Remark 1.3.3. A loop from Step 10 to Step 2 represents one iteration of the direction-finding procedure, i.e. a *null-step*: one subgradient is appended to the current approximation of $\partial_\varepsilon f(x^p)$. The detailed line-search is expanded between Steps 5 and 8. At each cycle, Step 7 decides whether

– to iterate the line-search, by a loop from Step 8 to Step 5,
– to iterate the direction-finding procedure,
– or to iterate the descent process, with an actual ε-descent obtained.

Now we have to prove convergence of this algorithm. First, we make sure that each line-search terminates.

Lemma 2.3.2 *Let $f : \mathbb{R}^n \to \mathbb{R}$ be convex. Suppose that the extrapolation and interpolation formulae in Step 8 of Algorithm 2.3.1 satisfy the safeguard-reduction Property II.3.1.3. Then, for each iteration (p, k), the number of loops from Step 8 to Step 5 is finite.*

PROOF. Suppose for contradiction that, for some fixed iteration (p, k), the line-search does not terminate. Suppose first that no $t_R > 0$ is ever generated. By construction, $\mu = 1$ and $s^\mu = s = s_L$ forever. One has therefore at each cycle

$$f(x^p + t_L d) + \varepsilon \geqslant f(x^p) \geqslant f(x^p + t_L d) - t_L \langle s^\mu, d \rangle,$$

where the first inequality holds because Step 7 never exits to Step 11; the second is because $s^\mu \in \partial f(x^p + t_L d)$. We deduce

$$t_L \langle s^\mu, d \rangle \geqslant -\varepsilon. \tag{2.3.2}$$

Now, the assumptions on the extrapolation formulae imply that $t_L \to \infty$. In view of the test Step 7 $\to$ Step 9, (2.3.2) cannot hold infinitely often.

Thus some $t_R > 0$ must eventually be generated, and Step 6 shows that, from then on, $\mu e_L + (1-\mu)e_R = \varepsilon$ at every subsequent cycle. This can be expressed as follows:
$$f(x^p) - \mu f(x^p + t_L d) - (1-\mu)f(x^p + t_R d)$$
$$= \varepsilon - t_L \mu \langle s_L, d\rangle - t_R(1-\mu)\langle s_R, d\rangle.$$

Furthermore, non-exit from Step 7 to Step 11 implies that the left-hand side above is smaller than ε, so
$$t_L \mu \langle s_L, d\rangle + t_R(1-\mu)\langle s_R, d\rangle \geq 0. \qquad (2.3.3)$$

By assumption on the interpolation formulae, t_L and t_R have a common limit $\bar{t} \geq 0$, and we claim that $\bar{t} > 0$. If not, the Lipschitz continuity of f in a neighborhood of x^p (Theorem IV.3.1.2) would imply the contradiction
$$0 < \varepsilon < e_R = f(x^p) - f(x^p + t_R d) + t_R \langle s_R, d\rangle \leq 2Lt_R \to 0$$
(L being a Lipschitz constant around x^p).

Thus, after division by $\bar{t} > 0$, (2.3.3) can be written as
$$\mu \langle s_L, d\rangle + (1-\mu)\langle s_R, d\rangle + \eta \geq 0$$
where the extra term
$$\eta := \frac{t_L - \bar{t}}{\bar{t}} \mu \langle s_L, d\rangle + \frac{t_R - \bar{t}}{\bar{t}}(1-\mu)\langle s_R, d\rangle$$
tends to 0 when the number of cycles tends to infinity. This is impossible because of the test Step 7 $\to$ Step 9. □

The rest is now a variant of Theorem 1.3.4.

Theorem 2.3.3 *The assumptions are those of Lemma* 2.3.2. *Then, either* $f(x^p) \to -\infty$, *or the stop of Step 3 occurs for some finite* p_*, *yielding* x^{p_*} *satisfying the approximate minimality condition* (2.3.1).

PROOF. In Algorithm 2.3.1, p cannot go to $+\infty$ if f is bounded from below. Consider the last iteration, say the $(p_*)^{\text{th}}$. From Lemma 2.3.2, each iteration k exits to Step 9. By construction, there holds for all k
$$s_{k+1} \in \partial_\varepsilon f(x^{p_*})$$
hence the sequence $\{s_k\}$ is bounded (x^{p_*} is fixed!), and
$$\langle s_{k+1}, d_k\rangle \geq -m' \|d_k\|^2.$$
We are in the situation of Lemma IX.2.1.1: k cannot go to $+\infty$ in view of Step 3. □

Remark 2.3.4 This proof reveals the necessity of taking the tolerance $m' > 0$ to stop each line-search. If m' were set to 0, we would obtain a non-implementable algorithm of the type IX.1.6.

Other tolerances can equally be used. In the next section, precisely, we will consider variations around the stopping criterion of the line-search. Here, we propose the following exercise: reproduce Lemma 2.3.2 with $m' = 0$, but with the descent criterion in Step 7 replaced by
$$\text{If } f < f_0 - \varepsilon' \text{ go to Step 11}$$
where ε' is fixed in $]0, \varepsilon[$. □

3 Putting the Algorithm in Perspective

As already stated in Remark 1.1.6, we are not interested in giving a specific choice for ε in algorithms such as 2.3.1. It is important for numerical efficiency only; but precisely, we believe that these algorithms cannot be made numerically efficient under their present form. We nevertheless consider two variants, obtained by giving ε rather special values. They are interesting for the sake of curiosity; furthermore, they illustrate an aspect of the algorithm not directly related to optimization, but rather to separation of closed convex sets – see §IX.3.3.

3.1 A Pure Separation Form

Suppose that $\bar{f} := \inf\{f(y) : y \in \mathbb{R}^n\} > -\infty$ is known and set

$$\bar{\varepsilon} := f(x) - \bar{f}, \qquad (3.1.1)$$

$x = x^1$ being the starting point of the ε-descent Algorithm 2.3.1. If we take this value $\bar{\varepsilon}$ for ε in that algorithm, the exit from Step 7 to Step 11 will never occur. The line-searches will be made along a sequence of successive directions, all issuing from the same starting x.

Then, we could let the algorithm run with $\varepsilon = \bar{\varepsilon}$, simply as stated in 2.3.1 (the exit-test from Step 7 to Step 11 – and Step 11 itself – could simply be suppressed with no harm). However, this does no good in terms of minimizing f: when the algorithm stops, we will learn that the approximate minimality condition (2.3.1) holds with $x^{p*} = x = x^1$ and $\varepsilon = \bar{\varepsilon}$; but we knew this before, even with $\delta = 0$.

Fortunately, there are better things to do. Observe first that, instead of minimizing f within $\bar{\varepsilon}$ (which has already been done), the idea of the algorithm is now to solve the following problem: by suitable calls to the black box (U1), compute $\bar{\varepsilon}$-subgradients at the given x, so as to obtain 0 – or at least a vector of small norm – in their convex hull. Even though the definition of $\bar{\varepsilon}$ implies $0 \in \partial_{\bar{\varepsilon}} f(x)$, the only constructive information concerning this latter set is the black box (U1), and 0 might never be produced by (U1). In other words, we must try to *separate* 0 from $\partial_{\bar{\varepsilon}} f(x)$. We will fail but the point is to explain this failure.

Remark 3.1.1 The above problem is not a theoretical pastime but is of great interest in Lagrangian relaxation. Suppose that (U1) is a "Lagrange black box", as discussed in §XII.1.1: each subgradient is of the form $s = -c(u)$, u being a primal variable. When we solve the above separation problem, we construct primal points $u_1, \ldots, u_k$ and associated convex multipliers $\alpha_1, \ldots, \alpha_k$ such that

$$\sum_{i=1}^{k} \alpha_i c(u_i) = 0.$$

In favourable situations – when some convexity is present – the corresponding combination $\sum_{i=1}^{k} \alpha_i u_i$ is the primal solution that was sought in the first place; see Theorem XII.2.3.4, see also Theorem XII.4.2.5. □

Thus, we are precisely in the framework of §IX.3.3: rather than a minimization algorithm, Algorithm 2.3.1 becomes a separation one, namely a particular instance of Algorithm IX.3.3.1. Using the notation $S := \partial_{\bar{\varepsilon}} f(x)$, what the latter algorithm needs in its Step 2 is a mechanism for computing the support function $\sigma_S(d)$ of S at any given $d \in \mathbb{R}^n$, together with a solution $s(d)$ in the exposed face $F_S(d)$. The comparison becomes perfect if we interpret the line-search of Algorithm 2.3.1 precisely as this mechanism.

So, upon exit from Step 7 to Step 9, we still want s^μ to be an $\bar{\varepsilon}$-subgradient at x; but we also want
$$\langle s^\mu, d \rangle \geq \langle s, d \rangle \quad \text{for all } s \in S,$$
in order to have
$$\langle s_{k+1}, d_k \rangle \ [= \langle s^\mu, d \rangle] \ = \sigma_S(d) = f'_{\bar{\varepsilon}}(x, d_k).$$

The next result shows that, once again, our task is to minimize $r_{\bar{\varepsilon}}$: this s_{k+1} must be some $s \in \partial f(x + 1/u\, d)$, with u yielding $0 \in \partial r_{\bar{\varepsilon}}(u)$.

Theorem 3.1.2 *Let x, $d \neq 0$ and $\varepsilon \geq 0$ be given. Suppose that $s \in \mathbb{R}^n$ satisfies one of the following two properties:*

(i) *either, for some $t > 0$, $s \in \partial f(x + td)$ and $e(x, x + td, s) = \varepsilon$;*

(ii) *or s is a cluster point of a sequence $\{s(t) \in \partial f(x + td)\}_t$ with $t \to +\infty$ and*
$$e(x, x + td, s(t)) \leq \varepsilon \quad \text{for all } t \geq 0.$$

Then
$$s \in \partial_\varepsilon f(x) \quad \text{and} \quad \langle s, d \rangle = f'_\varepsilon(x, d).$$

PROOF. The transportation formula (Proposition XI.4.2.2) directly implies that the s in (i) and the $s(t)$ in (ii) are in $\partial_\varepsilon f(x)$. Invoking the closedness of $\partial_\varepsilon f(x)$ for case (ii), we see that $s \in \partial_\varepsilon f(x)$ in both cases.

Now take an arbitrary $s' \in \partial_\varepsilon f(x)$, satisfying in particular
$$f(x + td) \geq f(x) + t \langle s', d \rangle - \varepsilon \quad \text{for all } t > 0. \tag{3.1.2}$$

[*Case (i)*] Replacing ε in (3.1.2) by $e(x, x + td, s)$, we obtain
$$0 \geq t\langle s', d \rangle - t\langle s, d \rangle.$$
Divide by $t > 0$: $\langle s, d \rangle = \sigma_{\partial_\varepsilon f(x)}(d)$ since s' was arbitrary.

[*Case (ii)*] Add the subgradient inequality $f(x) \geq f(x + td) - t\langle s(t), d \rangle$ to (3.1.2):
$$0 \geq t\langle s', d \rangle - t\langle s(t), d \rangle - \varepsilon.$$
Divide by $t > 0$ and let $t \to +\infty$ to see that $\langle s, d \rangle \geq \langle s', d \rangle$. □

In terms of Algorithm 2.3.1, we realize that the s^μ computed in Step 7 precisely aims at satisfying the conditions in Theorem 3.1.2:

(i) Suppose first that some $t_R > 0$ is generated (which implies $T_\varepsilon \neq \emptyset$). When sufficiently many interpolations are made, t_L and t_R become close to their common limit, say t_ε. Because $\partial f(\cdot)$ is outer semi-continuous, s_L and s_R are both close to $\partial f(x + t_\varepsilon d)$. Their convex combination s^μ is also close to the convex set $\partial f(x + t_\varepsilon d)$. Finally, $e(x, x + t_\varepsilon d, s^\mu)$ is close to $\mu e_L + (1 - \mu)e_R = \varepsilon$, by continuity of f. In other words, s^μ almost satisfies the assumptions of Theorem 3.1.2(i). This is illustrated in Fig. 3.1.1.

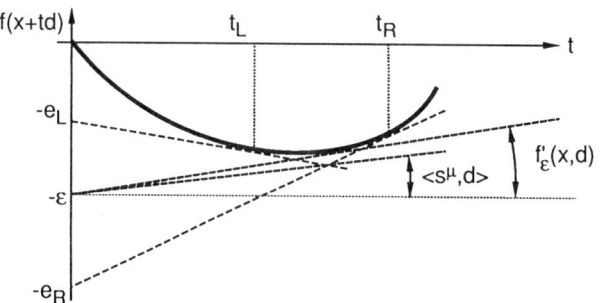

Fig. 3.1.1. Identifying the approximate directional derivative

(ii) If no $t_R > 0$ is generated and $t_L \to \infty$ (which happens when $T_\varepsilon = \emptyset$), it is case (ii) that is satisfied by the corresponding $s_L = s^\mu$.

In a word, Algorithm 2.3.1 becomes a realization of Algorithm IX.3.3.1 if we let the interpolations [resp. the extrapolations] drive $t_R - t_L$ small enough [resp. t_L large enough] before exiting from Step 7 to Step 9. It is then not difficult to adapt the various tolerances so that the approximation is good enough, namely $\langle s^\mu, d_k \rangle$ is close enough to $\sigma_S(d_k)$.

These ideas have been used for the numerical experiments of Figs. IX.3.3.1 and IX.3.3.2. It is the above variant that was the first form of Algorithm IX.3.3.1, alluded to at the end of §IX.3.3. In the case of TR48 (Example IX.2.2.6), $\partial_{\bar\varepsilon} f(x)$ was identified at $x = 0$, a significantly non-optimal point: $f(0) = -464816$, while $\bar f = -638565$. All generated s_k had norms roughly comparable. In the example MAXQUAD of VIII.3.3.3, this was far from being the case: the norms of the subgradients were fairly dispersed, remember Table VIII.3.3.1.

Remark 3.1.3 An interesting question is whether this variant produces a minimizing sequence. The answer is no: take the example of (IX.1.1), where the objective function was

$$(\xi, \eta) \mapsto f(\xi, \eta) := \max\{\xi + 2\eta, \xi - 2\eta, -\xi\}.$$

Start from the initial $x = (2, 1)$ and take $\varepsilon = \bar\varepsilon = 4$. We leave it as an exercise to see how Algorithm 1.3.1 proceeds: the first iteration, along the direction $(-1, -2)$, produces y_2 anywhere on the half-line $\{(2 - t, 1 - 2t) : t \geq 1/2\}$ and s_2 is unambiguously $(1, -2)$. Then, d_2 is collinear to $(-1, 0)$, $y_3 = (-1, 1)$ and $s_3 = (-1/3, 2/3)$ (for this last calculation, take $s_3 = \alpha(-1, 0) + (1 - \alpha)(1, 2)$, a convex combination of subgradients at y_3, and adjust α so that the corresponding linearization error $e(x, y_3, s_3)$ is equal to $\varepsilon = 4$). Finally $d_3 = 0$, although neither y_2 nor y_3 is even close to the minimum $(0, 0)$. These operations are illustrated by Fig. 3.1.2.

Note in this example that $d_3 = 0$ is a convex combination of s_2 and s_3: s_1 plays no role. Thus 0 is on the boundary of $\mathrm{co}\{s_1, s_2, s_3\} = \partial_4 f(x)$. If ε is decreased, s_2 and s_3 are slightly perturbed and $0 \notin \mathrm{co}\{s_1, s_2, s_3\}$: d_3 is no longer zero (although it stays small). The property $0 \in \mathrm{bd}\, \partial_4 f(x)$ does not hold by chance but was proved in Proposition XI.1.3.5: $\varepsilon = 4$ is the critical value of ε and the normal cone to $\partial_4 f(x)$ at 0 is not reduced to $\{0\}$. □

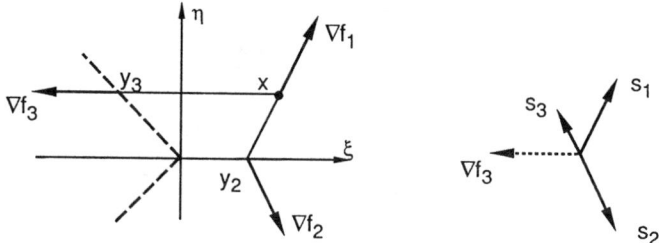

Fig. 3.1.2. Three iterations of the separating algorithm

3.2 A Totally Static Minimization Algorithm

Our second variant is also a sort of separation Algorithm IX.3.3.1. It also performs one single descent iteration, the starting point of the line-searches never being updated. It differs from the previous example, though: it is really aimed at minimizing f, and ε is no longer fixed but depends on the direction-index k. Specifically, for fixed $x = x^1$ and $d \neq 0$ in Algorithm 2.3.1, we set

$$\varepsilon = \varepsilon(x, d) := f(x) - \inf_{t>0} f(x + td). \tag{3.2.1}$$

A second peculiarity of our variant is therefore that ε is not known; and also, it may be 0 (for simplicity, we assume $\varepsilon(x, d) < +\infty$). Nevertheless, the method can still be made implementable in this case. In fact, a careful study of Algorithm 2.3.1 shows that the actual value of ε is needed in three places:

– for the exit-test from Step 7 to Step 11; with the ε of (3.2.1), this test is never passed and can be suppressed, just as in §3.1;
– in Step 6, to dispatch t as a t_L or t_R; Proposition 3.2.1 below shows that the precise value (3.2.1) is useless for this particular operation;
– in Step 6 again, to compute μ; it is here that lack of knowledge of ε has the main consequences.

Proposition 3.2.1 *For given x, $d \neq 0$ in $\mathbb{R}^n$ and ε of (3.2.1), the minimizers of $t \mapsto f(x + td)$ over $\mathbb{R}^+$ are also minimizers of q_ε – knowing that q_0 is the ordinary difference quotient, with $q_0(0) = f'(x, d)$. The converse is true if $\varepsilon > 0$.*

PROOF. Let $\bar{t}$ minimize $f(x+\cdot d)$ over $\mathbb{R}^+$. If $\bar{t} = 0$, then $\varepsilon(x, d) = 0$ and $q_{\varepsilon(x,d)}(0) = f'(x, d) \leqslant q_{\varepsilon(x,d)}(t)$ for all $t \geqslant 0$ (monotonicity of the difference quotient); the proof is finished.

If $\bar{t} > 0$, the definition

$$f(x + \bar{t}d) = f(x) - \varepsilon(x, d)$$

combined with the minimality condition

$$0 \in \langle \partial f(x + \bar{t}d), d \rangle$$

shows that $\bar{t}$ satisfies the minimality condition of Lemma 1.2.2(i) for $q_{\varepsilon(x,d)}$.

Now let $\{\tau_k\} \subset \mathbb{R}^+$ be a minimizing sequence for $f(x + \cdot d)$:

$$f(x + \tau_k d) \to f(x) - \varepsilon,$$

where ε is the $\varepsilon(x, d)$ of (3.2.1). If $\varepsilon > 0$, then 0 cannot be a cluster point of $\{\tau_k\}$: we can write

$$\frac{f(x + \tau_k d) - f(x) + \varepsilon}{\tau_k} \to 0,$$

and this implies $f'_\varepsilon(x, d) \leqslant 0$. If t_ε minimizes q_ε, we have

$$\frac{f(x + t_\varepsilon d) - f(x) + \varepsilon}{t_\varepsilon} = f'_\varepsilon(x, d) \leqslant 0.$$

Thus

$$f(x + t_\varepsilon d) \leqslant f(x) - \varepsilon,$$

t_ε does minimize $f(x + \cdot d)$ (and $f'_\varepsilon(x, d)$ is actually zero). □

It is therefore a good idea to minimize $f(x + td)$ with respect to $t \geqslant 0$: the implicitly given function $q_{\varepsilon(x,d)}$ will by necessity be minimized – and it is a much more natural task. A first consequence is that the decision to dispatch t is even simpler than before, namely:

$$\text{if } \langle s, d \rangle < 0, t \text{ must become a } t_L \quad (f \text{ seems to be decreasing}), \tag{3.2.2}$$

$$\text{if } \langle s, d \rangle > 0, t \text{ must become a } t_R \quad (f \text{ seems to be increasing}), \tag{3.2.3}$$

knowing that $\langle s, d \rangle = 0$ terminates the line-search, of course.

However, the object of the line-search is now to obtain t and $s_+ \in \partial f(x + td)$ satisfying $\langle s_+, d \rangle = 0$; the property $s_+ \in \partial_{\varepsilon(x,d)} f(x)$ will follow automatically. The choice of μ in Step 6 can then be adapted accordingly. The rationale for μ in Algorithm 2.3.1 was based on forcing s^μ to be a priori in $\partial_\varepsilon f(x)$, and it was only eventually that $\langle s^\mu, d \rangle$ became close to zero, i.e. large enough in that context. Here, it is rather the other way round: we do not know in what $\varepsilon(x, d)$-subdifferential the subgradient s^μ should be, but we do know that we want to obtain $\langle s^\mu, d \rangle = 0$. Accordingly, it becomes suitable to compute μ in Step 6 by

$$\mu \langle s_L, d \rangle + (1 - \mu) \langle s_R, d \rangle = 0 \tag{3.2.4}$$

when $t_R > 0$ (otherwise, take $\mu = 1$ as before). Observe that, fortunately, this is in harmony with the dispatching (3.2.2), (3.2.3): provided that s_R exists, $\mu \in [0, 1]$.

It remains to find a suitable stopping criterion for the line-search.

3 Putting the Algorithm in Perspective   221

– First, if no $t_R > 0$ is ever found (f decreasing along d), there is nothing to change: in view of Lemma 1.2.2(ii), $s^\mu = s(x + t_L d) \in \partial_{\varepsilon(x,d)} f(x)$ for all t_L; furthermore, Lemma 1.2.4(jj) still applies and eventually, $\langle s^\mu, d \rangle \geq -m' \|d\|^2$.
– The other case is when a true bracket $[t_L, t_R]$ is eventually found; then Lemma 2.2.3 implies $s^\mu \in \partial_{\bar{e}} f(x)$ where

$$\bar{e} := \mu e_L + (1 - \mu) e_R \qquad (3.2.5)$$

and μ is defined by (3.2.4). As for $\varepsilon(x, d)$, we obviously have

$$\varepsilon(x, d) \geq \underline{e} := \max \{f(x) - f(x + t_L), f(x) - f(x + t_R d)\}. \qquad (3.2.6)$$

Continuity of f ensures that $\bar{e}$ and $\underline{e}$ have the common limit $\varepsilon(x, d)$ when t_L and t_R both tend to the optimal (and finite) $t \geq 0$. Then, because of outer semi-continuity of $\varepsilon \mapsto \partial_\varepsilon f(x)$, s^μ is close to $\partial_{\varepsilon(x,d)} f(x)$ when t_R and t_L are close together.

In a word, without entering into the hair-splitting details of the implementation, the following variant of Algorithm 2.3.1 is reasonable.

Algorithm 3.2.2 (Totally Static Algorithm – Sketch) The initial point x is given, together with the tolerances $\eta > 0$, $\delta > 0$, $m' \in]0, 1[$. Set $k = 1$, $s_1 = s(x)$.

STEP 1. Compute $d_k := -\text{Proj } 0 / \{s_1, \ldots, s_k\}$.
STEP 2. If $\|d_k\| \leq \delta$ stop.
STEP 3. By an approximate minimization of $t \mapsto f(x + t d_k)$ over $t \geq 0$, find $t_k > 0$ and s_{k+1} such that

$$-m' \|d_k\|^2 \leq \langle s_{k+1}, d_k \rangle \leq 0 \qquad (3.2.7)$$

and

$$s_{k+1} \in \partial_{e_k} f(x),$$

where

$$e_k := f(x) - f(x + t_k d_k) + \eta.$$

STEP 4. Replace k by $k + 1$ and loop to Step 1. □

Here, there is no longer any descent test, but a new tolerance η is introduced, to make sure that $\bar{e} - \underline{e}$ of (3.2.5), (3.2.6) is small. Admitting that Step 3 terminates for each k, this algorithm is convergent:

Theorem 3.2.3 *In Algorithm 3.2.2, let*

$$x_k = x + t_k d_k \in \text{Argmin} \{f(x + t_i d_i) : i = 1, \ldots, k\}$$

be one of the best points found during the successive iterations. If f is bounded from below, the stop of Step 2 occurs for some finite k, and then

$$f(y) \geq f(x_k) - \eta - \delta \|y - x_k\| \quad \text{for all } y \in \mathbb{R}^n. \qquad (3.2.8)$$

PROOF. Corresponding to the best point x_k, set

$$\varepsilon_k := f(x) - f(x_k) + \eta. \qquad (3.2.9)$$

The sequence $\{\varepsilon_k\}$ is increasing. By construction, $e_k \leqslant \varepsilon_k$ and it follows that $s_{k+1} \in \partial_{\varepsilon_k} f(x)$ for each k.

If $\varepsilon_k \to +\infty$ when $k \to +\infty$, then $f(x_k) \to -\infty$. Otherwise, $\{\varepsilon_k\}$ is bounded, $\{s_{k+1}\}$ is therefore bounded as well. In view of (3.2.7), Lemma IX.2.1.1 applies: the algorithm must stop for some iteration k. At this k,

$$s_i \in \partial_{\varepsilon_i} f(x) \subset \partial_{\varepsilon_k} f(x) \quad \text{for } i = 1, \ldots, k$$

hence $d_k \in \partial_{\varepsilon_k} f(x)$. Using (3.2.9), we have for all $y \in \mathbb{R}^n$:

$$f(y) \geqslant f(x) + \langle d_k, y - x \rangle - \varepsilon_k = f(x_k) + \langle d_k, y - x \rangle - \eta. \qquad \square$$

From a practical point of view, this algorithm is of course not appropriate: it is not a sound idea to start each line-search from the same initial x. Think of a ballistic comparison, where each direction is a gun: to shoot towards a target $\bar{x}$, it should make sense to place the gun as close as possible to $\bar{x}$ – and not at the very first iterate, probably the worst of all! This is reflected in the minimality condition: write (3.2.8) with an optimal $y = \bar{x}$ (the only interesting y, actually); admitting that $\|\bar{x} - x\|$ is unduly large, an unduly small value of δ – i.e. of $\|d_k\|$ – is required to obtain any useful information about optimality of x_k. Remembering our analysis in §IX.2.2, this means an unduly large number of iterations.

The first inequality in (3.2.7) would suffice to let the algorithm run. The reason for the second inequality is to illustrate another aspect of the approach. Consider again the value $\bar{\varepsilon} = f(x) - \bar{f}$ of (3.1.1). Obviously in Algorithm 3.2.2,

$$s_{k+1} \in \partial_{e_k} f(x) \subset \partial_{\bar{\varepsilon}+\eta} f(x) \simeq \partial_{\bar{\varepsilon}} f(x)$$

for each k. Coming back to the problem of separating $\partial_{\bar{\varepsilon}} f(x) =: S$ and $\{0\}$, we see that Algorithm 3.2.2 is the second form of Algorithm IX.3.3.1 that was used for Figs. IX.3.3.1 and IX.3.3.2: for each direction d, its black box generates $s_+ \in S$ with $\langle s_+, d \rangle \simeq 0$ (instead of computing the support function). It is therefore a convenient instrument for illustrating the point made at the end of §IX.3.3.

XIV. Dynamic Construction of Approximate Subdifferentials: Dual Form of Bundle Methods

Prerequisites. Basic concepts of numerical optimization (Chap. II); descent principles for nonsmooth minimization (Chaps. IX and XIII); definition and elementary properties of approximate subdifferentials (Chap. XI); and to a lesser extent: minimality conditions for simple minimization problems, and elements of duality theory (Chaps. VII and XII).

1 Introduction: The Bundle of Information

1.1 Motivation

The methods of ε-descent, which were the subject of Chap. XIII, present a number of deficiencies from the numerical point of view. Their rationale itself – to decrease by ε the objective function at each iteration – is suspect: the fact that ε can (theoretically) be chosen in advance, without any regard for the real behaviour of f, must have its price, one way or another.

(a) The Choice of ε. At first glance, an attractive idea in the algorithms of Chap. XIII is to choose ε fairly large, so as to reduce the number of outer iterations, indexed by p. Then, of course, a counterpart is that the complexity of each such iteration must be expected to increase: k will grow larger for fixed p. Let us examine this point in more detail.

Consider one single iteration, say $p = 1$, in the schematic algorithm of ε-descent XIII.1.3.1. Set the stopping tolerance $\delta = 0$ and suppose for simplicity that q_ε (or r_ε) can be minimized exactly at each subiteration k. In a word, consider the following algorithm:

Algorithm 1.1.1 Start from $x = x^1 \in \mathbb{R}^n$. Choose $\varepsilon > 0$.

STEP 0. Compute $s_1 \in \partial f(x)$, set $k = 1$.
STEP 1. Compute $d_k = -\operatorname{Proj} 0/\{s_1, \ldots, s_k\}$. If $d_k = 0$ stop.
STEP 2. By a search along d_k issuing from x, minimize q_ε of (XIII.1.2.2) to obtain $t_k > 0$, $y_{k+1} = x + t_k d_k$ and $s_{k+1} \in \partial f(y_{k+1})$ such that

$$e(x, y_{k+1}, s_{k+1}) := f(x) - f(y_{k+1}) + t_k \langle s_{k+1}, d_k \rangle = \varepsilon. \qquad (1.1.1)$$

STEP 3. If $f(y_{k+1}) \geq f(x) - \varepsilon$, replace k by $k+1$ and loop to Step 1. Otherwise stop. □

This algorithm has essentially two possible outputs:
- Either a stop occurs in Step 3 (which plays the role of updating p in Step 2 of Algorithm XIII.1.3.1). Then one has on hand y ($= y_{k+1}$) with $f(y) < f(x) - \varepsilon$.
- Or Step 3 loops to Step 1 forever. It is a result of the theory that $\{d_k\}$ then tends to 0, the key properties being that $\langle s_{k+1}, d_k \rangle \geq 0$ for all k, and that $\{s_k\}$ is bounded. This case therefore means that the starting x minimizes f within ε. Same comment for the extraordinary case of a stop via $d_k = 0$ in Step 1.

Consider Algorithm 1.1.1 as a mapping $\varepsilon \mapsto \tilde{f}(\varepsilon)$, where $\tilde{f}(\varepsilon)$ is the best f-value obtained by the algorithm; in other words, in case of a stop in Step 3, $\tilde{f}(\varepsilon)$ is simply the last computed value $f(y_{k+1})$ (then, $\tilde{f}(\varepsilon) < f(x) - \varepsilon$); otherwise,

$$\tilde{f}(\varepsilon) = \inf \{f(y_k) : k = 1, 2, \ldots\}.$$

With $\bar{f} := \inf_{\mathbb{R}^n} f$, consider the "maximal useful" value for ε

$$\bar{\varepsilon} := f(x) - \bar{f}.$$

- For $\varepsilon < \bar{\varepsilon}$, $0 \notin \partial_\varepsilon f(x)$ and $\{d_k\}$ cannot tend to 0; the iterations must terminate with a success: there are finitely many loops from Step 3 to Step 1. In other words

$$\varepsilon < \bar{\varepsilon} \implies [\bar{f} = f(x) - \bar{\varepsilon} \leq] \ \tilde{f}(\varepsilon) < f(x) - \varepsilon, \qquad (1.1.2)$$

which implies in particular

$$\tilde{f}(\varepsilon) \to \bar{f} \quad \text{if } \varepsilon \uparrow \bar{\varepsilon}.$$

- For $\varepsilon \geq \bar{\varepsilon}$, $0 \in \partial_\varepsilon f(x)$ and Algorithm 1.1.1 cannot stop in Step 3: it has to loop between Step 1 and Step 3 – or stop in Step 1. In fact, we obtain for $\varepsilon = \bar{\varepsilon}$ nothing but (a simplified form of) the algorithm of §XIII.3.1. As mentioned there, $\{y_k\}$ need not be a minimizing sequence; see more particularly Remark XIII.3.1.3. In other words, it may well be that $\tilde{f}(\bar{\varepsilon}) > \bar{f}$. Altogether, we see that the function $\tilde{f}$ has no reason to be (left-)continuous at $\bar{\varepsilon}$.

Conclusion: it is hard to believe that (1.1.2) holds *in practice*, because when numerical calculations are involved, the discontinuity of $\tilde{f}$ is likely not to be exactly at $\bar{\varepsilon}$. When ε is close to $\bar{\varepsilon}$, $\tilde{f}(\varepsilon)$ is likely to be far from $\bar{f}$, just as it is when $\varepsilon = \bar{\varepsilon}$.

Remark 1.1.2 It is interesting to see what really happens numerically. Toward this end, take again the examples MAXQUAD of VIII.3.3.3 and TR48 of IX.2.2.6. To each of them, apply Algorithm 1.1.1 with $\varepsilon = \bar{\varepsilon}$, and Algorithm XIII.3.2.2. With the stopping tolerance δ set to 0, the algorithms should run forever. Actually, a stop occurs in each case, due to some failure: in the quadratic program computing the direction, or in the line-search. We call again $\tilde{f}$ the best f-value obtained when this stop occurs, with either of the two algorithms.

Table 1.1.1 summarizes the results: for each of the four experiments, it gives the number of iterations, the total number of calls to the black box (U1) of Fig. II.1.2.1, and the improvement of the final f-value obtained, relative to the initial value, i.e. the ratio

1 Introduction: The Bundle of Information

$$\frac{\tilde{f}(\bar{\varepsilon}) - \bar{f}}{f(x) - \bar{f}}.$$

Observe that the theory is somewhat contradicted: if one wanted to compare Algorithm 1.1.1 (supposedly non-convergent) and Algorithm XIII.3.2.2 (proved to converge), one should say that the former is better. We add that each test in Table 1.1.1 was run several times, with various initial x; all the results were qualitatively the same, as stated in the table. The lesson is that one must be modest when applying mathematics to computers: proper assessment of an algorithm implies an experimental study, in addition to establishing its theoretical properties; see also Remark II.2.4.4.

Table 1.1.1. What is a convergent algorithm?

	MAXQUAD			TR48		
Algorithm 1.1.1	87	718	4.10^{-4}	141	892	3.10^{-2}
Algorithm XIII.3.2.2	182	1000	2.10^{-3}	684	1897	4.10^{-2}

The apparently good behaviour of Algorithm 1.1.1 can be explained by the rather slow convergence of $\{d_k\}$ to 0. The algorithm takes advantage of its large number of iterations to explore the space around the starting x, and thus locates the optimum more or less accurately. By contrast, the early stop occurring in the 2-dimensional example of Remark XIII.3.1.3 gives the algorithm no chance. □

Knowing that a large ε is not necessarily safe, the next temptation is to take ε small. Then, it is hardly necessary to mention the associated danger: the direction may become that of steepest-descent, which has such a bad reputation, even for smooth functions (remember the end of §II.2.2). In fact, if f is smooth (or mildly kinky), every anti-subgradient is a descent direction. Thus, if ε is really small, the ε-steepest descent Algorithm XIII.1.3.1 will simply reduce to

$$x^{p+1} = x^p - t^p s(x^p), \qquad (1.1.3)$$

at least until x^p becomes nearly optimal. Only then, will the bundling mechanism enter into play.

The delicate choice of ε (not too large, not too small ...) is a major drawback of methods of ε-descent.

(b) The Role of the Line-Search. In terms of diminishing f, the weakness of the direction is even aggravated by the choice of the stepsize, supposed to minimize over $\mathbb{R}^+$ not the function $t \mapsto f(x^p + td_k)$ but the perturbed difference quotient q_ε.

Firstly, trying to minimize q_ε along a given direction is not a sound idea: after all, it is f that we would like to reduce. Yet the two functions q_ε and $t \mapsto f(x+td)$ may have little to do with each other. Figure 1.1.1 shows T_ε of Lemma XIII.1.2.2 for two extreme values of ε: on the left [right] part of the picture, ε is too small [too large]: the stepsize sought by the ε-descent Algorithm XIII.1.3.1 is too small [too large], and in both cases, f cannot be decreased efficiently.

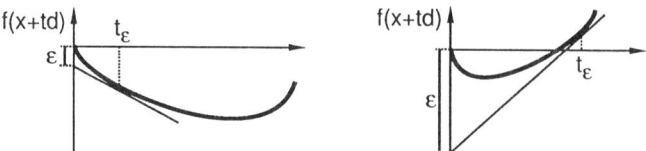

Fig. 1.1.1. Inefficient stepsizes in ε-descent methods

The second difficulty with the line-search is that, when d_k is not an ε-descent direction, the minimization of q_ε must be carried out rather accurately. In Table 1.1.1, the number of calls to the black box (U1) per iteration is in the range 5–10; recall that, in the smooth case, a line-search typically takes less than two calls per iteration. Even taking into account the increased difficulty when passing to a nonsmooth function, the present ratio seems hardly acceptable.

Still concerning the line-search, we mention one more problem: an initial stepsize is hard to guess. With methods of Chap. II, it was possible to initialize t by estimating the decrease of the line-search function $t \mapsto f(x + td)$ (Remark II.3.4.2). The behaviour of this function after changing x and d, i.e. after completing an iteration, could be reasonably predicted. Here the function of interest is q_ε of (XIII.1.2.2), which behaves wildly when x and d vary; among other things, it is infinite for $t = 0$, i.e. at x.

Remark 1.1.3 The difficulties illustrated by Fig. 1.1.1 are avoided by a further variant of §XIII.3.2: a descent test can be inserted in Step 4 of Algorithm XIII.3.2.2, updating x when $f(x + t_k d_k) < f(x) - \varepsilon$. Here $\varepsilon > 0$ is again chosen in advance, so this variant does not facilitate the choice of a suitable ε, nor of a suitable initial stepsize. Besides, the one-dimensional minimization of f must again be performed rather accurately. □

(c) Clumsy Use of the Information. Our last item in the list of deficiencies of ε-descent methods lies in their structure itself. Consider the two possible actions after terminating an iteration (p, k).

- (a_ε) In one case (ε-descent obtained), x^p is moved and all the past is forgotten. The polyhedron S_k is reinitialized to a singleton, which is the poorest possible approximation of $\partial_\varepsilon f(x^{p+1})$. This "Markovian" character is unfortunate: classically, all efficient optimization algorithms accumulate and exploit information about f from previous iterations. This is done for example by conjugate gradient and quasi-Newton methods of §II.2.
- (b_ε) In the second case (S_k enriched), x^p is left as it is, the information is now correctly accumulated. Another argument appears, however: the real problem of interest, namely to decrease f, is somewhat overlooked. Observe for example that no f-value is used to compute the directions d_k; more importantly, the iterate $x^p + t_k d_k$ produced by the line-search is discarded, even if it happens to be (possibly much) better than the current x^p.

In fact, the above argument is rather serious, although subtle, and deserves some explanation. The bundling idea is based on a certain separation process, whose speed of convergence

1 Introduction: The Bundle of Information

is questionable; see §IX.2.2. The (b_ε)-phase, which relies entirely and exclusively upon this process, has therefore a questionable speed of convergence as well. Yet the real problem, of minimizing a convex function f, is addressed by the (a_ε)-phase only. This sort of dissociation seems undesirable for performance.

In other words, there are two processes in the game: one is the convergence to 0 of some sequence of (approximate) subgradients $\hat{s}_k = -d_k$; the other is the convergence of $\{f(x^p)\}$ to $\bar{f}$. The first is theoretically essential – after all, it is the only way to get a minimality condition – but has fragile convergence qualities. One should therefore strive to facilitate its task with the help of the second process, by diminishing $\|d\|$- and f-values in an harmonious way. The overall convergence will probably not be improved *in theory*; but the algorithm must be given a chance to have a better behaviour *heuristically*.

(d) Conclusion. Let us sum up this Section 1.1: ε-descent methods have a number of deficiencies, and the following points should be addressed when designing new minimization methods.

(i) They should rely less on ε. In particular, the test for moving the current x^p should be less crude than simply decreasing f by an a priori value ε.
(ii) Their line-search should be based on reducing the natural objective function f. In particular, the general and sound principles of §II.3 should be applied.
(iii) They should take advantage as much as possible of all the information accumulated during the successive calls to (U1), perhaps from the very beginning of the iterations.
(iv) Their convergence should not be based entirely on that of $\{d_k\}$ to 0: they should take care as much as possible (on a heuristic basis) of the decrease of f to its minimal value.

In this chapter, we will introduce a method coping with (i) – (iii), and somehow with (iv) (to the extent that this last point can be quantified, i.e. very little). This method will retain the concept of keeping x^p fixed in some cases, in order to improve the approximation of $\partial_\varepsilon f(x^p)$ for some ε. In fact, this technique is mandatory when designing methods for nonsmooth optimization based on a descent principle. The direction will again be computed by projecting the origin onto the current approximation of $\partial_\varepsilon f(x^p)$, which will be substantially richer than in Chap. XIII. Also, the test for descent will be quite different: as in §II.3.2, it will require from f a definite relative decrease, rather than absolute. Finally, the information will be accumulated all along the iterations, and it is only for practical reasons, such as lack of memory, that this information will be discarded.

1.2 Constructing the Bundle of Information

Generally speaking, in most efficient minimization methods, the direction is computed at the current iterate in view of the information on hand concerning f. The only source for such information is the black box (U1) characterizing the problem to be solved; the following definition is therefore relevant.

Definition 1.2.1 At the current iteration of a minimization algorithm, call

$$\{y_i \in \mathbb{R}^n : i = 1, \ldots, L\}$$

the set of arguments at which (U1) has been called from the beginning. Then the *raw bundle* is the set of triples

$$\{y_i \in \mathbb{R}^n, \ f_i := f(y_i), \ s_i := s(y_i) \in \partial f(y_i) : i = 1, \ldots, L\} \qquad (1.2.1)$$

collecting all the information available. □

In (1.2.1), L increases at each iteration by the number of cycles needed by the line-search. The raw bundle is therefore an abstract object (unless L can be bounded a priori): no computer is able to store this potentially infinite amount of information. In order to make it tractable one must do *selection* and/or *compression* in the raw bundle.

– Selection means that only some designated triples $(y, f, s)_i$ are kept, L being replaced by a smaller value, say ℓ, which is kept under control, so as to respect the memory available in the computer.
– Compression means a transformation of the triples, so as to extract their most relevant characteristics.

Definition 1.2.2 The *actual bundle* (or simply "bundle", if no confusion is possible) is the bank of information that is

– explicitly obtained from the raw bundle,
– actually stored in the computer,
– and used to compute the current direction. □

There are several possibilities for selecting-compressing the raw bundle, depending on the role of the actual bundle in terms of the resulting minimization algorithm. Let us illustrate this point.

Example 1.2.3 (Quasi-Newton Methods) If one wanted to relate the present bundling concept to classical optimization methods, one could first consider the quasi-Newton methods of §II.2.3. There the aim of the bundle was to identify the second-order behaviour of f. The selection consisted of two steps:

(i) Only one triple (y, f, s) per iteration was "selected", namely $(x_k, f(x_k), s_k)$; any intermediate information obtained during each line-search was discarded: the actual number ℓ of elements in the bundle was the number k of line-searches.
(ii) Only the couple (y, s) was kept from each triple (y, f, s): f was never used.

Thus, the actual bundle was "compressed" to

$$\{(x_i, s_i) : i = 1, \ldots, k\} \qquad (1.2.2)$$

with i indexing iterations, rather than calls to (U1). This bundle (1.2.2), made of $2nk$ real numbers, was not explicitly stored. Rather, it was further "compressed" into a symmetric

$n \times n$ matrix – although, for $2nk \leq 1/2\, n(n+1)$, this could hardly be called a compression! – whose aim was to approximate $(\nabla^2 f)^{-1}$.

In summary, the actual bundle was a set of $1/2\, n(n+1)$ numbers, gathering the information contained in (1.2.2), and computed by recurrence formulae such as that of Broyden-Fletcher-Goldfarb-Shanno (II.2.3.9). □

The (i)-part of the selection in Example 1.2.3 is rather general in optimization methods. In practice, L in (1.2.1) is just the current number k of iterations, and i indexes line-searches rather than calls to (U1). In most optimization algorithms, the intermediate (f, s) computed during the line-searches are used only to update the stepsize, but not to compute subsequent directions.

Example 1.2.4 (Conjugate-Gradient Methods) Still along the same idea, consider now the conjugate-gradient method of §II.2.4. We have seen that its original motivation, based on the algebra of quadratic functions, was not completely clear. Rather, it could be interpreted as a way of smoothing out the sequence $\{d_k\}$ of successive directions (remember Fig. II.2.4.1).

There, we had a selection similar to (i) of Example 1.2.3; indeed the direction d_k was defined in terms of

$$\{s_i : i = 1, \ldots, k\}. \tag{1.2.3}$$

As for the compression, it was based on formulae – say of Polak-Ribière (II.2.4.8) – enabling the computation of d_k using only s_k and d_{k-1}, instead of the full set (1.2.3). In other words, the actual bundle was $\{d_{k-1}, s_k\}$, made up of two vectors. □

The above two examples are slightly artificial, in that the bundling concept is a fairly diverted way of introducing the straightforward calculations for quasi-Newton and conjugate-gradient methods. However, they have the merit of stressing the importance of the bundling concept: virtually all efficient minimization methods use it, one way or another. On the other hand, the cutting-plane algorithm of §XII.4.2 is an instance of an approach in which the bundling idea is blatant. Actually, that example is even too simple: no compression and no selection can be made, the raw bundle and actual bundle are there identical.

Example 1.2.5 (ε-Descent Methods) In Chap. XIII, the bundle was used to identify $\partial_\varepsilon f$ at the current iterate. To disclose the selection-compression mechanism, let us use the notation (p, k) of that chapter.

First of all, the directions depended only on s-values. When computing the current direction d_k issued from the current iterate x^p, one had on hand the set of subgradients (the raw bundle, in a way)

$$\left\{ s_1^1, s_2^1, \ldots, s_{k_1}^1;\ s_1^2, s_2^2, \ldots, s_{k_2}^2;\ \ldots;\ s_1^p, \ldots, s_k^p \right\},$$

where k_q, for $q = 1, \ldots, p-1$, was the total number of line-searches needed by the qth outer iteration (knowing that the pth is not complete yet).

In the selection, all the information collected from the previous descent-iterations (those having $q < p$) was discarded, as well as all f- and y-values. Thus, the actual bundle was essentially

$$\left\{ s_1^p, \ldots, s_k^p \right\}. \tag{1.2.4}$$

Furthermore, there was a compression mechanism, to cope with possibly large values of k. If necessary, an arbitrary subset in (1.2.4) was replaced by the convex combination

$$\hat{s}_k = \sum_{i=1}^{k} \alpha_i s_i^p \tag{1.2.5}$$

just computed in the previous iteration (remember Algorithm XIII.2.3.1). In a way, the *aggregate subgradient* (1.2.5) could be considered as synthesizing the most essential information from (1.2.4). □

The actual bundle in Example 1.2.5, with its s_k's and $\hat{s}_k$'s, is not so easy to describe with explicit formulae; a clearer view is obtained from inspecting step by step its evolution along the iterations. For this, we proceed to rewrite the schematic ε-descent Algorithm XIII.1.3.1 in a condensed form, with emphasis on this evolution. Furthermore, we give up the (p, k)-notation of Chap. XIII, to the advantage of the "standard" notation, used in Chap. II. Thus, the single index k will denote the number of iterations done from the very beginning of the algorithm. Accordingly, x_k will be the current iterate, at which the stopping criterion is checked, the direction d_k is computed, and the line-search along d_k is performed.

In terms of the methods of Chap. XIII, this change of notation implies a substantial change of philosophy: there is only one line-search per iteration (whereas a would-be outer iteration was a sequence of line-searches); and this line-search has two possible exits, corresponding to the two cases (a_ε) and (b_ε).

Call y_+ the final point tried by the current line-search: it is of the form $y_+ = x_k + t d_k$, for some stepsize $t > 0$. The two possible exits are:

(a_ε) When f has decreased by ε, the current iterate x_k is updated to $x_{k+1} = y_+ = x_k + t d_k$. The situation is rather similar to the general scheme of Chap. II and this is called a *descent-iteration*, or a *descent-step*. As for the bundle, it is totally reset to $\{s_{k+1}\} = \{s(x_{k+1})\} \subset \partial_\varepsilon f(x_{k+1})$. There is no need to use an additional index p.

(b_ε) When a new line-search must be performed from the same x, the trick is to set $x_{k+1} = x_k$, disregarding y_+. Here, we have a *null-iteration*, or a *null-step*. This is the new feature with respect to the algorithms of Chap. II: the current iterate is not changed, only the next direction is going to be changed.

With this notation borrowed from classical optimization (i.e. of smooth functions), Algorithm XIII.1.3.1 can now be described as follows. We neglect the irrelevant details of the line-search, but we keep the description of a possible deletion-compression mechanism.

Algorithm 1.2.6 (Algorithm of ε-Descent, Standard Notation) The starting point $x_1 \in \mathbb{R}^n$ is given; compute $s_1 \in \partial f(x_1)$. Choose the descent criterion $\varepsilon > 0$, the convergence parameter $\delta > 0$, the tolerance for the line-search $m' \in]0, 1[$ and the maximal bundle size $\bar{\ell} \geq 2$. Initialize the iteration index $k = 1$, the descent index $k_0 = 0$, and the bundle size $\ell = 1$.

1 Introduction: The Bundle of Information   231

STEP 1 (direction-finding and stopping criterion). Solve the minimization problem in the variable $\alpha \in \mathbb{R}^\ell$

$$\min \left\{ \tfrac{1}{2} \left\| \sum_{i=k_0+1}^{k_0+\ell} \alpha_i s_i \right\|^2 \; : \; \alpha_i \geq 0, \; \sum_{i=k_0+1}^{k_0+\ell} \alpha_i = 1 \right\}$$

and set $d_k := -\sum_{i=k_0+1}^{k_0+\ell} \alpha_i s_i$. If $\|d_k\| \leq \delta$ stop.

STEP 2 (line-search). Minimize r_ε along d_k accurately enough to obtain a stepsize $t > 0$ and a new iterate $y_+ = x_k + t d_k$ with $s_+ \in \partial f(y_+)$ such that

- either $f(y_+) < f(x_k) - \varepsilon$; then go to Step 3;
- or $\langle s_+, d_k \rangle \geq -m'\|d_k\|^2$ and $s_+ \in \partial_\varepsilon f(x_k)$; then go to Step 4.

STEP 3 (descent-step). Set $x_{k+1} = y_+$, $k_0 = k$, $\ell = 1$; replace k by $k+1$ and loop to Step 1.

STEP 4 (null-step). Set $x_{k+1} = x_k$, replace k by $k+1$. If $\ell < \bar{\ell}$, set $s_{k_0+\ell+1} = s_+$ from Step 2, replace ℓ by $\ell + 1$ and loop to Step 1. Else go to Step 5.

STEP 5 (compression). Delete at least two vectors from the list $\{s_{k_0+1}, \ldots, s_{k_0+\ell}\}$. With ℓ denoting the number of elements in the new list, set $s_{k_0+\ell+1} = -d_k$ from Step 1, set $s_{k_0+\ell+2} = s_+$ from Step 2; replace ℓ by $\ell + 2$ and loop to Step 1. □

In a way, we obtain something simpler than Algorithm XIII.1.3.1. A reader familiar with iterative algorithms and computer programming may even observe that the index k can be dropped: at each iteration, the old (x, d) can be overwritten; the only important indices are: the current bundle size ℓ, and the index k_0 of the last descent-iteration.

Observe the evolution of the bundle: it is totally refreshed after a descent-iteration (Step 3); it grows by one element at each null-iteration (Step 4), and it is compressed when necessary (Step 5). Indeed, the relative complexity of this flow-chart is mainly due to this compression mechanism. If it were neglected, the algorithm would become fairly simple: Step 5 would disappear and the number ℓ of elements in the bundle would just become $k - k_0$.

Recall that the compression in Step 5 is not fundamental, but is inserted for the sake of implementability only. By contrast, Definition 1.2.7 below will give a different type of compression, which is attached to the conception of the method itself. The mechanism of Step 5 will therefore be more suitably called *aggregation*, while "compression" has a more fundamental meaning, to be seen in (1.2.9).

Examining the ε-descent method in its form 1.2.6 suggests a weak point, not so visible in Chap. XIII: why, after all, should the actual bundle be reset entirely at k_0? Some of the elements $\{s_1, \ldots, s_{k_0}\}$ might well be in $\partial_\varepsilon f(x_k)$. Then why not use them to compute d_k? The answer, quite simple, is contained in the transportation formula of Proposition XI.4.2.2, extensively used in Chap. XIII.

Consider the number

$$e(x, y, s) := f(x) - [f(y) + \langle s, x - y \rangle] \quad \text{for all } x, y, s \text{ in } \mathbb{R}^n; \qquad (1.2.6)$$

we recall that, if $s \in \partial f(y)$, then $e(x, y, s) \geq 0$ and s is an $e(x, y, s)$-subgradient at x (this is just what the transportation formula says). The converse is also true:

for $s \in \partial f(y)$, $\quad s \in \partial_\varepsilon f(x) \iff e(x, y, s) \leqslant \varepsilon$. (1.2.7)

Note also that s returned from (U1) at y can be considered as a function of y alone (remember Remark VIII.3.5.1); so the notation $e(x, y)$ could be preferred to (1.2.6).

An important point here is that, for given x and y, this $e(x, y)$ can be *explicitly* computed and stored in the computer whenever $f(x)$, $f(y)$ and $s(y)$ are known. Thus, we have for example a straightforward test to detect whether a given s in the bundle can be used to compute the direction. Accordingly, we will use the following notation.

Definition 1.2.7 For each i in the raw bundle (1.2.1), the *linearization error* at the iterate x_k is the nonnegative number

$$e_i^k := f(x_k) - [f(y_i) + \langle s_i, x_k - y_i \rangle].$$ (1.2.8)

The abbreviated notation e_i will be used for e_i^k whenever convenient. □

Only the vectors s_i and the scalars e_i are needed by the minimization method of the present chapter. In other words, we are interested in the set

$$\left\{ s_i, e_i^k : i = 1, \ldots, k \right\}, \quad \text{where} \quad s_i \in \partial_{e_i^k} f(x_k);$$ (1.2.9)

and this is the actual bundle. Note: to be really accurate, the term "actual" should be reserved for the resulting bundle after aggregation of Step 5 in Algorithm 1.2.6; but we neglect this question of aggregation for the moment.

It is important to note here that the vectors y_i have been eliminated in (1.2.9); the memory needed to store (1.2.9) is $k(n+1)$, instead of $k(2n+1)$ for (1.2.1) (with $L = k$): we really have a compression here, and the information contained in (1.2.9) is definitely poorer than that in (1.2.1). However, (1.2.8) suggests that each e_i must be recomputed at each descent iteration; so, can we really maintain the bundle (1.2.9) without storing also the y's? The answer is yes, thanks to the following simple result.

Proposition 1.2.8 *For any two iterates k and k' and sampling point y_i, there holds*

$$e_i^{k'} = e_i^k + f(x_{k'}) - [f(x_k) + \langle s_i, x_{k'} - x_k \rangle].$$ (1.2.10)

PROOF. Just apply the definitions:

$$\begin{aligned} e_i^{k'} &= f(x_{k'}) - [f(y_i) + \langle s_i, x_{k'} - y_i \rangle] \\ e_i^k &= f(x_k) - [f(y_i) + \langle s_i, x_k - y_i \rangle] \end{aligned}$$

and obtain the result by mere subtraction. □

This result relies upon additivity of the linearization errors: for two different points x and x', the linearization error at x' is the sum of the one at x and of the error made when linearizing f from x to x' (with the same slope). When performing a descent step from x_k to $x_{k+1} \neq x_k$, it suffices to update all the linearization errors via (1.2.10), written with $k' = k + 1$.

Remark 1.2.9 An equivalent way of storing the relevant information is to replace (1.2.9) by

$$\{s_i, f^*(s_i) : i = 1, \ldots, k\}, \tag{1.2.11}$$

where we recognize in

$$f^*(s_i) := \langle s_i, y_i \rangle - f(y_i)$$

the value of the conjugate of f at s_i (Theorem X.1.4.1). Clearly enough, the linearization errors can then be recovered via

$$e_i^k = f(x_k) + f^*(s_i) - \langle s_i, x_k \rangle. \tag{1.2.12}$$

Compared to (1.2.8), formulae (1.2.11) and (1.2.12) have a more elegant and more symmetric appearance (which hides their nice geometric interpretation, though). Also they explain Proposition 1.2.8: y_i is eliminated from the notation. From a practical point of view, there are arguments in favor of each form, we will keep the form (1.2.8). It is more suggestive, and it will be more useful when we need to consider aggregation of bundle-elements. □

Let us sum up this section. The methods considered in this chapter compute the direction d_k by using the bundle (1.2.9) exclusively. This information is definitely poorer than the actual bundle (1.2.1), because the vectors y_i are "compressed" to the scalars e_i. Actually this bundle can be considered just as a bunch of e_i-subgradients at x_k, and the question is: are these approximate subgradients good for computing d_k? Here our answer will be: use them to approximate $\partial_\varepsilon f(x_k)$ for some chosen ε, and then proceed essentially as in Chap. XIII.

2 Computing the Direction

2.1 The Quadratic Program

In this section, we show how to compute the direction on the basis of our development in §1. The current iterate x_k is given, as well as a bundle $\{s_i, e_i\}$ described by (1.2.8), (1.2.9). Whenever possible, we will drop the iteration index k: in particular, the current iterate x_k will be denoted by x. The bundle size will be denoted by ℓ (which is normally equal to k, at least if no aggregation has been done yet), and we recall that the unit simplex of $\mathbb{R}^\ell$ is

$$\Delta_\ell := \left\{ \alpha \in \mathbb{R}^\ell : \sum_{i=1}^\ell \alpha_i = 1, \ \alpha_i \geq 0 \text{ for } i = 1, \ldots, \ell \right\}.$$

The starting idea for the algorithms in this chapter lies in the following simple result.

Proposition 2.1.1 *With $s_i \in \partial f(y_i)$ for $i = 1, \ldots, \ell$, e_i defined by (1.2.8) at $x_k = x$, and $\alpha = (\alpha_1, \ldots, \alpha_\ell) \in \Delta_\ell$, there holds*

$$f(z) \geq f(x) + \left\langle \sum_{i=1}^\ell \alpha_i s_i, z - x \right\rangle - \sum_{i=1}^\ell \alpha_i e_i \quad \text{for all } z \in \mathbb{R}^n.$$

In particular, for $\varepsilon \geqslant \min_i e_i$, the set
$$S(\varepsilon) := \left\{s = \sum_{i=1}^{\ell} \alpha_i s_i \, : \, \alpha \in \Delta_\ell, \, \sum_{i=1}^{\ell} \alpha_i e_i \leqslant \varepsilon\right\} \qquad (2.1.1)$$
is contained in $\partial_\varepsilon f(x)$.

PROOF. Write the subgradient inequalities
$$f(z) \geqslant f(y_i) + \langle s_i, z - y_i \rangle \quad \text{for } i = 1, \ldots, \ell$$
in the form
$$f(z) \geqslant f(x) + \langle s_i, z - x \rangle - e_i \quad \text{for } i = 1, \ldots, \ell$$
and obtain the results by convex combination. □

We will make constant use of the following notation: for $\alpha \in \Delta_\ell$, we set
$$s(\alpha) := \sum_{i=1}^{\ell} \alpha_i s_i, \quad e(\alpha) := \sum_{i=1}^{\ell} \alpha_i e_i, \qquad (2.1.2)$$
so that (2.1.1) can be written in the more compact form
$$S(\varepsilon) = \{s(\alpha) \, : \, \alpha \in \Delta_\ell, \, e(\alpha) \leqslant \varepsilon\}.$$

The above result gives way to a strategy for computing the direction, merely copying the procedure of Chap. XIII. There, we had a set of subgradients, call them $\{s_1, \ldots, s_\ell\}$, obtained via suitable line-searches; their convex hull was a compact convex polyhedron S included in $\partial_\varepsilon f(x)$. The origin was then projected onto S, so as to obtain the best hyperplane separating S from $\{0\}$. Here, Proposition 2.1.1 indicates that S can be replaced by the slightly more elaborate $S(\varepsilon)$ of (2.1.1): we still obtain an inner approximation of $\partial_\varepsilon f(x)$ and, with this straightforward substitution, the above strategy can be reproduced.

In a word, all we have to do is project the origin onto $S(\varepsilon)$: we choose $\varepsilon \geqslant 0$ and we compute the particular ε-subgradient
$$\hat{s} := \text{Proj}\, 0/S(\varepsilon) \qquad (2.1.3)$$
to obtain the direction $d = -\hat{s}$. After a line-search along this d is performed, we either move x to a better x_+ (descent-step) or we stay at x (null-step). A key feature, however, is that *in both cases*, the bundle size can be increased by one: as long as there is room in the computer, no selection has to be made from the raw bundle, which is thus enriched at *every iteration* (barring aggregation, which will be considered later). By contrast, the bundle size in the ε-descent Algorithm XIII.1.3.1 – or equivalently Algorithm 1.2.6 – was reset to 1 after each descent-step.

The decision "descent-step vs. null-step" will be studied in §3, together with the relevant algorithmic details. Here we focus our attention on the direction-finding problem, which is

2 Computing the Direction   235

$$\begin{vmatrix} \min \frac{1}{2} \left\| \sum_{i=1}^{\ell} \alpha_i s_i \right\|^2 & \alpha \in \mathbb{R}^{\ell}, & [\min \frac{1}{2} \|s(\alpha)\|^2] & \text{(i)} \\ \sum_{i=1}^{\ell} \alpha_i = 1, \ \alpha_i \geq 0 \text{ for } i = 1, \ldots, \ell, & [\alpha \in \Delta_{\ell}] & \text{(ii)} \\ \sum_{i=1}^{\ell} \alpha_i e_i \leq \varepsilon, & [e(\alpha) \leq \varepsilon] & \text{(iii)} \end{vmatrix} \quad (2.1.4)$$

which gives birth to the direction

$$d = -\hat{s} := -\hat{s}_\varepsilon := -s(\alpha) \quad \text{for some } \alpha \text{ solving (2.1.4).} \quad (2.1.5)$$

For a given bundle, the above direction-finding problem differs from the previous one — (XIII.1.3.1), (IX.1.8), or (VIII.2.1.6) — by the additional constraint (2.1.4)(iii) only.

The feasible set (2.1.4)(ii), (iii) is a compact convex polyhedron (possibly empty), namely a portion of the unit simplex Δ_{ℓ}, cut by a half-space. The set $S(\varepsilon)$ of (2.1.1) is its image under the linear transformation that, to $\alpha \in \mathbb{R}^{\ell}$, associates $s(\alpha) \in \mathbb{R}^n$ by (2.1.2). This confirms that $S(\varepsilon)$ too is a compact convex polyhedron (possibly empty). Note also that the family $\{S(\varepsilon)\}_{\varepsilon \geq 0}$ is nested and has a maximal element, namely the convex hull of the available subgradients:

$$\varepsilon \leq \varepsilon' \implies S(\varepsilon) \subset S(\varepsilon') \subset S(\infty) = \operatorname{co}\{s_1, \ldots, s_\ell\}.$$

Here $S(\infty)$ is a handy notation for "$S(\varepsilon)$ of a large enough ε, say $\varepsilon \geq \max_i e_i$".

Proposition 2.1.2 *The direction-finding problem (2.1.4) has an optimal solution if and only if*

$$\varepsilon \geq \min\{e_i : i = 1, \ldots, \ell\}, \quad (2.1.6)$$

in which case the direction d of (2.1.5) is well-defined, independently of the optimal α.

PROOF. Because all e_i and α_i are nonnegative, the constraints (2.1.4)(ii), (iii) are consistent if and only if (2.1.6) holds, in which case the feasible domain is nonempty and compact. The rest is classical. □

In a way, (2.1.3) is "equivalent" to (2.1.4), and the optimal set of the latter is the polyhedron

$$A_\varepsilon := \{\alpha \in \Delta_\ell : e(\alpha) \leq \varepsilon, \ s(\alpha) = \hat{s}_\varepsilon\}. \quad (2.1.7)$$

Uniqueness of $\hat{s}_\varepsilon$ certainly does not imply that A_ε is a singleton, though. In fact, consider the example with $n = 2$ and $\ell = 3$:

$$s_1 := (-1, 1), \ s_2 := (0, 1), \ s_3 := (1, 1); \quad e_1 = e_2 = 0, \ e_3 = \varepsilon = 1.$$

It is easy to see that $\hat{s} = s_2$ and that the solutions of (2.1.4) are those α satisfying (2.1.4)(ii) together with $\alpha_1 = \alpha_3$. Observe, correspondingly, that the set $\{e(\alpha) : \alpha \in A_\varepsilon\}$ is not a singleton either, but the whole segment $[0, \varepsilon/2]$.

Remark 2.1.3 Usually, $\min_i e_i = 0$ in (2.1.6), i.e. the bundle does contain some element of $\partial f(x_k)$ — namely $s(x_k)$. Then the direction-finding problem has an optimal solution for any $\varepsilon \geq 0$.

It is even often true that there is only one such minimal e_i, i.e.:

$\exists i_0 \in \{1, \ldots, \ell\}$ such that $e_{i_0} = 0$ and $e_i > 0$ for $i \neq i_0$.

This extra property is due to the fact that, usually, the only way to obtain a subgradient at a given x (here x_k) is to call the black box (U1) at this very x. Such is the case for example when f is strictly convex (Proposition VI.6.1.3):

$$f(y) > f(x) + \langle s, y - x \rangle \quad \text{for any } y \neq x \text{ and } s \in \partial f(x),$$

which implies in (1.2.6) that $e(x, y, s) > 0$ whenever $y \neq x$ and $s \in \partial f(y)$. On the other hand, the present extra property does not hold for some special objective functions, typically piecewise affine. □

Our new direction-finding problem (2.1.4) generalizes those seen in the previous chapters:

- When ε is large, say $\varepsilon \geqslant \max_i e_i$, there is no difference between $S(\varepsilon)$ and $S = S(\infty)$. The extra constraint (2.1.4)(iii) is inactive, we obtain nothing but the direction of the ε-descent Algorithm XIII.1.3.1.
- On the other hand, suppose ε lies at its minimal value of (2.1.6), say $\varepsilon = 0$ in view of Remark 2.1.3. Then any feasible α in (2.1.4)(ii), (iii) has to satisfy

$$\alpha_i > 0 \quad \Longrightarrow \quad e_i = \varepsilon \; [= \min_j e_j = 0].$$

Denoting by I_0 the set of indices with $e_i = \varepsilon$, the direction-finding problem becomes

$$\min \left\{ \tfrac{1}{2} \|s(\alpha)\|^2 \; : \; \alpha \in \Delta_\ell, \; \alpha_i = 0 \text{ if } i \notin I_0 \right\},$$

in which the extra constraint (2.1.4)(iii) has become redundant. In other words, the "minimal" choice $\varepsilon = \min_i e_i = 0$ disregards in the bundle the elements that are not in $\partial f(x_k)$; it is the direction d_k of Chap. IX that is obtained – a long step backward.
- Between these two extremes (0 and ∞), intermediate choices of ε allow some flexibility in the computation of the direction. When ε describes $\mathbb{R}^+$ the normalized direction describes, on the unit sphere of $\mathbb{R}^n$, a curve having as endpoints the directions of Chaps. IX and XIII.

Each linearization error e_i can also be considered as a *weight*, expressing how far the corresponding s_i is from $\partial f(x_k)$. The extra constraint (2.1.4)(iii) gives a preference to those subgradients with a smaller weight: in a sense, they are closer to $\partial f(x_k)$. With this interpretation in mind, the weights in Chap. XIII were rather 0–1: 0 for those subgradients obtained before the last outer iteration, 1 for those appearing during the present p^{th} iteration. Equivalently, each s_i in Algorithm 1.2.6 was weighted by 0 for $i \leqslant k_0$, and by 1 for $i > k_0$.

An index i will be called *active* if $\alpha_i > 0$ for some α solving (2.1.4). We will also speak of active subgradients s_i, and active weights e_i.

2.2 Minimality Conditions

We equip the α-space $\mathbb{R}^\ell$ with the standard dot-product. Then consider the function $\alpha \mapsto 1/2 \, \|s(\alpha)\|^2 =: \nu(\alpha)$ defined via (2.1.2). Observing that it is the composition

$$\mathbb{R}^\ell \ni \alpha \mapsto s = \sum_{i=1}^\ell \alpha_i s_i \in \mathbb{R}^n \mapsto \tfrac{1}{2}\|s\|^2 = v(\alpha) \in \mathbb{R},$$

standard calculus gives its partial derivatives:

$$\frac{\partial v}{\partial \alpha_j}(\alpha) = \langle s_j, s(\alpha) \rangle \quad \text{for } j = 1, \ldots, \ell. \tag{2.2.1}$$

Then it is not difficult to write down the minimality conditions for the direction-finding problem, based on the Lagrange function

$$\mathbb{R}^\ell \ni \alpha \mapsto \tfrac{1}{2}\|s(\alpha)\|^2 + \lambda \left(1 - \sum_{i=1}^\ell \alpha_i\right) + \mu[e(\alpha) - \varepsilon]. \tag{2.2.2}$$

Theorem 2.2.1 *We use the notation (2.1.2); α is a solution of (2.1.4) if and only if: it satisfies the constraints (2.1.4) (ii), (iii), and there is $\mu \geq 0$ such that*

$$\mu = 0 \quad \text{if } e(\alpha) < \varepsilon \quad \text{and}$$

$$\begin{array}{l}\langle s_i, s(\alpha) \rangle + \mu e_i \geq \|s(\alpha)\|^2 + \mu\varepsilon \quad \text{for } i = 1, \ldots, \ell \\ \qquad\qquad\qquad\qquad\qquad\qquad\text{with equality if } \alpha_i > 0.\end{array} \tag{2.2.3}$$

Besides, the set of such μ is actually independent of the particular solution α.

PROOF. Our convex minimization problem falls within the framework of §XII.5.3(c). All the constraints being affine, the weak Slater assumption holds and the Lagrangian (2.2.2) must be minimized with respect to α in the first orthant, for suitable values of λ and μ. Then, use (2.2.1) to compute the relevant derivatives and obtain the minimality conditions as in Example VII.1.1.6:

$$\mu[e(\alpha) - \varepsilon] = 0,$$

and, for $i = 1, \ldots, \ell$,

$$\langle s_i, s(\alpha) \rangle + \mu e_i - \lambda \geq 0,$$
$$\alpha_i[\langle s_i, s(\alpha) \rangle + \mu e_i - \lambda] = 0.$$

Add the last ℓ equalities to see that $\lambda = \|s(\alpha)\|^2 + \mu\varepsilon$, and recognize the Karush-Kuhn-Tucker conditions of Theorem VII.2.1.4; also, μ does not depend on α (Proposition VII.3.1.1). □

Thus, the right-hand side in (2.2.3) is the multiplier λ of the equality constraint in (2.1.4)(ii); it is an important number, and we will return to it later.

On the other hand, the extra constraint (2.1.4)(iii) is characteristic of the algorithm we have in mind; therefore, its multiplier μ is even more important than λ, and the rest of this section is devoted to its study. To single it out, we use the special Lagrangian

$$\mathbb{R}^\ell \ni \alpha \mapsto L_\mu(\alpha) := \tfrac{1}{2}\|s(\alpha)\|^2 + \mu[e(\alpha) - \varepsilon].$$

It must be minimized with respect to $\alpha \in \Delta_\ell$, to obtain the closed convex dual function

$$\mathbb{R} \ni \mu \mapsto \Theta(\mu) = -\min_{\alpha \in \Delta_\ell} L_\mu(\alpha). \tag{2.2.4}$$

Duality theory applies in a straightforward way: $L_\mu(\cdot)$ is a convex continuous function on the compact convex set Δ_ℓ. We know in advance that the set of dual solutions

$$M_\varepsilon = \operatorname{Argmin}\{\Theta(\mu) : \mu \geqslant 0\}$$

is nonempty: it is actually the set of those μ described by Theorem 2.2.1 for some solution $\alpha \in A_\varepsilon$ of (2.1.7). Then the minimality conditions have another expression:

Theorem 2.2.2 *We use the notation (2.1.2); $\alpha \in \Delta_\ell$ solves (2.1.4) if and only if, for some $\mu \geqslant 0$, it solves the minimization problem in (2.2.4) and satisfies*

$$e(\alpha) \leqslant \varepsilon \quad \text{with equality if } \mu \text{ is actually positive.} \tag{2.2.5}$$

PROOF. Direct application of Theorem VII.4.5.1, or also of §XII.5.3(c). □

We retain from this result that, to solve (2.1.4), one can equally solve (2.2.4) for some $\mu \geqslant 0$ (not known in advance!), and then check (2.2.5): if it holds, the α thus obtained is a desired solution, i.e. $s(\alpha) = \hat{s}_\varepsilon$. The next result gives a converse property:

Proposition 2.2.3 *For given $\mu \geqslant 0$, all optimal α in (2.2.4) make up the same $s(\alpha)$ in (2.1.2). If $\mu > 0$, they also make up the same $e(\alpha)$.*

PROOF. Let α and β solve (2.2.4). From convexity, $L_\mu(t\alpha + (1-t)\beta)$ is constantly equal to $\Theta(\mu)$ for all $t \in [0, 1]$, i.e.

$$t\{\langle s(\beta), s(\alpha) - s(\beta)\rangle + \mu[e(\alpha) - e(\beta)]\} + \tfrac{1}{2}t^2 \|s(\alpha) - s(\beta)\|^2 = 0.$$

The result follows by identifying to 0 the coefficients of t^2 and of t. □

Note: if we solve (2.2.5) with $\mu = 0$ and get a value $e(\alpha) > \varepsilon$, we cannot conclude that $s(\alpha) \neq \hat{s}_\varepsilon$, because another optimal α might give (2.2.5); only if $\mu > 0$, can we safely make this conclusion.

This point of view, in which $\mu \geqslant 0$ controls the value of $e(\alpha)$ coming out from the direction-finding problem, will be a basis for the next chapter. We also recall Everett's Theorem XII.2.1.1: if one solves (2.2.4) for some $\mu \geqslant 0$, thus obtaining some optimal α_μ, this α_μ solves a version of (2.1.4) in which the right-hand side of the extra constraint (iii) has been changed from ε to the a posteriori value $e(\alpha_\mu)$; this is a nonambiguous number, in view of Proposition 2.2.3. With the notation (2.1.2) and (2.1.7), $\alpha_\mu \in A_{e(\alpha_\mu)}$.

Another consequence of duality theory is the following important interpretation of M_ε:

Proposition 2.2.4 *The optimal value $v(\varepsilon) := 1/2 \|\hat{s}_\varepsilon\|^2$ in (2.1.4) is a convex function of ε, and its subdifferential is the negative of M_ε:*

$$\mu \in M_\varepsilon \quad \Longleftrightarrow \quad -\mu \in \partial v(\varepsilon).$$

PROOF. This is a direct consequence of, for example, Theorem VII.3.3.2.   □

Being the subdifferential of a convex function, the multifunction $\varepsilon \mapsto -M_\varepsilon$ is monotone, in the sense that

$$0 \leqslant \varepsilon < \varepsilon' \implies \mu \geqslant \mu' \text{ for all } \mu \in M_\varepsilon \text{ and } \mu' \in M_{\varepsilon'}. \qquad (2.2.6)$$

When ε diminishes, the multipliers can but increase, but they remain bounded:

Proposition 2.2.5 *With the notation* (2.1.1), *suppose* $S(0) \neq \emptyset$ (*i.e.* $\min_i e_i = 0$). *Then*

$$\mu \leqslant 2 \frac{K}{\underline{e}} \quad \text{for all } \varepsilon > 0 \text{ and } \mu \in M_\varepsilon,$$

where

$$K := \max \{ \|s_i\|^2 : i = 1, \ldots, \ell \}, \quad \underline{e} := \min \{ e_i : e_i > 0 \}.$$

PROOF. If $\mu = 0$, there is nothing to prove; if not, the optimal $e(\alpha)$ equals ε. Then take $\varepsilon \in \,]0, \underline{e}[$: there must be some j with $\alpha_j > 0$ and $e_j \geqslant \underline{e} > \varepsilon$ (the α's sum up to 1!). For this j, we can write

$$\|\hat{s}\|^2 + \mu \varepsilon = \langle s_j, \hat{s} \rangle + \mu e_j \geqslant \langle s_j, \hat{s} \rangle + \mu \underline{e}$$

hence

$$\mu(\underline{e} - \varepsilon) \leqslant \|\hat{s}\|^2 - \langle s_j, \hat{s} \rangle \leqslant 2K.$$

Divide by $\underline{e} - \varepsilon > 0$: our majorization holds in a neighborhood of 0^+. Because of (2.2.6), it holds for all $\varepsilon > 0$.   □

When $\varepsilon \to +\infty$, the extra constraint (2.1.4)(iii) eventually becomes inactive; μ reaches the value 0 and stays there. The threshold where this happens is

$$\bar{\varepsilon} := \min \{ e(\alpha) : \alpha \in \Delta_\ell, \ s(\alpha) = \hat{s}_\infty \} \qquad (2.2.7)$$

knowing that $\hat{s}_\infty = \text{Proj } 0/\{s_1, \ldots, s_\ell\}$. The function $\varepsilon \mapsto 1/2 \, \|\hat{s}_\varepsilon\|^2$ is minimal for $\varepsilon \in [\bar{\varepsilon}, +\infty[$ and we have

$$0 \in M_{\bar{\varepsilon}}; \quad \varepsilon > \bar{\varepsilon} \implies M_\varepsilon = \{0\}; \quad \varepsilon < \bar{\varepsilon} \implies 0 \notin M_\varepsilon.$$

Finally, an interesting question is whether M_ε is a singleton.

Proposition 2.2.6 *Suppose ε is such that, for some α solving* (2.1.4), *there is a j with $\alpha_j > 0$ and $e_j \neq \varepsilon$. Then M_ε consists of the single element*

$$\mu_\varepsilon = \frac{\langle \hat{s} - s_j, \hat{s} \rangle}{\varepsilon - e_j}.$$

PROOF. Immediate: for the j in question, we have

$$\langle s_j, \hat{s} \rangle + \mu e_j = \|\hat{s}\|^2 + \mu\varepsilon$$

and we can divide by $\varepsilon - e_j \neq 0$. □

- The situation described by this last result cannot hold when $\varepsilon = 0$: all the active weights are zero, then. Indeed consider the function v of Proposition 2.2.4; Theorem I.4.2.1 tells us that $M_0 = [\bar{\mu}, +\infty[$, where $\bar{\mu} = -D_+ v(0)$ is the common limit for $\varepsilon \downarrow 0$ of all numbers $\mu \in M_\varepsilon$. In view of Proposition 2.2.5, $\bar{\mu}$ is a finite number.
- On the other hand, for $\varepsilon > 0$, it is usually the case – at least if the bundle is rich enough – that (2.1.4) produces at least two active subgradients, with corresponding weights bracketing ε; when this holds, Proposition 2.2.6 applies.
- Yet, this last property need not hold: as an illustration in one dimension, take the bundle with two elements

$$s_1 = 1, \ s_2 = 2; \ e_1 = 1, \ e_2 = 0; \ \varepsilon = 1.$$

Direct calculations show that the set of multipliers μ in Theorem 2.2.1 is $[0, 1]$.

Thus, the graph of the multifunction $\varepsilon \longmapsto M_\varepsilon$ may contain vertical intervals other than $\{0\} \times M_0$. On the other hand, we mention that it cannot contain horizontal intervals other than $[\bar{\varepsilon}, +\infty[\times \{0\}$:

Proposition 2.2.7 *For $\varepsilon' < \varepsilon < \bar{\varepsilon}$ of (2.2.7), there holds*

$$\mu > \mu' \quad \text{for all } \mu \in M_\varepsilon \text{ and } \mu' \in M_{\varepsilon'}.$$

PROOF (sketch). A value of ε smaller than $\bar{\varepsilon}$ has positive multipliers μ; from Proposition 2.2.3, the corresponding $e(\alpha)$ is uniquely determined. Said in other terms, the dual function $\mu \mapsto \Theta(\mu)$ of (2.2.4) is differentiable for $\mu > 0$; its conjugate is therefore strictly convex on $]0, \bar{\varepsilon}[$ (Theorem X.4.1.3); but this conjugate is just the function $\varepsilon \mapsto 1/2 \|\hat{s}_\varepsilon\|^2$ (Theorem XII.5.1.1); with Proposition 2.2.4, this means that the derivative $-\mu$ of $\varepsilon \mapsto 1/2 \|\hat{s}_\varepsilon\|^2$ is strictly increasing. □

In summary, the function $\varepsilon \mapsto 1/2 \|\hat{s}_\varepsilon\|^2$ behaves as illustrated in Fig. 2.2.1 (which assumes $\min_i e_i = 0$), and has the following properties.

- It is a convex function with domain $[0, +\infty[$,
- with a finite right-slope at 0,
- ordinarily smooth for $\varepsilon > 0$,
- nowhere affine except for $\varepsilon \geqslant \bar{\varepsilon}$ of (2.2.7), where it is constant.
- We observe in particular that, because of convexity, the function is strictly decreasing on the segment $[0, \bar{\varepsilon}]$;
- and it reduces to a constant on $[0, +\infty[$ if $\bar{\varepsilon} = 0$, i.e. if $\hat{s}_\infty \in S(0)$.

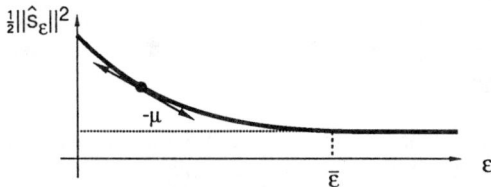

Fig. 2.2.1. Typical graph of the squared norm of the direction

2.3 Directional Derivatives Estimates

It is of interest to return to basics and remember why (2.1.4) is relevant for computing the direction.

- The reason for introducing (2.1.4) was to copy the development of Chap. XIII, with $S(\varepsilon)$ of (2.1.1) replacing the would-be $S = S_k = S(\infty)$.
- The reason for Chap. XIII was to copy Chap. IX, with $\partial_\varepsilon f(x)$ replacing $\partial f(x)$. The two approximating polyhedra S were essentially the same, as generated by calls to (U1); only their meaning was different: one was contained in $\partial_\varepsilon f(x)$, the other in $\partial f(x)$.
- Finally, Chap. IX was motivated by an implementation of the steepest-descent algorithm of Chap. VIII: the non-computable $\partial f(x)$ was approximated by S, directly available.

Now recall §VIII.1.1: what was needed there was a hyperplane separating $\{0\}$ and $\partial f(x)$ "best", i.e. minimizing $f'(x, \cdot)$ on some unit ball; we wanted

$$\min \{f'(x, d) : \|d\| = 1\} \tag{2.3.1}$$

and the theory of §VII.1.2 told us that, due to positive homogeneity, this minimization problem was in some sense equivalent to

$$\min \{\|s\|^* : s \in \partial f(x)\}. \tag{2.3.2}$$

Retracing our steps along the above path, we obtain:

- Passing from Chap. VIII to Chap. IX amounts to replacing in (2.3.1) and (2.3.2) $\partial f(x)$ by S and the corresponding support function $f'(x, \cdot)$ by σ_S. Furthermore, the norming is limited to $\|\cdot\| = \|\cdot\|^* = \|\cdot\|$.
- In Chap. XIII, S and σ_S become approximations of $\partial_\varepsilon f(x)$ and $f'_\varepsilon(x, \cdot)$ respectively.
- Finally the present chapter deals with $S(\varepsilon)$ and $\sigma_{S(\varepsilon)}$.

In summary, (2.1.4) is a realization of

$$\min \{\|s\|^* : s \in S(\varepsilon)\}, \tag{2.3.3}$$

which is in some sense equivalent to

$$\min \{\sigma_{S(\varepsilon)}(d) : \|d\| = 1\}, \tag{2.3.4}$$

knowing that we limit ourselves to the case $\|\cdot\| = \|\cdot\|^* = \|\cdot\|$.

Remark 2.3.1 We recall from §VIII.1.2 what the above "in some sense equivalent" means.

First of all, these problems are different indeed if the minimal value in (2.3.4) is non-negative, i.e. if the optimal $\hat{s}$ in (2.3.3) is zero. Then $0 \in S(\varepsilon)$, (2.1.4) produces no useful direction.

Second, the equivalence in question neglects the normalization. In (2.3.4), the right-hand side "1" of the constraint should be understood as any $\kappa > 0$. The solution-set stays collinear with itself when κ varies; remember Proposition VIII.1.1.5.

Passing between the two problems involves the inverse multifunctions

$$(\mathbb{R}^n, \|\cdot\|) \ni d \longmapsto \operatorname*{Argmax}_{\|s\|^* \leq 1} \langle s, d \rangle \subset (\mathbb{R}^n, \|\cdot\|^*)$$

and

$$(\mathbb{R}^n, \|\cdot\|^*) \ni s \longmapsto \operatorname*{Argmax}_{\|d\| \leq 1} \langle s, d \rangle \subset (\mathbb{R}^n, \|\cdot\|).$$

This double mapping simplifies when $\|\cdot\|$ is a Euclidean norm, say $\|d\|^2 = 1/2 \langle Qd, d \rangle$. Then the (unique) solutions $\hat{d}$ of (2.3.3) and $\hat{s}$ of (2.3.4) are linked by $\hat{d} = -Q^{-1}\hat{s}$ if we forget the normalization. In our present particular case of $\|\cdot\| = \|\cdot\|^* = \|\cdot\|$, this reduces to (2.1.5). We leave it as an exercise to copy §2.2 with an arbitrary norm. □

The following relation comes directly from the equivalence between (2.3.3) and (2.3.4).

Proposition 2.3.2 *With the notation* (2.1.1), (2.1.5), *there holds*

$$[f'_\varepsilon(x, -\hat{s}_\varepsilon) \geq] \quad \sigma_{S(\varepsilon)}(-\hat{s}_\varepsilon) = -\|\hat{s}_\varepsilon\|^2.$$

PROOF. This comes directly from (VIII.1.2.5), or can easily be checked via the minimality conditions. By definition, any $s \in S(\varepsilon)$ can be written $s = s(\alpha)$ with α feasible in (2.1.4)(ii), (iii); then (2.2.3) gives

$$\langle s(\alpha), -\hat{s} \rangle \leq -\|\hat{s}\|^2 + \mu[e(\alpha) - \varepsilon] \leq -\|\hat{s}\|^2$$

so $\sigma_{S(\varepsilon)}(-s) \leq -\|\hat{s}\|^2$; equality is obtained by taking α optimal, i.e. $s(\alpha) = \hat{s}$. □

Thus, the optimal value in the direction-finding problem (2.1.4) readily gives an (under-)estimate of the corresponding ε-directional derivative. In the ideal case $\partial_\varepsilon f(x) = S(\varepsilon)$, we would have $f'_\varepsilon(x, -\hat{s}) = -\|\hat{s}\|^2$.

Proposition 2.1.1 brings directly the following inequality: with the notation (2.1.5),

$$f(z) \geq f(x) + \langle \hat{s}_\varepsilon, z - x \rangle - e(\alpha) \geq f(x) - \|\hat{s}_\varepsilon\| \|z - x\| - \varepsilon \qquad (2.3.5)$$

for all $z \in \mathbb{R}^n$. This is useful for the stopping criterion: when $\|\hat{s}\|$ is small, ε can be decreased, unless it is small as well, in which case f can be considered as satisfactorily minimized. All this was seen already in Chap. XIII.

2 Computing the Direction

Remark 2.3.3 A relation with Remark VIII.1.3.7 can be established. Combining the bundle elements, we obtain for any $\alpha \in \Delta_\ell$ (not necessarily optimal) and $z \in \mathbb{R}^n$

$$f(z) \geq f(x) - \delta(\alpha)\|z - x\| - e(\alpha), \tag{2.3.6}$$

where we have set $\delta(\alpha) := \|s(\alpha)\|$. If $e(\alpha) \leq \varepsilon$, then $\delta(\alpha) \leq -\|\hat{s}\|$ by definition of $\hat{s}$ and we can actually write (2.3.5). In other words, $\|\hat{s}\|$ gives the most accurate inequality (2.3.6) obtainable from the bundle, when $e(\alpha)$ is restricted not to exceed ε. □

Having an interpretation of $\|\hat{s}\|^2$, which approximates the ε-directional derivative along the search direction $d = -\hat{s}$, we turn to the true directional derivative $f'(x, d)$. It is now the term $\|\hat{s}\|^2 + \mu\varepsilon$ that comes into play.

Proposition 2.3.4 *Suppose that*

(i) *the minimal value in (2.1.6) is zero: $e_j = 0$ for some $j \leq \ell$,*
(ii) *the corresponding α_j is positive for some solution of (2.1.4).*

Then, for any multiplier μ described by Theorem 2.2.1,

$$f'(x, -\hat{s}_\varepsilon) \geq -\|\hat{s}_\varepsilon\|^2 - \mu\varepsilon. \tag{2.3.7}$$

If, in addition, all the extreme points of $\partial f(x)$ appear among the ℓ elements of the bundle, then equality holds in (2.3.7).

PROOF. If $e_j = 0$, $s_j \in \partial f(x)$, so

$$f'(x, -\hat{s}) \geq \langle s_j, -\hat{s} \rangle$$

and if $\alpha_j > 0$, we obtain (2.3.7) with Theorem 2.2.1.

Furthermore, by the definition (1.2.6) of the weights e_j, any subgradient at x must have its corresponding $e = 0$. Then, from (2.2.3) written for each bundle element $(s_j, e_j = 0)$ extreme in $\partial f(x)$, we deduce

$$\langle s, -\hat{s} \rangle \leq -\|\hat{s}\|^2 - \mu\varepsilon \quad \text{for all } s \in \partial f(x)$$

and the second assertion follows from the first. □

Comparing with Proposition 2.3.2, we see that the approximation of $f'(x, d)$ is more fragile than $f'_\varepsilon(x, d)$. However, the assumptions necessary for equality in (2.3.7) are rather likely to hold in practice: for example when f has at x a gradient $\nabla f(x)$, which appears explicitly in the bundle (a normal situation), and which turns out to have a positive α in the composition of $\hat{s}$.

Remark 2.3.5 In the general case, there is no guarantee whether $-\|\hat{s}\|^2 - \mu\varepsilon$ is an over- or under-estimate of $f'(x, d)$: if no subgradient at x is active in the direction-finding problem, then (2.3.7) need not hold.

The assumptions (i), (ii) in Proposition 2.3.4 mean that $\partial f(x) \cap S(0)$ is nonempty and contains an element which is active for $\hat{s}$. The additional assumption means that $\partial f(x) \subset S(0)$;

244   XIV. Dual Form of Bundle Methods

and this implies equality of the two sets: indeed $\partial f(x) \supset S(0)$ because, as implied by Proposition XI.4.2.2, a subgradient s at some y with a positive weight $e(x, y, s)$ cannot be a subgradient at x. We emphasize: it is really $S(0)$ which is involved in the above relations, not $S(\varepsilon)$; for example, the property $S(\varepsilon) \cap \partial f(x) \neq \emptyset$ does not suffice for equality in (2.3.7). The following counter-example is instructive.

Take $\mathbb{R}^2 \ni (\xi, \eta) \mapsto f(\xi, \eta) = \xi^2 + \eta^2$ and $x = (1, 0)$. Suppose that the bundle contains just two elements, computed at $y_1 = (2, -1)$ and $y_2 = (-1, 2)$. In other words, $s_1 = (4, -2)$, $s_2 = (-2, 4)$. The reader can check that $e_1 = 2$, $e_2 = 8$. Finally take $\varepsilon = 4$; then it can be seen that

$$\nabla f(x) = (2, 0) = \tfrac{1}{3}(2s_1 + s_2) \in S(\varepsilon)$$

(but $\nabla f(x) \notin S(0) = \emptyset$!), and Proposition 2.3.4 does not apply. The idea of the example is illustrated by Fig. 2.3.1: $\hat{s}$ is precisely $\nabla f(x)$, so $f'(x, \hat{s}) = -\|\hat{s}\|^2$; yet, $\mu \neq 0$ (actually $\mu = 2$). Our estimate is therefore certainly not exact (indeed $f'(x, \hat{s}) = 4 < 12 = \|\hat{s}\|^2 + \mu\varepsilon$). □

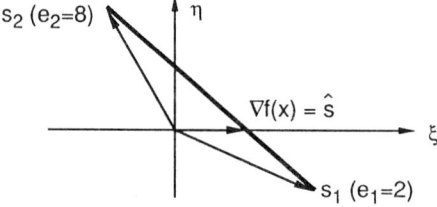

Fig. 2.3.1. A counter-example: $f'(x, d) \neq -\|s\|^2 - \mu\varepsilon$

2.4 The Role of the Cutting-Plane Function

Now we address the following question: is there a convex function having $S(\varepsilon)$ as approximate subdifferential? By construction, $d = -\hat{s}_\varepsilon$ would be an ε-descent direction for it: we would be very happy if this function were close to f.

The answer lies in the following concept:

Definition 2.4.1 The *cutting-plane function* associated with the actual bundle is

$$\check{f}(z) = f(x) + \max_{i=1,\ldots,\ell} [-e_i + \langle s_i, z - x \rangle]$$
$$= \max_{i=1,\ldots,\ell} [f(y_i) + \langle s_i, z - y_i \rangle] \qquad \text{for all } z \in \mathbb{R}^n. \qquad (2.4.1)$$

In the second expression above, y_i is the point at which (U1) has computed $f(y_i)$ and $s_i = s(y_i) \in \partial f(y_i)$; the first expression singles out the current iterate $x = x_k$. The equivalence between these two forms can readily be checked from the definition (1.2.8) of the weights e_i; see again Example VI.3.4. □

For $i = 1, \ldots, \ell$ and $z \in \mathbb{R}^n$, call

$$\bar{f}_i(z) := f(x) - e_i + \langle s_i, z - x \rangle = f(y_i) + \langle s_i, z - y_i \rangle \tag{2.4.2}$$

the linearization of f at y_i with slope $s_i \in \partial f(y_i)$; the subgradient inequality means $\bar{f}_i \leqslant f$ for each i, so the cutting-plane function

$$\check{f} = \max\{\bar{f}_i : i = 1, \ldots, \ell\}$$

is a piecewise affine *minorization* of f (the terminology "supporting plane" would be more appropriate than cutting plane: (2.4.2) defines a hyperplane supporting gr f).

Proposition 2.4.2 *With $\check{f}$ defined by (2.4.1) and $S(\varepsilon)$ by (2.1.1), there holds for all $\varepsilon \geqslant 0$*

$$\partial_\varepsilon \check{f}(x) = S(\varepsilon).$$

PROOF. Apply Example XI.3.5.3 to (2.4.1). □

Naturally, the constant term $f(x)$ could equally be dropped in (2.4.1) or (2.4.2), without changing $\partial_\varepsilon \check{f}(x)$. There are actually two reasons for introducing it:

– First, it allows the nice second form, already seen in the cutting-plane algorithm of §XII.4.2. Then we see the connection between our present approach and that of the cutting-plane algorithm: when we compute $\hat{s}_\varepsilon$, we are prepared to decrease the cutting-plane function $\check{f}$ by ε, instead of performing a full minimization of it.
– Second, $\check{f}(x) = f(x)$ if $\min_i e_i = 0$: $\check{f}$ can then be considered as an approximation of f near x, to the extent that $S(\varepsilon) = \partial_\varepsilon \check{f}(x)$ is an approximation of $\partial_\varepsilon f(x)$; remember Theorem XI.1.3.6, expressing the equivalence between the knowledge of f and of $\varepsilon \longmapsto \partial_\varepsilon f(x)$.

The case of a set M_ε not reduced to $\{0\}$ is of particular interest:

Theorem 2.4.3 *Let $\mu > 0$ be a multiplier described by Theorem 2.2.1. Then*

$$f(x) - \varepsilon - \tfrac{1}{\mu}\|\hat{s}_\varepsilon\|^2 = \check{f}(x - \tfrac{1}{\mu}\hat{s}_\varepsilon); \tag{2.4.3}$$

The above number is actually $\bar{f}_i(x - 1/\mu\, \hat{s}_\varepsilon)$, for each i such that equality holds in (2.2.3).

PROOF. For $0 < \mu \in M_\varepsilon$, divide (2.2.3) by μ and change signs to obtain for $i = 1, \ldots, \ell$ (remember that $s(\alpha)$, $\hat{s}$ and $\hat{s}_\varepsilon$ denote the same thing):

$$-\varepsilon - \tfrac{1}{\mu}\|\hat{s}\|^2 \geqslant -e_i + \langle s_i, -\tfrac{1}{\mu}\hat{s}\rangle = \bar{f}_i(x - \tfrac{1}{\mu}\hat{s}) - f(x);$$

and equality in (2.2.3) results in an equality above. On the other hand, there is certainly at least one $i \in \{1, \ldots, \ell\}$ such that equality holds in (2.2.3). Then the result follows from the definition of $\check{f}$. □

We give two interpretations of the above result:

– For an arbitrary $\alpha \in \Delta_\ell$, consider the affine function based on the notation (2.1.2):
$$\mathbb{R}^n \ni z \mapsto \bar{f}_\alpha(z) := f(x) - e(\alpha) + \langle s(\alpha), z - x \rangle \leqslant \check{f}(z),$$
where the last inequality comes directly by convex combination in (2.4.2) or (2.4.1). What (2.4.3) says is that, if α solves (2.1.4), then $\bar{f}_\alpha(x - 1/\mu\,\hat{s}_\varepsilon) = \check{f}(x - 1/\mu\,\hat{s}_\varepsilon)$ for any multiplier $\mu > 0$ (if there is one; incidentally, $e(\alpha)$ is then ε). In other words, this "optimal" $\bar{f}_\alpha$ gives one more support of epi f: it even gives a support of epi $\check{f}$ at $\left(x - 1/\mu\,\hat{s},\, \check{f}(x - 1/\mu\,\hat{s})\right)$.

– This was a primal interpretation; in the dual space, combine Propositions 2.3.2 and 2.4.2: $\check{f}'_\varepsilon(x, -\hat{s}_\varepsilon) = -\|\hat{s}_\varepsilon\|^2$, and the value (2.4.3)
$$\check{f}(x - \tfrac{1}{\mu}\hat{s}_\varepsilon) = -\varepsilon + \check{f}'_\varepsilon(x, -\tfrac{1}{\mu}\hat{s}_\varepsilon)$$
actually comes from §XI.2: $1/\mu$ is the optimal t in
$$\check{f}'_\varepsilon(x, -\hat{s}_\varepsilon) = \inf_{t > 0} \frac{\check{f}(x - t\hat{s}_\varepsilon) - \check{f}(x) + \varepsilon}{t}.$$

Figure 2.4.1, drawn in the one-dimensional graph-space along the direction $d = -\hat{s} = -\hat{s}_\varepsilon$, illustrates the above considerations. We have chosen an example where $\min_i e_i = 0$ and $-\|\hat{s}\|^2 - \mu\varepsilon > f'(x, d)$.

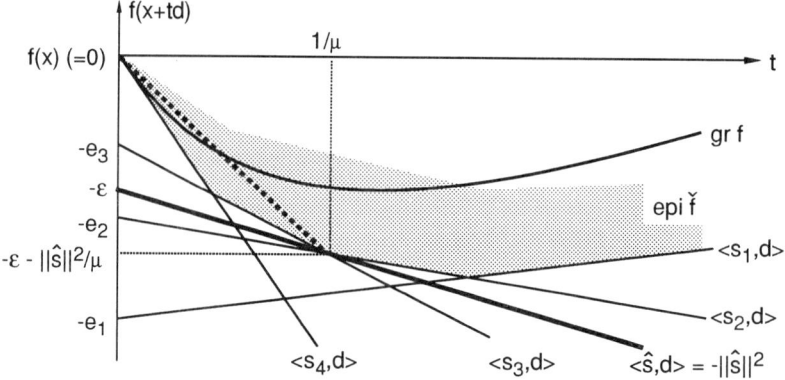

Fig. 2.4.1. The cutting-planes point of view

With respect to the environment space $\mathbb{R}^n \times \mathbb{R}$, the direction $-\hat{s}$ is defined as follows: a hyperplane in $\mathbb{R}^n \times \mathbb{R}$ passing through $(x, f(x) - \varepsilon)$ is characterized by some $s \in \mathbb{R}^n$, and is the set of points $(z, r) \in \mathbb{R}^n \times \mathbb{R}$ satisfying
$$r = r(z) = f(x) - \varepsilon + \langle s, z - x \rangle.$$
Such a hyperplane supports epi $\check{f}$ if and only if $s \in \partial_\varepsilon \check{f}(x) = S(\varepsilon)$. When a point in this hyperplane is moved so that its abscissa goes from x to $x + h$, its altitude varies by $\langle s, h \rangle$. For a normalized h, this variation is minimal (i.e. most negative) and equals $-\|s\|$ if $h = -s/\|s\|$. In words, $-s$ is the acceleration vector of a drop of water rolling on this hyperplane, and subject to gravity only. Now, the direction $-\hat{s}$ of Fig. 2.4.1 defines, among all the hyperplanes supporting

gr $\check{f}$, the one with the *smallest* possible acceleration. This is the geometric counterpart of the analytic interpretation of Remark 2.3.3.

The thick line in Fig. 2.4.1 is the intersection of the vertical hyperplane containing the picture with the "optimal" hyperplane thus obtained, whose equation is

$$r = \bar{f}^a(z) := f(x) - \varepsilon + \langle \hat{s}, z - x \rangle \quad \text{for all } z \in \mathbb{R}^n. \tag{2.4.4}$$

It represents an "aggregate" function $\bar{f}^a$, whose graph also supports gr f (and gr $\check{f}$), and which summarizes, in a way conditioned by ε, the linearizations $\bar{f}_i$ of (2.4.2). Using (2.4.3), we see that

$$\bar{f}^a(x - \tfrac{1}{\mu}\hat{s}) = f(x) - \varepsilon - \tfrac{1}{\mu}\|\hat{s}_\varepsilon\|^2 = \check{f}(x - \tfrac{1}{\mu}\hat{s}).$$

Thus, gr $\bar{f}^a$ touches epi $\check{f}$ at the point $\left(x - 1/\mu\,\hat{s},\ \check{f}(x - 1/\mu\,\hat{s})\right)$; hence $\hat{s} \in \partial \check{f}(x - 1/\mu\,\hat{s})$. Note in passing that the stepsize $t = 1/\mu$ thus plays a special role along the direction $-\hat{s}$. It will even become crucial in the next chapter, where these relations will be seen more thoroughly.

Remark 2.4.4 Suppose $\mu > 0$. Each linearization $\bar{f}_i$ in the bundle coincides with f and with $\check{f}$ at the corresponding y_i. The aggregate linearization $\bar{f}^a$ coincides with $\check{f}$ at $x - 1/\mu\,\hat{s}$; could $\bar{f}^a$ coincide with f itself at some y? Note in Fig. 2.4.1 that $y = x - 1/\mu\,\hat{s}$ would be such a coincidence point: $\hat{s}$ would be in $\partial f(x - 1/\mu\,\hat{s})$. In addition, for each i such that $\alpha_i > 0$ at some solution of (2.1.4), s_i would also be in $\partial f(x - 1/\mu\,\hat{s})$: this comes from the transportation formula used in

$$f(x - \tfrac{1}{\mu}\hat{s}) = \bar{f}_i(x - \tfrac{1}{\mu}\hat{s}) = f(y_i) + \langle s_i, x - \tfrac{1}{\mu}\hat{s} - y_i \rangle.$$

In Fig. 2.4.1, lift the graph of $\bar{f}^a$ as high as possible subject to supporting epi f. Analytically, this amounts to decreasing ε in (2.4.4) to the minimal value, say $\hat{\varepsilon}$, preserving the supporting property:

$$\hat{\varepsilon} = \inf\{\varepsilon\ :\ f(x) - \varepsilon + \langle \hat{s}, z - x \rangle \leqslant f(z) \text{ for all } z \in \mathbb{R}^n\} = f(x) + f^*(\hat{s}) - \langle \hat{s}, x \rangle.$$

On the other hand, use (1.2.12): with α solving (2.1.4),

$$\varepsilon = \sum_{i=1}^{\ell} \alpha_i e_i = f(x) + \sum_{i=1}^{\ell} \alpha_i f^*(s_i) - \langle \hat{s}, x \rangle.$$

The gap between gr $\bar{f}^a$ and epi f is therefore

$$\varepsilon - \hat{\varepsilon} = \sum_{i=1}^{\ell} \alpha_i f^*(s_i) - f^*(\hat{s}) \geqslant 0. \tag{2.4.5}$$

Another definition of $\hat{\varepsilon}$ is that it is the smallest e such that $\hat{s} \in \partial_e f(x)$ (transportation formula again). The aggregate bundle element $(\hat{s}, \varepsilon)$ thus appears as somehow corrupted, as compared to the original elements (s_i, e_i), which are "sharp" because $s_i \notin \partial_e f(x)$ if $e < e_i$. Finally, the gap (2.4.5) also explains our comment at the end of Remark 1.2.9: the use of f^* is clumsy for the aggregate element. □

To conclude, remember Example X.3.4.2: $\check{f}$ is the smallest convex function compatible with the information contained in the bundle; if g is a convex function satisfying

$$g(y_i) = f(y_i) \text{ and } s_i \in \partial g(y_i) \quad \text{for } i = 1, \ldots, \ell,$$

then $g \geqslant \check{f}$.

For a function g as above, $S(\varepsilon) = \partial_\varepsilon \check{f}(x) \subset \partial_\varepsilon g(x)$: in a sense, $S(\varepsilon)$ is the most pessimistic approximation of $\partial_\varepsilon f(x)$ – keeping in mind that we would like to stuff a convex polyhedron as large as possible into $\partial_\varepsilon f(x)$. However, $S(\varepsilon)$ is the largest possible set included *for sure* in $\partial_\varepsilon f(x)$, as we know from f only the information contained in the bundle. In other words, if $s \notin S(\varepsilon)$ (for some ε), then we might also have $s \notin \partial_\varepsilon f(x)$ (for the same ε); so would be the case if, for example, f turned out to be exactly $\check{f}$.

3 The Implementable Algorithm

The previous sections have laid down the basic ideas for constructing the algorithms we have in mind. To complete our description, there remains to specify the line-search, and to examine some implementation details.

3.1 Derivation of the Line-Search

In this section, we are given the current iterate $x = x_k$, and the current direction $d = d_k = -\hat{s}_\varepsilon$ obtained from (2.1.4). We are also given the current $\varepsilon > 0$ and the current multiplier $\mu \geqslant 0$ of the extra constraint (2.1.4)(iii). Our aim is to describe the line-search along d, which will produce a suitable $y_+ = x + td$ and its corresponding $s_+ \in \partial f(y_+)$ (note: we will often drop the subscript k and use "+" for the subscript $k+1$).

The algorithms developed in the previous chapters have clearly revealed the double role of the line-search, which must

(a) either find a *descent-step*, yielding a y_+ better than the current x, so that the next iterate x_+ can be set to this y_+,
(b) or find a *null-step*, with no better y_+, but with an s_+ providing a useful enrichment of the approximate subdifferential of f at x.

On the other hand, we announced in §1.2(b) that we want the line-search to follow as much as possible the principles exposed in §II.3. Accordingly, we must *test* each trial stepsize t, with three possible cases:

(0) this t is suitable and the line-search can be stopped;
(R) this t is not suitable and no suitable $\bar{t}$ should be searched on its right;
(L) this t is not suitable and no suitable $\bar{t}$ should be searched on its left.

Naturally, the (0)-clause is itself double, reflecting the alternatives (a) – (b): we must define what suitable descent- and null-steps are, respectively. The knowledge of

3 The Implementable Algorithm   249

$f'(x, d)$, and possibly of $f'(y_+, d)$, was required for this test in §II.3 – where they were called $q'(0)$ and $q'(t)$. Here, both numbers are unknown, but $\langle s_+, d \rangle$ is a reasonable estimate of the latter; and what we need is a reasonable estimate of the former. Thus, we suppose that a negative number is on hand, playing the role of $f'(x, d)$. We will see later how it can be computed, based on §2.3. For the moment, it suffices to call it formally $\tilde{v} < 0$.

A very first requirement for a descent-step is copied from (II.3.2.1): a coefficient $m \in {]}0, 1[$ is chosen and, if $t > 0$ does not satisfy the *descent test*

$$f(x + td) \leq f(x) + mt\tilde{v}, \tag{3.1.1}$$

it is declared "too large"; then case (R) occurs. On the other hand, a "too small" stepsize will be more conveniently defined after we see what a null-step is.

According to the general principles of Chap. XIII, a null-step is useful when f can hardly be decreased along d because $S(\varepsilon)$ of (2.1.1) approximates $\partial_\varepsilon f(x)$ poorly. The role of the line-search is then to improve this approximation with a new element (s_+, e_+) enriching the bundle (1.2.9). A new projection $\hat{s}_+$ will be computed, a new line-search will be performed from the same x, and here comes a crucial point: we must absolutely have

$$\langle s_+, \hat{s} \rangle + \mu e_+ < \|\hat{s}\|^2 + \mu\varepsilon. \tag{3.1.2}$$

Indeed, if (3.1.2) does not hold, just take some $\alpha \in \Delta_\ell$ solving (2.1.4), append the $(\ell + 1)^{\text{st}}$ multiplier $\alpha_+ := 0$ to it and realize that the old solution (α, μ) is again an optimal primal-dual pair: the minimization algorithm enters a disastrous loop. In a way, (3.1.2) plays the role of (XIII.2.1.1): it guarantees that the new element (s_+, e_+) will really define an $S_+(\varepsilon)$ richer than $S(\varepsilon)$; see also Remark IX.2.1.3.

– To keep the same spirit as in Chap. XIII, we ensure (3.1.2) by forcing first e_+ to be small. In addition to $\tilde{v}$, another parameter $\tilde{\varepsilon} > 0$ is therefore passed to the line-search which, if no descent is obtained, must at least produce an $s_+ \in \partial_{\tilde{\varepsilon}} f(x)$ to enrich the bundle. Admitting that $y_+ = x + td$ and $s_+ \in \partial f(y_+)$, this amounts to requiring

$$e_+ := e(x, y_+, s_+) = f(x) - f(y_+) + t\langle s_+, d \rangle \leq \tilde{\varepsilon}. \tag{3.1.3}$$

– For (3.1.2) to hold, not only e_+ but also $\langle s_+, \hat{s} \rangle = -\langle s_+, d \rangle$ must be small. This term represents a directional derivative, comparable to $\tilde{v}$ of (3.1.1); so we require

$$\langle s_+, d \rangle \geq m'\tilde{v} \tag{3.1.4}$$

for some coefficient m'; for convergence reasons, m' is taken in $]m, 1[$.

In summary, a null-iteration will be declared when (3.1.3) and (3.1.4) hold simultaneously: this is the second part of the (0)-clause, in which case t is accepted as a "suitable null-step".

Remark 3.1.1 Thus, we require the null-step to satisfy two inequalities, although the single (3.1.2) would suffice. In a way, this complicates the algorithm but our motivation is to keep closer to the preceding chapters: (3.1.3) is fully in the line of Chap. XIII, while (3.1.4) connotes Wolfe's criterion of (II.3.2.4).

Note that this strategy implies a careful choice of $\tilde{v}$ and $\tilde{\varepsilon}$: we have to make sure that [(3.1.4), (3.1.3)] does imply (3.1.2), i.e.

$$\langle s_+, \hat{s}\rangle \leqslant -m'\tilde{v} \text{ and } e_+ \leqslant \tilde{\varepsilon} \implies \langle s_+, \hat{s}\rangle + \mu e_+ < \|\hat{s}\|^2 + \mu\varepsilon.$$

Said otherwise, we want

$$\langle s_+, \hat{s}\rangle + \mu e_+ \leqslant -m'\tilde{v} + \mu\tilde{\varepsilon} \implies \langle s_+, \hat{s}\rangle + \mu e_+ < \|\hat{s}\|^2 + \mu\varepsilon.$$

This in turn will be ensured if

$$-m'\tilde{v} + \mu\tilde{\varepsilon} < \|\hat{s}\|^2 + \mu\varepsilon, \tag{3.1.5}$$

an inequality which we will have to keep in mind when choosing the tolerances. □

Finally, we come to the (L)-clause. From their very motivation, the concepts of null-step and of descent-step are mutually exclusive; and it is normal to reflect this exclusion in the criteria defining them. In view of this, we find it convenient to declare t as "not too small" when (3.1.3) does not hold: then t is accepted as a descent-step (if (3.1.1) holds as well), otherwise case (L) occurs. Naturally, observe that (3.1.3) holds for any t close enough to 0.

To sum up, the stopping criterion for the line-search will be as follows. The tolerances $\tilde{v} < 0, \tilde{\varepsilon} > 0$ are given (satisfying (3.1.5), but this is unimportant for the moment), as well as the coefficients $0 < m < m' < 1$. Accept the stepsize $t > 0$ with $y_+ = x + td$, $s_+ \in \partial f(y_+)$ and $e_+ = e(x, y_+, s_+)$ when:

$$f(y_+) \leqslant f(x) + mt\tilde{v} \text{ and } e_+ > \tilde{\varepsilon} \qquad \text{[descent-step]} \tag{3.1.6}$$

or

$$e_+ \leqslant \tilde{\varepsilon} \text{ and } \langle s_+, d\rangle \geqslant m'\tilde{v}. \qquad \text{[null-step]} \tag{3.1.7}$$

3.2 The Implementable Line-Search and Its Convergence

We are now in a position to design the line-search algorithm. It is based on the strategy of §II.3 (and more particularly Fig. II.3.3.1), suitably adapted to take §3.1 into account; its general organization is:

(0) t is convenient when it satisfies (3.1.6) or (3.1.7), with the appropriate exit-case;
(R) t is called t_R when (3.1.1) does not hold; subsequent trials will be smaller than t_R;
(L) t is called t_L in all other cases; subsequent trials will be larger than t_L.

Corresponding to these rules, the truth-value Table 3.2.1 specifies the decision made in each possible combination (T for true, F for false; a star* means an impossible case, ruled out by convexity). The line-search itself can be realized by the following algorithm.

Algorithm 3.2.1 (Line-Search, Nonsmooth Case) The initial point $x \in \mathbb{R}^n$ and stepsize $t > 0$, the direction $d \in \mathbb{R}^n$, the tolerances $\tilde{v} < 0, \tilde{\varepsilon} > 0$, and coefficients $m \in {]0, 1[}, m' \in {]m, 1[}$ are given. Set $t_L = 0, t_R = 0$.

Table 3.2.1. Exhaustive analysis of the test

(3.1.1)	(3.1.3)	(3.1.4)	decision
T	T	T	(0) (null)
T	T	F	(L)
T	F	T	(0) (descent)
T	F	F	(0) (descent)
F	T	T	(0) (null)
F	T	F	(R)*
F	F	T	(R)
F	F	F	(R)*

STEP 1. Compute $f(x+td)$ and $s = s(x+td)$; compute $e = e(x, x+td, s)$.

STEP 2 (test for a null-step). If (3.1.3) and (3.1.4) hold, stop the line-search with a null-step. Otherwise proceed to Step 3.

STEP 3 (test for large t). If (3.1.1) does not hold, set $t_R = t$ and go to Step 6. Otherwise proceed to Step 4.

STEP 4 (test for a descent-step; t is not too large). If (3.1.3) does not hold, stop the line-search with a descent-step. Otherwise set $t_L = t$ and proceed to Step 5.

STEP 5 (extrapolation). If $t_R = 0$, find a new t by extrapolation beyond t_L and loop to Step 1. Otherwise proceed to Step 6.

STEP 6 (interpolation). Find a new t by interpolation in $]t_L, t_R[$ and loop to Step 1. □

At each cycle, the extrapolation/interpolation formulae must guarantee a definite decrease of the bracket $[t_L, t_R]$, or a definite increase of t_L. This was called the safeguard-reduction property in §II.3.1. A simple way of ensuring it is, for example, to replace t by $10t$ in case of extrapolation and by $1/2 \, (t_L + t_R)$ in case of interpolation. More efficient formulae can be proposed, which were alluded to in Remark XIII.2.1.2. See also §II.3.4 for the forcing mechanism ensuring the safeguard-reduction property. Algorithm 3.2.1 is illustrated by the flow-chart of Fig. 3.2.1, which uses notation closer to that of §II.3.

Under these conditions, a simple adaptation of Theorem II.3.3.2 proves convergence of our line-search.

Theorem 3.2.2 *Let $f : \mathbb{R}^n \to \mathbb{R}$ be a convex function, and assume that the safeguard-reduction Property II.3.1.3 holds. Then Algorithm 3.2.1 either generates a sequence of stepsizes $t \to +\infty$ with $f(x + td) \to -\infty$, or terminates after a finite number of cycles, with a null- or a descent-step.*

PROOF. In what follows, we will use the notation $s_L := s(x + t_L d)$, $s_R := s(x + t_R d)$. Arguing by contradiction, we suppose that the stop never occurs, neither in Step 2, nor in Step 4. We observe that, at each cycle,

$$f(x + t_L d) \leq f(x) + m t_L \tilde{v}. \tag{3.2.1}$$

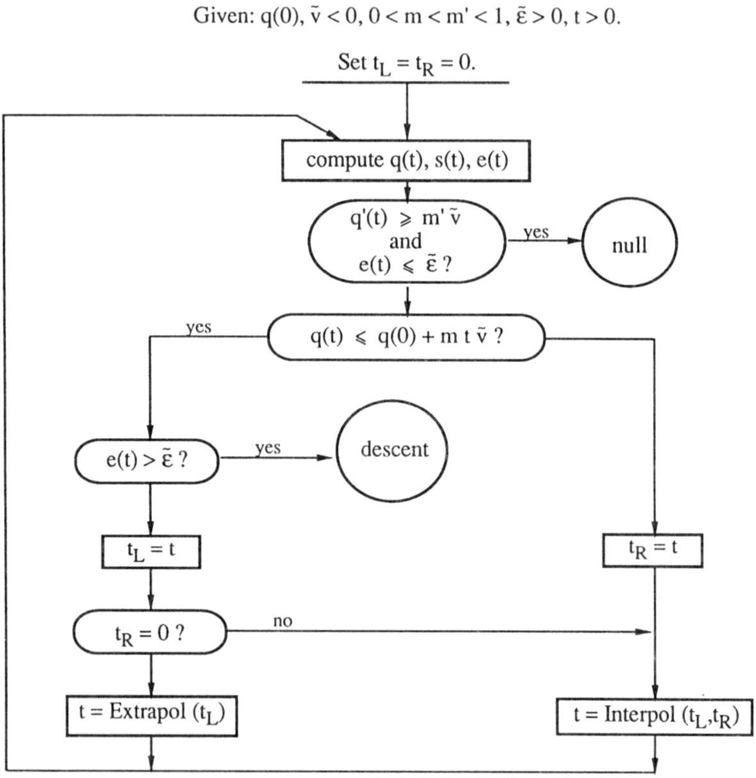

Fig. 3.2.1. Line-search with provision for null-step

[*Extrapolations*] Suppose first that Algorithm 3.2.1 loops indefinitely between Step 5 and Step 1: no interpolation is ever made. Then, by construction, every generated t is a t_L and tends to $+\infty$ by virtue of the safeguard-reduction property. Because $\tilde{v} < 0$, (3.2.1) shows that $f(x + t_L d) \to -\infty$.

[*Interpolations*] Thus, if $f(x + td)$ is bounded from below, t_R becomes positive at some cycle. From then on, the algorithm loops between Step 6 and Step 1. By construction, at each subsequent cycle,

$$f(x + t_R d) > f(x) + m t_R \tilde{v} ; \qquad (3.2.2)$$

the sequence $\{t_L\}$ is increasing, the sequence $\{t_R\}$ is decreasing, every t_L is smaller than every t_R, and the safeguard-reduction property implies that these two sequences have a common limit, say $\bar{t} \geqslant 0$.

By continuity, (3.2.1) and (3.2.2) imply

$$f(x + \bar{t} d) = f(x) + m \bar{t} \tilde{v} . \qquad (3.2.3)$$

[*Case* $\bar{t} = 0$] Compare (3.2.2) and (3.2.3) to see that $t_R > \bar{t}$; then pass to the limit in
$$\frac{f(x + t_R d) - f(x)}{t_R} > m\tilde{v}$$
to obtain
$$f'(x, d) \geqslant m\tilde{v} > m'\tilde{v}.$$
This is a contradiction: indeed, for $t_R \downarrow 0$,
$$\langle s_R, d \rangle \to f'(x, d) > m'\tilde{v} \quad \text{and} \quad e_R \to 0,$$
i.e. (3.1.3) and (3.1.4) are satisfied by t_R small enough; the stopping criterion of Step 2 must eventually be satisfied.

[*Case* $\bar{t} > 0$] Then t_L becomes positive after some cycle; write the subgradient inequality
$$f(x) \geqslant f(x + t_L d) + \langle s_L, x - x - t_L d \rangle$$
as
$$\langle s_L, d \rangle \geqslant \frac{f(x + t_L d) - f(x)}{t_L}$$
and pass to the limit: when the number of cycles goes to infinity, use (3.2.3) to see that the right-hand side tends to $m\tilde{v} > m'\tilde{v}$.

Thus, after some cycle, (3.1.4) is satisfied forever by $s_+ = s_L$. On the other hand, each t_L satisfies (3.1.3) (otherwise, a descent-step is found); we therefore obtain again a contradiction: t_L should be accepted as a null-step. □

As suggested in Remark 3.1.1, several strategies can be chosen to define a null-step. We have mentioned in §II.3.2 that several strategies are likewise possible to define a "too small" step. One specific choice cannot be made independently of the other, though: the truth-value table such as Table 3.2.1 must reflect the consistency property II.3.1.4. The reader may imagine a number of other possibilities, and study their consistency.

Remark 3.2.3 The present line-search differs from a "smooth" one by the introduction of the null-criterion, of course. With respect to Wolfe's line-search of §II.3.3, a second difference is the (L)-clause, which used to be "not (3.1.4)", and is replaced here by (3.1.3).

However, a slight modification in Algorithm 3.2.1 cancels this last difference. Keep the same definition for a null-step, but take Wolfe's definition of a descent-step; the test (0), (R), (L) becomes

$$(0) \left| \begin{array}{ll} f(y_+) \leqslant f(x) + mt\tilde{v} \quad \text{and} \quad \langle s_+, d \rangle \geqslant m'\tilde{v}, & \text{[descent-step]} \\ e_+ \leqslant \tilde{\varepsilon} \quad \text{and} \quad \langle s_+, d \rangle \geqslant m'\tilde{v}; & \text{[null-step]} \end{array} \right.$$
(R) $f(y_+) > f(x) + mt\tilde{v}$;
(L) all other cases.

Furthermore give priority to the null-criterion; in other words, in the ambiguous case when
$$f(y_+) \leqslant f(x) + mt\tilde{v} \quad \text{and} \quad \langle s_+, d \rangle \geqslant m'\tilde{v} \quad \text{and} \quad e_+ \leqslant \tilde{\varepsilon},$$
make a null-step.

The effect of this variant on the flow-chart 3.2.1 is to replace the box "$e(t) > \tilde{\varepsilon} \Rightarrow$ descent" by "$q'(t) > m'\tilde{v} \Rightarrow$ descent"; and the comparison with Fig. II.3.3.1 becomes quite eloquent. The only difference with (II.3.2.5) is now the insertion of the null-criterion in (0). The proof of Theorem 3.2.2 can be reproduced (easy exercise). Besides, a close look at the logic reveals that (3.1.4) still holds in case of a descent-step. In other words, the present variant is formally identical to its original: it still produces descent-steps satisfying (3.1.6), and null-steps satisfying (3.1.7).

Although not earth-shaking, this observation illustrates an important point, already alluded to in §VIII.3.3: when designing algorithms for nonsmooth minimization, one should keep as close as possible to the "smooth-philosophy" and extract its good features. In this sense, the algorithms of §XII.4 are somewhat suspect, as they depart too much from the smooth case. □

3.3 Derivation of the Descent Algorithm

Based on the previous developments, the general organization of the overall minimization algorithm becomes clear: at each iteration, having chosen ε, we solve (2.1.4) to obtain a primal-dual solution (α, μ), and $\hat{s}$ from (2.1.5). If $\|\hat{s}\|$ is small, (2.3.6) provides a minimality estimate. Otherwise the line-search 3.2.1 is done along $d = -\hat{s}$, to obtain either a descent-step as in §II.3, or a null-step obviating small moves. For a complete description of the algorithm, it remains to specify a few items: the line-search parameters $\tilde{v}$ and $\tilde{\varepsilon}$, the choice of ε (which plays such a central role), and the management of the bundle.

(a) Line-Search Parameters. The line-search needs two parameters $\tilde{v}$ and $\tilde{\varepsilon}$, for (3.1.6), (3.1.7). The status of $\tilde{\varepsilon}$ is rather clear: first, it is measured in the same units as ε, i.e. as the objective function. Second, it should not be larger than ε: the purpose of a null-step is to find $s_+ \in \partial_\varepsilon f(x_k)$. We therefore set

$$\tilde{\varepsilon} = \tilde{m}\varepsilon, \qquad (3.3.1)$$

for some fixed $\tilde{m} \in]0, 1]$. This coefficient allows some flexibility; there is no a priori reason to choose it very small nor close to 1.

On the other hand, we have two possible strategies concerning the $\tilde{v}$-question, respectively based on Chaps. II and IX.

– If we follow the ideas of §II.3.2, our aim is to decrease f by a fraction of its initial slope $f'(x, d)$ – see (3.1.1). This $f'(x, d)$ is unknown (and, incidentally, it may be nonnegative) but, from Proposition 2.3.4, it can conveniently be replaced by

$$\tilde{v} = -\|\hat{s}\|^2 - \mu\varepsilon. \qquad (3.3.2)$$

If this expression overestimates the real $f'(x, d)$, a descent-step will be easily found; otherwise the null-mechanism will play its role.

– If we decide to follow the strategy of a separation mechanism (§XIII.2.2 for example), the role of $\tilde{v}$ is different: it is aimed at forcing s_+ away from $S(\varepsilon)$ and should rather be $\sigma_{S(\varepsilon)}(d)$, given by Proposition 2.3.2 – see (3.1.4):

$$\tilde{v} = -\|\hat{s}\|^2. \qquad (3.3.3)$$

These choices are equally logical, we will keep both possibilities open (actually they make little difference, from theoretical as well as practical points of view).

(b) Stopping Criterion. The rationale for the stopping criterion is of course (2.3.5): when the projection $\hat{s}$ is close to 0, the current x minimizes f within the current ε, at least approximately. However, some complexity appears, due to the double role played by ε: it is not only useful to stop the algorithm, but it is also essential to compute the direction, via the constraint (2.1.4)(iii).

From this last point of view, ε must be reasonably large, subject to $S(\varepsilon)$ being a reasonable approximation of the ε-subdifferential of f at the current x. As a result, the minimization algorithm cannot be stopped directly when $\hat{s} \simeq 0$. One must still check that the (approximate) minimality condition is tight enough; if not, ε must be reduced and the projection recomputed. On the other hand, the partial minimality condition thus obtained is useful to safeguard from above the subsequent values of ε.

Accordingly, the general strategy will be as follows. First, a tolerance $\delta > 0$ is chosen, which plays the same role as in Algorithms II.1.3.1 and XIII.1.3.1. It has the same units as a subgradient-norm and the event "$\hat{s} \simeq 0$" is quantified by "$\|\hat{s}\| \leq \delta$". As for ε, three values are maintained throughout the algorithm:

(i) The current value, denoted by $\varepsilon = \varepsilon_k$, is used in (2.1.4)(iii) to solve the direction-finding problem.
(ii) The value $\underline{\varepsilon} > 0$ is the final tolerance, or the lowest useful value for ε_k: the user wishes to obtain a final iterate x satisfying

$$f(x) \leq f(y) + \underline{\varepsilon} + \delta \|y - x\| \quad \text{for all } y \in \mathbb{R}^n.$$

(iii) The value $\bar{\varepsilon}$ represents the best estimate

$$f(x) \leq f(y) + \bar{\varepsilon} + \delta \|y - x\| \quad \text{for all } y \in \mathbb{R}^n$$

obtained so far. It is decreased to the current ε each time (2.1.4) produces $\hat{s}$ with $\|\hat{s}\| \leq \delta$.

In practice, having the current $\bar{\varepsilon} \geq \underline{\varepsilon}$, the direction-finding problem is solved at each iteration with an ε chosen in $[\underline{\varepsilon}, \bar{\varepsilon}]$. The successive values $\bar{\varepsilon}$ form a decreasing sequence $\{\bar{\varepsilon}_k\}$, minorized by $\underline{\varepsilon}$, and $\bar{\varepsilon}_{k+1}$ is normally equal to $\bar{\varepsilon}_k$. In fact, an iteration $k + 1$ with $\bar{\varepsilon}_{k+1} < \bar{\varepsilon}_k$ is exceptional; during such an iteration, the direction-finding problem (2.1.4) has been solved several times, until a true direction $-\hat{s}$, definitely nonzero, has been produced. The final tolerance $\underline{\varepsilon}$ is fixed for the entire algorithm, which is stopped for good as soon as (2.1.4) is used with $\varepsilon = \underline{\varepsilon}$ and produces $\|\hat{s}\| \leq \delta$.

Knowing that $\underline{\varepsilon}$ and $\bar{\varepsilon}$ are managed in an automatic way, there remains to fix the current value ε of (i); we distinguish three cases.

– At the first iteration, the question is of minor importance in the sense that the first direction does not depend on ε (the bundle contains only one element, with $e_1 = 0$); it does play some role, however, via $\tilde{\varepsilon}$ of (3.3.1), in case the first direction $-s_1$ is not downhill. Choosing this initial ε is rather similar to choosing an initial stepsize.

- At an iteration following a null-step, the spirit of the method is to leave ε as it is: the direction is recomputed with the same ε – but with an additional element in the bundle. In a way, we are in the same situation as after an inner iteration $(p, k) \mapsto (p, k+1)$ of §XIII.2.3; also, note that (3.1.2) would become irrelevant if ε were changed.
- Admitting that null-steps occur only occasionally, as emergencies, the general case is when the current iteration follows a descent-step. It will be seen that the behaviour of the algorithm is drastically influenced by a proper choice of ε in this situation. Thus, after a descent-step, a new $\varepsilon = \varepsilon_{k+1} \in [\underline{\varepsilon}, \bar{\varepsilon}]$ is chosen to compute the direction d_{k+1} issuing from the new $x_{k+1} \neq x_k$. Although it is of *utmost importance* for efficient implementations, this choice of ε will not be specified here: we will state the algorithm independently of such a choice, thereby establishing convergence in a general situation.

(c) Management of the Bundle. At every "normal" iteration, a new element (s_+, e_+) is appended to the bundle. With $y_+ = x_k + t_k d_k$ and $s_+ \in \partial f(y_+)$ found by the kth line-search, there are two cases for e_+.

- If the line-search has produced a null-step, then $x_{k+1} = x_k$; hence

$$e_+ = e(x_k, y_+, s_+) = f(x_k) - f(y_+) + t_k \langle s_+, d_k \rangle .$$

- In the case of a descent-step, $x_{k+1} = y_+$ and $e_+ = e(x_{k+1}, y_+, s_+) = 0$. Because the current iterate is moved, we must also update the old linearization errors according to (1.2.10): for each index i in the old bundle, e_i is changed to

$$e_i + f(x_{k+1}) - f(x_k) - t_k \langle s_i, d_k \rangle .$$

Because a computer has finite memory, this appending process cannot go on forever: at some iteration, room must be made for the new element. In order to preserve convergence, room must also be made for an *aggregate* element, say $(s^{\text{a}}, e^{\text{a}})$, which we proceed to define (it has already been alluded to in §2.4, see Fig. 2.4.1).

The purpose of this aggregate element is to fit into the convergence theory of §IX.2.1, based on the fact that the present projection $\hat{s}$ belongs to the next polyhedron $S_+(\varepsilon)$ – at least in the case of a null-step. This leaves little choice: s^{a} must be $\hat{s}$ and e^{a} must not exceed ε.

On the other hand, the definition (1.2.9) of the bundle requires $s^{\text{a}} = \hat{s}$ to be an e^{a}-subgradient of f at x_k. Because $\hat{s} \in S(\varepsilon) \subset \partial_\varepsilon f(x_k)$, it suffices to take $e^{\text{a}} = \varepsilon$, or better

$$e^{\text{a}} = \hat{e} := \sum_{i=1}^{\ell} \alpha_i e_i \leqslant \varepsilon,$$

where ℓ is the current bundle size and $\alpha \in \Delta_\ell$ is an optimal solution of (2.1.4). Indeed,

$$\hat{s} \in \partial_{\hat{e}} f(x_k) \subset \partial_\varepsilon f(x_k) \tag{3.3.4}$$

and this is just what is needed for (1.2.9) to hold. Incidentally, $\hat{e} = \varepsilon$ if the multiplier μ is nonzero; otherwise we cannot even guarantee that $\hat{e}$ is well-defined, since the optimal α may not be unique.

Thus, when necessary, at least two elements are destroyed from the current bundle, whose size ℓ is therefore decreased to a value $\ell' \leq \ell - 2$. This makes room to append

$$(s_{\ell-1}, e_{\ell-1}) = (\hat{s}, \hat{e}) \quad \text{and} \quad (s_\ell, e_\ell) = (s_+, e_+) \tag{3.3.5}$$

(with possible renumbering).

Remark 3.3.1 If all the elements destroyed have $\alpha = 0$ from the quadratic problem (2.1.4), $(\hat{s}, \hat{e})$ brings nothing to the definition of $S(\varepsilon)$. In that case, no aggregation is necessary, only one element has to be destroyed. □

It is important to understand that the aggregate element is formally identical to any other "natural" element in (1.2.9), coming directly from (U1); and this is true despite the absence of any y such that $\hat{s} \in \partial f(y)$, thanks to the following property (in which the notation of §1.2 is used again, making explicit the dependence on $x = x_k$).

Lemma 3.3.2 *If* $(\hat{s}_k, \hat{e}^k) = (\hat{s}, \hat{e})$ *is the aggregate element at iteration* k, $\hat{s}_k \in \partial_{\hat{e}'} f(x')$ *for all* x', *where*

$$\hat{e}' := \hat{e}^k + f(x') - f(x_k) - \langle \hat{s}_k, x' - x_k \rangle .$$

PROOF. In view of (3.3.4), this is nothing but Proposition XI.4.2.4. □

At subsequent iterations, the aggregate weight $\hat{e} = \hat{e}^k$ will therefore be updated according to (1.2.10), just as the others; it may enter the composition of a further aggregation afterwards; this further aggregation will nevertheless be an appropriate ε-subgradient at an appropriate iterate. In a word, we can forget the notation $(\hat{s}, \hat{e})$; there is nothing wrong with the notation (3.3.5), in which an aggregate element becomes anonymous.

3.4 The Implementable Algorithm and Its Convergence

Now a detailed description of our algorithm can be given. We start with a remark concerning the line-search parameters.

Remark 3.4.1 As already mentioned in Remark 3.1.1, $\tilde{\varepsilon}$ (i.e. $\tilde{m}$) and $\tilde{v}$ are not totally independent of each other: they should satisfy (3.1.5); otherwise (3.1.2), which is essential in the case of a null-step, may not hold.

If $\tilde{v}$ is given by (3.3.3), we have

$$-m'\tilde{v} + \mu\tilde{\varepsilon} = m'\|\hat{s}\|^2 + \tilde{m}\varepsilon \leq \max\{m', \tilde{m}\}(-\|\hat{s}\|^2 + \mu\varepsilon)$$

and (3.1.5) – hence (3.1.2) – does hold if m and m' are simply smaller than 1. On the other hand, if $\tilde{v}$ is given by (3.3.2), we write

$$-m'\tilde{v} + \mu\tilde{\varepsilon} = m'\|\hat{s}\|^2 + (m' + \tilde{m})\mu\varepsilon$$

to see that the case $m' + \tilde{m} > 1$ is dangerous: the new subgradient may belong to the old $S(\varepsilon)$; in the case of a null-step, $\hat{s}$ and μ will stay the same – a disaster. Conclusion: if the value (3.3.2) is used for $\tilde{v}$, it is just safe to take

$$m' + \tilde{m} \leqslant 1. \tag{3.4.1}$$

□

The algorithm is then described as follows. Notes such as (^{1}) refer to explanations following the algorithm.

Algorithm 3.4.2 (Bundle Method in Dual Form) (^{1}) The data are: the initial x_1 and $\bar{\varepsilon} > 0$ (^{2}); the maximal bundle size $\bar{\ell} \geqslant 2$; the tolerances $\underline{\varepsilon} > 0$ and $\delta > 0$.

Fix the parameters m, m', $\tilde{m}$ such that $0 < m < m' < 1$ and $0 < \tilde{m} \leqslant 1$ (^{3}). Initialize the bundle with $s_1 = s(x_1)$, $e_1 = 0$, the iteration index $k = 1$, the bundle size $\ell = 1$; set $\varepsilon = \bar{\varepsilon}$.

STEP 1 (computing the direction). Replace ε by $\max\{\underline{\varepsilon}, \min(\varepsilon, \bar{\varepsilon})\}$ (^{4}). Solve (2.1.4) to obtain an optimal solution α and multiplier μ (^{5}). Set

$$\hat{s} := \sum_{i=1}^{\ell} \alpha_i s_i, \quad \hat{e} := \sum_{i=1}^{\ell} \alpha_i e_i.$$

STEP 2 (stopping criterion). If $\|\hat{s}\| > \delta$ (^{6}) go to Step 3. Otherwise:
  If $\hat{e} \leqslant \underline{\varepsilon}$ stop (^{7}); otherwise diminish $\bar{\varepsilon}$ (^{8}) and go to Step 1.

STEP 3 (line-search). Perform the line-search 3.2.1 issuing from x_k, along the direction $d = -\hat{s}$, with the descent-parameter $\tilde{v} < 0$ (^{9}) and the null-step parameter $\tilde{\varepsilon} = \tilde{m}\varepsilon$.

  Obtain a positive stepsize $t > 0$ and $s_+ = s(x_k + td)$, realizing either a descent-step or a null-step (^{10}).

STEP 4 (managing the bundle size). If $\ell = \bar{\ell}$: delete at least 2 elements from the bundle (^{11}); insert the element $(\hat{s}, \hat{e})$ coming from Step 1 (^{7}); call again $\ell < \bar{\ell}$ the length of the new list thus obtained.

STEP 5 (appending the new element). Append $s_{\ell+1} := s_+$ to the bundle.

  – In the case of a null-step, append $e_{\ell+1} := f(x_k) - f(x_k + td) + t\langle s_+, d\rangle$.
  – In the case of a descent-step, append $e_{\ell+1} := 0$ and, for all $i \in \{1, \ldots, \ell\}$, change each e_i to

$$e_i + f(x_k + td) - [f(x_k) + t\langle s_i, d\rangle].$$

Choose a new ε (^{12}), replace k by $k+1$, ℓ by $\ell+1$ and loop to Step 1. □

Notes 3.4.3

(^{1}) We use this name because the algorithm has been conceived via a development in the dual space, to construct approximate subdifferentials. The next chapter will develop similar algorithms, based on primal arguments only.

(^{2}) This is the same problem as the very first initialization of the stepsize (Remark II.3.4.2): use for $\bar{\varepsilon}$ an estimate of $f(x_1) - \bar{f}$, where $\bar{f}$ is the infimal value of f.

(^{3}) Suggested values are: $m = \tilde{m} = 0.1$, $m' = 0.2$. Remember Remark 3.4.1.

(^{4}) The aim of this operation is to make sure that $0 < \underline{\varepsilon} \leqslant \varepsilon \leqslant \bar{\varepsilon}$, whatever happens, with $\underline{\varepsilon} > 0$ fixed once and for all. Then (2.1.4) is consistent and ε cannot approach 0.

(^{5}) Our notation is that of §2: $\alpha \in \Delta_\ell \subset \mathbb{R}^\ell$ and $\mu \geqslant 0$ is the multiplier associated with the inequality constraint (2.1.4)(iii). For convenience, let us write again the minimality conditions (2.2.3):

$$\langle s_i, \hat{s} \rangle + \mu e_i \geqslant -\|\hat{s}\|^2 + \mu \hat{e} \quad \text{for } i \in \{1, \ldots, \ell\}, \qquad (3.4.2)$$
$$\text{with equality if } \alpha_i > 0.$$

(^{6}) We want to detect $\hat{e}$-optimality of the current iterate x. For this, it is a good idea to let δ vary with ε. In fact, we have

$$f(y) \geqslant f(x) - \hat{e} - \|\hat{s}\| \, \|y - x\|.$$

Suppose that an idea of the diameter of the picture is on hand: a bound M is known such that, for all $y \in \mathbb{R}^n$,

$$f(y) < f(x) - \hat{e} \implies \|y - x\| \leqslant M.$$

Then we have

$$f(y) \geqslant f(x) - \hat{e} - M \|\hat{s}\| \quad \text{for all } y \in \mathbb{R}^n,$$

which means roughly $\hat{e}$-optimality if $M \|\hat{s}\|$ is small compared to $\hat{e}$, say

$$M \|\hat{s}\| \leqslant 0.1 \hat{e}.$$

In a word, instead of fixing δ forever in the data, we can take at each iteration

$$\delta = \frac{0.1}{M} \max\{\underline{\varepsilon}, \hat{e}\}. \qquad (3.4.3)$$

(^{7}) In case $\mu = 0$, the value $\hat{e}$ is ambiguous. In view of the gap mentioned in Remark 2.4.4, it is useful to preserve as much accuracy as possible; the aggregate weight could therefore be taken as small as possible, and this means the optimal value in

$$\min \left\{ \sum_{i=1}^\ell \alpha_i e_i \ : \ \alpha \in \Delta_\ell, \ \sum_{i=1}^\ell \alpha_i s_i = \hat{s} \right\}.$$

(^{8}) The algorithm is a sequence of loops between Step 5 and Step 1, until $\hat{e}$-optimality is detected; and it will be shown below that this happens after finitely many such loops. Then $\bar{\varepsilon}$ must be decreased fast enough, so that the lower threshold $\underline{\varepsilon}$ is eventually reached. A sensible strategy is to take

$$\bar{\varepsilon} = \max\{0.1 \hat{e}, \underline{\varepsilon}\}.$$

(^{9}) We have seen that $\tilde{v}$ can have the two equally sensible values (3.3.2) or (3.3.3). We do not specify the choice in this algorithm, which is stated formally. Remember Remark 3.4.1, though.

(^{10}) We recall from §3.2 that, in the case of a descent-step,

$$f(x_k + td) \leqslant f(x) + mt\tilde{v} \quad \text{and} \quad e(x_k, x_k + td, s_+) > \tilde{m}\varepsilon$$

(meaning that t is neither too small nor too large) while, in the case of a null-step, t is small enough and s_+ is useful, i.e.:

$$e(x_k, x_k + td, s_+) \leqslant \tilde{m}\varepsilon \quad \text{and} \quad \langle s_+, d \rangle \geqslant m'\tilde{v}.$$

(^{11}) If a deletion is made after a null-step, if the gradient at the current x (i.e. the element with $e = 0$) is deleted, and if ε is subsequently decreased at Step 2, then (2.1.4) may become inconsistent; see (^{4}) above. To be on the safe side, this element should not be deleted, which means that the maximal bundle size $\bar{\ell}$ should actually be at least 3.

(^{12}) Here lies a key of the algorithm: between $\underline{\varepsilon}$ and $\bar{\varepsilon}$, a possibly wide range of values are available for ε. Each of them will give a different direction, thus conditioning the efficiency of the algorithm (remember from Chap. II that a good direction is a key to an efficient algorithm). A rough idea of sensible values is known, since they have the same units as the objective f. This, however, is not enough: full efficiency wants more accurate values. In a way, the requirement (i) in the conclusion of §1.1 is not totally fulfilled. More will be said on this point in §4. □

We now turn to convergence questions. Naturally, the relevant tools are those of Chaps. IX or XIII: the issue will be to show that, although the e's and the ε's are moving, the situation eventually stabilizes and becomes essentially as in §IX.2.1.

In what follows, we call "iteration" a loop from Step 5 to Step 1. During one such iteration, the direction-finding problem may be solved several times, for decreasing values of ε. We will denote by $\varepsilon_k, \hat{s}_k, \mu_k$ the values corresponding to the last such resolution, which are those actually used by the line-search. We will also assume that $\tilde{v}$ is given either by (3.3.2) or by (3.3.3). Our aim is to show that there are only finitely many iterations, i.e. the algorithm does stop in Step 2.

Lemma 3.4.4 *Let $f : \mathbb{R}^n \to \mathbb{R}$ be convex and assume that the line-search terminates at each iteration. Then, either $f(x_k) \to -\infty$, or the number of descent-steps is finite.*

PROOF. The descent test (3.1.1) and the definition (3.3.2) or (3.3.3) of $\tilde{v}$ imply

$$f(x_k) - f(x_{k+1}) \geq m t_k \|\hat{s}_k\|^2 = m \|\hat{s}_k\| \, \|x_{k+1} - x_k\| \geq m\delta \|x_{k+1} - x_k\| \quad (3.4.4)$$

at each descent-iteration, i.e. each time x_k is changed (here $\delta > 0$ is either fixed, or bounded away from 0 in case the refinement (3.4.3) is used). Suppose that $\{f(x_k)\}$ is bounded from below.

We deduce first that $\{x_k\}$ is bounded. Then the set $\cup_k \partial_{\bar{\varepsilon}} f(x_k)$ is bounded (Proposition XI.4.1.2), so $\{s_k\}$ is also bounded, say by L.

Now, we have from the second half "not (3.1.3)" of the descent-test (3.1.6):

$$\tilde{\varepsilon}_k = \tilde{m}\varepsilon_k < f(x_k) - f(x_{k+1}) + \langle s(x_{k+1}), x_{k+1} - x_k \rangle \leq 2L\|x_{k+1} - x_k\| .$$

Combining with (3.4.4), we see that f decreases at least by the fixed quantity $m \, \tilde{m} \, \delta \, \underline{\varepsilon}/(2L)$ each time it is changed: this process must be finite. □

When doing null-steps, we are essentially in the situation of §XIII.2.2, generating a sequence of nested polyhedra $\{S_k(\varepsilon)\}_k$ with ε fixed. If $\tilde{v}$ is given by (3.3.3), each new subgradient s_+ satisfies

$$\langle s_+, d \rangle \geq -m'\|d\|^2 = m'\sigma_{S(\varepsilon)}(d)$$

(remember Proposition 2.3.2) and the convergence theory of §IX.2.1 applies directly: the event "$\|\hat{s}_k\| \leq \delta$" eventually occurs. The case of $\tilde{v}$ given by (3.3.2) requires a slightly different argument, we give a result valid for both choices.

Lemma 3.4.5 *The assumptions are those of Lemma 3.4.4; assume also* $m' + \tilde{m} \leq 1$. *Then, between two descent-steps, only finitely many consecutive null-steps can be made without reducing* $\bar{\varepsilon}$ *in Step 2.*

PROOF. Suppose that, starting from some iteration k_0, only null-steps are made. From then on, x_k and ε_k remain fixed at some x and ε; because $\hat{e}_k \leq \varepsilon$, we have $\hat{s}_k \in S_{k+1}(\varepsilon) \subset \partial_\varepsilon f(x)$. The sequence $\{\|\hat{s}_k\|^2\}$ is therefore decreasing; with $\hat{s}_{k+1} = \text{Proj } 0/S_{k+1}(\varepsilon)$, we can write

$$\|\hat{s}_{k+1}\|^2 \leq \langle \hat{s}_{k+1}, \hat{s}_k \rangle \leq \|\hat{s}_{k+1}\| \, \|\hat{s}_k\| \leq \|\hat{s}_k\|^2$$

where we have used successively: the characterization of the projection $\hat{s}_{k+1}$, the Cauchy-Schwarz inequality, and monotonicity of $\{\|\hat{s}_k\|^2\}$. This implies the following convergence properties: when $k \to +\infty$,

$$\begin{array}{c} \|\hat{s}_k\|^2 \to \delta', \quad \langle \hat{s}_{k+1}, \hat{s}_k \rangle \to \delta', \\ \|\hat{s}_{k+1} - \hat{s}_k\|^2 = \|\hat{s}_{k+1}\|^2 - 2\langle \hat{s}_{k+1}, \hat{s}_k \rangle + \|\hat{s}_k\|^2 \to 0. \end{array} \quad (3.4.5)$$

Because of the stopping criterion, $\delta' \geq \delta^2 > 0$; and this holds even if the refinement (3.4.3) is used.

Now, we obtain from the minimality condition (3.4.2) at the $(k+1)^{\text{st}}$ iteration:

$$\|\hat{s}_{k+1}\|^2 + \mu_{k+1}\varepsilon \leq \langle \hat{s}_{k+1}, s_{k+1} \rangle + \mu_{k+1} e_{k+1} \leq \langle \hat{s}_{k+1}, s_{k+1} \rangle + \mu_{k+1} \tilde{m}\varepsilon,$$

which we write as

$$(1 - \tilde{m})\mu_{k+1}\varepsilon \leq \langle \hat{s}_k, s_{k+1} \rangle + \langle \hat{s}_{k+1} - \hat{s}_k, s_{k+1} \rangle - \|\hat{s}_{k+1}\|^2. \quad (3.4.6)$$

On the other hand, (3.1.4) gives

$$\langle \hat{s}_k, s_{k+1} \rangle \leq -m' \tilde{v}_k \leq m'(\|\hat{s}_k\|^2 + \mu_k \varepsilon),$$

the second inequality coming from the definition (3.3.2) or (3.3.3) of $\tilde{v} = \tilde{v}_k$. So, combining with (3.4.6):

$$(1 - \tilde{m})\mu_{k+1}\varepsilon \leq m' \mu_k \varepsilon + \delta_k \leq (1 - \tilde{m})\mu_k \varepsilon + \delta_k,$$

where

$$\delta_k := m' \|\hat{s}_k\|^2 + \langle \hat{s}_{k+1} - \hat{s}_k, s_{k+1} \rangle - \|\hat{s}_{k+1}\|^2.$$

In view of (3.4.5) and remembering that $\{s_{k+1}\} \subset \partial_{\tilde{m}\varepsilon} f(x)$ is bounded,

$$\delta_k \to (m' - 1)\delta' < 0.$$

Thus, when $k \to +\infty$, 0 cannot be a cluster point of $\{\delta_k\}$: at each k large enough, the relation

$$\mu_{k+1} \leq \mu_k + \frac{\delta_k}{(1 - \tilde{m})\varepsilon}$$

shows that μ_k diminishes at least by a fixed quantity. Since $\mu_k \geq 0$, this cannot go on forever. □

With these two results, convergence of the overall algorithm is easy to establish; but some care must be exercised in the control of the upper bound $\bar{\varepsilon}$: as mentioned in 3.4.3(^{8}), Step 2 should be organized in such a way that infinitely many decreases would result in an $\bar{\varepsilon}$ tending to 0. This explains the extra assumption made in the following result.

Theorem 3.4.6 *Let $f : \mathbb{R}^n \to \mathbb{R}$ be convex. Assume for example that $\bar{\varepsilon}$ is divided by 10 at each loop from Step 2 to Step 1 of Algorithm 3.4.2; if $\tilde{v}$ is given by (3.3.2), assume also $m' + \tilde{m} \leqslant 1$. Then:*

– *either $f(x_k) \to -\infty$ for $k \to +\infty$,*
– *or the line-search detects at some iteration k that f is unbounded from below,*
– *or the stop occurs in Step 2 for some finite k, at which there holds*

$$f(y) \geqslant f(x_k) - \underline{\varepsilon} - \delta \|y - x_k\| \quad \text{for all } y \in \mathbb{R}^n .$$

PROOF. Details are left to the reader. By virtue of Lemma 3.4.4, the sequence $\{x_k\}$ stops. Then, either Lemma IX.2.1.4 or Lemma 3.4.5 ensures that the event "$\|\hat{s}_k\| \leqslant \delta$" occurs in Step 2 as many times as necessary to reduce $\bar{\varepsilon}$ to its minimal value $\underline{\varepsilon}$. □

We conclude with some remarks.

(i) The technical reason for assuming dom $f = \mathbb{R}^n$ is to bound the sequence $\{s_k\}$ (an essential property). Such a bound may also hold under some weaker assumption, say for example if the sublevel-set $S_{f(x_1)}(f)$ is compact and included in the interior of dom f. Another reason is pragmatic: what could the black box (U1) produce, when called at some $y \notin$ dom f?

(ii) As in Chap. XIII, our convergence result is not classical: it does not establish that the algorithm produces $\{x_k\}$ satisfying $f(x_k) \to \inf f$. For this, $\underline{\varepsilon}$ and δ should be set to zero, thus destroying the proofs of Lemma 3.4.4 and Proposition 3.4.5. If δ is set to 0, one can still prove by contradiction that 0 adheres to $\{\hat{s}_k\}$. On the other hand, the role of $\underline{\varepsilon}$ is much more important: if 0 adheres to $\{\varepsilon_k\}$, d_k may be close to the steepest-descent direction, and we know from §VIII.2.2 that this is dangerous.

(iii) As already mentioned in §3.3(b), $\underline{\varepsilon}$ is ambiguous as it acts both as a stopping criterion and as a safeguard against bad directions. Note, however, that more refined controls of ε_k could also be considered: for example, a property like

$$\sum_{k=1}^{\infty} \varepsilon_k \|\hat{s}_k\| = +\infty$$

ensures a sufficient decrease in f at each descent-iteration, which still allows the last argument in the proof of Lemma 3.4.4.

(iv) Finally observe that Lemma 3.4.4 uses very little of the definition of $\hat{s}$. As far as descent-steps are concerned, other directions may also work well, provided that (3.1.6) holds (naturally, $\tilde{v}$ must then be given some suitable value, interpreted as a convergence parameter to be driven to 0). It is only when null-steps come into play that the bundling mechanism, i.e. (2.1.4), becomes important. Indeed, observe the similarity between the proofs of Lemma 3.4.4 and of Theorem II.3.3.6.

4 Numerical Illustrations

This section illustrates the numerical behaviour of the dual bundle Algorithm 3.4.2. In particular, we compare to each other various choices of the parameters appearing in the definition of the algorithm. Unless otherwise specified, the experiments below are generally conducted with the following values of the parameters:

$$\tilde{v} = -\|\hat{s}\|^2 - \mu\varepsilon, \quad m = \tilde{m} = 0.1, \quad m' = 0.2, \quad \bar{\ell} = 100,$$

and a computer working with 6 digit-accuracy; as for the stopping criterion, it uses the refinement (3.4.3) and generally $\underline{\varepsilon} = 10^{-3}(1 + |\bar{f}|)$, i.e. 3 digit-accuracy is required ($\bar{f}$ being the minimal value of f).

4.1 Typical Behaviour

First, we specify how ε can typically be chosen, in Step 5 of the algorithm; since it will enter into play at the next iteration, we call ε_{k+1} or ε_+ this value (as before, the index k is dropped whenever possible). The following result is useful for this matter.

Proposition 4.1.1 *Let the current line-search, made from x along $d = -\hat{s}$, produce $y_+ = x + td$, $s_+ \in \partial f(y_+)$, and $e_+ = e(x, y_+, s_+) = f(x) - f(y_+) + t\langle s_+, d\rangle$. Then there holds*

$$-\|\hat{s}\|^2 \leqslant f'_\varepsilon(x, d) \leqslant \langle s_+, d \rangle + \frac{\varepsilon - e_+}{t} = \frac{\varepsilon - \Delta}{t}, \quad (4.1.1)$$

with $\Delta := f(x) - f(y_+)$.

PROOF. The first inequality results from Propositions 2.1.1 and 2.3.2. The second was given in Remark XI.2.3.3 (see Fig. 4.1.1 if necessary): in fact,

$$f'_{e_+}(x, d) = \langle s_+, d \rangle, \qquad \text{[transportation formula XI.4.2.2]}$$
$$\tfrac{1}{t} \in \partial(-f'_\cdot)_{e_+}(x, d), \qquad \text{[Theorem XI.2.3.2]}$$

where the last set denotes the subdifferential of the convex function $\varepsilon \mapsto -f'_\varepsilon(x, d)$ at $\varepsilon = e_+$. □

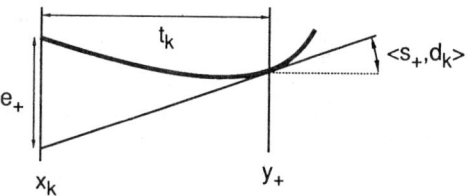

Fig. 4.1.1. The e_+-directional derivative and its super-derivative

See Fig. 4.1.2 for illustration; observe from the definitions that $\sigma_{S(\varepsilon)}$ is piecewise affine, and remember from Example XI.2.1.3 that the actual f'_ε behaves typically like $\varepsilon^{1/2}$ for $\varepsilon \downarrow 0$. This result can be exploited when a descent-step is made from $x_k = x$ to $x_{k+1} = y_+$. For example, the average

$$\sigma_k := \tfrac{1}{2}\left(-\|\hat{s}_k\|^2 + \frac{\varepsilon_k - \Delta_k}{t_k}\right)$$

is a possible estimate of the true $f'_{\varepsilon_k}(x, d)$; and then, we can for example use the following simple rules:

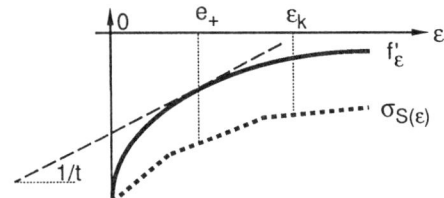

Fig. 4.1.2. Approximations of the approximate derivative

Strategy 4.1.2 (Standard ε-Strategy)
– If $\sigma_k > 0$, we were too optimistic when choosing ε_k: we decrease it.
– If $\sigma_k \leqslant 0$, ε can be safely increased.

In both cases, ε is multiplied by a specific factor not too far from 1. □

The detailed computation of the above factor is omitted. Specific rules for choosing it are more in the domain of cooking recipes than mathematics; so we skip them. Naturally, all the experiments reported below use the same such rules.

To give a rough idea of the result, take the test-problem MAXQUAD of VIII.3.3.3: we want to minimize

$$\max\left\{\tfrac{1}{2}x^\top A_j x + b_j^\top x + c_j \,:\, j = 1, \ldots, m\right\},$$

where $x \in \mathbb{R}^{10}$, $m = 5$, and each A_j is positive definite. The minimal value is known to be 0. The top-part of Fig. 4.1.3 gives the decrease of f (in logarithmic scale) as a function of the number of calls to the black box (U1); the stopping criterion $\underline{\varepsilon} = 10^{-4}$ was obtained after 143 such calculations, making up a total of 33 descent- and 26 null-iterations. Observe the evolution of f: it suggests long levels during which the ε-subdifferential is constructed, and which are separated by abrupt decreases, when it is suitably identified.

The lower part of Fig. 4.1.3 represents the corresponding evolution of $\|\hat{s}_k\|^2$. Needless to say, the abrupt increases happen mainly when the partial stopping criterion of Step 2 is obtained and $\bar{\varepsilon}$ is decreased. Indeed, Strategy 4.1.2 is of little influence in this example, and most of the iterations have $\varepsilon = \bar{\varepsilon}$. Table 4.1.1 displays the information concerning this partial stopping criterion, obtained 7 times during the run (the 7th time being the last).

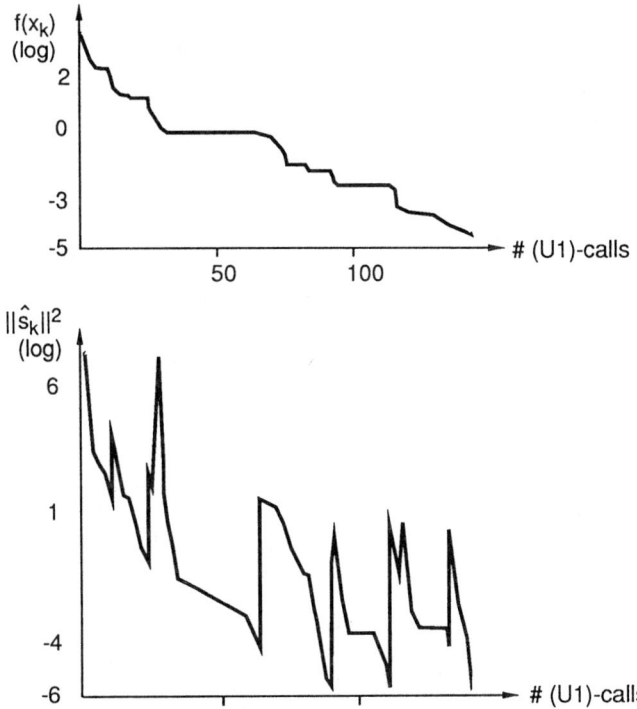

Fig. 4.1.3. A possible behaviour of the dual bundle algorithm

The test-problem TR48 of IX.2.2.6 further illustrates how the algorithm can behave. With $\varepsilon = 50$ (remembering that the optimal value is -638565, this again means 4-digit accuracy), $\bar{\varepsilon}$ is never decreased; the stopping criterion $\|\hat{s}\|^2 = 10^{-13}$ is obtained after 86 descent- and 71 null-iterations, for a total of 286 calls to (U1). Figure 4.1.4 shows the evolution of $f(x_k)$, again as a function of the number of calls to (U1); although slightly smoothed, the curve gives a fair account of reality; ε_k follows the same pattern. The behaviour of $\|\hat{s}_k\|^2$ is still as erratic as in Fig. 4.1.3; same remark for the multiplier μ_k of the constraint (2.1.4)(iii).

Table 4.1.1. The successive partial stopping criteria

k	#(U1)-calls	$f(x_k)$	$f(x_k) - \varepsilon_k$	$\|\hat{s}_k\|^2$
5	10	119.	-72.	13.
13	24	7.7	-8.3	1.10^{-2}
27	65	0.353	-1.	6.10^{-5}
37	91	0.009	-0.08	4.10^{-7}
47	113	0.003	-0.003	0.
55	135	0.0002	-0.0002	3.10^{-14}
59	143	0.00003	-0.0001	3.10^{-14}

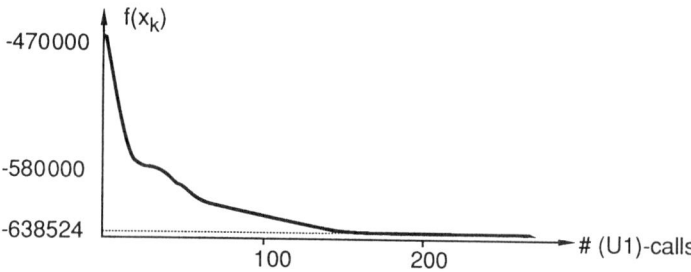

Fig. 4.1.4. Evolution of the objective with TR48

4.2 The Role of ε

To give an idea of how important the choice of ε_k can be, run the dual bundle Algorithm 3.4.2 with the two examples above, using various strategies for ε_k. Table 4.2.1 records the number of iterations required to reach 3-digit accuracy. We recall from Remark II.3.2.1 that the most important column is the third.

Table 4.2.1. Performance with various ε-strategies

	MAXQUAD			TR48		
$\varepsilon_{k+1} =$	#desc	#null	#(U1)-calls	#desc	#null	#(U1)-calls
standard 4.1.2	25	24	118	63	46	176
$f(x_k) - \bar{f}$	20	0	46	58	42	167
$0.1[f(x_k) - \bar{f}]$	33	1	65	162	37	326
ε_k	30	20	112	83	81	262
$f(x_k) - f(x_{k+1})$	24	0	52	351	55	612
$-t_k \tilde{v}_k$	29	5	73	305	43	530

The two strategies using $f(x_k) - \bar{f}$ are of course rarely possible, as they need a knowledge of the minimal value. They are mentioned mainly to illustrate the range for sensible values of ε; and they suggest that these values should be rather optimistic. From the motivation of the method itself (to decrease f by ε at each iteration), one could think that $\varepsilon_{k+1} = 0.1[f(x_k) - \bar{f}]$ should be more realistic; but this is empirically contradicted.

The last strategy in Table 4.2.1 is based on a (too!) simple observation: if everything were harmonious, we would have $\varepsilon \simeq f(x) - f(x_+) \simeq -t\tilde{v}$, a reasonable value for ε_+. The strategy $\varepsilon_+ = f(x) - f(x_+)$ is its logical predecessor.

Remark 4.2.1 (ε-Instability) Observe the relative inefficiency of these last two strategies; in fact, they illustrate an interesting difficulty. Suppose that, for some reason, ε is unduly small at some iteration. The resulting direction is bad (remember steepest descent!), so the algorithm makes little progress; as a result, ε_+ is going to be even smaller, thus amplifying the bad choice of ε. This is exactly what happens with TR48: in the last two experiments in Table 4.2.1, ε_k reaches its lower level $\underline{\varepsilon}$ fairly soon, and stays there forever.

The interesting point is that this difficulty is hard to avoid: to choose ε_+, the only available information is the observed behaviour of the algorithm during the previous iteration(s). The problem is then to decide whether this behaviour was "normal" or not; if not, the reason is still to be determined: was it because ε was too large, or too small? We claim that here lies a *key* of all kinds of bundle methods. □

These two series of results do not allow clear conclusions about which implementable strategies seem good. We therefore supplement the analysis with the test-problem TSP442, given in IX.2.2.7. The results are now those of Table 4.2.2; they indicate the importance of ε much more clearly. Observe, naturally, the high proportion of null-steps when ε is constant (i.e. too large); a contrario, the last two strategies have ε too small (cf. Remark 4.2.1), and have therefore a small proportion of null-steps. This table assesses the standard strategy 4.1.2, which had a mediocre behaviour on MAXQUAD.

Table 4.2.2. Performance on TSP442

$\varepsilon_{k+1} =$	#desc	#null	#(U1)-calls
standard 4.1.2	38	66	164
$f(x_k) - \bar{f}$	47	41	126
$0.1[f(x_k) - \bar{f}]$	182	32	275
ε_k	80	329	640
$f(x_k) - f(x_{k+1})$	340	19	418
$-t_k \tilde{v}_k$	292	24	378

Another series of experiments test the initial value for ε, which has a direct influence on the standard Strategy 4.1.2. The idea is to correct the value of ε_1, so as to reduce hazard. The results are illustrated by Table 4.2.3, in which the initialization is: $\varepsilon_2 = \kappa[f(x_2) - f(x_1)]$ for $\kappa = 0.1, 1, 5, 10$; in the last line, $\varepsilon_4 = f(x_4) - f(x_1)$ – we admit that ε_1, ε_2 and ε_3 have little importance. From now on, we will call "standard" the ε-Strategy 4.1.2 supplemented with this last initialization.

Table 4.2.3. Influence of the initial ε

	MAXQUAD			TR48			TSP442		
κ	desc	null	calls	desc	null	calls	desc	null	calls
0.1	25	15	88	295	51	*502	34	93	215
1	24	9	63	65	47	181	71	125	292
5	25	16	97	57	95	283	663	695	2310
10	25	16	97	433	148	*1021	231	543	1466
ε_4	16	15	82	93	34	224	36	36	102

A *star in the table means that the required 3-digit accuracy has not been reached: because of discrepancies between f- and s-values (due to roundoff errors), the linesearch has failed before the stopping criterion could be met.

Finally, we recall from §VIII.3.3 that smooth but stiff problems also provide interesting applications. Precisely, the line "nonsmooth" in Table VIII.3.3.2 was obtained by the present algorithm, with the ε-strategy $-t_k \tilde{v}_k$. When the standard strategy is used, the performances are comparable, as recorded in Table 4.2.4 (to be read just as Table VIII.3.3.2; note that $\pi = 0$ corresponds to Table 4.2.3).

Table 4.2.4. The standard strategy applied to "smooth" problems

π	100	10	1	10^{-1}	10^{-2}	10^{-3}	0
iter(#U1)	6(11)	9(20)	12(23)	28(56)	28(57)	33(67)	31(82)

4.3 A Variant with Infinite ε : Conjugate Subgradients

Among the possible ε-strategies, a simple one is $\varepsilon_k \equiv +\infty$, i.e. very large. The resulting direction is then (opposite to) the projection of the origin onto the convex hull of all the subgradients in the bundle – insofar as $\bar{\varepsilon}$ is also large. Naturally, $\tilde{\varepsilon} = \tilde{m}\varepsilon$ is then a clumsy value: unduly many null-steps will be generated. In this case, it is rather $\tilde{\varepsilon}$ that must be chosen via a suitable strategy.

We have applied this idea to the three test-problems of Table 4.2.3 with $\tilde{\varepsilon} = \tilde{m}[f(x_k) - \bar{f}]$ (in view of Tables 4.2.1 and 4.2.2, this choice is a sensible one, if available). The results, displayed in Table 4.3.1, show that the strategy is not too good. Observe for example the very high proportion of null-steps with TSP442: they indicate that the directions are generally bad. We also mention that, when $\tilde{\varepsilon}$ is computed according to the standard strategy, the instability 4.2.1 appears, and the method fails to reach the required 3-digit accuracy (except for MAXQUAD, in which the standard strategy itself generates large ε's).

Table 4.3.1. Performance of conjugate subgradients

	#desc	#null	#(U1)-calls
MAXQUAD	28	15	97
TR48	79	93	294
TSP442	79	1059	*1858

These results show that large values for ε are inconvenient. Despite the indications of Tables 4.2.1 and 4.2.2, one should not be "overly optimistic" when choosing ε: in fact, since d is computed in view of $S(\varepsilon)$, this latter set should not be "too much larger" than the actual $\partial_\varepsilon f(x)$. Figure 4.3.1 confirms this fact: it displays the evolution of $f(x_k)$ for TR48 in the experiment above, with convergence pushed to 4-digit accuracy. Vertical dashed lines indicate iterations at which the partial stopping criterion is satisfied; then $\tilde{\varepsilon}$ is reduced and forces ε_k down to a smaller value; after a while, this value becomes again large in terms of $f(x_k) - \bar{f}$, the effect of $\tilde{\varepsilon}$ vanishes and the performance degrades again. We should mention that this reduction mechanism is triggered by the δ-strategy of (3.4.3); without it, the convergence would be much slower.

Remark 4.3.1 We see that choosing sensible values for ε is a real headache:
– If ε_k is too small, d_k becomes close to the disastrous steepest-descent direction. In the worst case, the algorithm has to stop because of roundoff errors, namely when: d_k is not numer-

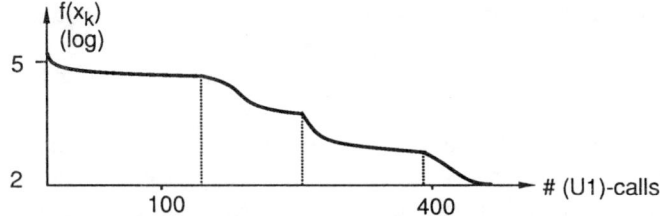

Fig. 4.3.1. Evolution of f in conjugate subgradients; TR48

ically downhill, but nevertheless (3.1.3) cannot be fulfilled in practice, $\tilde{\varepsilon}$ being negligible compared to f-values. This is the cause of *stars in Tables 4.2.3 and 4.3.1.
- If ε_k is too large, $S(\varepsilon_k)$ is irrelevant and d_k is not good either.
- Of these two extremes, the second is probably the less dangerous, but Remark 4.2.1 tells us that the first is precisely the harder to avoid. Furthermore, Fig. 4.3.1 suggests that an appropriate δ-strategy is then compulsory, although hard to adjust.

From our experience, we even add that, when the current ε is judged inconvenient, no available information can help to decide whether it is too small or too large. Furthermore, a decision made at a given iteration does not yield immediate effects: it takes a few iterations for the algorithm to recover from the old bad choice of ε. All these reasons make it advisable to bias the strategy (4.1.2 or any other) towards large ε-values. □

The interest of the present variant is mainly historical: it establishes a link with conjugate gradients of §II.2.4; and for this reason, it used to be called the *conjugate subgradient* method. In fact, suppose that f is convex and quadratic, and that the line-searches are exact: at each iteration, $f(x_k + td_k)$ is minimized with respect to t, and no null-step is ever accepted. Then the theory of §II.2.4 can easily be reproduced to realize that:

- the gradients are mutually orthogonal: $\langle s_i, s_j \rangle = 0$ for all $i \neq j$;
- the direction is actually the projection of the origin onto the *affine* hull of the gradients;
- this direction is just the same as in the conjugate-gradient method;
- these properties hold for any $\bar{\ell} \geqslant 2$.

Remember in particular Remark II.2.4.6. For this interpretation to hold, the assumptions "f convex quadratic and line-searches exact" are essential. From the experience reported above, we consider this interpretation as too thin to justify the present variant in a non-quadratic context.

4.4 The Role of the Stopping Criterion

(a) Stopping Criterion as a Safeguard. The stopping criterion is controlled by two parameters: δ and $\underline{\varepsilon}$. As already mentioned (see (iii) at the end of §3.4, and also Remark 4.3.1 above), $\underline{\varepsilon}$ is not only a tolerance for stopping, but also a safeguard against dangerous steepest-descent directions. As such, it may have a drastic influence on the behaviour of the algorithm, possibly from very early iterations. When $\underline{\varepsilon}$ is small, the algorithm may become highly inefficient: the directions may become close to steepest descent, and also $\tilde{\varepsilon}$ may become small, making it hard to find null-steps when needed. Furthermore, this phenomenon is hard to avoid, remember Remark 4.2.1.

In Table 4.2.3, for example, this is exactly what happens on the first line with TR48, in which ε_k is generally too small simply because it is initialized on a small value. Actually, the phenomenon also occurs on each "bad" line, even when ε could be thought of as too large. Take for example the run TSP442 with $\kappa = 5$: Fig. 4.4.1 shows, in logarithmic scale and relative values, the simultaneous evolution of $f(x_k) - \bar{f}$ and ε_k (remember from Tables 4.2.1 and 4.2.2 that, in a harmonious run, the two curves should evolve together). It blatantly suffers the instability phenomenon of Remark 4.2.1: in this run, ε_1 is too large; at the beginning, inefficient null-steps are taken; then 4.1.2 comes into play and ε is reduced down to $\underline{\varepsilon}$; and this reduction goes much faster than the concomitant reduction of f. Yet, the ε-strategy systematically takes ε_{k+1} in $[0.9\varepsilon_k, 4\varepsilon_k]$ – i.e. it is definitely biased towards large ε-values.

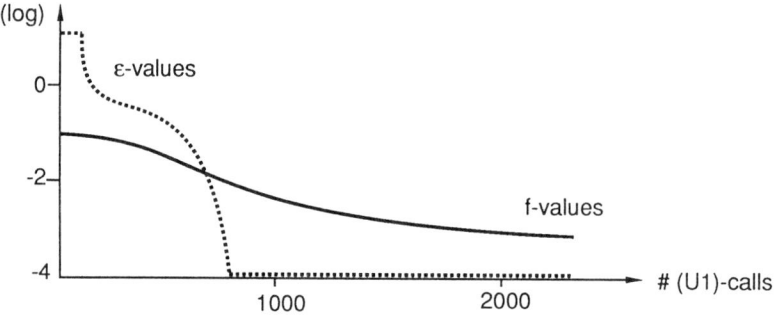

Fig. 4.4.1. TSP442 with a large initial ε

(b) The Tail of the Algorithm. Theorem 3.4.6 says that $\varepsilon_k \leqslant \bar{\varepsilon}$ eventually stabilizes at the threshold $\underline{\varepsilon}$. During this last phase, the algorithm becomes essentially that of §XIII.3.1, realizing the separation process of §IX.3.3. Then, of course, f decreases little (but descent-steps do continue to appear, due to the value $\tilde{m} = 0.1$) and the relevant convergence is rather that of $\{\|\hat{s}_k\|^2\}$ to 0.

For a more precise study of this last phase, we have used six more TSP-problems of the type IX.2.2.7, with respectively 14, 29, 100, 120, 614 and 1173 (dual) variables. For each such problem, Table 4.4.1 singles out the iteration k_0 where ε reaches its limit $\underline{\varepsilon}$ (which is here $10^{-4} \bar{f}$), and stays there till the final iteration k_f, where the convergence criterion $\|\hat{s}\|^2 \leqslant \delta^2 = 10^{-7}$ is met. All these experiments have been performed with $\bar{\ell} = 400$, on a 15-digit computer; the strategy $\varepsilon_k = f(x_k) - \bar{f}$ has been used, in order for k_0 to be well-defined. The last column in Table 4.4.1 gives the number of active subgradients in the final direction-finding problem. It supposedly gives a rough idea of the dimensionality of $S(\underline{\varepsilon})$, hence of the $\underline{\varepsilon}$-subdifferential at the final iterate.

Table 4.4.1. The f-phase and s-phase in TSP

TSP	k_0	k_f	$k_f - k_0$	$\|\hat{s}_{k_0}\|^2$	$\|\hat{s}_{k_f}\|^2$	#-active
14	14	14	0	0.4	0.	7
29	30	32	2	0.2	10^{-29}	7
100	45	71	26	0.7	10^{-29}	28
120	107	218	111	0.2	10^{-7}	74
442	300	2211	1911	0.04	10^{-7}	194
614	247	1257	1010	0.2	10^{-7}	141
1173	182	1315	1133	0.5	10^{-6}	204

Remark 4.4.1 The speed of convergence is illustrated by Fig. 4.4.2, which shows the evolution of $\|\hat{s}\|^2$ during the 1911 iterations forming the last phase in TSP442. Despite the theoretical predictions of §IX.2.2(b), it does suggest that, in practice, $\{\hat{s}_k\}$ converges to 0 at a rate definitely better than sublinear (we believe that the rate – linear – is measured by the first part of the curve, the finitely many possible $s(x)$ explaining the steeper second part, starting roughly at iteration 1500). This observation supplements those in §IX.2.2(c). □

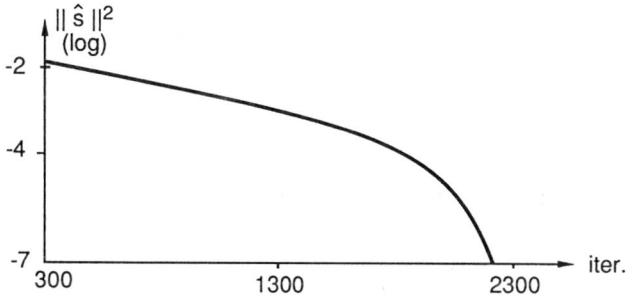

Fig. 4.4.2. Practical convergence of the separation process; TSP442

To measure the influence of $\underline{\varepsilon}$ on each of the two phases individually, Table 4.4.2 reports on the same experiments, made with TSP442, but for varying $\underline{\varepsilon}$. Full accuracy was required, in the sense that δ was set to 10^{-27} (in view of roundoff errors, this appeared to be a smallest allowable value, below which the method got lost in the line-search or in the quadratic program). The column Δf represents the final relative error $[f(x_f) - \bar{f}]/\bar{f}$; observe that it is of order $\tilde{m}\underline{\varepsilon}$.

The positive correlation between $\underline{\varepsilon}$ and the number of active subgradients is easy to understand: when a set decreases, its dimensionality can but decrease. We have no explanation for the small value of $k_f - k_0$ in the line "10^{-2}".

4.5 The Role of Other Parameters

An important item is the maximal length $\bar{\ell}$ of the bundle: it is reasonable to expect that the algorithm goes faster when $\bar{\ell}$ is larger, and reasonable values for $\bar{\ell}$ should

272   XIV. Dual Form of Bundle Methods

Table 4.4.2. Full accuracy in the s-phase with TSP442

$\bar{\varepsilon}/|\bar{f}|$	k_0	k_f	$k_f - k_0$	$\|\hat{s}_{k_0}\|^2$	$\|\hat{s}_{k_f}\|^2$	#-active
10^{-2}	21	1495	1474	6.	7.10^{-4}	357
10^{-3}	100	3202	3102	0.4	6.10^{-5}	248
10^{-4}	330	1813	1483	0.04	5.10^{-6}	204
10^{-5}	536	1601	1065	0.02	1.10^{-6}	185
10^{-6}	710	1828	1118	0.03	4.10^{-7}	170

be roughly comparable to the dimensionality of the subdifferential near a minimum point.

Table 4.5.1 (number of iterations and number of calls to (U1) to reach 4-digit accuracy) illustrates this point, with the following bundle-compression strategy: when $\ell = \bar{\ell}$, all nonactive elements in the previous quadratic problem are destroyed from the bundle; if no such element is found, a full compression is done, keeping only the aggregate element $(\hat{s}, \hat{e})$. This strategy is rather coarse and drastically stresses the influence of $\bar{\ell}$.

Table 4.5.1. Influence of $\bar{\ell}$ on TR48

$\bar{\ell}$	30	50	70	700	150	171
iter(#U1)	7193(13762)	821(1714)	362(693)	216(464)	201(372)	171(313)

Table 4.5.2. Influence of $\bar{\ell}$ on TSP1173

$\bar{\ell}$	2	5	10	100
iter(#U1)	101(184)	117(194)	106(186)	137(203)

The same experiment, conducted with TSP1173, is reported in Table 4.5.2. The results are paradoxical, to say the least; they suggest once more that theoretical predictions in numerical analysis are suspect, as long as they are not checked experimentally.

Remark 4.5.1 One more point concerning $\bar{\ell}$: as already mentioned on several occasions, most of the computing time is spent in (U1). If $\bar{\ell}$ becomes really large, however, solving the quadratic problem may become expensive. It is interesting to note that, in most of our examples (with a large number of variables, say beyond 10^2), the operation that becomes most expensive for growing $\bar{\ell}$ is not the quadratic problem itself, but rather the computation of the scalar products $\langle s_+, s_i \rangle$, $i = 1, \ldots, \ell$. All this indicates the need for a careful choice of $\bar{\ell}$-values, and of compression strategies. □

The parameter $\tilde{m}$ is potentially important, via its direct control of null-steps. Comparative experiments are reported in Table 4.5.3, which just reads as Table 4.2.3 – in which $\tilde{m}$ was 0.1. Note from (3.4.1) that the largest value allowed is $\tilde{m} = 0.8$.

According to these results, $\tilde{m}$ may not be crucial for performances, but it does play a role. The main reason is that, when $\tilde{m}$ is small, e_+ can be much smaller than ε; then the approximation (4.1.1) becomes hazardous and the standard ε-strategy may present difficulties. This, for example, causes the failure in the line "0.01" of TR48 via the instability 4.2.1.

Table 4.5.3. Influence of $\widetilde{m}$

$\widetilde{m}$	MAXQUAD			TR48			TSP442		
	desc	null	calls	desc	null	calls	desc	null	calls
0.8	18	40	98	35	167	368	34	93	215
0.5	19	26	117	57	163	411	71	125	292
0.1	16	15	82	118	47	295	36	36	102
0.01	32	7	90	357	13	*468	236	1	242

Finally, we study the role of $\tilde{v}$, which can take either of the values (3.3.2) or (3.3.3). To illustrate their difference, we have recorded at each iteration of the run "TSP442 + standard algorithm" the ratio

$$\frac{\|\hat{s}\|^2 + \mu\varepsilon}{\|\hat{s}\|^2}.$$

Figure 4.5.1 gives the histogram: it shows for example that, for 85% of the iterations, the value (3.3.2) was not larger than 3 times the value (3.3.3). The ratio reached a maximal value of 8, but it was larger than 5 in only 5% of the iterations. The statistics involved 316 iterations and, to avoid a bias, we did not count the tail of the algorithm, in which μ was constantly 0.

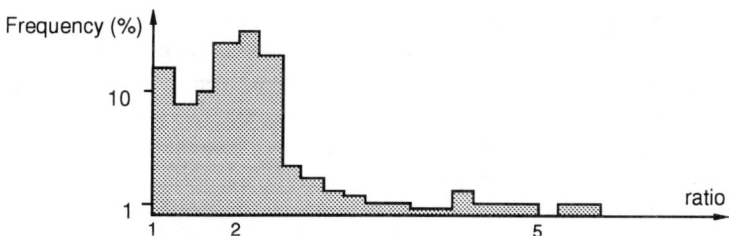

Fig. 4.5.1. Comparing two possible values of $\tilde{v}$

Thus, the difference between (3.2.2) and (3.2.3) can be absorbed by a proper redefinition of the constants m and m'. We mention that experiments conducted with (3.3.3) give, with respect to the standard strategy (3.3.2), differences which may be non-negligible, but which are certainly not significant: they never exceed 20 calls to (U1), for any required accuracy and any test-problem. This also means that the role of m and m' is minor (a fact already observed with classical algorithms for smooth functions).

4.6 General Conclusions

A relevant question is now: have we fulfilled the requirements expressed in the conclusion of §1.1? Judging from our experiments, several observations can be made.

(i) The algorithms of this chapter do enter the general framework of Chap. II. A comparison of the line-searches (Figs. II.3.3.1 and 3.2.1) makes this point quite clear. These algorithms can be viewed as robust minimization methods, aimed at solving ill-conditioned problems – compare Table 4.2.4 with the "smooth" methods in Table VIII.3.3.2. Also, they use as much as possible of the information accumulated along the iterations, the only limitations coming from the memory available in the computer.

(ii) On the other hand, the need for a parameter hard to choose (namely ε) has not been totally eliminated: the standard Strategy 4.1.2 does lack full reliability, and we do not know any definitely better one. The performances can sometimes rely heavily on the accuracy required by the user; and in extreme cases, roundoff errors may cause failure far from the sought minimum point.

(iii) The convergence of $\{\hat{s}_k\}$ to 0 – measured by $k_f - k_0$ in Tables 4.4.1 and 4.4.2 – is a slow process, indeed; and it gets a lot slower when the number of variables increases – or rather the dimension of the relevant subdifferential. This confirms that constructing a convex set may be a difficult task, and algorithms relying upon it may perform poorly.

(iv) On the positive side, f converges dramatically faster during the first phase of the descent algorithm, measured by k_0 in the tables of §4.4. Of course, this convergence is more painful when the required accuracy becomes more strict, but in acceptable proportions; and also, it does not depend so much on the number of variables; but once again, it heavily depends on the ε-strategy.

We believe that this last point (iv) is very important, in view of point (iv) in the conclusion of §1.1: minimizing a convex function is theoretically equivalent to separating $\{0\}$ from some subdifferential; but in practice, the situation is quite different: forcing f to decrease to $\bar{f}$ may be a lot easier than forcing $-\|\hat{s}\|^2$ to decrease to 0. Note, however, that this comment can be but heuristic: remember Remark IX.2.2.5.

A consequence of the above observation is that finding an approximate minimum may be much simpler than checking how approximate it is: although very attractive, the idea of having a stopping criterion of the type $f(x) \leqslant \bar{f} + \underline{\varepsilon}$ appears as rather disappointing. Consider for example Table 4.4.2: on the first line, it took 1474 iterations to realize that the 21th iterate was accurate within 10^{-2} (here, the question of an adequate δ is of course raised – but unsolved). These 1474 "idle" iterations were not totally useless, though: they simultaneously improved the 19th iterate and obtained a 10^{-3}-optimum. Unfortunately, the line "10^{-5}" did 10 times better in only 536 iterations! and this is normal: §4.3 already warned us that keeping ε_k constant is a bad idea.

In summary, the crucial problem of finding an adequate ε is hard, mainly because of instabilities. Safeguarding it from below is not a real cure; it may even be a bad idea in practice, insofar as it prevents regular decreases of the objective function.

XV. Acceleration of the Cutting-Plane Algorithm: Primal Forms of Bundle Methods

Prerequisites. Chapter VI (subdifferentials and their elementary properties) is essential and Chap. VII or XII (minimality conditions and elementary duality for simple convex minimization problems) will suffice for a superficial reading. However, practically all the book is useful for a complete understanding of every detail; in particular: Chap. II (basic principles for minimization algorithms), Chap. IV (properties of perspective-functions), Chap. IX (bundling mechanism for descent algorithms), Chap. X (conjugacy, smoothness of the conjugate), Chap. XI (approximate subdifferentials, infimal convolution), Chap. XII (classical algorithms for convex minimization, duality schemes in the convex case), Chap. XIV (general principles of dual bundle methods).

Introduction. In Chap. XII, we sketched two numerical algorithms for convex minimization: subgradients and cutting planes. Apparently, they have nothing to do with each other; one has a *dual* motivation, in the sense that it uses a subgradient (a dual object) as a direction of motion from the current iterate; the second is definitely *primal*: the objective function is replaced by an approximation which is minimized to yield the next iterate.

Here we study more particularly the cutting-plane method, for which we propose a number of accelerated versions. We show that these versions are primal adaptations of the dual bundle methods of Chap. XIV. They define a sort of continuum having two endpoints: the algorithms of subgradients and of cutting planes; a link is thus established between these two methods.

Throughout this chapter,

$$\boxed{f : \mathbb{R}^n \to \mathbb{R} \text{ is convex}}$$

and we want to minimize f. As always when dealing with numerical algorithms, we assume the existence of a black box (U1) which, given $x \in \mathbb{R}^n$, computes $f(x)$ together with some subgradient $s(x) \in \partial f(x)$.

1 Accelerating the Cutting-Plane Algorithm

We have seen in §XII.4.2 that a possible algorithm for minimizing f is the *cutting-plane* algorithm, and we briefly recall how it works. At iteration k, suppose that the iterates $y_1, \ldots, y_k$ have been generated and the corresponding (bundle of) information $f(y_1), \ldots, f(y_k)$, $s_1 = s(y_1), \ldots, s_k = s(y_k)$ has been collected. The cutting-plane approximation of f, associated with the sampling points $y_1, \ldots, y_k$, is the piecewise affine function of Definition XIV.2.4.1:

$$\mathbb{R}^n \ni y \mapsto \check{f}_k(y) := \max\{f(y_i) + \langle s_i, y - y_i\rangle \,:\, i = 1, \ldots, k\}. \qquad (1.0.1)$$

There results immediately from convexity that this function is an under-estimate of f, which is exact at each sampling point:

$$\check{f}_k \leqslant f \quad \text{and} \quad \check{f}_k(y_i) = f(y_i) \text{ for } i = 1, \ldots, k.$$

The idea of the cutting-plane algorithm is to minimize $\check{f}_k$ at each iteration. More precisely, some compact convex set C is chosen, so that the problem

$$y_{k+1} \in \operatorname*{Argmin}_{y \in C} \check{f}_k(y) \qquad (1.0.2)$$

does have a solution, which is thus taken as the next iterate. We recall from Theorem XII.4.2.3 the main convergence properties of this algorithm: denoting by $\bar{f}_C$ the minimal value of f over C, we have

$$\check{f}_k(y_{k+1}) \leqslant \check{f}_{k'}(y_{k'+1}) \leqslant \bar{f}_C \leqslant f(y_\ell) \quad \text{for all } k \leqslant k' \text{ and all } \ell,$$
$$f(y_k) \to \bar{f}_C \text{ and } \check{f}_k(y_{k+1}) \to \bar{f}_C \text{ when } k \to +\infty.$$

To make sure that our original problem is really solved, C should contain at least one minimum point of f; so finding a convenient C is not totally trivial. Furthermore, it is widely admitted that the numerical performance of the cutting-plane algorithm is intolerably low. Both questions are addressed in the present chapter.

1.1 Instability of Cutting Planes

Consider the simple example illustrated by Fig. 1.1.1: with $n = 1$, take $f(x) = 1/2 \, x^2$ and start with two iterates $y_1 = 1$, $y_2 = -\varepsilon < 0$. Then y_3 is obviously the solution of $\check{f}_1(y) = \check{f}_2(y)$, i.e.

$$\tfrac{1}{2} + y - 1 = \tfrac{1}{2}(-\varepsilon)^2 + (-\varepsilon)(y + \varepsilon),$$

and $y_3 = 1/2 - 1/2\,\varepsilon$. If ε gets smaller, y_2 increases, and y_3 increases as well. In algorithmic terms, we say: if the current iterate is better (y_2 comes closer to the solution 0), the next iterate is worse (y_3 goes further from this solution); the algorithm is *unstable*.

Remark 1.1.1 We mention a curious consequence of this phenomenon. Forgetting the artificial set C, consider the linear program expressing (1.0.2):

$$\inf_{y,r}\{r \,:\, f(y_i) + \langle s_i, y - y_i \rangle \leqslant r \text{ for } i = 1, \ldots, k\}.$$

Taking k Lagrange multipliers $\alpha_1, \ldots, \alpha_k$, its dual is (see §XII.3.3 if necessary, $\Delta_k \subset \mathbb{R}^k$ is the unit simplex):

1 Accelerating the Cutting-Plane Algorithm

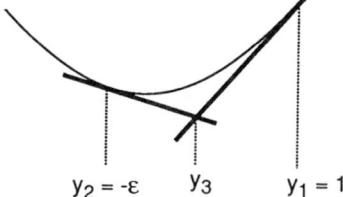

$y_2 = -\varepsilon$ y_3 $y_1 = 1$

Fig. 1.1.1. Instability of cutting planes

$$\sup \sum_{i=1}^{k} \alpha_i [f(y_i) - \langle s_i, y_i \rangle], \quad \alpha \in \Delta_k, \quad \sum_{i=1}^{k} \alpha_i s_i = 0.$$

Now assume that f is quadratic; then the gradient mapping $s = \nabla f$ is affine. It follows that, if α is an optimal solution of the dual problem above, we have:

$$\nabla f \left(\sum_{i=1}^{k} \alpha_i y_i \right) = \sum_{i=1}^{k} \alpha_i \nabla f(y_i) = \sum_{i=1}^{k} \alpha_i s_i = 0.$$

The point $y^* := \sum_{i=1}^{k} \alpha_i y_i$ minimizes f.

More generally, suppose that f is a differentiable convex function and assume that the iterates y_i cluster together; by continuity, $\nabla f(y^*) \simeq 0$: the point y^* is nearly optimal and can be considered as a good next iterate y_{k+1}. Unfortunately this idea is killed by the instability of the cutting planes: because $y_{k+1} = y^*$ is good, the next iterates will be bad. □

In our example of Fig. 1.1.1, the instability is not too serious; but it can become disastrous in less naive situations. Indeed the next example cooks a black box (U1) for which reducing the initial gap $f(y_1) - \bar{f}_C$ by a factor $\varepsilon < 1$ requires some $(1/\varepsilon)^{(n-2)/2}$ iterations: with 20 variables, a billion iterations are needed to obtain just one digit accuracy!

Example 1.1.2 We use an extra variable η, which plays a role for the first two iterations only. Given $\varepsilon \in]0, 1/2[$, we want to minimize the function

$$\mathbb{R}^n \times \mathbb{R} \ni (y, \eta) \mapsto f(y, \eta) = \max\{|\eta|, -1 + 2\varepsilon + \|y\|\}$$

on the unit ball of $\mathbb{R}^n \times \mathbb{R}$: C in (1.0.2) is therefore taken as this unit ball. The optimal value is obviously 0, obtained for $\eta = 0$ and y anywhere in the ball $B(0, 1-2\varepsilon) \subset \mathbb{R}^n$. Starting the cutting-plane algorithm at $(0, 1) \in \mathbb{R}^n \times \mathbb{R}$, which gives the first objective-value 1, the question is: how many iterations will be necessary to obtain an objective-value of at most ε?

The first subgradient is $(0, 1)$ and the second iterate is $(0, -1)$, at which the objective-value is again 1 and the second subgradient is $(0, -1)$. The next cutting-plane problem is then

$$\min_{y, \eta, r} \{r : r \geq |\eta|, \ \|y\|^2 + \eta^2 \leq 1\}. \tag{1.1.1}$$

Its minimal value is 0, obtained at all points of the form $(y, 0)$ with y describing the unit ball B of $\mathbb{R}^n$. The constraints of (1.1.1) can thus be formulated as

$$r \geq 0, \quad \|y\|^2 \leq 1 \text{ (i.e. } y \in B) \quad \text{and} \quad \eta = 0;$$

in other words, the variable η is dropped, we are now working in $B \subset \mathbb{R}^n$. The above constraints will be present in all the subsequent cutting-plane problems, whose minimal values will be nonnegative (because of the permanent constraint $r \geq 0$), and in fact exactly 0 (because $\check{f}_k \leq f$).

We adopt the convention that, if the kth cutting-plane problem has some solution of norm 1, then it produces such a solution for y_{k+1}, rather than an interior point of its optimal set.

For example, the third iterate y_3 has norm 1, its objective-value is 2ε, and the third cutting-plane problem is

$$\min_{y,r} \{r : r \geq 0, \ y \in B, \ r \geq 2\varepsilon + \langle y_3, y - y_3\rangle\}.$$

Look at Fig. 1.1.2(a): the minimal value is still 0 and the effect of the above third constraint is to cut from the second optimal set (B itself) the portion defined by

$$\langle y_3, y \rangle > 1 - 2\varepsilon.$$

More generally, as long as the kth optimal set $B_k \subset B$ contains a vector of norm 1, the $(k+1)^{\text{st}}$ iterate will have an objective-value of 2ε, and will cut from B_k a similar portion obtained by rotation. As a result, no ε-optimal solution can be produced before all the vectors of norm 1 are eliminated by these successive cuts. For this, k must be so large that $k - 2$ times the area $S(\varepsilon)$ of the cap

$$\{y \in \mathbb{R}^n : \|y\| = 1, \ \langle v, y \rangle > 1 - 2\varepsilon\}$$

is at least equal to the area S_n of the boundary of B; see Fig. 1.1.2(a), where the thick line represents $S(\varepsilon)$, v has norm 1 and stands for $y_3, y_4, \ldots$).

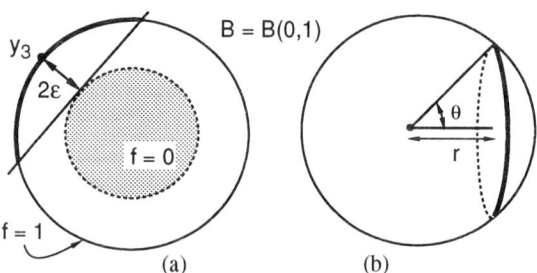

Fig. 1.1.2. Cuts and surfaces in the unit ball

It is known that the area of the boundary of $B(0, r)$ in $\mathbb{R}^n$ is $r^{n-1} S_n$. The area of the infinitesimal ring displayed in Fig. 1.1.2(b), at distance r of the origin, is therefore

$$S_{n-1}\sqrt{1-r^2}^{n-2}\,dr = S_{n-1}\sin^{n-1}\theta\,d\theta\,;\qquad(1.1.2)$$

hence, setting $\theta_\varepsilon := \cos^{-1}(1-2\varepsilon)$,

$$S(\varepsilon) = S_{n-1}\int_0^{\theta_\varepsilon}\sin^{n-1}\theta\,d\theta \leqslant S_{n-1}\int_0^{\theta_\varepsilon}\theta^{n-1}\,d\theta = S_{n-1}\tfrac{1}{n}(\theta_\varepsilon)^n\,.$$

Using (1.1.2) again:

$$S_n = 2S_{n-1}\int_0^{\pi/2}\sin^{n-1}\theta\,d\theta \geqslant 2S_{n-1}\int_0^{\pi/2}\sin^{n-1}\theta\cos\theta\,d\theta = \tfrac{2}{n}S_{n-1}\,.$$

Thus, the required number of iterations $S_n/S(\varepsilon)$ is at least $2/(\theta_\varepsilon)^n$. Knowing that $\theta_\varepsilon \simeq 2\sqrt{\varepsilon}$, we see the disaster. □

It is interesting to observe that the instability demonstrated above has the same origin as the need to introduce C in (1.0.2): the minima of $\check{f}_k$ are hardly controlled, in extreme cases they are "at infinity". The artificial set C had primarily a theoretical motivation: to give a meaning to the problem of minimizing $\check{f}_k$; but this appears to have a pragmatic supplement: to prevent a wild behaviour of $\{y_k\}$. Accordingly, we can even say that C (which after all is not so artificial) should be "small", which also implies that it should be "appropriately located" to catch a minimum of f. Actually, one should rather have a varying set C_k, with diameter shrinking to 0 as the cutting-plane iterations proceed. Such was not the case in our counter-example above: the "stability set" C remained equal to the entire $B(0,1)$ for an enormous number of iterations. This seemingly innocent remark can be considered as a starting point for the methods in this chapter.

1.2 Stabilizing Devices: Leading Principles

Our stability issue existed already in Chap. II: in §II.2.1, we wanted to minimize a certain directional derivative $\langle s_k, \cdot \rangle$, which played the role of $\check{f}_k$ – both functions are identical for $k = 1$. Due to positive homogeneity, a normalization had to be introduced; the situation was the same for the steepest-descent problem of §VIII.1.

We choose to follow the same idea here, and we describe abstractly the kth iteration of a stabilized algorithm as follows:

(i) we have a *model*, call it φ_k, supposed to represent f;
(ii) we choose a *stability center*, call it x_k;
(iii) we choose a *norming*, call it $\|\cdot\|_k$;
(iv) then we compute a next iterate y_{k+1} realizing a compromise between diminishing the model:

$$\varphi_k(y_{k+1}) < \varphi_k(x_k)\,,$$

and keeping close to the stability center:

$$\|y_{k+1} - x_k\|_k \quad \text{small}\,.$$

In Chap. II or VIII, φ_k approximated f to first order and was *positively homogeneous*; we wanted to solve

$$\min\{\varphi_k(y) \,:\, \|y - x_k\|_k = \kappa\}$$

with $\varphi_k(y) = f'(x_k, y - x_k)$. As shown in §VIII.1.2, this was "equivalent" to

$$\min\{\varphi_k(y) \,:\, \|y - x_k\|_k \leqslant \kappa\}$$

or also to

$$\min[\varphi_k(y) + \mu\|y - x_k\|_k]$$

(see (II.2.3.1), (II.2.3.2), for example), and the resulting direction $y_{k+1} - x_k$ was independent of $\kappa > 0$ or $\mu > 0$. Here the model $\varphi_k = \check{f}_k$ is no longer positively homogeneous; we see immediately an important consequence: the parameter κ or μ will have a decisive influence on the resulting y_{k+1}.

Just as for the first-order models, we emphasize that this stabilizing trick is able to cope simultaneously with the two main deficiencies of the pure cutting-plane algorithm: the need for a compact set C is totally eliminated, and the stability question will be addressed by a proper management of x_k and/or $\|\cdot\|_k$.

Remark 1.2.1 Mixing the cutting-plane model with a norming is arguably a rather ambiguous operation:

– As already observed, some flexibility is added via κ or μ (with the first-order models of §II.2.2, these parameters had an trifling role). Because varying κ or μ amounts to dilating $\|\cdot\|_k$, we can also say that the *size* of the normalizing unit ball is important, in addition to its shape.

– Again because of the lack of positive homogeneity, the solution y_{k+1} of the problem in κ-form should not be viewed as giving a direction issuing from x_k. As a result, the concept of line-search is going to be inappropriate.

– The cutting-plane function has a *global* character, which hardly fits with the localizing effect of the normalization: $\check{f}_k$ cannot be considered as a local approximation of f near a distinguished point like x_k. Actually, unless f happens to be very smooth, an accurate approximation φ_k of f is normally out of reach, in contrast to what we saw in Chap. II. see the Newton-like models of §II.2.3.

All in all, one might see more negative than positive aspects in our technique; but nothing better is available yet. On the other hand, several useful interpretations will be given in the rest of the chapter, which substantially increase the value of the idea. For the moment, let us say that the above remarks must be kept in mind when specifying our stabilizing devices. □

Let us now review the list of items (i) – (iv) of the beginning of this Section 1.2.

(i) The model φ_k, i.e. the cutting-plane function $\check{f}_k$ of (1.0.1), can be enriched by one affine piece for example every time a new iterate y_{k+1} is obtained from the model-problem in (iv). Around this general philosophy, several realizations are possible. Modelling is actually just the same as the bundling concept of §XIV.1.2; in particular, an important ingredient will be the aggregation technique already seen on several occasions.

(ii) The stability center should approximate a minimum point of f, as reasonably as possible; using objective-values to define the word "reasonably", the idea will be to take for x_k one of the sampling points $y_1, \ldots, y_k$ having the best f-value. Actually, a recursive strategy will be used: at each iteration, the stability center is either left as it is, or moved to the new iterate y_{k+1}, depending on how good $f(y_{k+1})$ is.

(iii) Norming is quite an issue, our study will be limited to multiples of a fixed norm. Said otherwise, our attention will be focused on the κ or the μ considered above, and $\|\cdot\|_k$ will be kept fixed (normally to the Euclidean norm).

(iv) Section 2 will be devoted to the stabilized problem; three conceptually equivalent possibilities will be given, which can be viewed as primal interpretations of the dual bundle method of Chap. XIV.

In fact, the whole stabilizing idea is primarily characterized by (ii), which relies on the following crucial technique. In addition to the next iterate y_{k+1}, the stabilized problem yields a "nominal decrease" for f; this is a nonnegative number δ_k, giving an idea of the gain $f(x_k) - f(y_{k+1})$ to be expected from the current stability center x_k to the next iterate y_{k+1}; see Example 1.2.3 below. Then x_k is set to y_{k+1} if the actual gain is at least a fraction of the "ideal" gain δ_k. With emphasis put on the management of the stability center, the resulting algorithm looks like the following:

Algorithm 1.2.2 (Schematic Stabilized Algorithm) Start from some $x_1 \in \mathbb{R}^n$; choose a descent coefficient $m \in \,]0, 1[$ and a stopping tolerance $\underline{\delta} \geqslant 0$. Initialize $k = 1$.

STEP 1. Choose a convex model-function φ_k (for example $\check{f}_k$) and a norming $\|\cdot\|_k$ (for example a multiple of the Euclidean norm).

STEP 2. Solve the stabilized model-problem, whatever it is, to obtain the next iterate y_{k+1}. Upon observation of $\varphi_k(x_k) - \varphi_k(y_{k+1})$, choose a nominal decrease $\delta_k \geqslant 0$ for f.
  If $\delta_k \leqslant \underline{\delta}$ stop

STEP 3. If
$$f(x_k) - f(y_{k+1}) \geqslant m\delta_k, \qquad (1.2.1)$$
set $x_{k+1} = y_{k+1}$; otherwise set $x_{k+1} = x_k$. Replace k by $k+1$ and loop to Step 1. □

Apart from moving the stability center, several other decisions can be based on the descent-test (1.2.1), or on more sophisticated versions of it: essentially the model and/or the norming may or may not be changed. Classical algorithms of Chap. II can be revisited in the light of the above approach, and this is not a totally frivolous exercise.

Example 1.2.3 (Line-Search) At the given x_k, take the first-order approximation

$$y \mapsto \varphi_k(y) = f(x_k) + \langle s_k, y - x_k \rangle \qquad (1.2.2)$$

as model and choose a symmetric positive definite operator Q to define the norming.

(a) Steepest Descent, First Order.

Let the stabilized problem be

$$\min \left\{ \varphi_k(y) : \tfrac{1}{2}\langle Q(y - x_k), y - x_k \rangle \leq \tfrac{1}{2}\kappa \right\} \quad (1.2.3)$$

for some radius $\kappa > 0$; stabilization is thus forced via an explicit *constraint*. From the minimality conditions of Theorem VII.2.1.4, there is a multiplier $\mu \geq 0$ such that

$$s_k = -\mu Q(y - x_k). \quad (1.2.4)$$

From there, we get

$$\begin{aligned}\langle s_k, y - x_k \rangle &= -\mu \langle Q(y - x_k), y - x_k \rangle \\ &= -\mu\kappa \quad \text{[transversality condition]} \\ &= -\sqrt{\kappa \langle s_k, Q^{-1} s_k \rangle}\end{aligned}$$

(to obtain the last equality, take the scalar product of (1.2.4) with $\kappa Q^{-1} s_k$). We interpret these relations as follows: between the stability center and a solution of the stabilized problem, the model decreases by

$$\varphi_k(x_k) - \varphi_k(y_{k+1}) = \sqrt{\kappa \langle s_k, Q^{-1} s_k \rangle} = \mu\kappa =: \delta_k \geq 0.$$

Knowing that $\varphi_k(x_k) = f(x_k)$ and $\varphi_k \simeq f$, this δ_k can be viewed as a "nominal decrease" for f. Then Algorithm 1.2.2 corresponds to the following strategy:

(i) If δ_k is small, stop the algorithm, which is justified to the extent that $\|s_k\|$ is small. This in turn is the case if κ is not too small, and the maximal eigenvalue of Q is not too large: then the original problem, or its first-order approximation, is not too affected by the stabilization constraint.

(ii) If δ_k is far from zero, the next iterate $y_{k+1} = x_k - 1/\mu_k Q^{-1} s_k$ is well defined; setting $d_k := -Q^{-1} s_k$ (a direction) and $t_k := 1/\mu_k$ (a stepsize), (1.2.1) can be written

$$f(y_{k+1}) \leq f(x_k) - m\delta_k = f(x_k) - mt_k \langle s_k, d_k \rangle.$$

We recognize the descent test (II.3.2.1), universally used for classical line-searches. This test is thus encountered once more, just as in §XIV.3.1. With Remark II.3.2.3 in mind, observe the interpretative role of δ_k in terms of the initial derivative of a line-search function:

$$[tq'(0) = t \langle s_k, d_k \rangle =] \; \delta_k = \varphi_k(y_{k+1}) - \varphi_k(x_k).$$

(ii$_1$) Moving x_k in Step 4 of Algorithm 1.2.2 can be interpreted as stopping a line-search; this decision is thus made whenever "the stepsize is not too large": the concept of a "stepsize too small" has now disappeared.

(ii$_2$) What should be done in the remaining cases, when y_{k+1} is not good? This is not specified in Algorithm 1.2.2, except perhaps an allusion in Step 1. In a classical line-search, the model would be left unchanged, and the norming would be reinforced: (1.2.3) would be solved again with the same φ_k but with a smaller κ, i.e. a larger μ, or a smaller stepsize $1/\mu$.

(b) Steepest Descent, Second Order. With the same φ_k of (1.2.2), take the stabilized problem as
$$\min \{\varphi_k(y) + \tfrac{1}{2}\mu\langle Q(y - x_k), y - x_k\rangle : y \in \mathbb{R}^n\} \tag{1.2.5}$$
for some *penalty* coefficient $\mu > 0$. The situation is quite comparable to that in (a) above; a slight difference is that the multiplier μ is now explicitly given, as well as the solution $x_k - 1/\mu\, Q^{-1}s_k$.

A more subtle difference concerns the choice of δ_k: the rationale for (a) was to approximate f by φ_k, at least on some restricted region around x_k. Here we are bound to consider that f is approximated *by the minimand* in (1.2.5); otherwise, why solve (1.2.5) at all? It is therefore more logical to set
$$\delta_k = \varphi_k(x_k) - \varphi_k(y_{k+1}) - \tfrac{1}{2}\mu\langle Q(y_{k+1} - x_k), y_{k+1} - x_k\rangle = \tfrac{1}{2\mu}\langle s_k, Q^{-1}s_k\rangle.$$
With respect to (a), the nominal decrease of f is divided by two; once again, remember Remark II.3.2.3, and also Fig. II.3.2.2. □

Definition 1.2.4 We will say that, when the stability center is changed in Algorithm 1.2.2, we make a *descent-step*; otherwise we make a *null-step*.
The set of descent iterations is denoted by $K \subset \mathbb{N}$:
$$[k \in K \iff x_{k+1} = y_{k+1}] \quad \text{and} \quad [k \notin K \iff x_{k+1} = x_k]. \qquad \square$$

The above terminology suggests a link between the bundle methods of the preceding chapters and the line-searches of Chap. II; at the same time it points out the main difference: in case of a null-step, bundle methods enrich the piecewise affine model φ_k, while line-searches act on the norming (shortening κ, increasing μ) and do not change the model.

1.3 A Digression: Step-Control Strategies

The ambiguity between cases (a) and (b) in Example 1.2.3 suggests a certain inconsistency in the line-search principle. Let us come back to the situation of Chap. II, with a smooth objective function f, and let us consider a Newton-type approach: we have on hand a symmetric positive definite operator Q representing, possibly equal to, the Hessian $\nabla^2 f(x_k)$. Then the model (quadratic but not positively homogeneous)
$$y \mapsto \varphi_k(y) = f(x_k) + \langle s_k, y - x_k\rangle + \tfrac{1}{2}\langle Q(y - x_k), y - x_k\rangle \tag{1.3.1}$$
is supposedly a good approximation of f near x_k, viewed as a stability center. Accordingly, we compute
$$\bar{y} = \arg\min \varphi_k \quad \text{and} \quad d_k = \bar{y} - x_k,$$
hoping that the Newtonian point $\bar{y}$ is going to be the next iterate.

Then the line-search strategy of §II.3 is as follows: check if the stepsize $t = 1$ in the update $x_{k+1} = x_k + t d_k$ is suitable; if not, use the actual behaviour of f, as described by the black box (U1), to *search* the next iterate along the half-*line* $x_k + \mathbb{R}^+ d_k$. Now we ask an embarrassing question: what is so magic with this half-line? If the descent test is not passed by our Newtonian point $\bar{y}$, why in the world should x_{k+1} lie on the half-line issuing from x_k and pointing to it? Needless to say, this half-line has nothing magic: it is present for historical reasons only, going back to 1847 when A. Cauchy invented the steepest-descent method.

The stabilizing concepts developed in §1.2 suggest a much more natural idea. In fact, with the model φ_k of (1.3.1), take for example a stabilized problem in constrained form

$$\min \left\{\varphi_k(y) \,:\, \tfrac{1}{2}\|y - x_k\|^2 \leqslant \tfrac{1}{2}\kappa\right\} \tag{1.3.2}$$

and call $y(\kappa)$ its solution. Considering $\kappa > 0$ as an unknown parameter, the answer of the black box (U1) at $y(\kappa)$ can be used to adjust κ, *instead of* a stepsize along the direction $\bar{y} - x_k$. When κ describes $\mathbb{R}^+$, $y(\kappa)$ describes a certain curve parameterized by κ; we still can say that κ is a stepsize, but along the curve in question, instead of a half-line. If we insisted on keeping traditional terminology, we could say that we have a line-search where the direction depends on the stepsize. Officially, the present approach is rather called the *trust-region* technique: the constraint in (1.3.2) defines a region, in which the model φ_k is expectedly trustful.

To exploit this idea, it suffices to design a test (0), (R), (L) as in §II.3.1, using the "curved-search function" $q(\kappa) := f(y(\kappa))$ to *search* $y(\kappa)$ along the *curve* implicitly defined by (1.3.2). The tests seen in §II.3.2 use the initial derivative $q'(0)$ which is unknown but this difficulty has an easy solution: it is time to remember Remark II.3.2.3, which related the use of $q'(0)$ to the concept of a model. Just as in §1.2, the nominal decrease $\varphi_k(x_k) - \varphi_k(y(\kappa))$ can be substituted for the would-be $-tq'(0)$.

To give a specific illustration, let us adapt the criterion of Goldstein and Price to the case of a "curved-search".

Algorithm 1.3.1 (Goldstein and Price Curved-Search) The data are: the current iterate x_k, the model φ_k, the descent coefficients $m \in \,]0, 1[$ and $m' \in \,]m, 1[$. Set $\kappa_L = 0$ and $\kappa_R = 0$; take an initial $\kappa > 0$.

STEP 0. Solve (1.3.2) to obtain $y(\kappa)$ and set $\delta := f(x_k) - \varphi_k(y(\kappa))$.
STEP 1 (Test for large κ). If $f(y(\kappa)) > f(x_k) - m\delta$, set $\kappa_R = \kappa$ and go to Step 4.
STEP 2 (κ is not too large). If $f(y(\kappa)) \geqslant f(x_k) - m'\delta$, stop the curved-search with $x_{k+1} = y(\kappa)$.
   Otherwise set $\kappa_l = \kappa$ and go to Step 3.
STEP 3 (Extrapolation). If $\kappa_R > 0$ go to Step 4.
   Otherwise find a new κ by extrapolation beyond κ_L and loop to Step 0.
STEP 4 (Interpolation). Find a new κ by interpolation in $]\kappa_L, \kappa_R[$ and loop to Step 0.
□

If φ_k does provide a relevant Newtonian point $\bar{y}$, the initial κ should be chosen large enough to produce it. Afterwards, each new κ could be computed by polynomial interpolation, as in §II.3.4; this implies a parametric study of (1.3.2), so as to get the derivative of the curved-search function $\kappa \mapsto f(y(\kappa))$. Note also that the same differential information would be needed if we wanted to implement a "Wolfe curved-search".

Remark 1.3.2 This trust-region technique was initially motivated by non-convexity. Suppose for example that Q in (1.3.1) is indefinite, as might well happen with $Q = \nabla^2 f(x_k)$ if f is not convex. Then the Newtonian point $\bar{y} = x_k - Q^{-1} s_k$ becomes suspect (although it is still of interest for solving the equation $\nabla f(x) = 0$); it may not even exist if Q is degenerate. Furthermore, a line-search along the associated direction $d_k = \bar{y} - x_k$ may be disastrous because $f'(x_k, d_k) = \langle s_k, d_k \rangle = -\langle Q d_k, d_k \rangle$ need not be negative.

Indeed, Newton's method with line-search has little relevance in a non-convex situation. By contrast, the solution $y(\kappa)$ of (1.3.2) makes a lot of sense in terms of minimizing f:

- It always exists, since the trust-region is compact. Much more general models φ_k could even be handled; the only issue would be the actual computation of $y(\kappa)$.
- To the extent that φ_k really approximates f to second order near x_k, (1.3.2) is consistent with the original problem.
- If Q happens to be positive definite, the Newtonian point is recovered, provided that κ is chosen large enough.
- If, for some reason, the curved-search produces small values of κ, the move from x_k to $y(\kappa)$ is made roughly along the steepest-descent direction; and this is good: the steepest-descent direction, precisely, is steepest for small moves. □

Admitting that a model such as φ_k is trusted around x_k only, the trust-region technique is still relevant even when φ_k is nicely convex. In this case, the stabilized problem can also be formulated as

$$\min \left\{ \varphi_k(y) + \tfrac{1}{2}\mu \|y - x_k\|^2 \ :\ y \in \mathbb{R}^n \right\}$$

instead of (1.3.2) (Proposition VII.3.1.4). The parameter μ will represent an alternative curvilinear coordinate, giving the explicit solution $x_k - (Q + \mu I)^{-1} s_k$ if the model is (1.3.1).

The interest of our digression is to view the stabilization introduced in §1.2 as some form of trust-region technique. Indeed consider one iteration of Algorithm 1.2.2, and suppose that a null-step is made. Suppose that Step 1 then keeps the same model φ_k but enhances the norming by taking a larger μ, or a smaller κ. This corresponds to Algorithm 1.3.1 with no Step 2: $y(\kappa)$ is accepted as soon as a sufficient decrease is obtained in Step 1.

2 A Variety of Stabilized Algorithms

We give in this section several algorithms realizing the general scheme of §1.2. They use conceptually equivalent stabilized problems in Step 2 of Algorithm 1.2.2. Three of them are formulated in the primal space $\mathbb{R}^n$; via an interpretation in the dual space, they are also conceptually equivalent to the bundle method of Chap. XIV. The notations are those of §1: $\check{f}_k$ is the cutting-plane function of (1.0.1); the Euclidean norm $\|\cdot\|$ is assumed for the stabilization (even though our development is only descriptive, and could accommodate more general situations).

286   XV. Primal Forms of Bundle Methods

2.1 The Trust-Region Point of View

The first idea that comes to mind is to force the next iterate to be a priori in a ball associated with the given norming, centered at the given stability center, and having a given radius κ. The sequence of iterates is thus defined by

$$y_{k+1} \in \mathrm{Argmin}\,\{\check{f}_k(y) \,:\, \|y - x_k\| \leqslant \kappa\}.$$

This approach has the same rationale as Example 1.2.3(a). The original model $\check{f}_k$ is considered as a good approximation of f in $B(x_k, \kappa)$, a *trust-region* drawn around the stability center. Accordingly, $\check{f}_k$ is minimized in this trust-region, any point outside it being disregarded. The resulting algorithm in its crudest form is then as follows.

Algorithm 2.1.1 (Cutting Planes with Trust Region) The initial point x_1 is given, together with a stopping tolerance $\underline{\delta} \geqslant 0$. Choose a trust-region radius $\kappa > 0$ and a descent coefficient $m \in {]}0, 1[$. Initialize the descent-set $K = \emptyset$, the iteration-counter $k = 1$ and $y_1 = x_1$; compute $f(y_1)$ and $s_1 = s(y_1)$.

STEP 1. Define the model

$$y \mapsto \check{f}_k(y) := \max\,\{f(y_i) + \langle s_i, y - y_i \rangle \,:\, i = 1, \ldots, k\}.$$

STEP 2. Compute a solution y_{k+1} of

$$\min\,\left\{\check{f}_k(y) \,:\, \tfrac{1}{2}\|y - x_k\|^2 \leqslant \tfrac{1}{2}\kappa^2\right\} \tag{2.1.1}$$

and set

$$\delta_k := f(x_k) - \check{f}_k(y_{k+1}) \geqslant 0.$$

STEP 3. If $\delta_k \leqslant \underline{\delta}$ stop.
STEP 4. Compute $f(y_{k+1})$ and $s_{k+1} = s(y_{k+1})$. If

$$f(y_{k+1}) \leqslant f(x_k) - m\delta_k,$$

set $x_{k+1} = y_{k+1}$ and append k to the set K (descent-step). Otherwise set $x_{k+1} = x_k$ (null-step).
  Replace k by $k + 1$ and loop to Step 1.   □

Here K represents the set of descent iterations, see Definition 1.2.4 again. Its role is purely notational and will appear when we study convergence. The algorithm could well be described without any reference to K. Concerning the nominal decrease δ_k, it is useful to understand that $f(x_k) = \check{f}_k(x_k)$ by construction. When a series of null-steps is taken, the cutting-plane algorithm is applied within the stability set $C := B(x_k, \kappa)$, which is changed at every descent-step. Such a change happens possibly long before f is minimized on C, and this is crucial for efficiency: minimizing f accurately over C is a pure waste of time if C is far from a minimum of f.

Remark 2.1.2 An efficient solution scheme of the stabilized problem (2.1.1) can be developed, based on duality: for $\mu > 0$, the Lagrange function

$$y \mapsto L(y, \mu) := \check{f}_k(y) + \tfrac{1}{2}\mu \left(\|y - x_k\|^2 - \kappa^2 \right)$$

has a unique minimizer $y(\mu)$, which can be computed exactly. It suffices to find $\mu > 0$ solving the equation $\|y(\mu) - x_k\| = \kappa$, if there is one; otherwise $\check{f}_k$ has an unconstrained minimum in $B(x_k, \kappa)$. Equivalently, the concave function $\mu \mapsto L(y(\mu), \mu)$ must be minimized over $\mu \geqslant 0$. □

The solution-set of (2.1.1) has a qualitative description.

Proposition 2.1.3 *Denote by $\kappa_\infty \geqslant 0$ the distance from x_k to the minimum-set of $\check{f}_k$ (with the convention $\kappa_\infty = +\infty$ if Argmin $\check{f}_k = \emptyset$). For $0 \leqslant \kappa \leqslant \kappa_\infty$, (2.1.1) has a unique solution, which lies at distance κ from x_k. For $\kappa \geqslant \kappa_\infty$, the solution-set of (2.1.1) is Argmin $\check{f}_k \cap B(x_k, \kappa)$.*

PROOF. The second statement is trivial (when applicable, i.e. when $\kappa_\infty < +\infty$).
Now take $\eta \geqslant 0$ such that the sublevel-set

$$S := \{x : \check{f}_k(x) \leqslant f(x_k) - \eta\}$$

is nonempty, and let x^* be the projection of x_k onto S. If $\kappa \leqslant \|x^* - x_k\| \leqslant \kappa_\infty$, any solution $\bar{y}$ of (2.1.1) must be at a distance κ from x_k: otherwise $\bar{y}$, lying in the interior of $B(x_k, \kappa)$, would minimize $\check{f}_k$ locally, hence globally, and the property $\|\bar{y} - x_k\| < \kappa \leqslant \|x^* - x_k\|$ would contradict the definition of x^*. The solution-set of (2.1.1) is therefore a convex set on the surface of a Euclidean ball: it is a singleton. □

Now we turn to a brief study of convergence, which uses typical arguments. First of all, $\{f(x_k)\}$ is a decreasing sequence, which has a limit in $\mathbb{R} \cup \{-\infty\}$. If $\{f(x_k)\}$ is unbounded from below, f has no minimum and $\{x_k\}$ is a "minimizing" sequence. The only interesting case is therefore

$$f_* := \lim_{k \to +\infty} f(x_k) > -\infty. \tag{2.1.2}$$

Lemma 2.1.4 *With the notation (2.1.2), and m denoting the descent coefficient in Step 4 of Algorithm 2.1.1, there holds $\sum_{k \in K} \delta_k \leqslant [f(x_1) - f_*]/m$.*

PROOF. Take $k \in K$, so that the descent test gives

$$\delta_k \leqslant \frac{f(x_k) - f(x_{k+1})}{m}.$$

Let k' be the successor of k in K. Because the stability center is not changed after a null-step, $f(x_{k+1}) = \cdots = f(x_{k'})$ and we have

$$\delta_{k'} \leqslant \frac{f(x_{k'}) - f(x_{k'+1})}{m} = \frac{f(x_{k+1}) - f(x_{k'+1})}{m}.$$

The recurrence is thus established: sum over K to obtain the result. □

This shows that, if no null-step were made, the method would converge rather fast: $\{\delta_k\}$ would certainly tend to 0 faster than $\{k^{-1}\}$ (to be compared with the speed $k^{-2/n}$ of Example 1.1.2). In particular, the stopping criterion would act after finitely many iterations. Note that, when the method stops, we have by construction

$$f(x) \geq f(x_k) - \underline{\delta} \quad \text{for all } x \in B(x_k, \kappa).$$

Using convexity, an approximate optimality condition is derived for x_k; it is this idea that is exploited in Case 2 of the proof below.

Theorem 2.1.5 *Let Algorithm 2.1.1 be used with fixed $\kappa > 0$, $m \in]0, 1[$ and $\underline{\delta} = 0$. Then $\{x_k\}$ is a minimizing sequence.*

PROOF. We take (2.1.2) into account and we distinguish two cases.

[*Case 1*: K is an infinite set of integers] Suppose for contradiction that there are $\bar{x} \in \mathbb{R}^n$ and $\eta > 0$ such that

$$f(\bar{x}) \leq f(x_k) - \eta \quad \text{for } k = 1, 2, \ldots$$

Lemma 2.1.4 tells us that $\lim_{k \in K} \delta_k = 0$. Hence, for k large enough in K:

$$f(x_k) - \tfrac{\eta}{2} \leq \check{f}_k(y_{k+1}) \leq \check{f}_k(y) \leq f(y) \quad \text{for all } y \in B(x_k, \kappa)$$

and $B(x_k, \kappa)$ cannot contain $\bar{x}$. Then consider the construction of Fig. 2.1.1: z_k is between x_k and $\bar{x}$, at a distance κ from x_k, and we set $r_k := \|\bar{x} - z_k\|$. We have

$$z_k = \frac{r_k x_k + \kappa \bar{x}}{r_k + \kappa};$$

from convexity,

$$f(z_k) \leq \frac{r_k f(x_k) + \kappa f(\bar{x})}{r_k + \kappa} \leq \frac{r_k f(x_k) + \kappa [f(x_k) - \eta]}{r_k + \kappa} = f(x_k) - \frac{\eta \kappa}{r_k + \kappa}.$$

In other words

$$\frac{\eta \kappa}{r_k + \kappa} \leq f(x_k) - f(z_k) \leq f(x_k) - \check{f}_k(z_k) \leq \delta_k.$$

Write this inequality for each k (large enough) in K and sum up. By construction, $\|x_{k+1} - x_k\| \leq \kappa$ and the triangle inequality implies that $r_{k+1} \leq r_k + \kappa$ for all $k \in K$: the left-hand side forms a divergent series; but this is impossible in view of Lemma 2.1.4.

[*Case 2*: K is finite] In this second case, only null-steps are taken after some iteration k^*:

$$f(y_{k+1}) > m \check{f}_k(y_{k+1}) + (1 - m) f(x_{k^*}) \quad \text{for all } k \geq k^*. \tag{2.1.3}$$

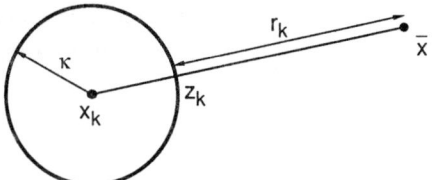

Fig. 2.1.1. Majorization outside the trust-region

From then on, Algorithm 2.1.1 reduces to the ordinary cutting-plane algorithm applied on the (fixed) compact set $C = B(x_{k*}, \kappa)$. It is convergent: when $k \to +\infty$, Theorem XII.4.2.3 tells us that $\{\check{f}_k(y_{k+1})\}$ and $\{f(y_{k+1})\}$ tend to the minimal value of f on $B(x_{k*}, \kappa)$, call it $\bar{f}^*$. Passing to the limit in (2.1.3),

$$(1-m)\bar{f}^* \geqslant (1-m)f(x_{k*}).$$

Because $m < 1$, we conclude that x_{k*} minimizes f on $B(x_{k*}, \kappa)$, hence on the whole space. □

Note in the above proof that null-steps and descent-steps call for totally different arguments. In one case, similar to §XII.4.2 or §XIV.3.4, the key is the accumulation of the information into the successive models $\check{f}_k$. The second case uses the definite decrease of f at each descent iteration, and this connotes the situation in §II.3.3. Another remark, based upon this proof, is that the stop does occur at some finite k if $\underline{\delta} > 0$: either because of Lemma 2.1.4, or because $\check{f}_k(y_{k+1}) \uparrow f(x_{k*})$ in Case 2.

We thus have a convergent stabilization of the cutting-plane algorithm. However, this algorithm is too simple: even though the number of descent-steps is relatively small (Lemma 2.1.4), many null-steps might still be necessary. In fact, Case 2 in the proof of Theorem 2.1.5 does suggest that we have not reached our goal. To obtain something really efficient, the radius κ of the trust-region should *tend to* 0 when $k \to \infty$. Otherwise, the disastrous Counter-example 1.1.2 would again crop up.

2.2 The Penalization Point of View

The trust-region technique of §2.1 was rather abrupt: the next iterate was controlled by a mere *switch*: "on" inside the trust-region, "off" outside it. Something more flexible is obtained if the distance from the stability center acts as a *weight*.

Here we choose a coefficient $\mu > 0$ (the strength of a spring) and our model is

$$y \mapsto \check{f}_k(y) + \tfrac{1}{2}\mu\|y - x_k\|^2;$$

needless to say, pure cutting planes would be obtained with $\mu = 0$. This strategy is in the spirit of Example 1.2.3(b): it is the model itself that is given a stabilized form; its unconstrained minimization will furnish the next iterate, and its decrease will give the nominal decrease for f.

Remark 2.2.1 This stabilizing term has been known for a long time in the framework of least-squares calculations. Suppose we have to minimize the function

$$\mathbb{R}^n \ni x \mapsto f(x) := \frac{1}{2} \sum_{j=1}^m f_j^2(x),$$

where each f_j is smooth. According to the general principles of §II.2, a second-order model of f is desirable.

Observe that (we assume the dot-product for simplicity)

$$\nabla^2 f(x) = Q(x) + \sum_{j=1}^m f_j(x) \nabla^2 f_j(x) \quad \text{with} \quad Q(x) := \sum_{j=1}^m \nabla f_j(x) [\nabla f_j(x)]^\top;$$

$Q(x)$ is thus a reasonable approximation of $\nabla^2 f(x)$, at least if the $f_j(x)$'s or the $\nabla^2 f_j(x)$'s are small. The so-called method of *Gauss-Newton* exploits this idea, taking as next iterate a minimum of the corresponding model

$$y \mapsto f(x_k) + [\nabla f(x_k)]^\top (y - x_k) + \tfrac{1}{2} (y - x_k)^\top Q(x_k)(y - x_k).$$

An advantage is that only first-order derivatives are required; furthermore $Q(x_k)$ is automatically positive semi-definite.

However, trouble will appear when $Q(x_k)$ is singular or ill-conditioned. Adding to it a multiple of the identity matrix, say μI, is then a good idea: this is just our present stabilization by penalty. In a Gauss-Newton framework, μ is traditionally called the coefficient of Levenberg and Marquardt. □

Just as in the previous section, we give the resulting algorithm in its crudest form:

Algorithm 2.2.2 (Cutting Planes with Stabilization by Penalty) The initial point x_1 is given, together with a stopping tolerance $\underline{\delta} \geq 0$. Choose a spring-strength $\mu > 0$ and a descent-coefficient $m \in \,]0, 1[$. Initialize the descent-set $K = \emptyset$, the iteration-counter $k = 1$, and $y_1 = x_1$; compute $f(y_1)$ and $s_1 = s(y_1)$.

STEP 1. With $\check{f}_k$ denoting the cutting-plane function (1.0.1), compute the solution y_{k+1} of

$$\min \left[\check{f}_k(y) + \tfrac{1}{2} \mu \| y - x_k \|^2 \right] \quad (2.2.1)$$

and set

$$\delta_k := f(x_k) - \check{f}_k(y_{k+1}) - \tfrac{1}{2} \mu \| y_{k+1} - x_k \|^2 \geq 0.$$

STEP 2. If $\delta_k \leq \underline{\delta}$ stop.
STEP 3. Compute $f(y_{k+1})$ and $s_{k+1} = s(y_{k+1})$. If

$$f(y_{k+1}) \leq f(x_k) - m\delta_k$$

set $x_{k+1} = y_{k+1}$ and append k to the set K (descent-step). Otherwise set $x_{k+1} = x_k$ (null-step).

Replace k by $k + 1$ and loop to Step 1. □

With respect to the trust-region variant, a first obvious difference is that the stabilized problem is easier: it is the problem with quadratic objective and affine constraints

$$\left| \begin{array}{ll} \min \left[r + \frac{1}{2}\mu \|y - x_k\|^2 \right] & (y, r) \in \mathbb{R}^n \times \mathbb{R} \\ f(y_i) + \langle s_i, y - x_i \rangle \leqslant r & \text{for } i = 1, \ldots, k. \end{array} \right. \quad (2.2.2)$$

A second difference is the value of δ_k: here the nominal decrease for f is more logically taken as the decrease of its piecewise quadratic model; this is comparable to the opposition (a) vs. (b) in Example 1.2.3. We will see later that the difference is actually of little significance: Algorithm 2.2.2 would be almost the same with $\delta_k = f(x_k) - \check{f}_k(y_{k+1})$.

In fact, neglecting any δ_k-detail, the two variants are conceptually equivalent in the sense that they produce the same $(k + 1)^{\text{st}}$ iterate, provided that κ and μ are properly chosen:

Proposition 2.2.3

(i) *For any $\kappa > 0$, there is $\mu \geqslant 0$ such that any solution of the trust-region problem (2.1.1) also solves the penalized problem (2.2.1).*

(ii) *Conversely, for any $\mu \geqslant 0$, there is $\kappa \geqslant 0$ such that any solution of (2.2.1) (assumed to exist) also solves (2.1.1).*

PROOF. [(i)] When $\kappa > 0$, the results of Chap. VII can be applied to the convex minimization problem (2.1.1): Slater's assumption is satisfied, the set of multipliers is nonempty, and there is a $\mu \geqslant 0$ such that the solutions of (2.1.1) can be obtained by unconstrained minimization of the Lagrangian

$$y \mapsto \check{f}_k(y) + \tfrac{1}{2}\mu(\|y - x_k\|^2 - \kappa^2)$$

(Proposition VII.3.1.4).

[(ii)] Suppose that (2.2.1) has a solution y_{k+1}. For any $y \in B(x_k, \|y_{k+1} - x_k\|)$,

$$\check{f}_k(y) + \tfrac{1}{2}\mu \|y_{k+1} - x_k\|^2 \geqslant \check{f}_k(y) + \tfrac{1}{2}\mu \|y - x_k\|^2 \geqslant \check{f}_k(y_{k+1}) + \tfrac{1}{2}\mu \|y_{k+1} - x_k\|^2$$

and we see that y_{k+1} solves (2.1.1) for $\kappa = \|y_{k+1} - x_k\|$. □

The relation between our two forms of stabilized problems can be made more precise. See again Proposition 2.1.3: when κ increases from 0 to the value κ_∞, the unique solution of (2.1.1) describes a curve linking x_k to the minimum of $\check{f}_k$ that is closest to x_k; and if $\check{f}_k$ is unbounded from below, this curve is unbounded (recall from the end of §V.3.4, or of §VIII.3.4, that the piecewise affine function $\check{f}_k$ attains its infimum when it is bounded from below).

Now, for $\mu > 0$, call $y(\mu)$ the unique solution of (2.2.1). Proposition 2.2.3 tells us that, when μ describes $\mathbb{R}_*^+$, $y(\mu)$ describes the same curve as above, neglecting some possible problem at $\kappa \downarrow 0$ (which corresponds to $\mu \to +\infty$). This establishes a mapping

$$\mu \mapsto \kappa = \kappa(\mu) \quad \text{from }]0, +\infty[\text{ onto }]0, \kappa_\infty[$$

defined by $\kappa(\mu) = \|y(\mu) - x_k\|$ for $\mu > 0$. Note that this mapping is not invertible: the μ of Proposition 2.2.3(i) need not be unique.

Beware that the equivalence between (2.1.1) and (2.2.1) is not of a practical nature. It simply means that

- choosing κ and then solving the trust-region problem,
- or choosing μ and then solving the penalized problem

are *conceptually* equivalent. Nevertheless, there is no *explicit* relation giving a priori one parameter as a function of the other.

Section 3 will be entirely devoted to this second variant, including a study of its convergence.

2.3 The Relaxation Point of View

The point made in the previous section is that the second term $1/2\,\mu\|y - x_k\|^2$ in (2.2.1) can be interpreted as the dualization of a certain constraint $\|y - x_k\| \leq \kappa$, whose *right-hand side* $\kappa = \kappa(\mu)$ becomes a function of its *multiplier* μ. Likewise, we can interpret the first term $\check{f}_k(y)$ as the dualization of a constraint $\check{f}_k(y) \leq \ell$, whose right-hand side $\ell = \ell(\mu)$ will be a function of its multiplier $1/\mu$.

In other words, a third possible stabilized problem is

$$\left| \begin{array}{l} \min \tfrac{1}{2}\|y - x_k\|^2, \\ \check{f}_k(y) \leq \ell \end{array} \right. \tag{2.3.1}$$

for some *level* ℓ. We leave it to the reader to adapt Proposition 2.2.3, thereby studying the equivalence of (2.3.1) with the previous variants. A difficulty has now appeared, though: what if (2.3.1) has an empty feasible set, i.e. if $\ell < \inf \check{f}_k$? By contrast, the parameter κ or μ of the previous sections could be given arbitrary positive values throughout the iterations, even if a fixed κ led to possible inefficiencies. The present variant thus needs some more sophistication.

On the other hand, the level used in (2.3.1) suggests an obvious nominal decrease for f: it is natural to set $x_{k+1} = y_{k+1}$ if this results in a definite objective-decrease from $f(x_k)$ towards ℓ. Then the resulting algorithm has the following form, in which we explicitly let the level depend on the iteration.

Algorithm 2.3.1 (Cutting Planes with Level-Stabilization) The initial point x_1 is given, together with a stopping tolerance $\underline{\delta} \geq 0$. Choose a descent-coefficient $m \in]0, 1[$. Initialize the descent-set $K = \emptyset$, the iteration counter $k = 1$, and $y_1 = x_1$; compute $f(y_1)$ and $s_1 = s(y_1)$.

STEP 1. Choose a level $\ell = \ell_k$ satisfying $\inf \check{f}_k \leq \ell < f(x_k)$; perform the stopping criterion.

STEP 2. Compute the solution y_{k+1} of (2.3.1).

STEP 3. Compute $f(y_{k+1})$ and $s_{k+1} = s(y_{k+1})$. If

$$f(y_{k+1}) \leq f(x_k) - m[f(x_k) - \ell] \tag{2.3.2}$$

set $x_{k+1} = y_{k+1}$ and append k to the set K (descent-step). Otherwise set $x_{k+1} = x_k$ (null-step).

Replace k by $k + 1$ and loop to Step 1. □

2 A Variety of Stabilized Algorithms   293

To explain the title of this subsection, consider the problem of solving a system of inequalities: given an index-set J and smooth functions f_j for $j \in J$, find $x \in \mathbb{R}^n$ such that

$$f_j(x) \leq 0 \quad \text{for all } j \in J. \tag{2.3.3}$$

It may be a difficult one, especially when J is a large set (possibly infinite), and/or the f_j's are not affine. Several techniques are known for a numerical treatment of this problem.

- The *relaxation* method addresses large sets J and consists of a dynamic selection of appropriate elements in J. For example: at the current iterate x_k we take a most violated inequality, i.e. we compute an index j_k such that

$$f_{j_k}(x_k) \geq f_j(x_k) \quad \text{for all } j \in J;$$

then we solve

$$\min \left\{ \tfrac{1}{2} \|x - x_k\|^2 \;:\; f_{j_k}(x) \leq 0 \right\}.$$

Unless x_k is already a solution of (2.3.3), we certainly obtain $x_{k+1} \neq x_k$. Refined variants exist, in which more than one index is taken into account at each iteration.

- Newton's principle (§II.2.3) can also be used: each inequality can be linearized at x_k, (2.3.3) being replaced by

$$f_j(x_k) + \langle \nabla f_j(x_k), x - x_k \rangle \leq 0 \quad \text{for all } j \in J.$$

- When combining the two techniques, we obtain an implementable method, in which a sequence of quadratic programming problems are solved.

Now take our original problem of minimizing f, and assume that the optimal value $\bar{f} = \inf f$ is known. We must clearly solve $f(x) \leq \bar{f}$, to which the above technique can be applied: linearizing f at the current x_k, we are faced with

$$\min \left\{ \tfrac{1}{2} \|x - x_k\|^2 \;:\; f(x_k) + \langle s(x_k), x - x_k \rangle \leq \bar{f} \right\}.$$

Remark 2.3.2 The solution of this projection problem can be computed explicitly: the minimality conditions are

$$x - x_k + \mu s(x_k) = 0, \quad \mu[f(x_k) + \langle s(x_k), x - x_k \rangle - \bar{f}] = 0.$$

Assuming $s(x_k) \neq 0$ and $\bar{f} < f(x_k)$ (hence $x \neq x_k$, $\mu > 0$), the next iterate is

$$x_k - \frac{f(x_k) - \bar{f}}{\|s(x_k)\|^2} s(x_k).$$

This is just the subgradient Algorithm XII.4.1.1, in which the knowledge of $\bar{f}$ is exploited to provide a special stepsize. □

When formulating our minimization problem as the single inequality $f(x) \leq \bar{f}$, the essential resolution step is the linearization of f (Newton's principle). Another possible formulation is via infinitely many cutting-plane inequalities: f can be expressed as a supremum of affine functions,

$$f(x) = \sup \{ f(y) + \langle s(y), y - x \rangle \;:\; y \in \mathbb{R}^n \}.$$

To minimize f is therefore to find x solving the system of affine inequalities

$$f(y) + \langle s(y), x - y \rangle \leqslant \bar{f} \quad \text{for all } y \in \mathbb{R}^n,$$

where the indices $i \in J$ of (2.3.3) are rather denoted by $y \in \mathbb{R}^n$. We can apply to this problem the relaxation principle and *memorize* the inequalities visited during the successive iterations. At the given iterate x_k, we recover a problem of the form (2.3.1).

Let us conclude this comparison: if the infimum of f is known, then Algorithm 2.3.1 with $\ell_k \equiv \bar{f}$ is quite a suitable variant. In other cases, the whole issue for this algorithm will be to identify the unknown value $\bar{f}$, thus indicating suitable rules for the management of $\{\ell_k\}$.

We just give an example of a convergence proof, limited to the case of a known $\bar{f}$.

Theorem 2.3.3 *Let f have a minimum point $\bar{x}$. Then Algorithm 2.3.1, applied with $\ell_k \equiv f(\bar{x}) =: \bar{f}$ and $\underline{\delta} = 0$, generates a minimizing sequence $\{x_k\}$.*

PROOF. The proof technique is similar to that of the trust-region variant.

[*Case 1*: K is infinite] At each descent iteration, we have by construction

$$f(x_{k+1}) - \bar{f} \leqslant f(x_k) + m[\bar{f} - f(x_k)] - \bar{f} = (1-m)[f(x_k) - \bar{f}].$$

Reasoning as in Lemma 2.1.4,

$$f(x_k) - \bar{f} \leqslant (1-m)^{\nu(k)}[f(x_1) - \bar{f}],$$

if $\nu(k)$ denotes the number of descent-steps that have been made prior to the kth iteration. If the algorithm performs infinitely many descent-steps, $f(x_k) - \bar{f} \to 0$ and we are done.

[*Case 2*: K is finite] Suppose now that the sequence $\{x_k\}$ stops at some iteration k^*, so that $x_k = x_{k^*}$ for all $k \geqslant k^*$. We proceed to prove that $\bar{f}$ is a cluster value of $\{f(y_k)\}$.

First, $\check{f}_k(\bar{x}) \leqslant f(\bar{x}) = \bar{f}$ for all k: $\bar{x}$ is feasible in each stabilized problem, so $\|y_k - x_{k^*}\| \leqslant \|\bar{x} - x_{k^*}\|$ by construction. We conclude that $\{y_k\}$ is bounded; from their definition, all the model-functions $\check{f}_k$ have a fixed Lipschitz constant L, namely a Lipschitz constant of f on $B(x_{k^*}, \|\bar{x} - x_{k^*}\|)$ (Theorem IV.3.1.2).

Then take k and k' arbitrary with $k^* \leqslant k \leqslant k'$, observe that $\check{f}_{k'}(y_k) = f(y_k)$ by definition of $\check{f}_{k'}$, so

$$f(y_k) \leqslant \check{f}_{k'}(y_{k'+1}) + L\|y_k - y_{k'+1}\| \leqslant \bar{f} + L\|y_k - y_{k'+1}\|, \tag{2.3.4}$$

where the last inequality comes from the definition of $y_{k'+1}$.

Because $\{y_k\}$ is bounded, $\|y_k - y_{k'+1}\|$ cannot stay away from 0 when k and k' go independently to infinity: for any $\varepsilon > 0$, we can find large enough k and k' such that $\|y_k - y_{k'+1}\| \leqslant \varepsilon$. Thus, $\liminf f(y_k) = \bar{f}$ and failure to satisfy of the descent test (2.3.2) implies that x_{k^*} is already optimal. □

The comments following Theorem 2.1.5 are still valid, concerning the proof technique; it can even be said that, after a descent-step, all the linearizations appearing in $\check{f}_k$ can be refreshed, thus reinitializing $\check{f}_{k+1}$ to the affine function $f(x_{k+1}) + \langle s_{k+1}, \cdot - x_{k+1} \rangle$. This modification does not affect Case 2; and Case 1 uses no memory mechanism.

Remark 2.3.4 Further consideration of (2.3.4) suggests another interesting technical comment: suppose in Algorithm 2.3.1 that the level ℓ is fixed but the descent test is ignored: only null-steps are taken. Then, providing that $\{y_k\}$ is bounded, $\{f(y_k)\}$ reaches any level above ℓ. If $\ell = \bar{f}$, this has several implications:

– First, as far as proving convergence is concerned, the concept of descent-step is useless: we could just set $m = 1$ in the algorithm; the stability center would remain fixed throughout, the minimizing sequence would be $\{y_k\}$.
– Even further: the choice of the stability center, to be projected onto the sublevel-set of $\check{f}_k$, is moderately important. Technically, its role is limited to preserving the boundedness of $\{y_k\}$.
– Existence of a minimum of f is required to guarantee this boundedness: in a way, our present algorithm is weaker than the trust-region form, which did not need this existence. However, we leave it as an exercise to reproduce the proof of Theorem 2.3.3 for a variant of Algorithm 2.3.1 using a more conservative choice of the level:

$$\ell_k = f(x_k) - m'[f(x_k) - \bar{f}] \quad \text{with} \quad m' \in \,]m, 1[\,.$$

– Our proof of Theorem 2.3.3 is qualitative; compare it with §IX.2.1(a). A quantitative argument could also be conceived of, as in §IX.2.1(b). □

To conclude, let us return to our comparison with the inequality-solving problem (2.3.3), assumed to have a nonempty solution-set. For this problem, the simplest Newton algorithm takes x_{k+1} as the unique solution of

$$\min \left\{ \tfrac{1}{2}\|x - x_k\|^2 \,:\, f_i(x) + \langle \nabla f_i(x_k), x - x_k \rangle \leqslant 0 \text{ for } i \in J \right\}. \quad (2.3.5)$$

This makes numerical sense if J is finite (and not large): we have a quadratic program to solve at each iteration. Even in this case, however, the algorithm may not converge because a Newton method is only locally convergent.

To get global convergence, line-searching is a natural technique: we can take a next iterate along the half-line pointing towards the solution of (2.3.5) and decreasing "substantially" the natural objective function $\max_{i \in J} f_i$; we are right in the framework of §II.3. The message of the present section is that another possible technique is to *memorize* the linearizations; as already seen in §1.2, a null-step resembles one cycle in a line-search. This technique enables the resolution of (2.3.3) with an infinite index-set J, and also with nonsmooth functions f_i. The way is open to minimization methods with nonsmooth constraints. Compare also the present discussion with §IX.3.2.

2.4 A Possible Dual Point of View

Consider again (2.2.1), written in expanded form:

$$\left| \begin{array}{l} \min \left[r + \tfrac{1}{2}\mu\|y - x_k\|^2 \right] \quad (y, r) \in \mathbb{R}^n \times \mathbb{R}, \\ f(y_i) + \langle s_i, y - y_i \rangle \leqslant r \quad \text{for } i = 1, \ldots, k. \end{array} \right. \quad (2.4.1)$$

The dual of this quadratic program can be formulated explicitly, yielding very instructive interpretations. In the result below, Δ_k is the unit simplex of $\mathbb{R}^k$ as usual; the coefficients

$$e_i := e(x_k, y_i, s_i) := f(x_k) - f(y_i) - \langle s_i, x_k - y_i \rangle \quad \text{for } i = 1, \ldots, k \quad (2.4.2)$$

are the linearization errors between y_i and x_k, already encountered in several previous chapters (see Definitions XI.4.2.3 and XIV.1.2.7).

Lemma 2.4.1 *For $\mu > 0$, the unique solution of the penalized problem (2.2.1) = (2.4.1) is*

$$y_{k+1} = x_k - \frac{1}{\mu} \sum_{i=1}^{k} \alpha_i s_i, \quad (2.4.3)$$

where $\alpha \in \mathbb{R}^k$ solves

$$\min_{\alpha \in \Delta_k} \left[\tfrac{1}{2} \left\| \sum_{i=1}^{k} \alpha_i s_i \right\|^2 + \mu \sum_{i=1}^{k} \alpha_i e_i \right]. \quad (2.4.4)$$

Furthermore, there holds

$$\check{f}_k(y_{k+1}) = f(x_k) - \sum_{i=1}^{k} \alpha_i e_i - \tfrac{1}{\mu} \left\| \sum_{i=1}^{k} \alpha_i s_i \right\|^2. \quad (2.4.5)$$

PROOF. This is a direct application of Chap. XII (see §XII.3.4 if necessary). Take k nonnegative dual variables $\alpha_1, \ldots, \alpha_k$ and form the Lagrange function which, at $(y, r, \alpha) \in \mathbb{R}^n \times \mathbb{R} \times \mathbb{R}^k$, has the value

$$r + \tfrac{1}{2}\mu \|y - x_k\|^2 + \sum_{i=1}^{k} \alpha_i \langle s_i, y \rangle + \sum_{i=1}^{k} \alpha_i [f(y_i) - \langle s_i, y_i \rangle - r].$$

Its minimization with respect to the primal variables (y, r) implies first the condition $\sum_{i=1}^{k} \alpha_i = 1$ (otherwise we get $-\infty$), and results in y given as in (2.4.3). Plugging this value back into the Lagrangian, we obtain the dual problem associated with (2.4.1):

$$\max_{\alpha \in \Delta_k} \left\{ -\tfrac{1}{2\mu} \left\| \sum_{i=1}^{k} \alpha_i s_i \right\|^2 + \sum_{i=1}^{k} \alpha_i [f(y_i) + \langle s_i, x_k - y_i \rangle] \right\},$$

in which the notation (2.4.2) can be used. Equating the primal and dual objective-values gives (2.4.5) directly.

Finally, the solution-set of the dual problem is not changed if we multiply its objective function by $\mu > 0$, and add the constant term $f(x_k) = \sum_{i=1}^{k} \alpha_i f(x_k)$. This gives (2.4.4). □

With the form (2.4.4) of stabilized problem, we can play the same game as in the previous subsections: the linear term in α can be interpreted as the dualization of a constraint whose right-hand side, say ε, is a function of its multiplier μ. In other words: given $\mu > 0$, there is an ε such that the solution of (2.2.1) is given by (2.4.3), where α solves

$$\left| \begin{array}{l} \min \tfrac{1}{2} \left\| \sum_{i=1}^{k} \alpha_i s_i \right\|^2, \quad \alpha \in \Delta_k, \\ \sum_{i=1}^{k} \alpha_i e_i \leqslant \varepsilon. \end{array} \right. \quad (2.4.6)$$

Conversely, let the constraint in this last problem have a positive multiplier μ; then the associated y_{k+1} of (2.4.3) is also the unique solution of (2.2.1) with this μ.

A thorough observation of our notation shows that

$$e_i \geq 0 \text{ for } i = 1, \ldots, k \quad \text{and} \quad e_j = 0 \text{ for some } j \leq k.$$

It follows that the correspondence $\varepsilon \leftrightarrows \mu$ involves nonnegative values of ε only. Furthermore, we will see that the values of ε that are relevant for convergence must depend on the iteration index. In summary, our detour into the dual space has revealed a fourth conceptually equivalent stabilized algorithm:

Algorithm 2.4.2 (Cutting Planes with Dual Stabilization) The initial point x_1 is given, together with a stopping tolerance $\underline{\delta} \geq 0$. Choose a descent-coefficient $m \in \,]0,1[\,$. Initialize the descent-set $K = \emptyset$, the iteration counter $k = 1$, and $y_1 = x_1$; compute $f(y_1)$ and $s_1 = s(y_1)$.

STEP 1. Choose $\varepsilon \geq 0$ such that the constraint in (2.4.6) has a positive multiplier μ (for simplicity, we assume this is possible).

STEP 2. Solve (2.4.6) to obtain an optimal $\alpha \in \Delta_k$ and a multiplier $\mu > 0$. Unless the stopping criterion is satisfied, set

$$\hat{s} := \sum_{i=1}^{k} \alpha_i s_i, \quad y_{k+1} = x_k - \tfrac{1}{\mu}\hat{s}.$$

STEP 3. Compute $f(y_{k+1})$ and $s_{k+1} = s(y_{k+1})$. If

$$f(y_{k+1}) \leq f(x_k) - m\left(\varepsilon + \tfrac{1}{2\mu}\|\hat{s}\|^2\right)$$

set $x_{k+1} = y_{k+1}$ and append k to the set K (descent-step). Otherwise set $x_{k+1} = x_k$ (null-step).

Replace k by $k+1$ and loop to Step 1. □

The interesting point about this interpretation is that (2.4.6) is just the direction-finding problem (XIV.2.1.4) of the *dual bundle methods*. This allows some useful comparisons:

(i) A stabilized cutting-plane method is a particular form of bundle method, in which the "line-search" chooses systematically $t = 1/\mu$ (the inverse multiplier of (2.4.6), which must be positive). The privileged role played by this particular stepsize was already outlined in §XIV.2.4; see Fig. XIV.2.4.1 again.

(ii) Alternatively, a dual bundle method can be viewed as a stabilized cutting-plane method, in which a line-search is inserted. Since Remark 1.2.1 warns us against such a mixture, it might be desirable to imagine something different.

(iii) The stopping criterion in Algorithm 2.4.2 is not specified but we know from the previous chapters that x_k is approximately optimal if both ε and $\|\hat{s}\|$ are small; see for example (XIV.2.3.5). This interpretation is also useful for the stopping criteria used in the previous subsections. Indeed (2.4.5) reveals two distinct components in the nominal decrease $\check{f}_k(y_{k+1}) - f(x_k)$. One, ε, is directly comparable to f-values; the other, $\|\hat{s}\|^2/\mu$, connotes shifts in x via the "rate" of decrease $\|\hat{s}\|$; see the note XIV.3.4.3([6]).

(iv) In §XIV.3.3(a), we mentioned an ambiguity concerning the descent test used by a dual bundle method. If the value (XIV.3.3.2) is used for the line-search parameter $\tilde{v}$, the dual bundle algorithm uses the descent test

$$f(x_k - t\hat{s}) \leq f(x_k) - mt\mu \left(\varepsilon + \tfrac{1}{\mu}\|\hat{s}\|^2\right).$$

In view of (2.4.5), we recognize in the last parenthesis the nominal decrease $\check{f}_k(x_k - t\hat{s}) - f(x_k)$, recommended for the trust-region variant.

(v) In their dual form, bundle methods lend themselves to the aggregation technique. A solution α of (2.4.6) not only yields the aggregate subgradient $\hat{s}$, but also an aggregate linearization error $\hat{e} := \sum_{i=1}^{k} \alpha_i e_i$. As seen in §XIV.2.4, this corresponds to the aggregate linearization

$$y \mapsto f(x_k) - \hat{e} + \langle \hat{s}, y - x_k \rangle$$

which minorizes f, simply because it minorizes $\check{f}_k$. The way is thus open to "economic" cutting-plane algorithms (possibly stabilized), in which the above aggregate linearization can take the place of one or more direct linearizations, thereby diminishing the complexity of the cutting-plane function.

Recall from §XIV.3.4 that a rather intricate control of ε_k is necessary, just to ensure convergence of Algorithm 2.4.2. Having said enough about this problem in Chap. XIV, we make here a last remark. To pass from the primal stabilized μ-problem to its dual (2.4.4) or (2.4.6), we applied Lagrange duality to the developed form (2.4.1); and to make the comparison more suggestive, we introduced the linearization errors (2.4.2). The same interpretative work can be done by applying Fenchel duality (§XII.5.4) directly to (2.2.1). This gives something more abstract, but also more intrinsic, involving the conjugate function of $\check{f}_k$:

Proposition 2.4.3 *For $\mu > 0$, the unique solution y_{k+1} of (2.2.1) is*

$$y_{k+1} = x_k - \tfrac{1}{\mu}\hat{s},$$

where $\hat{s} \in \partial \check{f}_k(y_{k+1})$ is the unique minimizer of the closed convex function

$$\mathbb{R}^n \ni s \mapsto \check{f}_k^*(s) - \langle s, x_k \rangle + \tfrac{1}{2\mu}\|s\|^2. \qquad (2.4.7)$$

PROOF. Set $g_1 := \check{f}_k$, $g_2 := 1/2\mu \|\cdot - x_k\|^2$ and apply Proposition XII.5.4.1: because g_2 is finite everywhere, the qualification assumption (XII.5.4.3) certainly holds. The conjugate of g_2 can be easily computed, and minimizing the function of (2.4.7) is exactly the dual problem of (2.2.1). Denoting its solution by $\hat{s}$, the solution of (2.2.1) is then the (therefore nonempty) set $\partial \check{f}_k^*(\hat{s}) \cap \{-1/\mu\, \hat{s} + x_k\}$. Then there holds

$$y_{k+1} = x_k - \tfrac{1}{\mu}\hat{s} \in \partial \check{f}_k^*(\hat{s}), \quad \text{i.e.} \quad \hat{s} \in \partial \check{f}_k(y_{k+1}). \qquad \square$$

Needless to say, (2.4.7) is simply a compact form for the objective function of (2.4.4). To see this, use either of the following two useful expressions for the cutting-plane model:

$$\begin{aligned}\check{f}_k(y) &= f(x_k) + \max\{-e_i + \langle s_i, y - x_k \rangle : i = 1, \ldots, k\} \\ &= \max\{-f^*(s_i) + \langle s_i, y \rangle : i = 1, \ldots, k\}.\end{aligned}$$

Its conjugate can be computed with the help of various calculus rules from Chap. X (see in particular §X.3.4); we obtain a convex hull of needles:

$$\begin{aligned}\check{f}_k^*(s) &= -f(x_k) + \langle s, x_k\rangle + \min\left\{\sum_{i=1}^k \alpha_i e_i \,:\, \alpha \in \Delta_k,\ \sum_{i=1}^k \alpha_i s_i = s\right\} \\ &= \min\left\{\sum_{i=1}^k \alpha_i f^*(s_i) \,:\, \alpha \in \Delta_k,\ \sum_{i=1}^k \alpha_i s_i = s\right\}.\end{aligned}$$

Plugging the first expression into (2.4.7) yields the minimization problem (2.4.4); and with the second value, we obtain an equivalent form:

$$\min_{\alpha \in \Delta_k}\left\{\tfrac{1}{2}\left\|\sum_{i=1}^k \alpha_i s_i\right\|^2 + \mu \sum_{i=1}^k \alpha_i [f^*(s_i) - \langle s_i, x_k\rangle]\right\}.$$

2.5 Conclusion

This Section 2 has reviewed a number of possible algorithms for minimizing a (finite-valued) convex function, based on two possible motivations:

- Three of them work in the primal space. They start from the observation that the cutting-plane algorithm is unstable: its next iterate is "too far" from the current one, and should be pulled back.
- One of them works in the dual space. It starts from the observation that the steepest-descent algorithm is sluggish: the next iterate is "too close" to the current one, and should be pushed forward.

In both approaches, the eventual aim is to improve the progress towards an optimal solution, as measured in terms of the objective function.

Remark 2.5.1 We mention here that our list of variants is not exhaustive.

- When starting from the μ-problem (2.2.1), we introduced a trust-region constraint, or a level constraint, ending up with two more primal variants.
- When starting from the same μ-problem in its dual form (2.4.4), we introduced a linearization-error constraint, which corresponded to a level point of view in the dual space. Altogether, we obtained the four possibilities, reviewed above.
- We could likewise introduce a trust-region point of view in the dual space: instead of (2.4.6), we could formulate

$$\min_{\alpha \in \Delta_k}\left\{\sum_{i=1}^k \alpha_i e_i \,:\, \left\|\sum_{i=1}^k \alpha_i s_i\right\|^2 \leqslant \sigma\right\},$$

for some dual radius σ (note: $-\sigma$ represents the value of a certain support function, remember Proposition XIV.2.3.2).
- Another idea would be to take the dual of (2.1.1) or (2.3.1), and then to make analogous changes of parameter.

We do not know what would result from these various exercises. □

All these algorithms follow essentially the same strategy: solve a quadratic program depending on a certain internal parameter: κ, μ, ℓ, ε, or whatever. They are conceptually equivalent, in the sense that they generate the same iterates, providing that their respective parameters are linked by a certain a posteriori relation; and they are characterized precisely by the way this parameter is chosen.

Figure 2.5.1 represents the relations connecting all these parameters; it plots the two model-functions, $\hat{f}_k$ and its piecewise quadratic perturbation, along $\hat{s}$ interpreted as a direction. The graph of f lies somewhere above gr $\hat{f}_k$ and meets the two model-graphs at $(x_k, f(x_k))$. The dashed line represents the trace along $x_k + \mathbb{R}\hat{s}$ of the affine function

$$y \mapsto f(x_k) - \varepsilon + \langle \hat{s}, y - x_k \rangle .$$

If the variant mentioned in Remark 2.5.1 is used, take the graph of an affine function of slope $-\sigma$, and lift it as high as possible so as to support gr $\check{f}_k$: this gives the other parameters.

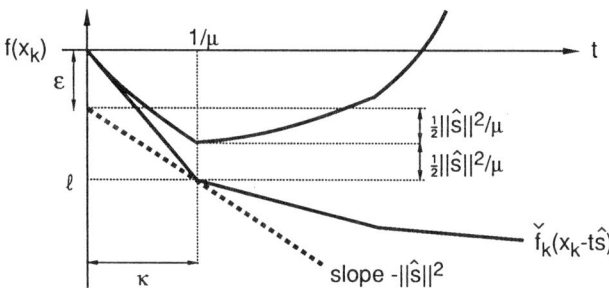

Fig. 2.5.1. A posteriori relations

Remark 2.5.2 This picture clearly confirms that the nominal decreases for the trust-region and penalized variants are not significantly different: in view of (2.4.5), they are respectively

$$\delta_{TR} = \varepsilon + \tfrac{1}{\mu}\|\hat{s}_k\|^2 \quad \text{and} \quad \delta_P = \varepsilon + \tfrac{1}{2\mu}\|\hat{s}_k\|^2 .$$

Thus $\delta_{TR} \in [\delta_P, 2\delta_P]$. □

So far, our review has been purely descriptive; no argument has been given to prefer any particular variant. Yet, the numerical illustrations in §XIV.4.2 have clearly demonstrated the importance of a proper choice of the associated parameter (be it ε, κ, μ, ℓ, or whatever). The selection of a particular variant should therefore address two questions, one theoretical, one practical:

– specific rules for choosing the parameter efficiently, in terms of speed of convergence;
– effective resolution of the stabilized problem, which is extremely important in practice: being routinely executed, it must be fast and fail-safe.

With respect to the second criterion, the four variants are approximately even, with a slight advantage to the ε-form: some technical details in quadratic programming make its resolution process more stable than the other three.

The first criterion is also the more decisive; but unfortunately, it does not yield a clear conclusion. We refer to the difficulties illustrated in §XIV.4.2 for efficient choices of ε; and apart from the situation of a known minimal value $\bar{f}$ (in which case the level-strategy becomes obvious), little can be said about appropriate choices of the other parameters.

It turns out, however, that the variant by penalization has a third possible motivation, which gives it a definite advantage. It will be seen in §4 that it is intimately related to the Moreau-Yosida regularization of Example XI.3.4.4. The next section is therefore devoted to a thorough study of this variant.

3 A Class of Primal Bundle Algorithms

In this section, we study more particularly primal bundle methods in penalized form, introduced in §2.2. Because they work directly in the primal space, they can handle possible constraints rather easily. We therefore assume that the problem to solve is actually

$$\inf \{f(x) : x \in C\}; \qquad (3.0.1)$$

f is still a convex function (finite everywhere), and now C is a closed convex subset of $\mathbb{R}^n$. The only restriction imposed on C is of a practical nature, namely each stabilized problem (2.2.1) must still be solvable, even if the constraint $y \in C$ is added. In practice, this amounts to assuming that C is a closed convex polyhedron:

$$C = \{x \in \mathbb{R}^n : \langle a_j, x \rangle \leqslant b_j \text{ for } j = 1, \ldots, m\}, \qquad (3.0.2)$$

$(a_j, b_j) \in \mathbb{R}^n \times \mathbb{R}$ being given for $j = 1, \ldots, m$. As far as this chapter is concerned, the only effect of introducing C is a slight complication in the algebra; the reader may take $C = \mathbb{R}^n$ throughout, if this helps him to follow our development more easily. Note, anyway, that (3.0.1) is the unconstrained minimization of $g := f + I_C$; because f is assumed finite everywhere, $\partial g = \partial f + N_C$ (Theorem XI.3.1.1) and little is essentially changed with respect to our previous developments.

3.1 The General Method

The model will not be exactly $\check{f}_k$ of (1.0.1), so we prefer to denote it abstractly by φ_k, a convex function finite everywhere. Besides, the iteration index k is useless for the moment; the stabilized problem is therefore denoted by

$$\min \left\{ \varphi(y) + \tfrac{1}{2}\mu \|y - x\|^2 : y \in C \right\}, \qquad (3.1.1)$$

x and μ being the (given) k^{th} stability center and penalty coefficient respectively. Once again, this problem is assumed numerically solvable; such is the case for φ piecewise affine and C polyhedral.

The model φ will incorporate the *aggregation* technique, already seen in previous chapters, which we proceed to describe in a totally primal language. First we reproduce Lemma 2.4.2.

Lemma 3.1.1 *With $\varphi : \mathbb{R}^n \to \mathbb{R}$ convex, $\mu > 0$ and C closed convex, (3.1.1) has a unique solution y_+, characterized by the formulae*

$$y_+ = x - \tfrac{1}{\mu}(\hat{s} + \hat{p}), \quad \hat{s} \in \partial\varphi(y_+), \quad \hat{p} \in N_C(y_+). \qquad (3.1.2)$$

Furthermore
$$\varphi(y) \geq f(x) + \langle \hat{s}, y - x \rangle - \hat{e} \quad \text{for all } y \in \mathbb{R}^n,$$
where
$$\hat{e} := f(x) - \varphi(y_+) - \tfrac{1}{\mu}\langle \hat{s}, \hat{s} + \hat{p}\rangle \geq 0. \tag{3.1.3}$$

PROOF. The assumptions clearly imply that (3.1.1) has a unique solution. Using the geometric minimality condition VII.1.1.1 and some subdifferential calculus, this solution is seen to be the unique point y_+ satisfying
$$0 \in \partial\varphi(y_+) + \mu(y_+ - x) + N_C(y_+),$$
which is just (3.1.2).

We thus have
$$\varphi(y) \geq \varphi(y_+) + \langle \hat{s}, y - y_+\rangle \quad \text{for all } y \in \mathbb{R}^n.$$

Using, as in Proposition XI.4.2.2, a "transportation trick" from y_+ to x, this can be written
$$\varphi(y) \geq f(x) + \langle \hat{s}, y - x\rangle - f(x) + \varphi(y_+) + \langle \hat{s}, x - y_+\rangle.$$

In view of (3.1.2), we recognize the expression (3.1.3) of $\hat{e}$. □

Proposition 3.1.2 *With the notation of Lemma 3.1.1, take an arbitrary function $\psi : \mathbb{R}^n \to \mathbb{R} \cup \{+\infty\}$ satisfying*
$$\psi(y) \geq f(x) - \hat{e} + \langle \hat{s}, y - x\rangle =: \bar{f}^a(y) \quad \text{for all } y \in \mathbb{R}^n, \tag{3.1.4}$$
with equality at $y = y_+$. Then y_+ minimizes on C the function
$$C \ni y \mapsto \tilde{\psi}(y) := \psi(y) + \tfrac{1}{2}\mu\|y - x\|^2.$$

PROOF. Use again the same transportation trick. using (3.1.3) and (3.1.2), the relations defining ψ can be written
$$\psi(y) \geq \varphi(y_+) + \langle \hat{s}, y - y_+\rangle,$$
with equality at $y = y_+$. Adding the term $1/2\,\mu\|y - x\|^2$ to both sides,
$$\tilde{\psi}(y) \geq \varphi(y_+) + \langle \hat{s}, y - y_+\rangle + \tfrac{1}{2}\mu\|y - x\|^2,$$
again with equality at $y = y_+$. Then it suffices to observe from (3.1.2) that the function of the right-hand side is minimized over C at $y = y_+$: indeed its gradient at y_+ is
$$\hat{s} + \mu(y_+ - x) = -\hat{p} \in -N_C(y_+)$$
and the geometric minimality condition VII.1.1.1(iii) is satisfied. □

The affine function $\bar{f}^a$ appearing in (3.1.4) is the *aggregate* linearization of f, already seen in §XIV.2.4. It minorizes φ (Lemma 3.1.1) and can also be written

$$\mathbb{R}^n \ni y \mapsto \bar{f}^a(y) = \varphi(y_+) + \langle \hat{s}, y - y_+ \rangle. \tag{3.1.5}$$

Proposition 3.1.2 tells us in particular that the next iterate y_+ would not be changed if, instead of φ, the model were any convex function ψ sandwiched between $\bar{f}^a$ and φ.

Note that the aggregate linearization concerns exclusively f, the function that is modeled by φ. On the other hand, the indicator part of the (unconstrained) objective function $f + I_C$ is treated directly, without any modelling; so aggregation is irrelevant for it.

Remark 3.1.3 We could take for example $\psi = \bar{f}^a$ in Proposition 3.1.2. In relation to Fig. 2.5.1, this suggests another construction, illustrated by Fig. 3.1.1. For given x and $\mu > 0$, assume $\hat{p} = 0$ (so as to place ourselves in the framework of §2); with y of the form $x - t\hat{s}$, draw the parabola of equation

$$r = r(y) = r_0 - \tfrac{1}{2}\mu \|y - x\|^2,$$

where r_0 is lifted as high as possible to support the graph of the model $\check{f}_k = \varphi$; there is contact $r(y) = \varphi(y)$ at the unique point $(y_+, \varphi(y_+))$. Because y_+ minimizes the quadratic function $\bar{f}^a + 1/2\,\mu \|\cdot - x\|^2$, the dashed line gr $\bar{f}^a$ of Fig. 3.1.1 is tangent to our parabola; the value $\bar{f}^a(x)$ unveils the key parameter $\hat{e}$ (the would-be ε of Fig. 2.5.1), from which the whole picture can be reconstructed.

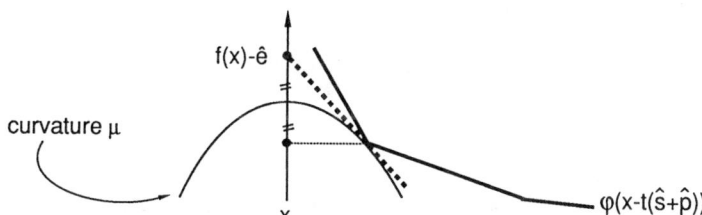

Fig. 3.1.1. Supporting a convex epigraph with a parabola

The above observation, together with the property of parabolas given in Fig. II.3.2.2, illustrates once more the point made in Remark 2.5.2 concerning nominal decreases. □

Keeping in mind the strategy used in the previous chapters, the above two results can be used as follows. When (3.1.1) is solved, a new affine piece $(y_+, f(y_+), s_+)$ will be introduced to give the new model

$$\varphi_+(y) = \max\{\varphi(y), f(y_+) + \langle s_+, y - y_+ \rangle\}.$$

Before doing so, we may wish to "simplify" φ to some other ψ, in order to make room in the computer and/or to simplify the next quadratic program. For example, we may wish to discard some old affine pieces; this results in $\psi \leqslant \varphi$. After such an operation, we must incorporate the affine piece $\bar{f}^{\mathrm{a}}$ into the definition of the simpler ψ, so that we will have $\bar{f}^{\mathrm{a}} \leqslant \psi \leqslant \varphi$. This *aggregation* operation has been seen already in several previous chapters and we can infer that it will not impair convergence. Besides, the simpler function ψ is piecewise affine if φ is such, and the operation can be repeated at any later iteration.

In terms of the objective function f, the aggregate linearization $\bar{f}^{\mathrm{a}}$ is not attached to any $\hat{y}$ such that $\hat{s} \in \partial f(\hat{y})$, so the notation (1.0.1) is no longer correct for the model. For the same reason, characterizing the model in terms of triples $(y, f, s)_i$ is clumsy: we need only *couples* $(s, r)_i \in \mathbb{R}^n \times \mathbb{R}$ characterizing affine functions. In addition to its slope s_i, each such affine function could be characterized by its value at 0 but we prefer to characterize it by its value at the current stability center x; furthermore we choose to call $f(x) - e_i$ this value. Calling ℓ the total number of affine pieces, all the necessary information is then characterized by a *bundle* of couples

$$(s_i, e_i) \in \mathbb{R}^n \times \mathbb{R}^+ \quad \text{for } i = 1, \ldots, \ell$$

and the model is

$$\mathbb{R}^n \ni y \mapsto \varphi(y) = f(x) + \max_{i=1,\ldots,\ell} [-e_i + \langle s_i, y - x \rangle].$$

We recognize notation already introduced in §XIV.1.2. The slopes s_i are either direct subgradients of f computed by the black box (U1) at some sampling points y_i, or vectors of the form $\hat{s}$ obtained after one aggregation or more. As for the linearization errors e_i, they can be computed recursively: when the stability center x is changed to x_+, each e_i must be changed to the difference e_i^+ between $f(x_+)$ and the value at x_+ of the ith linearization. In other words,

$$f(x_+) - e_i^+ = f(x) - e_i + \langle s_i, x_+ - x \rangle,$$

hence

$$e_i^+ = e_i + f(x_+) - f(x) - \langle s_i, x_+ - x \rangle.$$

Finally, we prefer to work with the inverse of the penalty parameter; its interpretation as a stepsize is much more suggestive in a primal context.

In summary, re-introducing the iteration index k, our stabilized problem to be solved at the kth iteration is

$$\left| \begin{array}{ll} \min \left[r + \frac{1}{2t_k} \|y - x_k\|^2 \right] & (y, r) \in C \times \mathbb{R}, \\ r \geqslant f(x_k) - e_i + \langle s_i, y - x_k \rangle & \text{for } i = 1, \ldots, \ell, \end{array} \right. \tag{3.1.6}$$

a quadratic program if C is a closed convex polyhedron; the extra variable r stands for φ-values. It can easily be seen that (3.1.6) has a unique solution (y_{k+1}, r_{k+1}) with $\varphi(y_{k+1}) = r_{k+1}$.

The precise algorithm can now be stated, with notations combining those of Chap. XIV and §2.2.

Algorithm 3.1.4 (Primal Bundle Method With Penalty) The initial point x_1 is given, together with a stopping tolerance $\underline{\delta} \geqslant 0$ and a maximal bundle size $\bar{\ell}$. Choose a descent-coefficient $m \in \,]0,1[$, say $m = 0.1$. Initialize the descent-set $K = \emptyset$, the iteration counter $k = 1$ and the bundle size $\ell = 1$. Compute $f(x_1)$ and $s_1 = s(x_1)$. Set $e_1 = 0$, corresponding to the initial bundle (s_1, e_1), and the initial model

$$y \mapsto \varphi_1(y) := f(x_1) + \langle s_1, y - x_1 \rangle.$$

STEP 1 (main computation and stopping test). Choose a "stepsize" $t_k > 0$ and solve (3.1.6). As stated in Lemma 3.1.1, its unique solution is $y_{k+1} = x_k - t_k(\hat{s}_k + \hat{p}_k)$, with $\hat{s}_k \in \partial \varphi_k(y_{k+1})$ and $\hat{p}_k \in N_C(y_{k+1})$. Set

$$\hat{e}_k := f(x_k) - \varphi_k(y_{k+1}) - t_k \langle \hat{s}_k, \hat{s}_k + \hat{p}_k \rangle,$$
$$\delta_k := f(x_k) - \varphi_k(y_{k+1}) - \tfrac{1}{2} t_k \|\hat{s}_k + \hat{p}_k\|^2.$$

If $\delta_k \leq \underline{\delta}$ stop.

STEP 2 (descent test). Compute $f(y_{k+1})$ and $s(y_{k+1})$; if the descent test

$$f(y_{k+1}) \leq f(x_k) - m \delta_k \tag{3.1.7}$$

is not satisfied declare "null-step" and go to Step 4.

STEP 3 (descent-step). Set $x_{k+1} = y_{k+1}$. Append k to the set K; for $i = 1, \ldots, \ell$, change e_i to

$$e_i + f(x_{k+1}) - f(x_k) - \langle s_i, x_{k+1} - x_k \rangle;$$

change also $\hat{e}_k$ similarly.

STEP 4 (managing the bundle size). If $\ell = \bar{\ell}$ then: delete at least 2 elements from the bundle and insert the element $(\hat{s}_k, \hat{e}_k)$.

Call again $(s_i, e_i)_{i=1,\ldots,\ell}$ the new bundle thus obtained (note: $\ell < \bar{\ell}$).

STEP 5 (loop). Append $(s_{\ell+1}, e_{\ell+1})$ to the bundle, where $e_{\ell+1} = 0$ in case of descent-step and, in case of null-step:

$$e_{\ell+1} = f(x_k) - [f(y_{k+1}) + \langle s_{\ell+1}, x_k - y_{k+1} \rangle].$$

Replace ℓ by $\ell + 1$ and define the model

$$y \mapsto \varphi_{k+1}(y) = f(x_{k+1}) + \max_{i=1,\ldots,\ell} [-e_i + \langle s_i, y - x_{k+1} \rangle].$$

Replace k by $k+1$ and loop to Step 1. □

This algorithm is directly comparable to Algorithm XIV.3.4.2; observe, however, how much simpler it is – hardly more complex than its schematic version 2.2.2. Let us add some notes playing the role of those in XIV.3.4.3.

Notes 3.1.5

(i) The initialization t_1 can use Remark II.3.4.2 if we have an estimate Δ of the total decrease $f(x_1) - \bar{f}$. Then the initial t_1 can be obtained from the formula $t_1 \|s_1\|^2 = 2\Delta$. Not surprisingly, we then have $\delta_1 = \Delta$.

(ii) Lemma 3.1.1 essentially says that $\hat{s}_k \in \partial_{\hat{e}_k} f(x_k)$ and our convergence analysis will establish that $\hat{s}_k + \hat{p}_k \in \partial_{\varepsilon_k}(f + I_C)(x_k)$, with $\varepsilon_k \geq 0$ given in Lemma 3.2.1 below. Because the objective function is $f + I_C$, the whole issue will be to show that $\varepsilon_k \to 0$ and $\hat{s}_k + \hat{p}_k \to 0$. Accordingly, it may be judged convenient to split the stopping tolerance δ into two terms: stop when

$$\varepsilon_k \leq \varepsilon \quad \text{and} \quad \|\hat{s}_k + \hat{p}_k\| \leq \delta,$$

for given tolerances ε and δ, respectively homogeneous to objective-values and norms of subgradients.

(iii) See Remark 2.5.2 for the nominal decrease, which could also be set to $f(x_k) - \varphi_k(y_{k+1})$; the descent parameter m can absorb the difference between the two possible δ_k's. Note also that we subtract the optimal value in the stabilized problem from $f(x_k)$, not from $\varphi_k(x_k)$. This is because the aggregation may make $\varphi_k(x_k) < f(x_k)$; a detail which will be important when we establish convergence.

(iv) This algorithm is "semi-abstract", to the extent that the important choice of t_k is left unspecified. No wonder: we have not made the least progress towards intelligent choices of the stability parameter, be it μ, κ, ℓ, ε or whatever (see the end of §2.5). □

This algorithm requires access individually to $\hat{s}_k$ (the aggregate slope) and $\hat{p}_k$ (to compute y_{k+1} knowing $\hat{s}_k$), which are of course given from the minimality conditions in the stabilized problem. Thus, instead of (3.1.1) = (3.1.6), one might prefer to solve a dual problem.

(a) Abstractly: formulate (3.1.1) as

$$\min_{y \in \mathbb{R}^n} \left[\varphi_k(y) + \tfrac{1}{2t_k} \|y - x_k\|^2 + I_C(y) \right].$$

The conjugates of the three terms making up the above objective function are respectively φ_k^*, $1/2\, t_k \|\cdot\|^2 + \langle \cdot, x_k\rangle$ and σ_C. The dual problem (for example Fenchel's dual of §XII.5.4) is then the minimization with respect to $(s, p) \in \mathbb{R}^n \times \mathbb{R}^n$ of

$$\varphi_k^*(s) + \tfrac{1}{2}t_k \|-s - p\|^2 - \langle s + p, x_k\rangle + \sigma_C(p)$$

or equivalently

$$\varphi_k^*(s) + \tfrac{1}{2}t_k \|s + p - \mu_k x_k\|^2 + \sigma_C(p).$$

Remark 3.1.6 Note in passing the rather significant role of $\mu_k = 1/t_k$, which appears as more than a coefficient needed for numerical efficiency: multiplication by μ_k actually sends a vector of $\mathbb{R}^n$ to a vector of the *dual* space $(\mathbb{R}^n)^*$.

Indeed, it is good practice to view μ_k as the operator $\mu_k I$, which could be more generally a symmetric operator $Q: \mathbb{R}^n \to (\mathbb{R}^n)^*$. The same remark is valid for the ordinary gradient method of Chap. II: writing $y = x - ts(x)$ should be understood as $y = x - Q^{-1}s(x)$. This mental operation is automatic in the Newton method, in which $Q = \nabla^2 f(x)$. □

(b) More concretely: consider (3.1.6), where C is a closed convex polyhedron as described in (3.0.2). The corresponding conjugate functions φ_k^* and σ_C could be computed to specify the dual problem of (a). More simply, we can also follow the example of §XII.3.4: formulate (3.0.2), (3.1.6) as

$$\left|\begin{array}{l} \min\left[r + \frac{1}{2t_k}\|y - x_k\|^2\right] \\ r \geqslant f(x_k) - e_i + \langle s_i, y - x_k \rangle \quad \text{for } i = 1, \ldots, \ell, \\ \langle a_j, y - x_k \rangle \leqslant b_j - \langle a_j, x_k \rangle \quad \text{for } j = 1, \ldots, m, \end{array}\right.$$

which uses the variable $y - x_k$. Taking ℓ multipliers α_i and m multipliers γ_j, we set

$$s(\alpha) := \sum_{i=1}^{\ell} \alpha_i s_i, \quad a(\gamma) := \sum_{j=1}^{m} \gamma_j a_j,$$
$$e(\alpha) := \sum_{i=1}^{\ell} \alpha_i e_i, \quad b(\gamma) := \sum_{j=1}^{m} \gamma_j [b_j - \langle a_j, x_k \rangle]$$

and we form the Lagrange function

$$L(y, \alpha, \gamma) = \frac{1}{2t_k}\|y - x_k\|^2 + \langle s(\alpha) + a(\gamma), y - x_k \rangle - e(\alpha) - b(\gamma) + $$
$$+ \left(1 - \sum_{i=1}^{\ell} \alpha_i\right) + \sum_{i=1}^{\ell} \alpha_i f(x_k).$$

Its minimization with respect to (y, r) gives the dual problem

$$\left|\begin{array}{l} \min\left[\frac{1}{2}t_k \|s(\alpha) + a(\gamma)\|^2 + e(\alpha) + b(\gamma)\right], \\ \sum_{i=1}^{\ell} \alpha_i = 1, \quad \alpha_i \geqslant 0 \text{ for } i = 1, \ldots, \ell, \\ \gamma_j \geqslant 0 \quad \text{for } j = 1, \ldots, m, \end{array}\right.$$

from which we obtain $\hat{s}_k = \sum_{i=1}^{\ell} \alpha_i s_i$ and $\hat{p}_k = \sum_{j=1}^{m} \gamma_j a_j$, as well as $\hat{e}_k = \sum_{i=1}^{\ell} \alpha_i e_i$. We obtain also a term not directly used by the algorithm:

$$\hat{b}_k := \sum_{j=1}^{m} \gamma_j [b_j - \langle a_j, x_k \rangle];$$

from the transversality conditions,

$$\hat{b}_k = \langle \hat{p}_k, y_{k+1} - x_k \rangle = -t_k \langle \hat{p}_k, \hat{s}_k + \hat{p}_k \rangle \geqslant 0.$$

Knowing that the essential operation is conjugation of the polyhedral function $\varphi_k + I_C$, compare the above dual problem with Example X.3.4.3. Our "bundle" could be viewed as made up of two parts: $(s_i, e_i)_{i=1,\ldots,\ell}$, as well as $(a_j, b_j - \langle a_j, x_k \rangle)_{j=1,\ldots,m}$; the multipliers of the second part vary in the nonnegative orthant, instead of the unit simplex.

3.2 Convergence

Naturally, convergence of Algorithm 3.1.4 cannot hold for arbitrary choice of the stepsizes t_k. When studying convergence, two approaches are therefore possible: either give specific rules to define t_k – and establish convergence of the corresponding implementation – or give abstract conditions on $\{t_k\}$ for which convergence can be established. We choose the second solution here, even though our conditions on $\{t_k\}$ lack implementability. In fact, our aim is to demonstrate a technical framework, rather than establish a particular result.

The two observations below give a feeling for "reasonable" abstract conditions. Indeed:

(i) A small t_k, or a large μ_k, is dangerous: it might over-emphasize the role of the stabilization, resulting in unduly small moves from x_k to y_{k+1}. This is suggested by analogy with Chap. II: the descent-test in Step 2 takes care of large stepsizes but so far, nothing prevents them from being too small.

(ii) It is dangerous to make a null-step with large t_k. Here, we are warned by the convergence theory of Chap. IX: what we need in this situation is $\hat{s}_k + \hat{p}_k \to 0$, and this depends crucially on the boundedness of the successive subgradients s_k, which in turn comes from the boundedness of the successive iterates y_k. A large t_k gives a y_{k+1} far away.

First of all, we fix the point made in Note 3.1.5(ii).

Lemma 3.2.1 *At each iteration of Algorithm 3.1.4, $\varphi_k \leqslant f$ and there holds*

$$f(y) \geqslant f(x_k) + \langle \hat{s}_k + \hat{p}_k, y - x_k \rangle - \varepsilon_k \quad \text{for all } y \in C,$$

with

$$\varepsilon_k := \hat{e}_k - t_k \langle \hat{p}_k, \hat{s}_k + \hat{p}_k \rangle = f(x_k) - \varphi_k(y_{k+1}) - t_k \|\hat{s}_k + \hat{p}_k\|^2.$$

PROOF. At the first iteration, φ_1 is the (affine) cutting-plane function of (1.0.1): $\varphi_1 \leqslant f$. Assume recursively that $\varphi_k \leqslant f$. Keeping Lemma 3.1.2 in mind, the compression of the bundle at Step 4 replaces φ_k by some other convex function $\psi \leqslant f$. Then, when appending at Step 5 the new piece

$$f(x_k) - e_{\ell+1} + \langle s_{\ell+1}, y - x_k \rangle =: \bar{f}_{\ell+1}(y) \leqslant f(y) \quad \text{for all } y \in \mathbb{R}^n,$$

the model becomes $\varphi_{k+1} = \max\{\psi, \bar{f}_{\ell+1}\} \leqslant f$: the required minoration does hold for all k.

Then add the inequalities

$$\begin{aligned} f(y) &\geqslant \varphi_k(y) \geqslant f(x_k) + \langle \hat{s}_k, y - x_k \rangle - \hat{e}_k \quad &\text{[from Lemma 3.1.1]} \\ 0 &\geqslant \langle \hat{p}_k, y - y_{k+1} \rangle = \langle \hat{p}_k, y - x_k \rangle + t_k \langle \hat{p}_k, \hat{p}_k + \hat{s}_k \rangle \end{aligned}$$

to obtain the stated value of ε_k. □

In terms of the nominal decrease, we can also write (see Fig. 2.5.1 again)

$$\varepsilon_k = \delta_k - \tfrac{1}{2} t_k \|\hat{s}_k + \hat{p}_k\|^2. \tag{3.2.1}$$

If both $\{\varepsilon_k\}$ and $\{\hat{s}_k + \hat{p}_k\}$ tend to zero, any accumulation point of $\{x_k\}$ will be optimal (Proposition XI.4.1.1: the graph of $(\varepsilon, x) \mapsto \partial_\varepsilon(f + I_C)(x)$ is closed). However, convergence can be established even if $\{x_k\}$ is unbounded, with a more subtle argument, based on inequality (3.2.3) below.

As before, we distinguish two cases: if there are infinitely many descent steps, the objective function decreases "sufficiently", thanks to the successive descent tests (3.1.7); and if the sequence $\{x_k\}$ stops, it is the bundling mechanism that does the job.

Theorem 3.2.2 *Let Algorithm 3.1.4 be applied to the minimization problem (3.0.1), with the stopping tolerance $\underline{\delta} = 0$. Assume that K is an infinite set.*

(i) If
$$\sum_{k \in K} t_k = +\infty, \qquad (3.2.2)$$

then $\{x_k\}$ is a minimizing sequence.

(ii) If, in addition, $\{t_k\}$ has an upper bound on K, and if (3.0.1) has a nonempty set of solutions, then the whole sequence $\{x_k\}$ converges to such a solution.

PROOF. [*Preliminaries*] Let $k \in K$, so $x_{k+1} = y_{k+1}$. With y arbitrary in C, write

$$\|y - x_{k+1}\|^2 = \|y - x_k\|^2 + 2\langle y - x_k, x_k - x_{k+1}\rangle + \|x_k - x_{k+1}\|^2.$$

In view of (3.1.2), the cross-product can be bounded above using Lemma 3.2.1 and, with (3.2.1), we obtain

$$\|y - x_{k+1}\|^2 \leq \|y - x_k\|^2 + 2t_k[f(y) - f(x_k) + \delta_k]. \qquad (3.2.3)$$

Now $\{f(x_k)\}$ has a limit $f_* \in [-\infty, +\infty[$. If $f_* = -\infty$, the proof is finished. Otherwise the descent test (3.1.7) implies that

$$\sum_{k \in K} \delta_k \leq \frac{1}{m} \sum_{k=1}^{\infty} [f(x_k) - f(x_{k+1})] = \frac{f(x_1) - f_*}{m} < +\infty, \qquad (3.2.4)$$

where we have used the fact that $f(x_{k+1}) - f(x_k) = 0$ if $k \notin K$ (the same argument was used in Lemma 2.1.4).

[*(i)*] Assume for contradiction that there are $y \in C$ and $\eta > 0$ such that

$$f(y) \leq f(x_k) - \eta \quad \text{for all } k \in K,$$

so that (3.2.3) gives with this y:

$$\|y - x_{k+1}\|^2 \leq \|y - x_k\|^2 + 2t_k(\delta_k - \eta).$$

Because of (3.2.4), $\lim_{k \in K} \delta_k = 0$; there is a k_0 such that

$$\|y - x_{k+1}\|^2 \leq \|y - x_k\|^2 - t_k \eta \quad \text{for } k_0 \leq k \in K.$$

Sum these inequalities over $k \in K$. Remembering again that $x_{k+1} = x_k$ if $k \notin K$, the terms $\|y - x_k\|^2$ cancel out and we obtain the contradiction

$$0 \leq \|y - x_{k_0}\|^2 - \eta \sum_{k_0 \leq k \in K} t_k = -\infty.$$

[*(ii)*] Now let $\bar{x}$ minimize f over C and use $y = \bar{x}$ in (3.2.3):

$$\|\bar{x} - x_{k+1}\|^2 \leq \|\bar{x} - x_k\|^2 + 2t_k \delta_k \quad \text{for all } k \in K. \qquad (3.2.5)$$

If $\{t_k\}$ is bounded on K, (3.2.4) implies that the series $\sum_{k \in K} t_k \delta_k$ is convergent; once again, sum the inequalities (3.2.5) over $k \in K$ to see that $\{x_k\}$ is bounded. Extract some cluster point; in view of (i), this cluster point minimizes f on C and can legitimately be called $\bar{x}$.

Given an arbitrary $\eta > 0$, take some k_1 large enough in K so that

$$\|\bar{x} - x_{k_1}\|^2 \leq \eta/2 \quad \text{and} \quad \sum_{k_1 \leq k \in K} t_k \delta_k \leq \eta/4.$$

Perform once more on (3.2.5) the summation process, with k running in K from k_1 to an arbitrary $k_2 \geq k_1$ ($k_2 \in K$). We conclude:

$$\|\bar{x} - x_{k_2+1}\|^2 \leq \|\bar{x} - x_{k_1}\|^2 + 2 \sum_{k \in K \cap \{k_1, \ldots, k_2\}} t_k \delta_k \leq \eta. \qquad \square$$

It is interesting to note the similarity between (3.2.2) and the condition (XII.4.1.3), establishing convergence of the basic subgradient algorithm (although the latter used normalized directions). This rule is motivated by the case of an affine objective function: if $f(x) = r + \langle s, x \rangle$, we certainly have

$$x_{k+1} = x_k - t_k s = x_1 - \left(\sum_{i=1}^{k} t_i \right) s.$$

To obtain $f(x_k) \to -\infty$, we definitely need the "cumulated stepsizes" to be unbounded.

Apart from the conditions on the stepsize, the only arguments needed for Theorem 3.2.2 are: the definition (3.1.2) of the next iterate, the inequalities stated in Lemma 3.2.1, and the descent test (3.1.7); altogether, the bundling mechanism is by no means involved. Thus, consider a variant of the algorithm in which, when a descent has been obtained, Step 4 flushes the bundle entirely: ℓ is set to 1 and the aggregate piece $(\hat{s}_k, \hat{e}_k)$ is simply left out. Theorem 3.2.2 still applies to such an algorithm. Numerically, the idea is silly, though: it uses the steepest-descent direction whenever possible, and we have insisted again and again that something worse can hardly be invented.

To establish convergence of Algorithm 3.1.4, it remains to fix the case of infinitely many consecutive null-steps: what happens if the stability center stops at some x_{k_0}? The basic argument is then as in other bundle methods. The bundling mechanism forces the nominal decreases $\{\delta_k\}$ to tend to 0 at a certain speed. In view of (3.2.1), $\varepsilon_k \to 0$ and, providing that $\{t_k\}$ does not tend to 0 too fast, $\hat{s}_k + \hat{p}_k \to 0$; then Lemma 3.2.1 proves optimality of x_{k_0}. The key is to transmit the information contained in the aggregate linearization $\bar{f}^a$, which therefore needs to be indexed by k.

Lemma 3.2.3 *Denote by*

$$\mathbb{R}^n \ni y \mapsto \bar{f}_k^a(y) := \varphi_k(y_{k+1}) + \langle \hat{s}_k, y - y_{k+1} \rangle$$

the aggregate linearization obtained at the kth iteration of Algorithm 3.1.4. For all $y \in C$, there holds

$$\bar{f}_k^a(y) + \tfrac{1}{2t_k} \|y - x_k\|^2 \geq f(x_k) - \delta_k + \tfrac{1}{2t_k} \|y - y_{k+1}\|^2. \qquad (3.2.6)$$

PROOF. Consider the quadratic function $\tilde{\psi} := \bar{f}_k^a + 1/2 t_k \| \cdot - x_k \|^2$; it is strongly convex of modulus $1/t_k$ (in the sense of Proposition IV.1.1.2), and has at y_{k+1} the gradient $\hat{s}_k + (y_{k+1} - x_k)/t_k = -\hat{p}_k$. From Theorem VI.6.1.2,

$$\tilde{\psi}(y) \geq \tilde{\psi}(y_{k+1}) - \langle \hat{p}_k, y - y_{k+1} \rangle + \frac{1}{2t_k}\|y - y_{k+1}\|^2 \quad \text{for all } y \in \mathbb{R}^n.$$

Because $\hat{p}_k \in N_C(y_{k+1})$, the second term can be dropped whenever $y \in C$. On the other hand, using various definitions, we have

$$\tilde{\psi}(y_{k+1}) = \varphi_k(y_{k+1}) + \frac{1}{2t_k}\|y_{k+1} - x_k\|^2 = f(x_k) - \delta_k ;$$

the result follows. □

Theorem 3.2.4 *Consider Algorithm 3.1.4 with the stopping tolerance $\underline{\delta} = 0$. Assume that K is finite: for some k_0, each iteration $k \geq k_0$ produces a null-step. If*

$$t_k \leq t_{k-1} \quad \text{for all } k > k_0 \tag{3.2.7}$$

$$\sum_{k > k_0} \frac{t_k^2}{t_{k-1}} = +\infty, \tag{3.2.8}$$

then x_{k_0} minimizes f on C.

PROOF. In all our development below, $k > k_0$ and we use the notation of Lemma 3.2.3.
[*Step 1*] Write the definition of the kth nominal decrease:

$$\begin{aligned}
f(x_{k_0}) - \delta_k &= \varphi_k(y_{k+1}) + \frac{1}{2t_k}\|y_{k+1} - x_{k_0}\|^2 \\
&\geq \varphi_k(y_{k+1}) + \frac{1}{2t_{k-1}}\|y_{k+1} - x_{k_0}\|^2 && \text{[because of (3.2.7)]} \\
&\geq \bar{f}_{k-1}^a(y_{k+1}) + \frac{1}{2t_{k-1}}\|y_{k+1} - x_{k_0}\|^2 && \text{[Step 4 implies } \varphi_k \geq \bar{f}_{k-1}^a] \\
&\geq f(x_{k_0}) - \delta_{k-1} + \frac{1}{2t_{k-1}}\|y_{k+1} - y_k\|^2 . && \text{[from (3.2.6)]}
\end{aligned}$$

Thus we have proved

$$\delta_k + \frac{1}{2t_{k-1}}\|y_{k+1} - y_k\|^2 \leq \delta_{k-1} \quad \text{for all } k > k_0 . \tag{3.2.9}$$

[*Step 2*] In particular, $\{\delta_k\}$ is decreasing. Also, set $y = x_k = x_{k_0}$ in (3.2.6): knowing that $\varphi_k \leq f$ (Lemma 3.2.1), we have

$$f(x_{k_0}) \geq \varphi_k(x_{k_0}) \geq \bar{f}_k^a(x_{k_0}) \geq f(x_{k_0}) - \delta_k + \frac{1}{2t_k}\|x_{k_0} - y_{k+1}\|^2 .$$

Hence

$$\|y_{k+1} - x_{k_0}\|^2 \leq 2t_k \delta_k \leq 2t_{k_0} \delta_{k_0} =: R^2 ,$$

the sequence $\{y_k\}$ is bounded; L will be a Lipschitz constant for f and each φ_k on $B(x_{k_0}, R)$ and will be used to bound from below the decrease from δ_k to δ_{k+1}. Indeed, Step 5 of the algorithm forces $f(y_k) = \varphi_k(y_k)$, therefore

$$f(y_{k+1}) - \varphi_k(y_{k+1}) = f(y_{k+1}) - f(y_k) + \varphi_k(y_k) - \varphi_k(y_{k+1}) \leqslant$$
$$\leqslant 2L \|y_{k+1} - y_k\|. \quad (3.2.10)$$

[*Step 3*] On the other hand, the descent test (3.1.7) is not satisfied:

$$f(y_{k+1}) > f(x_{k_0}) - m\delta_k;$$

subtract the inequality

$$\varphi_k(y_{k+1}) = f(x_{k_0}) - \delta_k - \tfrac{1}{2t_k}\|y_{k+1} - x_{k_0}\|^2 \leqslant f(x_{k_0}) - \delta_k$$

and combine with (3.2.10):

$$(1-m)\delta_k < f(y_{k+1}) - \varphi_k(y_{k+1}) \leqslant 2L\|y_{k+1} - y_k\|.$$

Insert this in (3.2.9):

$$C \frac{\delta_k^2}{t_{k-1}} < \delta_{k-1} - \delta_k \quad \text{for all } k > k_0, \quad (3.2.11)$$

where we have set $C := 1/2\,(1-m)^2/(2L)^2$.

[*Epilogue*] Summing the inequalities (3.2.11):

$$C \sum_{k > k_0} \frac{\delta_k^2}{t_{k-1}} < \delta_{k_0} < +\infty.$$

In view of (3.2.7), (3.2.1), this implies $\varepsilon_k \to 0$, and also

$$\sum_{k > k_0} \frac{t_k^2}{t_{k-1}} \|\hat{s}_k + \hat{p}_k\|^4 < +\infty.$$

It remains to use (3.2.8): $\liminf \|\hat{s}_k + \hat{p}_k\|^4 = 0$. Altogether, Lemma 3.2.1 tells us that $0 \in \partial(f + I_C)(x_{k_0})$. □

Remark 3.2.5 In this proof, as well as that of Theorem 3.2.2, the essential argument is that $\delta_k \to 0$. The algorithm does terminate if the stopping tolerance $\underline{\delta}$ is positive; upon termination, Lemma 3.2.1 then gives an approximate minimality condition, depending on the magnitude of t_k. Thus, a proof can also be given in the spirit of those in the previous chapters, showing how the actual implementation behaves in the computer.

Note also that the proof gives an indication of the speed at which $\{\delta_k\}$ tends to 0: compare (3.2.11) with Lemma IX.2.2.1. Numerically speaking, it is a nice property that the two components of the "compound" convergence parameter

$$\delta_k = \varepsilon_k + \tfrac{1}{2}t_k\|\hat{s}_k + \hat{p}_k\|^2$$

are simultaneously driven to 0. See again Fig. XIV.4.1.3 and §XIV.4.4(b).

We mentioned in Remark IX.2.1.3 that a tolerance m'' could help an approximate resolution of the quadratic program for dual bundle methods. In an actual implementation of the present algorithm, a convenient stopping criterion should also be specified for the quadratic solver. We omit such details here, admitting that exact arithmetic is used for an exact resolution of (3.1.6). □

Both conditions (3.2.2), (3.2.8) rule out small stepsizes and connote (XII.4.1.3), used to prove convergence of the basic subgradient algorithm; note for example that (3.2.8) holds if $t_k = 1/k$. Naturally, the comments of Remark XII.4.1.5 apply again, concerning the *practical irrelevance* of such conditions. The situation is even worse here: because we do not know in advance whether $k \in K$, the possibility of checking (3.2.2) becomes even more remote. Our convergence results 3.2.2 and 3.2.4 are more interesting for their proofs than for their statements. To guarantee (3.2.2), (3.2.8), a simple strategy is to bound t_k from below by a positive threshold $\underline{t} > 0$. In view of (3.2.7), it is even safer to impose a fixed stepsize $t_k \equiv t > 0$; then we get Algorithm 2.2.2, with a fixed penalty parameter $\mu = 1/t$. However, let us say again that no reasonable value for t is a priori available; the corresponding algorithm is hardly efficient in practice.

On the other hand, (3.2.7) is numerically meaningful and says: when a null-step has been made, do not increase the stepsize. This has a practical motivation. Indeed suppose that a null-step has been made, so that the only change for the next iteration is the new piece (s_{k+1}, e_{k+1}) appearing in the bundle. Increasing the stepsize might well reproduce the iterate $y_{k+2} = y_{k+1}$, in which case the next call to the black box (U1) would be redundant.

Remark 3.2.6 We have already mentioned that (3.0.1) is the problem of minimizing the sum $f + I_C$ of two closed convex functions. More generally, let a minimization problem be posed in the form

$$\inf\{f(x) + g(x) : x \in \mathbb{R}^n\},$$

with f and g closed and convex. Several bundling patterns can be considered:

(i) First of all, $f + g$ can be viewed as one single objective function h, to which the general bundle method can be applied. Provided that h is finite everywhere, there is nothing new so far. We recall here that the local Lipschitz continuity of the objective function is technically important, as it guarantees the boundedness of the sequence $\{s_k\}$.

(ii) A second possibility is to apply the bundling mechanism to f alone, keeping g as it is; this is what we have done here, keeping the constraint $y \in C$ explicitly in each stabilized problem. Our convergence results state that this approach is valid if f is finite everywhere, while g is allowed the value $+\infty$.

(iii) When both f and g are finite everywhere, a third possibility is to manage two "decomposed" bundles separately. Indeed suppose that the black box (U1) of Fig. II.1.2.1 is able to answer individual objective-values $f(y)$ and $g(y)$ and subgradient-values, say $s(y)$ and $p(y)$ – rather than the sums $f(y) + g(y)$ and $s(y) + p(y)$. Then f and g can be modelled separately:

$$\check{f}_k(y) := \max_{i=1,\ldots,k}[f(y_i) + \langle s(y_i), y - y_i\rangle] \leq f(y),$$
$$\check{g}_k(y) := \max_{i=1,\ldots,k}[g(y_i) + \langle p(y_i), y - y_i\rangle] \leq g(y).$$

The resulting model is more accurate than the "normal" model

$$\check{h}_k(y) := \max_{i=1,\ldots,k}[(f+g)(y_i) + \langle (s+p)(y_i), y - y_i\rangle] \leq f(y) + g(y),$$

simply because $\check{h}_k \leq \check{f}_k + \check{g}_k$ (this is better seen if two different indices i and j are used in the definition of $\check{f}$ and $\check{g}$: the maximum $\check{h}_k$ of a sum is smaller than the sum $\check{f}_k + \check{g}_k$ of maxima).

Exploiting this idea costs in memory (two additional subgradients are stored at each iteration) but may result in faster convergence. ☐

3.3 Appropriate Stepsize Values

Each iteration of Algorithm 3.1.4 offers a continuum of possibilities, depending on the particular value chosen for the stepsize t_k. As already said on several occasions, numerical efficiency requires a careful choice of such values. We limit ourselves to some general ideas here, since little is known on this question.

In this subsection, we simplify the notation, returning to the unconstrained framework of §2 and dropping the (fixed) iteration index k. The current iterate x is given, as well as the bundle $\{(s_i, e_i)\}$, with $e_i \geq 0$ and $s_i \in \partial_{e_i} f(x)$ for $i = 1, \ldots, \ell$. We denote by $y(t)$ ($t > 0$) the unique solution of the stabilized problem:

$$y(t) := \operatorname*{argmin}_{y} \left[\varphi(y) + \tfrac{1}{2t} \|y - x\|^2 \right]$$

where φ is the model

$$y \mapsto \varphi(y) = f(x) + \max_{i=1,\ldots,\ell} [-e_i + \langle s_i, y - x \rangle].$$

Recall the fundamental formulae:

$$\begin{aligned} y(t) &= x - t\hat{s}(t), \quad \hat{s}(t) \in \partial_{\varepsilon(t)} f(x), \\ \varepsilon(t) &= f(x) - \varphi(y(t)) - \langle \hat{s}(t), x - y(t) \rangle, \end{aligned} \quad (3.3.1)$$

where $\hat{s}(t) = \sum_{i=1}^{\ell} \alpha_i s_i$ can be obtained from the dual problem: $\Delta_\ell \subset \mathbb{R}^\ell$ being the unit simplex,

$$\min_{\alpha \in \Delta_\ell} \left[\tfrac{1}{2} t \| \textstyle\sum_{i=1}^{\ell} \alpha_i s_i \|^2 + \textstyle\sum_{i=1}^{\ell} \alpha_i e_i \right]. \quad (3.3.2)$$

(a) Small Stepsizes. When $t \downarrow 0$, the term $\sum_{i=1}^{\ell} \alpha_i e_i$ in (3.3.2) is pushed down to its minimal value. Actually, this minimal value is even attained for finitely small $t > 0$:

Proposition 3.3.1 *Assume that $e_i = 0$ for some i. Then there exists $\underline{t} > 0$ such that, for $t \in]0, \underline{t}]$, $\hat{s}(t)$ solves*

$$\min \left\{ \tfrac{1}{2} \| \textstyle\sum_{i=1}^{\ell} \alpha_i s_i \|^2 \, : \, \alpha \in \Delta_\ell, \, \textstyle\sum_{i=1}^{\ell} \alpha_i e_i = 0 \right\}. \quad (3.3.3)$$

EXPLANATION. We do not give a formal proof, but we link this result to the direction-finding problem for dual bundle methods. Interpret $1/t =: \mu$ in (3.3.2) as a multiplier of the constraint in the equivalent formulation (see §2.4)

$$\min_{\alpha \in \Delta_\ell} \left\{ \tfrac{1}{2} \| \textstyle\sum_{i=1}^{\ell} \alpha_i s_i \|^2 \, : \, \textstyle\sum_{i=1}^{\ell} \alpha_i e_i \leq \varepsilon \right\}.$$

When $\varepsilon \downarrow 0$ we know from Proposition XIV.2.2.5 that such a multiplier is bounded, say by $\underline{\mu} =: 1/\underline{t}$. ☐

The $\hat{s}(t)$ obtained from (3.3.3) is nothing but the substitute for steepest-descent, considered in Chap. IX. Note also that we have $\hat{s}(t) \in \partial f(x)$: small stepsizes in Algorithm 3.1.4 mimic the basic subgradient algorithm of §XII.4.1. Proposition 3.3.1 tells us that this sort of steepest-descent direction is obtained when t is small. We know already from Chap. II or VIII that such directions are dangerous, and this explains why $t = t_k$ should not be small in Algorithm 3.1.4: not only for theoretical convergence but also for numerical efficiency.

(b) Large Stepsizes. The case $t \to +\infty$ represents the other extreme, and a first situation is one in which φ is bounded from below on $\mathbb{R}^n$. In this case,

- being piecewise affine, φ has a nonempty set of minimizers (§3.4 in Chap. V or VIII);
- when $t \to +\infty$, $y(t)$ has a limit which is the minimizer of φ that is closest to x; this follows from Propositions 2.2.3 and 2.1.3. We will even see in Remark 4.2.6 that this limit is attained for finite t, a property which relies upon the piecewise affine character of φ.

Thus, we say: among the minimizers of the cutting-plane model (if there are any), there is a distinguished one which can be reached with a large stepsize in Algorithm 3.1.4. It can also be reached by the trust-region variant, with $\kappa = \kappa_\infty$ of Proposition 2.1.3.

On the other hand, suppose that φ is not bounded from below; then there is no cutting-plane iterate, and no limit for $y(t)$ when $t \to +\infty$. Nevertheless, minimization of the cutting-plane model can give something meaningful in this case too. In the next result, we use again the notation Proj for the projection onto the closed convex hull of a set.

Proposition 3.3.2 *For $\hat{s}(t)$ given by (3.3.2), there holds*

$$\hat{s}(t) \to \hat{s}(\infty) = \text{Proj}\, 0/\{s_1, \ldots, s_\ell\} \quad \text{when } t \to +\infty.$$

PROOF. With $t > 0$, write the minimality conditions of (3.3.2) (see for example Theorem XIV.2.2.1):

$$\langle s_i, \hat{s}(t) \rangle + \tfrac{1}{t} e_i \geq \|\hat{s}(t)\|^2 + \tfrac{1}{t} \varepsilon(t) \quad \text{for } i = 1, \ldots, \ell$$

and let $1/t \downarrow 0$. Being convex combinations of the bundle elements, $\hat{s}(t)$ and $e(t)$ are bounded; $e(t)/t \to 0$ and any cluster point $\hat{s}(\infty)$ of $\hat{s}(t)$ satisfies

$$\langle s_i, \hat{s}(\infty) \rangle \geq \|\hat{s}(\infty)\|^2 \quad \text{for } i = 1, \ldots, \ell.$$

Also, as $\hat{s}(\infty)$ is in the convex hull of $\{s_1, \ldots, s_\ell\}$, the only possibilty for $\hat{s}(\infty)$ is to be the stated projection (see for example Proposition VIII.1.3.4). □

Thus, assume $\hat{s}(\infty) \neq 0$ in the above result. When $t \to +\infty$, $y(t)$ is unbounded but $[y(t) - x]/t = \hat{s}(t)$ converges to a nonzero *direction* $\hat{s}(\infty)$, the solution of

$$\min_{\alpha \in \Delta_\ell} \tfrac{1}{2} \left\| \sum_{i=1}^{\ell} \alpha_i s_i \right\|^2. \tag{3.3.4}$$

We recognize the direction-finding problem of a dual bundle method, with ε set to a large value, say $\varepsilon \geqslant \max_i e_i$. The corresponding algorithm was called conjugate subgradient in §XIV.4.3.

To sum up, consider the following hybrid way of computing the next iterate y_+ in Algorithm 3.1.4.

– If the solution $\hat{s}(\infty)$ of (3.3.4) is nonzero, make a line-search along $-\hat{s}(\infty)$.
– If $\hat{s}(\infty) = 0$, take the limit as $t \to +\infty$ of $y(t)$ described by (3.3.1); in this case too, a line-search can be made to cope with a $y(\infty)$ which is very far, and hence has little value.

The essence of this algorithm is to give a meaning to the cutting-plane iteration as much as possible. Needless to say, it should not be recommended.

Remark 3.3.3 Having thus established a connection between conjugate subgradients and cutting planes, another observation can be made. Suppose that f is quadratic; we saw in §XIV.4.3 that, taking the directions of conjugate subgradients, and making exact line-searches along each successive such direction, we obtained the ordinary conjugate-gradient algorithm of §II.2.4. Keeping in mind Remark 1.1.1, we are bound to notice mysterious relationships between piecewise affine and quadratic models of convex functions. □

(c) On-Line Control of the Stepsize. Our development (a) – (b) above clearly shows the difficulty of choosing the stepsize in Algorithm 3.1.4: small [resp. large] values result in some form of steepest descent [resp. cutting plane]; both are disastrous in practice. Once again, this difficulty is not new, it was with us all the way through §XIV.4. To guide our choice, an on-line strategy can be suggested, based on the trust-region principles explained in §1.3, and combining the present primal motivations with the dual viewpoint of Chap. XIV. For this, it suffices to follow the general scheme of Fig. XIV.3.2.1, say, but with the direction recomputed each time the stepsize is changed.

Thus, the idea is to design a test (0), (R), (L) which, upon observation of the actual objective function at the solution $y(t)$ of (3.3.1), (3.3.2), decides:

(0_d) This solution is convenient for a descent-step,
(0_n) or this solution is convenient for a null-step.
(R) This solution corresponds to a t too large.
(L) This solution corresponds to a t too small.

No matter how this test is designed, it will result in the following pattern:

Algorithm 3.3.4 (Curved-Search Pattern, Nonsmooth Case) The data are the initial x, the model φ, and an initial $t > 0$. Set $t_L = t_R = 0$.

STEP 1. Solve (3.3.1), (3.3.2) and compute $f(y(t))$ and $s(y(t)) \in \partial f(y(t))$. Apply the test (0), (R), (L).

STEP 2 (Dispatching). In case (0) stop the line-search, with either a null- or a descent-step. In case (L) [resp. (R)] set $t_L = t$ [resp. $t_R = t$] and proceed to Step 3.

STEP 3 (Extrapolation). If $t_R > 0$ go to Step 4. Otherwise find a new t by extrapolation beyond t_L and loop to Step 1.

STEP 4 (Interpolation). Find a new t by interpolation in $]t_L, t_R[$ and loop to Step 1. □

Of course, the stabilizing philosophy of the present chapter should be kept. This means that the descent test

$$f(y(t)) \leqslant f(x) - m\left[\varepsilon(t) + \tfrac{1}{2}t\|\hat{s}(t)\|^2\right] \qquad (3.3.5)$$

should be a basis for the test (0), (R), (L):

(i) if it is satisfied, another test should tell us if we are in case (0_d) or (L);
(ii) if it is not satisfied, another test should tell us if we are in case (0_n) or (R).

Note that, in cases (R) and (L), the next t will work with the same model φ. Then, computing some (generalized) derivative $y'(t)$ by a parametric study of (3.3.2), the way is open to interpolation formulae as in §II.3.4, for convenient computation of the next t.

By contrast, the model will change in case (0); then this parametric study is no longer relevant: the question of choosing t_{k+1} remains intact. In addition, just how to design the additional tests (i), (ii) above is not totally clear. We will therefore not elaborate on this approach any longer. It is again impeded by the ever present question in bundle methods: when the current iterate is not suitable, and in particular when (3.3.5) does not hold, should we change the value of the parameter (be it t, μ, κ, ℓ, ε or whatever), and if so, how; or should we enrich the model, or do both things at the same time?

The end of this chapter is rather devoted to an interesting theoretical aspect of bundle methods.

4 Bundle Methods as Regularizations

Consider one iteration of the primal bundle method of §3. Given the current iterate, model and stepsize, we minimize

$$\varphi(y) + \tfrac{1}{2t}\|y - x\|^2$$

with respect to $y \in \mathbb{R}^n$ (assuming an unconstrained situation for simplicity). Now the above optimal value can be viewed as a *function* of $x \in \mathbb{R}^n$, and we recognize in this function the Moreau-Yosida regularization of φ, already seen on several occasions.

In fact, bundle methods and Moreau-Yosida regularization are intimately related and the aim of this section is to explore this relation.

4.1 Basic Properties of the Moreau-Yosida Regularization

In this subsection, we collect and complete some results previously given concerning the Moreau-Yosida regularization, seen from the point of view of convex minimization. In contrast with the previous sections, we consider now a general closed convex function: in what follows,

$$\boxed{f \in \overline{\text{Conv}}\,\mathbb{R}^n \text{ and } M \text{ is a symmetric positive definite operator.}}$$

The function

$$\mathbb{R}^n \ni x \mapsto f_M(x) := \min_{y \in \mathbb{R}^n} \left[f(y) + \tfrac{1}{2}\langle M(y-x), y-x \rangle \right] \tag{4.1.1}$$

is the *Moreau-Yosida regularization* of f, associated with M. Thus we allow quadratic perturbations slightly more general than the mere $\|\cdot\|^2 = \langle \cdot, \cdot \rangle$ of, say, Example XI.3.4.4. It will sometimes be convenient to call

$$\mathbb{R}^n \times \mathbb{R}^n \ni (x,y) \mapsto g(x,y) := f(y) + \tfrac{1}{2}\langle M(y-x), y-x \rangle$$

the function appearing in (4.1.1).

Lemma 4.1.1 *The minimization problem in (4.1.1) has a unique solution, characterized as the unique point $y \in \mathbb{R}^n$ satisfying*

$$M(x-y) \in \partial f(y). \tag{4.1.2}$$

PROOF. For each x, the minimand $g(x, \cdot)$ is a strictly convex function; as such, it has at most one minimum. On the other hand, f is minorized by some affine function; $g(x, \cdot)$ is therefore 1-coercive and, being also closed, it does have one minimum.

Now, because the quadratic term in g is finite everywhere, the calculus rule XI.3.1.1 on the sum of two convex functions applies, and the subdifferential of $g(x, \cdot)$ at y is

$$\partial_y g(x,y) = \partial f(y) + M(y-x)$$

(with the convention $\emptyset + \{s\} = \emptyset$ for all s). Thus (4.1.2) represents the necessary and sufficient minimality condition $0 \in \partial_y g(x,y)$, and has a solution which is the unique minimizer of $g(x, \cdot)$. □

Note here that the convexity of f is important but not its closedness. The function $g(x, \cdot)$ would still be strongly convex even if f were not closed; and all minimizing sequences would have the same unique limit point. Then nothing would be essentially changed if f were replaced by cl f, in (4.1.1) and (4.1.2) as well.

Definition 4.1.2 (Proximal Point) We will extensively use the following system of notation:

$$p_M(x) := \operatorname*{argmin}_y \left[f(y) + \tfrac{1}{2}\langle M(y-x), y-x \rangle \right]$$

is called the *proximal point* of x (associated with f and M); x can be called the proximal center;

$$s_M(x) := M[x - p_M(x)] \in \partial f(p_M(x)) \tag{4.1.3}$$

is the particular subgradient of f at $p_M(x)$ defined via Lemma 4.1.1; we set

$$W := M^{-1},$$

so that there holds

$$p_M(x) = x - W s_M(x). \tag{4.1.4}$$

□

It is important to understand here that $s_M(x)$ of (4.1.3) is a *distinguished* subgradient of f at $p_M(x)$, which comes from the calculation in (4.1.1); it must not be confused with the *arbitrary* subgradient $s(p_M(x))$ which would come from a black box (U1).

Interpretation 4.1.3 The operator M defines the scalar product $\langle\langle x, x'\rangle\rangle := \langle Mx, x'\rangle$, with its associated norm $\|x\| := \sqrt{\langle\langle x, x\rangle\rangle}$. For this norm, $p_M(x)$ is the projection of x onto a certain sublevel-set of f, namely the one at the level $f(p_M(x))$. Indeed take y such that $f(y) \leq f(p_M(x))$. Combining with the subgradient inequality

$$f(p_M(x)) + \langle s_M(x), y - p_M(x)\rangle \leq f(y) \leq f(p_M(x))$$

and using (4.1.3), we obtain

$$\langle\langle x - p_M(x), y - p_M(x)\rangle\rangle \leq 0 \quad \text{for all } y \in S_{f(p_M(x))}(f).$$

Since $p_M(x)$ is obviously in $S_{f(p_M(x))}(f)$, we recognize the characterization of the asserted projection (Theorem III.3.1.1).

Geometrically, Fig. 4.1.1 illustrates the construction with $M = I$. Let a level ℓ be decreasing from the value $f(x)$ and, for each ℓ, take the projection y_ℓ of x onto the sublevel-set $S_\ell(f)$. This y_ℓ is characterized by the property $x - y_\ell \in N_{S_\ell(f)}(y_\ell)$. Depending on the value of ℓ, there are a number of possibilities:

- $\partial f(y_\ell)$ may be empty; $S_\ell(f)$ may also be empty;
- in a "normal" case ($\partial f(y_\ell) \neq \emptyset$ and ℓ strictly larger than the infimum of f), there is some $s_\ell \in \partial f(y_\ell)$ which is collinear with $x - y_\ell$, say $s_\ell = t(x - y_\ell)$ with $t \geq 0$ (see Fig. VI.1.3.2);
- when $t = 1$, we have just picked the correct level and $y_\ell = p_M(x)$;
- a consequence of Lemma 4.1.1 is that there is exactly one ℓ-value for which $t = 1$.

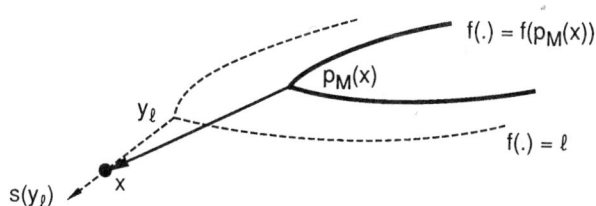

Fig. 4.1.1. Projecting x onto a sublevel-set

Note the similarity between this construction and the level-variant of §2.3. Note also from Lemma 4.1.1 that $p_M(x)$ may be on the boundary of dom f but, even in this case, f has a nonempty subdifferential at $p_M(x)$. □

From now on, we will denote by

$$\lambda_{\max}(Q) := \lambda_1(Q) \geq \cdots \geq \lambda_n(Q) =: \lambda_{\min}(Q) > 0$$

the eigenvalues of a symmetric positive definite operator Q, and we recall that W is the inverse of the given M.

Theorem 4.1.4 *The function f_M of (4.1.1) is finite everywhere, convex and differentiable; its gradient is*

$$\nabla f_M(x) = s_M(x) = M[x - p_M(x)]. \tag{4.1.5}$$

Its conjugate is

$$\mathbb{R}^n \ni s \mapsto f_M^*(s) = f^*(s) + \tfrac{1}{2}\langle s, Ws\rangle. \tag{4.1.6}$$

Furthermore, there holds for all x and x' in $\mathbb{R}^n$:

$$\langle s_M(x) - s_M(x'), W[s_M(x) - s_M(x')]\rangle \leq \langle s_M(x) - s_M(x'), x - x'\rangle \tag{4.1.7}$$

and

$$\|\nabla f_M(x) - \nabla f_M(x')\| \leq \lambda_{\max}(M)\|x - x'\|.$$

PROOF. Take an arbitrary $x_0 \in \text{dom } f$; we clearly have for all $x \in \mathbb{R}^n$

$$f_M(x) \leq f(x_0) + \tfrac{1}{2}\langle M(x_0 - x), x_0 - x\rangle < +\infty;$$

indeed f_M is the inf-convolution of f with the quadratic function $x \mapsto 1/2\langle Mx, x\rangle$, differentiable and finite everywhere. Then f_M is convex; its gradient is given by (XI.3.4.4) and its conjugate is a sum (Corollary X.2.1.3).

Now take the scalar product of $W[s_M(x) - s_M(x')]$ with both sides of

$$s_M(x) - s_M(x') = M(x - x') - M[p_M(x) - p_M(x')];$$

use the monotonicity of ∂f in (4.1.3) to obtain (4.1.7).

Finally, (4.1.7) directly gives

$$\lambda_{\min}(W)\|s_M(x) - s_M(x')\| \leq \|x - x'\|$$

and (4.1.5) completes the proof. □

Note that f_M can also be viewed as a marginal function associated with the minimand g (convex in (x, y) and differentiable in x); ∇f_M can therefore be derived from (XI.3.3.7); see also Corollary VI.4.5.3 and the comments following it. The global Lipschitz property of ∇f_M can also be proved with Theorem X.4.2.1.

Thus, we definitely have a *regularization* of f, let us show that we have an *approximation* as well; in words, the minimization (4.1.1) with a "big" operator M results in $p_M(x)$ close to x.

Proposition 4.1.5 *When $\lambda_{\min}(M) \to +\infty$ the following convergence properties hold:*

$$f_M(x) \to f(x) \quad \text{for all } x \in \mathbb{R}^n, \tag{4.1.8}$$
$$p_M(x) \to x \quad \text{for all } x \in \text{dom } f. \tag{4.1.9}$$

PROOF. There exists an affine function minorizing f: for some $(s_0, r_0) \in \mathbb{R}^n \times \mathbb{R}$,

$$\langle s_0, y \rangle - r_0 \leqslant f(y) \quad \text{for all } y \in \mathbb{R}^n.$$

Use this inequality to minorize the left-hand side in

$$f(p_M(x)) + \tfrac{1}{2}\langle M[p_M(x) - x], p_M(x) - x \rangle = f_M(x) \leqslant f(x), \quad (4.1.10)$$

so that

$$\langle s_0, p_M(x) \rangle - r_0 + \tfrac{1}{2}\lambda_{\min}(M)\|p_M(x) - x\|^2 \leqslant f_M(x).$$

With some algebraic manipulations, we obtain

$$\tfrac{1}{2}\lambda_{\min}(M)\left\| p_M(x) - x + \tfrac{1}{\lambda_{\min}(M)} s_0 \right\|^2 + r(M) \leqslant f_M(x), \quad (4.1.11)$$

where we have set

$$r(M) := \tfrac{-1}{2\lambda_{\min}(M)}\|s_0\|^2 + \langle s_0, x \rangle - r_0.$$

Note: $\{r(M)\}$ is bounded when $\lambda_{\min}(M) \to +\infty$.

Let $x \in \operatorname{dom} f$. Combine (4.1.11) with the inequality in (4.1.10) and divide by $\lambda_{\min}(M)$ to see that $p_M(x) \to x$ when $\lambda_{\min}(M) \to +\infty$; (4.1.9) is proved and the lower semi-continuity of f implies that $\liminf f(p_M(x)) \geqslant f(x)$. Since there obviously holds

$$f(p_M(x)) \leqslant f_M(x) \leqslant f(x), \quad (4.1.12)$$

(4.1.8) is also proved for $x \in \operatorname{dom} f$.

Now let $x \notin \operatorname{dom} f$; we must prove that $f_M(x) \to +\infty$. Assume for contradiction that there are a sequence $\{M_k\}$ with $\lim_{k \to +\infty} \lambda_{\min}(M_k) = +\infty$, and a number R such that

$$f_{M_k}(x) \leqslant R \quad \text{for } k = 1, 2, \ldots$$

Then (4.1.11) shows that $p_{M_k}(x) \to x$ as before. The lower semi-continuity of f and (4.1.12) imply

$$+\infty = f(x) \leqslant \liminf_{k \to +\infty} f(p_{M_k}(x)) \leqslant \liminf_{k \to +\infty} f_{M_k}(x) \leqslant R,$$

which is the required contradiction. □

Now we pass to properties relating the minimization of f with that of f_M. We start with a result useful for numerical purposes: it specifies the decrease of f when its perturbation $g(x, \cdot)$ of (4.1.1) is minimized, instead of f itself.

Lemma 4.1.6 *For all $x \in \mathbb{R}^n$,*

$$f_M(x) + \tfrac{1}{2}\langle s_M(x), W s_M(x) \rangle = f(p_M(x)) + \langle s_M(x), W s_M(x) \rangle \leqslant f(x). \quad (4.1.13)$$

PROOF. Use (4.1.4): the first relation is the definition of f_M; the second is the subgradient inequality

$$f(x) \geqslant f(p_M(x)) + \langle s_M(x), x - p_M(x) \rangle. \quad \Box$$

Theorem 4.1.7 *Minimizing f and f_M are equivalent problems, in the sense that*

$$\inf_{x \in \mathbb{R}^n} f_M(x) = \inf_{x \in \mathbb{R}^n} f(x) \qquad (4.1.14)$$

(an equality in $\mathbb{R} \cup \{-\infty\}$), and that the following statements are equivalent:

(i) x *minimizes* f;
(ii) $p_M(x) = x$;
(iii) $s_M(x) = 0$;
(iv) x *minimizes* f_M;
(v) $f(p_M(x)) = f(x)$;
(vi) $f_M(x) = f(x)$.

PROOF. Observe from (4.1.6) that $-f_M^*(0) = -f^*(0)$; this is just (4.1.14). To prove our chain of equivalences, remember first that M and W are invertible.

When (i) holds, $y = x$ minimizes simultaneously f and the quadratic term in (4.1.1): (ii) holds. On the other hand, (ii) $\Leftrightarrow$ (iii) because of (4.1.4), and (iii) $\Leftrightarrow$ (iv) because $\nabla f_M(x) = s_M(x)$. Now, if (iv) = (ii) holds, (v) is trivial; conversely, (4.1.13) shows that (v) implies (iii) and (vi).

In summary, we have proved

$$(i) \implies (ii) \iff (iii) \iff (iv) \iff (v) \implies (vi).$$

Finally assume that (vi) holds; use (4.1.13) again to see that (iii) = (ii) holds; then

$$0 = s_M(x) \in \partial f(p_M(x)) = \partial f(x). \qquad \square$$

4.2 Minimizing the Moreau-Yosida Regularization

Theorem 4.1.7 gives a number of equivalent formulations for the problem of minimizing f. Among them, (ii) shows that we must find a fixed point of the mapping $x \mapsto p_M(x)$, which resembles a projection: see Remark 4.1.3. As such, p_M is nonexpansive (for the norm associated with $\langle\!\langle \cdot, \cdot \rangle\!\rangle$) and the iteration formula $x_{k+1} = p_M(x_k)$ appears as a reasonable proposal.

This is known as the *proximal point* algorithm. Note that this algorithm can be formulated with a varying operator M, say $x_{k+1} = p_{M_k}(x_k)$. In terms of minimizing f, this still makes sense, even though the Moreau-Yosida interpretation disappears.

Algorithm 4.2.1 (Proximal Point Algorithm) Start with an initial point $x_1 \in \mathbb{R}^n$ and an initial symmetric positive definite operator M_1. Set $k = 1$.

STEP 1. Compute the unique solution $y = x_{k+1}$ of

$$\min_{y \in \mathbb{R}^n} \left[f(y) + \tfrac{1}{2}\langle M_k(y - x_k), y - x_k \rangle \right].$$

STEP 2. If $x_{k+1} = x_k$ stop.

STEP 3. Choose a new symmetric positive definite operator M_{k+1}. Replace k by $k+1$ and loop to Step 1. □

Naturally, this is only a conceptual algorithm because we do not specify how Step 1 can be performed. In addition, it may sound silly to minimize the perturbation $g(x_k, \cdot)$ instead of f itself; said otherwise, the best idea should be to take $M_1 = 0$: then no iteration would be needed. For the moment, our development is purely theoretical and the next subsection will address numerical aspects.

We set $W_k := M_k^{-1}$ and we recall from Lemma 4.1.1 that each iterate is characterized by the relations

$$x_{k+1} = x_k - W_k \hat{s}_k, \quad \hat{s}_k \in \partial f(x_{k+1}); \qquad (4.2.1)$$

here the notation $\hat{s}_k$ replaces the former s_{M_k}. Furthermore we have from (4.1.13)

$$f(x_{k+1}) + \langle W_k \hat{s}_k, \hat{s}_k \rangle \leq f(x_k)$$

and we can write

$$f(x_k) \to f_* \in \mathbb{R} \cup \{-\infty\} \quad \text{when } k \to +\infty.$$

For future use, we introduce the number

$$\delta_k := f(x_k) - f_{M_k}(x_k) = f(x_k) - f(x_{k+1}) - \tfrac{1}{2}\langle \hat{s}_k, W_k \hat{s}_k \rangle \geq 0. \qquad (4.2.2)$$

According to Theorem 4.1.7, $\delta_k = 0$ if and only if x_k minimizes f (or f_{M_k}); a key issue is therefore to establish the property $\delta_k \to 0$.

In view of (4.1.9), taking "big" operators M_k is dangerous: then the iterates x_k will bog down. On the other hand, "small" M_k's seem safe, since the zero operator yields immediately a minimum of f. This explains the convergence condition

$$\sum_{k=1}^{\infty} \lambda_{\min}(W_k) = +\infty. \qquad (4.2.3)$$

It rules out unduly small [resp. large] eigenvalues of W_k [resp. M_k].

Lemma 4.2.2 *Assume that f_* is a finite number. Then $\sum_{k=1}^{\infty} \delta_k < +\infty$ and (4.2.3) implies that 0 is a cluster point of $\{\hat{s}_k\}$.*

PROOF. From (4.2.2), we have for all k

$$\delta_k + \tfrac{1}{2}\lambda_{\min}(W_k)\|\hat{s}_k\|^2 \leq \delta_k + \tfrac{1}{2}\langle \hat{s}_k, W_k \hat{s}_k \rangle \leq f(x_k) - f(x_{k+1})$$

and by summation

$$\sum_{k=1}^{\infty} [\delta_k + \tfrac{1}{2}\lambda_{\min}(W_k)\|\hat{s}_k\|^2] \leq f(x_1) - f_*.$$

If the right-hand side is finite, the two series $\sum \delta_k$ and $\sum \lambda_{\min}(W_k)\|\hat{s}_k\|^2$ are convergent. If the convergence condition (4.2.3) holds, $\|\hat{s}_k\|^2$ cannot stay away from zero. □

Theorem 4.2.3 *In Algorithm 4.2.1, assume that the convergence condition (4.2.3) holds. If $\{x_k\}$ is bounded, all its cluster points are minimizers of f.*

PROOF. Because f is minorized by an affine function, boundedness of $\{x_k\}$ implies $f_* > -\infty$. In view of Lemma 4.2.2, we can take a subsequence for which $\hat{s}_k \to 0$; the corresponding subsequence $\{x_{k+1}\}$ is bounded and we can extract from it a further subsequence tending to some limit x_*. Then (4.2.1) implies $0 \in \partial f(x_*)$ (Proposition VI.6.2.1). Thus f_* is the minimal value of f, and the result follows since f_* is also the objective-value at any other cluster point of $\{x_k\}$. □

Among other things, boundedness of $\{x_k\}$ implies existence of a minimizer of f. If we call t_k the minimal eigenvalue of W_k, the convergence condition (4.2.3) has a familiar flavour: see Sections 3.2 and XII.4.1. However, it is not clear whether this condition suffices to guarantee that f_* is the infimum of f even when $\{x_k\}$ is unbounded. For this, we need to take M_k proportional to a fixed operator.

Theorem 4.2.4 *Assume that $M_k = \mu_k M$, with $\mu_k > 0$ for all k, and M symmetric positive definite.*

(i) *If the convergence condition (4.2.3) holds, i.e. if*

$$\sum_{k=1}^{\infty} \frac{1}{\mu_k} = +\infty,$$

then $\{x_k\}$ is a minimizing sequence.

(ii) *If, in addition, $\{\mu_k\}$ has a positive lower bound, and if f has a nonempty set of minimum points, then the entire sequence $\{x_k\}$ converges to such a point.*

PROOF. Our proof is rather quick because the situation is in fact as in Theorem 3.2.2. The usual transportation trick in the subgradient inequality

$$f(y) \geqslant f(x_{k+1}) + \langle \hat{s}_k, y - x_{k+1} \rangle$$

gives

$$f(y) \geqslant f(x_k) + \langle M_k(x_k - x_{k+1}), y - x_k \rangle - \varepsilon_k \qquad (4.2.4)$$

where, using (4.2.2),

$$\varepsilon_k := f(x_k) - f(x_{k+1}) - \langle \hat{s}_k, W_k \hat{s}_k \rangle = \delta_k - \frac{1}{2\mu_k} \langle \hat{s}_k, M^{-1} \hat{s}_k \rangle.$$

Denoting by $\langle\!\langle u, v \rangle\!\rangle$ the scalar product $\langle Mu, v \rangle$ ($|\!|\!| \cdot |\!|\!|$ will be the associated norm, note that both are independent of k), we write (4.2.4) as

$$f(y) \geqslant f(x_k) + \tfrac{1}{2}\mu_k \langle\!\langle x_k - x_{k+1}, y - x_k \rangle\!\rangle - \delta_k.$$

Use this to majorize the cross-product in the development

$$|\!|\!| y - x_{k+1} |\!|\!|^2 = |\!|\!| y - x_k |\!|\!|^2 + 2\langle\!\langle x_k - x_{k+1}, y - x_k \rangle\!\rangle + |\!|\!| x_k - x_{k+1} |\!|\!|^2$$

and obtain

$$\|y - x_{k+1}\|^2 \leq \|y - x_k\|^2 + 2\frac{f(y) - f(x_k) + \delta_k}{\mu_k} \quad \text{for all } y \in \mathbb{R}^n.$$

Now, if $f_* = -\infty$, the proof is finished. Otherwise, the above inequality establishes with Lemma 4.2.2 inequalities playing the role of (3.2.3) and (3.2.4); then copy the proof of Theorem 3.2.2 with $C = \mathbb{R}^n$ and $t_k = 1/\mu_k$; the descent-set K is now the whole of $\mathbb{N}$, which simplifies some details. □

Another interesting particular case is when the proximal point Algorithm 4.2.1 does terminate at Step 2 with $\hat{s}_k = 0$ and an optimal x_k.

Proposition 4.2.5 *Assume that f has a finite infimum and satisfies the following property:*

$$\exists \eta > 0 \quad \text{such that} \quad \partial f(x) \cap B(0, \eta) \neq \emptyset \implies x \text{ minimizes } f. \quad (4.2.5)$$

If the convergence condition (4.2.3) holds, the stop in Algorithm 4.2.1 occurs for some k.

PROOF. Lemma 4.2.2 guarantees that the event $\|\hat{s}_k\| \leq \eta$ eventually occurs, implying optimality of x_{k+1}. From Theorem 4.1.7, the algorithm will stop at the next iteration. □

Observe the paradoxical character of the above statement: *finite* termination does not match *divergence* of the series $\sum \lambda_{\min}(W_k)$! Remember that the correct translation of (4.2.3) is: the matrices W_k are computed in such a way that

$$\forall R \geq 0, \exists k(R) \in \mathbb{N}_* \quad \text{such that} \quad \sum_{k=1}^{k(R)} \lambda_{\min}(W_k) \geq R.$$

We have purposely used a somewhat sloppy statement, in order to stress once more that properties resembling (4.2.3) have little relevance in reality.

Remark 4.2.6 The meaning of (4.2.5) deserves comment. In words, it says that all subgradients of f at all nonoptimal points are uniformly far from 0. In particular, use the notation $\hat{s}(x) := \text{Proj} \, 0/\partial f(x)$ for the subgradient of f at x that has least Euclidean norm. If f satisfies (4.2.5), $\hat{s}(x)$ is either 0 (x optimal) or larger than η in norm (x not optimal). In the latter case, f can be decreased locally from x at a rate at least η: simply move along the steepest-descent direction $-\hat{s}(x)$. We say in this case that f has a *sharp* set of minima.

Suppose for example that f is piecewise affine:

$$f(x) = \max\{\langle s_j, x \rangle - b_j : j = 1, \ldots, m\} \quad \text{for all } x \in \mathbb{R}^n.$$

Taking an arbitrary nonempty subset $J \subset \{1, \ldots, m\}$, define

$$\hat{s}^J := \text{Proj} \, 0/\{s_j : j \in J\}$$

to obtain $2^m - 1$ vectors $\hat{s}^J$; all the possible $\hat{s}(x)$ are taken from this finite list. Setting

$$\eta := \min\left\{\|\hat{s}^J\| \,:\, \hat{s}^J \neq 0\right\},$$

we therefore see that (4.2.5) holds for this particular $\eta > 0$. Conclusion: when (4.2.3) holds, the proximal point algorithm terminates for piecewise affine functions that are bounded from below. □

Returning to our original problem of minimizing f, the proximal point algorithm is based on the formulation (ii) in Theorem 4.1.7; but formulations in terms of convex minimization can also be used. First of all, observe from (4.2.1) that Algorithm 4.2.1 is an *implicit* gradient method (preconditioned by W_k, and with unit stepsizes) to minimize f.

Formulation 4.1.7(iv) is another alternative. To use it conveniently, take a fixed operator $M_k = M$ in Algorithm 4.2.1; in view of (4.1.5), the update formula written as

$$x_{k+1} = x_k - M^{-1}\nabla f_M(x_k)$$

shows that we also have an *explicit* gradient method (preconditioned by M^{-1}, and with unit stepsizes) to minimize f_M.

On the other hand, the Moreau-Yosida regularization is very smooth: its gradient is Lipschitzian; so any method of Chap. II can be chosen, in particular the powerful quasi-Newton methods of §II.2.3. This is attractive but not quite straightforward:

- f_M cannot be computed explicitly, so actual implementations can only mimic such methods, with suitable approximations of f_M and ∇f_M;
- the possibility of having a varying operator M_k opens the way to intriguing interpretations, in which the objective function becomes f_{M_k} and depends on the iteration index.

These ideas will not be developed here: we limit our study to a possible algorithm computing f_M.

4.3 Computing the Moreau-Yosida Regularization

In this subsection, x and M are fixed; they can be for example x_k and M_k at the current iteration of the proximal point Algorithm 4.2.1. We address the problem of minimizing the perturbed objective function of (4.1.1)

$$\mathbb{R}^n \ni y \mapsto g(x,y) := f(y) + \tfrac{1}{2}\langle M(y-x), y-x\rangle$$

to obtain the optimal $y = p_M(x)$. Since we turn again to numerical algorithms, we assume that our convex function f is finite everywhere; the above g is therefore a convex finite-valued function as well. The minimization of g would not necessitate particular comments, except for the fact that our eventual aim is to minimize f, really. Then several possibilities are available:

- The basic subgradient algorithm of §XII.4.1 is not very attractive: the only difference between f and g is the 1-coercivity of the latter, a property which does not help this method much. In other words, a subgradient algorithm should be more fruitfully applied directly to f.

4 Bundle Methods as Regularizations

- The basic cutting-plane algorithm of §XII.4.2 is still impeded by the need for compactness.
- Bundle methods will take cutting-plane approximations for $g(x, \cdot)$, introduce a stabilizing quadratic term, and solve the resulting quadratic program. This is somewhat redundant: a quadratic perturbation has already been introduced when passing from f to $g(x, \cdot)$.
- Then a final possibility suggests itself, especially if we remember Remark 3.2.6(ii): apply the cutting-plane mechanism to f only, obtaining a "hybrid" bundle method in penalized form.
- This last technique can also be seen as an "auto-stabilized" cutting-plane method, in which the quadratic term in the objective function $g(x, \cdot)$ is kept as a stabilizer (preconditioned by M). Because no artificial stabilizer is introduced, the stability center x must not be moved and the descent test must be inhibited, so as to take enough "null-steps", until $f_M(x) = \min g(x, \cdot)$ is reached. Actually, the quadratic term in g plays the role of the artificial C introduced for (1.0.2).

For consistency with §3, the index k will still denote the current iteration of the resulting algorithm. The key is then to replace f in the proximal problem (4.1.1) by a model-function φ_k satisfying

$$\varphi_k \leqslant f ; \tag{4.3.1}$$

also, φ_k is piecewise affine, but this is not essential. Then the next iterate is the proximal point of x associated with φ_k:

$$y_{k+1} := \operatorname{argmin}\left[\varphi_k(y) + \tfrac{1}{2}\langle M(y-x), y-x\rangle\right]. \tag{4.3.2}$$

According to Lemma 4.1.1, we have (once again, $W = M^{-1}$)

$$y_{k+1} = x - W\hat{s}_k \quad \text{with} \quad \hat{s}_k \in \partial\varphi_k(y_{k+1}) \tag{4.3.3}$$

and the aggregate linearization

$$y \mapsto \bar{f}_k^{\mathrm{a}}(y) := \varphi_k(y_{k+1}) + \langle \hat{s}_k, y - y_{k+1}\rangle \tag{4.3.4}$$

will be useful for the next model φ_{k+1}. With respect to §4.2, beware that $\hat{s}_k$ does not refer to the proximal point of some varying x_k associated with f: here the proximal center x is fixed, it is the function φ_k that is varying.

We obtain an algorithm which is implementable, in the sense that it needs only a black box (U1) that, given $y \in \mathbb{R}^n$, computes $f(y)$ and $s(y) \in \partial f(y)$.

Algorithm 4.3.1 (Hybrid Bundle Method) The preconditioner M and the proximal center x are given. Choose the initial model $\varphi_1 \leqslant f$, a convex function which can be for example

$$y \mapsto \varphi_1(y) = f(x) + \langle s(x), y - x\rangle,$$

and initialize $k = 1$.

STEP 1. Compute y_{k+1} from (4.3.2) or (4.3.3).

STEP 2. If $f(y_{k+1}) = \varphi_k(y_{k+1})$ stop.
STEP 3. Update the model to any convex function φ_{k+1} satisfying (4.3.1), as well as

$$\varphi_{k+1} \geqslant \bar{f}_k^a \tag{4.3.5}$$

and

$$\varphi_{k+1}(y) \geqslant f(y_{k+1}) + \langle s(y_{k+1}), y - y_{k+1}\rangle \quad \text{for all } y \in \mathbb{R}^n. \tag{4.3.6}$$

Replace k by $k+1$ and loop to Step 1. □

This algorithm is of course just a form of Algorithm 3.1.4 with a few modifications:
- $C = \mathbb{R}^n$; a simplification which yields $\hat{p}_k = 0$.
- The notation is "bundle-free": no list of affine functions is assumed, and the successive models are allowed more general forms than piecewise affine; only the essential properties (4.3.1), (4.3.5), (4.3.6) are retained.
- The stabilizing operator M is fixed but is not proportional to the identity (a negligible generalization).
- The proximal center $x_k = x$ is never updated; this can be simulated by taking a very large $m > 0$ in (3.1.7).
- The stopping criterion in Step 2 has no reason to ever occur; in view of (4.3.1), it means that f coincides with its model φ_k at the "trial proximal point" y_{k+1}.

Proposition 4.3.2 *If Algorithm 4.3.1 terminates at Step 2, $y_{k+1} = p_M(x)$.*

PROOF. When the stop occurs, y_{k+1} satisfies by construction

$$g(x, y_{k+1}) = \varphi_k(y_{k+1}) + \tfrac{1}{2}\langle M(y_{k+1} - x), y_{k+1} - x\rangle \leqslant$$
$$\leqslant \varphi_k(y) + \tfrac{1}{2}\langle M(y - x), y - x\rangle \leqslant g(x, y) \quad \text{for all } y \in \mathbb{R}^n. \qquad \square$$

A more realistic stopping criterion could be

$$[0 \leqslant] \; f(y_{k+1}) - \varphi_k(y_{k+1}) \leqslant \underline{\delta} \tag{4.3.7}$$

for some positive tolerance $\underline{\delta}$. The whole issue for convergence is indeed the property $f(y_{k+1}) - \varphi_k(y_{k+1}) \to 0$, which we establish with the tools of §3.2. First of all, remember that nothing is changed if the objective function of (4.3.2) is a posteriori replaced by

$$y \mapsto \gamma_k(y) := \bar{f}_k^a(y) + \tfrac{1}{2}\langle M(y - x), y - x\rangle.$$

In fact, γ_k is strongly convex and the definition (4.3.4) of $\bar{f}_k^a$ gives $0 = \nabla \gamma_k(y_{k+1})$.

Lemma 4.3.3 *For all k and all $y \in \mathbb{R}^n$, there holds*

$$\gamma_k(y) = \gamma_k(y_{k+1}) + \tfrac{1}{2}\langle M(y - y_{k+1}), y - y_{k+1}\rangle.$$

4 Bundle Methods as Regularizations 329

PROOF. Use the simple identity, valid for all u, v in $\mathbb{R}^n$ and symmetric M:

$$\tfrac{1}{2}\langle M(u+v), u+v\rangle = \tfrac{1}{2}\langle Mu, u\rangle + \langle Mu, v\rangle + \tfrac{1}{2}\langle Mv, v\rangle.$$

Take $u = y_{k+1} - x$, $v = y - y_{k+1}$ and add to the equality (4.3.4) to obtain the result. □

This equality plays the role of (3.2.6); it allows us to reproduce Steps 1 and 2 in the proof of Theorem 3.2.4.

Theorem 4.3.4 *For $f : \mathbb{R}^n \to \mathbb{R}$ convex and M symmetric positive definite, the sequence $\{y_k\}$ generated by Algorithm 4.3.1 satisfies*

$$f(y_{k+1}) - \varphi_k(y_{k+1}) \to 0, \tag{4.3.8}$$

$$y_k \to p_M(x).$$

PROOF. We assume that the stop never occurs, otherwise Proposition 4.3.2 applies. In view of (4.3.5),

$$\bar{f}^a_{k-1}(y_{k+1}) \leq \varphi_k(y_{k+1}) \quad \text{for all } k > 1.$$

Add $1/2\,\langle M(y_{k+1} - x), y_{k+1} - x\rangle$ to both sides and use Lemma 4.3.3 with k replaced by $k-1$ to obtain

$$\gamma_{k-1}(y_k) + \tfrac{1}{2}\langle M(y_{k+1} - y_k), y_{k+1} - y_k\rangle \leq \gamma_k(y_{k+1}) \leq f(x) \quad \text{for } k > 1.$$

The sequence $\{\gamma_k(y_{k+1})\}$ is increasing and $y_{k+1} - y_k \to 0$.

Now set $y = x$ in Lemma 4.3.3:

$$\tfrac{1}{2}\langle M(x - y_{k+1}), x - y_{k+1}\rangle \leq \gamma_k(x) - \gamma_k(y_{k+1}) \leq f(x) - \gamma_1(y_2)$$

so that $\{y_k\}$ is bounded. Then use (4.3.6) at the $(k-1)^{\text{st}}$ iteration:

$$\varphi_k(y_{k+1}) \geq f(y_k) + \langle s(y_k), y_{k+1} - y_k\rangle;$$

subtracting $f(y_{k+1})$ from both sides and using a Lipschitz constant L for f around x,

$$0 \geq \varphi_k(y_{k+1}) - f(y_{k+1}) \geq -2L\|y_{k+1} - y_k\|;$$

(4.3.8) follows.

Finally extract a convergent subsequence from $\{y_k\}$; more precisely, let $K \subset \mathbb{N}$ be such that $y_{k+1} \to y_*$ for $k \to +\infty$ in K. Then $\varphi_k(y_{k+1}) \to f(y_*)$ and, passing to the limit in the subgradient inequality

$$\varphi_k(y_{k+1}) + \langle M(x - y_{k+1}), y - y_{k+1}\rangle \leq \varphi_k(y) \leq f(y) \quad \text{for all } y \in \mathbb{R}^n$$

shows that $M(x - y_*) \in \partial f(y_*)$. Because of Lemma 4.1.1, $y_* = p_M(x)$. □

We terminate with an important comment concerning the stopping criterion. Remember that, for our present concern, $p_M(x)$ is computed only to implement an "outer" algorithm minimizing f, as in §4.2. Then it is desirable to stop the "inner" Algorithm 4.3.1 early if x is far from a minimizer of f (once again, why waste time in minimizing the perturbed function $g(x, \cdot)$ instead of f itself?). This implies an ad hoc value δ in (4.3.7).

Adding and subtracting $f(x)$, write (4.3.7) as

$$f(y_{k+1}) \leqslant f(x) - \{[f(x) - \varphi_k(y_{k+1})] - \underline{\delta}\}$$

which can be viewed as a *descent* criterion. Under these circumstances, we can think of a $\underline{\delta}$ depending on k, and comparable to $f(x) - \varphi_k(y_{k+1})$ (a number which is available when Step 2 is executed). For example, with some coefficient $m \in \,]0, 1[$, we can take

$$\underline{\delta} = (1-m)[f(x) - \varphi_k(y_{k+1})],$$

in which case (4.3.7) becomes

$$f(y_{k+1}) \leqslant f(x) - m[f(x) - \varphi_k(y_{k+1})].$$

We recognize a descent test used in bundle methods: see more particularly the trust-region Algorithm 2.1.1. The descent iteration of such a bundle method will then update the current stability center $x = x_k$ to this "approximate proximal point" y_{k+1}.

Conclusion: seen from the proximal point of view of this Section 4, bundle methods provide three ingredients:

– an inner algorithm to compute a proximal point, based on cutting-plane approximations of f;
– a way of stopping this internal algorithm dynamically, when a sufficient decrease is obtained for the original objective function f;
– an outer algorithm to minimize f, using the output of the inner algorithm to mimic the proximal point formula.

Furthermore, each inner algorithm can use for its initialization the work performed during the previous outer iteration: this is reflected in the initial φ_1 of Algorithm 4.3.1, in which all the cutting planes computed from the very beginning of the iterations can be accumulated.

This explains our belief expressed at the very end of §2.5: the penalized form is probably the most interesting variant among the possible bundle methods.

Bibliographical Comments

Just as we did with the first volume, let us repeat that [159] is a must for convex analysis in finite dimension. On the other hand, we recommend [89] for an exhaustive account of bundle methods, with the most refined techniques concerning convergence, and also various extensions.

Chapter IX. The technique giving birth to the bundling mechanism can be contrasted with an old separation algorithm going back to [1]. The complexity theory that is alluded to in §2.2(b) comes from [133] and our Counter-example 2.2.4 is due to A.S. Nemirovskij.

It is important to know that this bundling mechanism can be extended to nonconvex functions without major difficulty (at least theoretically). The approach, pioneered in [55], goes as follows. First, one considers locally Lipschitzian functions; denote by $\partial f(x)$ the *Clarke generalized gradient* of such a function f at x ([36, 37]). This is just the set co $\gamma f(x)$ of §VI.6.3(a) and would be the subdifferential of f at x if f were convex. Assume that some $s(y) \in \partial f(y)$ can be computed at each y, just as in the convex case. Given a direction $d\ (= d_k)$, the line-search 1.2 is performed and the interesting situation is when $t \downarrow 0$, with no descent obtained; this may happen when

$$\limsup_{t \downarrow 0} \frac{f(x+td) - f(x)}{t} \geq 0. \qquad (*)$$

As explained in §1.3(a), the key is then to produce a cluster point s^* of $\{s(x+td)\}_t$ satisfying $\langle s^*, d \rangle \geq 0$; to be on the safe side, we need

$$\liminf_{t \downarrow 0} \langle s(x+td), d \rangle \geq 0. \qquad (**)$$

If f is convex, $(*)$ automatically implies $(**)$. If not, the trick is simply to set the property "$(*) \Rightarrow (**)$" as an *axiom* restricting the class of Lipschitz functions that can be minimized by this approach.

A possible such axiom is for example

$$\limsup_{t \downarrow 0} \frac{f(x+td) - f(x)}{t} \leq \liminf_{t \downarrow 0} \langle s(x+td), d \rangle,$$

resulting in the rather convenient classes of *semi-smooth* functions of R. Mifflin [122]. On the other hand, a minimal requirement allowing the bundling mechanism to work is obtained if one observes that, in the actual line-search, the sequences $\{t_k\}$ giving $f(x + t_k d)$ in (∗) and $s(x + t_k d)$ in (∗∗) are just the same. What is needed is simply the axiom defined in [24]:

$$\limsup_{t \downarrow 0} \left[\frac{f(x+td) - f(x)}{t} - \langle s(x+td), d \rangle \right] \leqslant 0.$$

Convex quadratic programming problems are usually solved by one universal *pivoting* technique, see [23]. The starting idea is that a solution can be explicitly computed if the set of active constraints is known (solve a system of linear equations). The whole technique is then iterative:

– at each iteration, a subset J of the inequality constraints is selected;
– each constraint in J is replaced by an equality, the other constraints being discarded;
– this produces a point x_J, whose optimality is tested;
– if x_J is not optimal, J is updated and the process is repeated until a correct J^* is identified, yielding a solution of the original problem.

Chapter X. As suggested from the introduction to this chapter, the transformation $f \mapsto f^*$ has its origins in a publication of A. Legendre (1752-1833), dated from 1787; remember also Young's inequality in one dimension (see the bibliographical comments on Chap. I). Since then, this transformation has received a number of names in the literature: conjugate, polar, maximum transformation, etc. However, it is now generally agreed that an appropriate terminology is *Legendre-Fenchel transform*, which we precisely adopted in this chapter. Let us mention that W. Fenchel (1905-1988) wrote in a letter to C. Kiselman, dated March 7, 1977: "I do not want to add a new name, but if I had to propose one now, I would let myself by guided by analogy and the relation with polarity between convex sets (in dual spaces) and I would call it for example *parabolic polarity*". Fenchel was influenced by his geometric (projective) approach, and also by the fact that the "parabolic" function $f : x \mapsto f(x) = 1/2 \|x\|^2$ is the only one that satisfies $f^* = f$. In our present finite-dimensional context, it is mainly Fenchel who studied "his" transformation, in papers published between 1949 and 1953.

The essential part of §1.5 (Theorem 1.5.6) is new. The close-convexification, or biconjugacy, of a function arises naturally in variational problems ([49, 81]): minimizing an objective function like $\int_a^b \ell(t, x(t), \dot{x}(t))dt$ is related to the minimization of the "relaxed" form $\int_a^b \overline{\text{co}}\, \ell(t, x(t), \dot{x}(t))dt$, where $\overline{\text{co}}\, \ell$ denotes the closed convex hull of the function $\ell(t, x, \cdot)$. The question leading to Corollary 1.5.2 was answered in [43], in a calculus of variations framework; a short and pedagogical proof can be found in [74]. Lemma 1.5.3 is due to M. Valadier [180, p. 69]; the proof proposed here is more legible and detailed. The results of Proposition 1.5.4 and Theorem 1.5.5 appeared in [64].

The calculus rules of §2 are all rather classical; two of them deserve a particular comment: post-composition with an increasing convex function, and maximum of functions. The first is treated in [94] for vector-valued functions. As for max-functions, we limited ourselves here to finitely many functions but more general cases are treated similarly. In fact, consider the following situation (cf. §VI.4.4): T is compact in some metric space; $f : T \times \mathbb{R}^n \to \mathbb{R}$ satisfies

$f(t, \cdot) =: f_t$ is a convex function from $\mathbb{R}^n$ to $\mathbb{R}$ for all $t \in T$
$f(\cdot, x)$ is upper semi-continuous on T for all $x \in \mathbb{R}^n$;
hence $f(x) := \max_{t \in T} f_t(x) < +\infty$ for all $x \in \mathbb{R}^n$.

As already said in §VI.4.4, f is then jointly upper semi-continuous on $T \times \mathbb{R}^n$, so that $f^* : (t, s) \mapsto f^*(t, s) := (f_t)^*(s)$ is jointly lower semi-continuous and it follows that $\varphi := \min_{t \in T} f^*(t, \cdot)$ is lower semi-continuous; besides, φ is also 1-coercive since $\varphi^* = f$ is finite everywhere. Altogether, the results of §1.5 can be applied to φ: with the help of Proposition 1.5.4, $f^* = \overline{\text{co}}\,\varphi = \text{co}\,\varphi$ can be expressed in a way very similar to Theorem 2.4.7. We have here an alternative technique to compute the subdifferential of a max-function (Theorem VI.4.4.2).

The equivalence (i) $\Leftrightarrow$ (ii) of Theorem 3.2.3 is a result of T.S. Motzkin (1935); our proof is new. Proposition 3.3.1 was published in [120, §II.3]. Our Section 4.2(b) is partly inspired from [63], [141]: Corollary 4.2.9 can be found in [63], while J.-P. Penot was more concerned in [141] by the "one-sided" aspect suitable for the unilateral world of convex analysis, even though all his functions φ_s were convex quadratic forms. For Corollary 4.2.10, we mention [39]; see also [170], where this result is illustrated in various domains of mathematics.

Chapter XI. Approximate subgradients appeared for the first time in [30, §3]. They were primarily motivated by topological considerations (related to their regularization properties), and the idea of using them for algorithmic purposes came in [22]. In [137, 138, 139], E.A. Nurminskii used (1.3.5) to design algorithms for convex minimization; see also [111], and the considerations at the end of [106]. Approximate subgradients also have intriguing applications in *global optimization*. Let f and g be two functions in $\overline{\text{Conv}}\,\mathbb{R}^n$. Then x minimizes (globally) the difference $f - g$ if and only if $\partial_\varepsilon g(x) \subset \partial_\varepsilon f(x)$ for all $\varepsilon > 0$. This result was published in [75] and, to prove its sufficiency part, one can for example start from (1.3.7).

The characterization (1.3.4) of the support function of the graph of the multifunction $\varepsilon \mapsto \partial_\varepsilon f(x)$ also appeared in [75]. The representation of closed convex functions via approximate directional derivatives (Theorem 1.3.6) was published independently in [73, §2] and [111, §1]. As for the fundamental expression (2.1.1) of the approximate directional derivative, it appeared in [131, p. 67] and [159, pp. 219–220], but with different proofs. The various properties of approximate difference quotients, detailed in our §2.2, come from [73, §2].

Concerning approximate subdifferential calculus (§3), let us mention that general results, with vector-valued functions, had been announced in [95]; the case of convex functions with values in $\mathbb{R} \cup \{+\infty\}$ is detailed in the survey paper [72], which inspired our exposition. Qualification assumptions can be avoided to develop calculus rules of a different nature, using the (richer) information contained in $\{\partial_\eta f_j(x) : 0 < \eta \leq \bar{\eta}\}$, instead of the mere $\partial f_j(x)$. For example, with f_1 and f_2 in $\overline{\text{Conv}}\,\mathbb{R}^n$,

$$\partial(f_1 + f_2)(x) = \bigcap_{0 < \eta \leq \bar{\eta}} \text{cl}\left[\partial_\eta f_1(x) + \partial_\eta f_2(x)\right], \quad \text{where } \bar{\eta} > 0 \text{ is arbitrarily small.}$$

This formula emphasizes once more the smoothing or "viscosification" effect of approximate subdifferentials. For a more concrete development of approximate minimality conditions, alluded to in §3.1 and the end of §3.6, see [175].

The local Lipschitz property of $\partial_\varepsilon f(\cdot)$, with fixed $\varepsilon > 0$, was observed for the first time in [136]; the overall formalization of this result, and the proof used here, were published in [71]. Passing from a globally Lipschitzian convex function to an arbitrary convex function was anyway the motivation for introducing the regularization-approximation technique via the inf-convolution with $c\|\cdot\|$ ([71, §2]). Theorem 4.2.1 comes from [30]. The transportation formula (4.2.2), and the neighborhoods of (4.2.3), were motivated by the desire to make the algorithm of [22] implementable (see [100]).

Chapter XII. References to duality without convexity assumptions are not so common; we can mention [32] (especially for §3.1), [52], [65]; see also [21]. However, there is a wealth of works dealing with Lagrangian relaxation in integer programming; a milestone in this subject was [68], which gave birth to the test-problems TSP of IX.2.2.7.

The subgradient method of §4.1 was discovered by N.Z. Shor in the beginning of the sixties, and its best account is probably that of [145]; for more recent developments, and in particular accelerations by dilation of the space, see [171]. Our proof of convergence is copied from [134]. The original references to cutting planes are the two independent papers [35] and [84]; their column-generation variant in a linear-programming context appeared in [40].

The views expressed in §5 go along those of [163]. The most complete theory of augmented Lagrangians is given in the works of D.P. Bertsekas, see for example [20]. Dualization schemes when contraints take their values in a cone are also explained in [114]. Several authors have related Fenchel and Lagrange duality schemes. Our approach is inspired from [115].

Chapters XIII and XIV. The ε-descent algorithm, going back to [100], is the ancestor of bundle methods. It was made possible thanks to the work [22], which was done at the same time as the speculations detailed in Chap. IX; the latter were motivated by a particular economic lotsizing problem of the type XII.1.2(b), coming from the glass industry. This observation points out once more how applied mathematics can be a delicate subject: the publication of purely theoretical papers like [22] may sometimes be necessary for the resolution of apparently innocent applied problems. Using an idea of [116], the method can be extended to the resolution of variational inequalities.

Then came the *conjugate-subgradient* form (§XIV.4.3) in [101] and [188], and Algorithm XIV.3.4.2 was introduced soon after in [102]. From then on, the main works concerning these methods dealt with generalizations to the nonconvex case and the treatment of constraints, in particular by R. Mifflin [123]. At that time, the similarity with conjugate gradients (Remark II.2.4.6) was felt as a key in favour of conjugate subgradients; but we now believe that this similarity is incidental. We mention here that conjugate gradients might well become obsolete for "smooth" optimization anyway, see [61, 113] among others. In fact, it is probably with other interior point methods, as in [62], that the most promising connections of bundle methods remain to be explored.

The special algorithm for nonsmooth univariate minimization, alluded to in Remark XIII.2.1.2, can be found in [109]. It is probably the only existing algorithm that converges globally and superlinearly in this framework. Defining algorithms having such qualities for several variables is a real challenge; it probably requires first an appropriate definition of second-order differentiation of a convex function. These questions have been pending for the last two decades.

Chapter XV. Primal bundle methods appeared in [103], after it was realized from [153] that (dual) bundle methods were connected with sequential quadratic programming (see the bibliographical comments of Chap. VIII). R. Mifflin definitely formalized the approach in [124], with an appropriate treatment of non-convexity. In [88], K.C. Kiwiel gave the most refined proof of convergence and proved finite termination for piecewise affine functions. Then he proposed a wealth of adaptations to various situations: noisy data, sums of functions, ... see for example his review [90]. J. Zowe contributed a lot to the general knowledge of bundle methods [168]. E.A. Nurminskii gave an interesting interpretation in [137, 138, 139], which can be sketched as follows: at iteration k,
- choose a distance in the dual graph space;
- choose a certain point of the form $P_k = (0, r_k)$ in this same space;
- choose an approximation E_k of epi f^*; more specifically, take the epigraph of $(\check{f}_k)^*$;
- project P_k onto E_k in the sense of the distance chosen (which is not necessarily a norm); this gives a vector (s_k, θ_k), and $-s_k$ can be used for a line-search.

The counter-example in §1.1 was published in [133, §4.3.6]; for the calculations that it needs in Example 1.1.2, see [18] for example. In classical "smooth" optimization, there is a strong tendency to abandon line-searches, to the advantage of the trust-region technique alluded to in §1.3 and overviewed in [127].

The trust-region variant of §2.1 has its roots in [117]. The level variant of §2.3 is due to A.S. Nemirovskij and Yu. Nesterov, see [110]. As for the relaxation variant, it strongly connotes and generalizes [146]; see [46] for an account of the Gauss-Newton and Levenberg-Marquardt methods.

Our convergence analysis in §3 comes from [91]. Several researchers have felt attracted by the connection between quadratic models and cutting planes approximations (Remark 3.3.3): for example [185], [147], [92].

The Moreau-Yosida regularization is due to K. Yosida for maximal monotone operators, and was adapted in [130] to the case of a subgradient mapping. The idea of exploiting it for numerical purposes goes back to [14, Chap.V] for solving ill-posed systems of linear equations. This was generalized in [119] for the minimization of convex functions, and was then widely developed in primal-dual contexts: [161] and its derivatives. The connection with bundle methods was realized in the beginning of the eighties: [59], [11].

References

1. Aizerman, M.A., Braverman, E.M., Rozonoer, L.I.: The probability problem of pattern recognition learning and the method of potential functions. Automation and Remote Control **25**,9 (1964) 1307–1323.
2. Alexeev, V., Galeev, E, Tikhomirov, V.: *Recueil de Problèmes d'Optimisation*. Mir, Moscow (1984).
3. Alexeev, V., Tikhomirov, V., Fomine, S.: *Commande Optimale*. Mir, Moscou (1982).
4. Anderson Jr., W.N., Duffin, R.J.: Series and parallel addition of matrices. J. Math. Anal. Appl. **26** (1969) 576–594.
5. Artstein, Z.: Discrete and continuous bang-bang and facial spaces or: look for the extreme points. SIAM Review **22**,2 (1980) 172–185.
6. Asplund, E.: Differentiability of the metric projection in finite-dimensional Euclidean space. Proc. Amer. Math. Soc. **38** (1973) 218–219.
7. Aubin, J.-P.: *Optima and Equilibria: An Introduction to Nonlinear Analysis*. Springer, Berlin Heidelberg (1993).
8. Aubin, J.-P.: *Mathematical Methods of Game and Economic Theory*. North-Holland (1982) (revised edition).
9. Aubin, J.-P., Cellina, A.: *Differential Inclusions*. Springer, Berlin Heidelberg (1984).
10. Auslender, A.: *Optimisation, Méthodes Numériques*. Masson, Paris (1976).
11. Auslender, A.: Numerical methods for nondifferentiable convex optimization. In: *Nonlinear Analysis and Optimization*. Math. Prog. Study **30** (1987) 102–126.
12. Barbu, V., Precupanu, T.: *Convexity and Optimization in Banach Spaces*. Sijthoff & Noordhoff (1982).
13. Barndorff-Nielsen, O.: *Information and Exponential Families in Statistical Theory*. Wiley & Sons (1978).
14. Bellman, R.E., Kalaba, R.E., Lockett, J.: *Numerical Inversion of the Laplace Transform*. Elsevier (1966).
15. Ben Tal, A., Ben Israel, A., Teboulle, M.: Certainty equivalents and information measures: duality and extremal principles. J. Math. Anal. Appl. **157** (1991) 211–236.
16. Berger, M.: *Geometry I, II (Chapters 11, 12)*. Springer, Berlin Heidelberg (1987).
17. Berger, M.: Convexity. Amer. Math. Monthly **97**,8 (1990) 650–678.
18. Berger, M., Gostiaux, B.: *Differential Geometry: Manifolds, Curves and Surfaces*. Springer, New York (1990).
19. Bertsekas, D.P.: Necessary and sufficient conditions for a penalty method to be exact. Math. Prog. **9** (1975) 87–99.
20. Bertsekas, D.P.: *Constrained Optimization and Lagrange Multiplier Methods*. Academic Press (1982).

21. Bertsekas, D.P.: Convexification procedures and decomposition methods for nonconvex optimization problems. J. Optimization Th. Appl. **29**,2 (1979) 169–197.
22. Bertsekas, D.P., Mitter, S.K.: A descent numerical method for optimization problems with nondifferentiable cost functionals. SIAM J. Control **11**,4 (1973) 637–652.
23. Best, M.J.: Equivalence of some quadratic programming algorithms. Math. Prog. **30**,1 (1984) 71–87.
24. Bihain, A.: Optimization of upper semi-differentiable functions. J. Optimization Th. Appl. **4** (1984) 545–568.
25. Bonnans, J.F: Théorie de la pénalisation exacte. Modélisation Mathématique et Analyse Numérique **24**,2 (1990) 197–210.
26. Borwein, J.M.: A note on the existence of subgradients. Math. Prog. **24** (1982) 225–228.
27. Borwein, J.M., Lewis, A.: *Convexity, Optimization and Functional Analysis*. Wiley Interscience – Canad. Math. Soc. (in preparation).
28. Brenier, Y.: Un algorithme rapide pour le calcul de transformées de Legendre-Fenchel discrètes. Note aux C.R. Acad. Sci. Paris **308** (1989) 587–589.
29. Brøndsted, A.: *An Introduction to Convex Polytopes*. Springer, New York (1983).
30. Brøndsted, A., Rockafellar, R.T.: On the subdifferentiability of convex functions. Proc. Amer. Math. Soc. **16** (1965) 605–611.
31. Brousse, P.: *Optimization in Mechanics: Problems and Methods*. North-Holland (1988).
32. Cansado, E.: Dual programming problems as hemi-games. Management Sci. **15**,9 (1969) 539–549.
33. Castaing, C., Valadier, M.: *Convex Analysis and Measurable Multifunctions*. Lecture Notes in Mathematics, vol. 580. Springer, Berlin Heidelberg (1977).
34. Cauchy, A.: Méthode générale pour la résolution des systèmes d'équations simultanées. Note aux C. R. Acad. Sci. Paris **25** (1847) 536–538.
35. Cheney, E.W., Goldstein, A.A.: Newton's method for convex programming and Tchebycheff approximation. Numer. Math. **1** (1959) 253–268.
36. Clarke, F.H.: Generalized gradients and applications. Trans. Amer. Math. Soc. **205** (1975) 247–262.
37. Clarke, F.H.: *Optimization and Nonsmooth Analysis*. Wiley & Sons (1983), reprinted by SIAM (1990).
38. Crandall, M.G., Ishii, H., Lions, P.-L.: User's guide to viscosity solutions of second order partial differential equations. Bull. Amer. Math. Soc. **27**,1 (1992) 1–67.
39. Crouzeix, J.-P.: A relationship between the second derivative of a convex function and of its conjugate. Math. Prog. **13** (1977) 364–365.
40. Dantzig, G.B. Wolfe, P.: A decomposition principle for linear programs. Oper. Res. **8** (1960) 101–111.
41. Davidon, W.C.: Variable metric method for minimization. AEC Report ANL5990, Argonne National Laboratory (1959).
42. Davidon, W.C.: Variable metric method for minimization. SIAM J. Optimization **1** (1991) 1–17.
43. Dedieu, J.-P.: Une condition nécessaire et suffisante d'optimalité en optimisation non convexe et en calcul des variations. Séminaire d'Analyse Numérique, Univ. Paul Sabatier, Toulouse (1979–80).
44. Demjanov, V.F.: Algorithms for some minimax problems. J. Comp. Syst. Sci. **2** (1968) 342–380.
45. Demjanov, V.F., Malozemov, V.N.: *Introduction to Minimax*. Wiley & Sons (1974).

46. Dennis, J., Schnabel, R.: *Numerical Methods for Constrained Optimization and Nonlinear Equations*. Prentice Hall (1983).
47. Dubois, J.: Sur la convexité et ses applications. Ann. Sci. Math. Quebec **I**,1 (1977) 7–31.
48. Dubuc, S.: *Problèmes d'Optimisation en Calcul des Probabilités*. Les Presses de l'Université de Montréal (1978).
49. Ekeland, I., Temam, R.: *Convex Analysis and Variational Problems*. North-Holland, Amsterdam (1976).
50. Eggleston, H.G.: *Convexity*. Cambridge University Press, London (1958).
51. Ellis, R.S.: *Entropy, Large Deviations and Statistical Mechanics*. Springer, New York (1985).
52. Everett III, H.: Generalized Lagrange multiplier method for solving problems of optimum allocation of resources. Oper. Res. **11** (1963) 399–417.
53. Fenchel, W.: Convexity through the ages. In: *Convexity and its Applications* (P.M. Gruber and J.M. Wills, eds.). Birkhäuser, Basel (1983) 120–130.
54. Fenchel, W.: Obituary for the death of –. Det Kongelige Danske Videnskabernes Selskabs Aarbok (Oversigten) [Yearbook of the Royal Danish Academy of Sciences] (1988–89) 163–171.
55. Feuer, A.: An implementable mathematical programming algorithm for admissible fundamental functions. PhD. Thesis, Columbia Univ. (1974).
56. Fletcher, R.: *Practical Methods of Optimization*. Wiley & Sons (1987).
57. Fletcher, R., Powell, M.J.D.: A rapidly convergent method for minimization. The Computer Journal **6** (1963) 163–168.
58. Flett, T.M.: *Differential Analysis*. Cambridge University Press (1980).
59. Fukushima, M.: A descent algorithm for nonsmooth convex programming. Math. Prog. **30**,2 (1984) 163–175.
60. Geoffrion, A.M.: Duality in nonlinear programming: a simplified application-oriented development. SIAM Review **13**,11 (1971) 1–37.
61. Gilbert, J.C., Lemaréchal,C.: Some numerical experiments with variable-storage quasi-Newton algorithms. Math. Prog. **45** (1989) 407–435.
62. Goffin, J.-L., Haurie, A., Vial, J.-Ph.: Decomposition and nondifferentiable optimization with the projective algorithm. Management Sci. **38**,2 (1992) 284–302.
63. Gorni, G.: Conjugation and second-order properties of convex functions. J. Math. Anal. Appl. **158**,2 (1991) 293–315.
64. Griewank, A., Rabier, P.J.: On the smoothness of convex envelopes. Trans. Amer. Math. Soc. **322** (1990) 691–709.
65. Grinold, R.C.: Lagrangian subgradients. Management Sci. **17**,3 (1970) 185–188.
66. Gritzmann, P., Klee, V.: Mathematical programming and convex geometry. In: *Handbook of Convex Geometry* (Elsevier, North-Holland, to appear).
67. Gruber, P.M.: History of convexity. In: *Handbook of Convex Geometry* (Elsevier, North-Holland, to appear).
68. Held, M., Karp, R.M.: The traveling-salesman problem and minimum spanning trees. Math. Prog. **1**,1 (1971) 6–25.
69. Hestenes, M.R., Stiefel, M.R.: Methods of conjugate gradients for solving linear systems. J. Res. NBS **49** (1959) 409–436.
70. Hiriart-Urruty, J.-B.: Extension of Lipschitz functions. J. Math. Anal. Appl. **77** (1980) 539–554.
71. Hiriart-Urruty, J.-B.: Lipschitz r-continuity of the approximate subdifferential of a convex function. Math. Scand. **47** (1980) 123–134.

72. Hiriart-Urruty, J.-B.: ε-subdifferential calculus. In: *Convex Analysis and Optimization* (J.-P. Aubin and R. Vinter, eds.). Pitman (1982), pp. 43–92.
73. Hiriart-Urruty, J.-B.: Limiting behaviour of the approximate first order and second order directional derivatives for a convex function. Nonlinear Anal. Theory, Methods & Appl. **6**,12 (1982) 1309–1326.
74. Hiriart-Urruty, J.-B.: When is a point x satisfying $\nabla f(x) = 0$ a global minimum of f? Amer. Math. Monthly **93** (1986) 556–558.
75. Hiriart-Urruty, J.-B.: Conditions nécessaires et suffisantes d'optimalité globale en optimisation de différences de fonctions convexes. Note aux C.R. Acad. Sci. Paris **309**, I (1989) 459–462.
76. Hiriart-Urruty, J.-B., Ye, D.: Sensitivity analysis of all eigenvalues of a symmetric matrix. Preprint Univ. Paul Sabatier, Toulouse (1992).
77. Holmes, R.B.: *A Course on Optimization and Best Approximation*. Lecture Notes in Mathematics, vol. 257. Springer, Berlin Heidelberg (1972).
78. Holmes, R.B.: *Geometrical Functional Analysis and its Applications*. Springer, Berlin Heidelberg (1975).
79. Hörmander, L.: Sur la fonction d'appui des ensembles convexes dans un espace localement convexe. Ark. Mat. **3**,12 (1954) 181–186.
80. Ioffe, A.D., Levin, V.L.: Subdifferentials of convex functions. Trans. Moscow Math. Soc. **26** (1972) 1–72.
81. Ioffe, A.D., Tikhomirov, V.M.: *Theory of Extremal Problems*. North-Holland (1979).
82. Israel, R.B.: *Convexity in the Theory of Lattice Gases*. Princeton University Press (1979).
83. Karlin, S.: *Mathematical Methods and Theory in Games, Programming and Economics*. Mc Graw-Hill, New York (1960).
84. Kelley, J.E.: The cutting plane method for solving convex programs. J. SIAM **8** (1960) 703–712.
85. Kim, K.V., Nesterov, Yu.E., Cherkassky, B.V.: The estimate of complexity of gradient computation. Soviet Math. Dokl. **275**,6 (1984) 1306–1309.
86. Kiselman, C.O.: How smooth is the shadow of a smooth convex body? J. London Math. Soc. (2) **33** (1986) 101–109.
87. Kiselman, C.O.: Smoothness of vectors sums of plane convex sets. Math. Scand. **60** (1987), 239–252.
88. Kiwiel, K.C.: An aggregate subgradient method for nonsmooth convex minimization. Math. Prog. **27** (1983) 320–341.
89. Kiwiel, K.C.: *Methods of Descent for Nondifferentiable Optimization*. Lecture Notes in Mathematics, vol. 1133. Springer, Berlin Heidelberg (1985).
90. Kiwiel, K.C.: A survey of bundle methods for nondifferentiable optimization. In: Proceedings, XIII. International Symposium on Mathematical Programming, Tokyo (1988).
91. Kiwiel, K.C.: Proximity control in bundle methods for convex nondifferentiable minimization. Math. Prog. **46**,1 (1990) 105–122.
92. Kiwiel, K.C.: A tilted cutting plane proximal bundle method for convex nondifferentiable optimization. Oper. Res. Lett. **10** (1991) 75–81.
93. Kuhn, H.W.: Nonlinear programming: a historical view. SIAM-AMS Proceedings **9** (1976) 1–26.
94. Kutateladze, S.S.: Changes of variables in the Young transformation. Soviet Math. Dokl. **18**,2 (1977) 545–548.
95. Kutateladze, S.S.: Convex ε-programming. Soviet Math. Dokl. **20** (1979) 391–393.
96. Kutateladze, S.S.: ε-subdifferentials and ε-optimality. Sib. Math. J. (1981) 404–411.

97. Laurent, P.-J.: *Approximation et Optimisation*. Hermann, Paris (1972)
98. Lay, S.R.: *Convex Sets and their Applications*. Wiley & Sons (1982).
99. Lebedev, B.Yu.: On the convergence of the method of loaded functional as applied to a convex programming problem. J. Num. Math. and Math. Phys. **12** (1977) 765–768.
100. Lemaréchal, C.: An algorithm for minimizing convex functions. In: Proceedings, IFIP74 (J.L. Rosenfeld, ed.). Stockholm (1974), pp. 552–556.
101. Lemaréchal, C.: An extension of Davidon methods to nondifferentiable problems. In: *Nondifferentiable Optimization* (M.L. Balinski, P. Wolfe, eds.). Math. Prog. Study **3** (1975) 95–109.
102. Lemaréchal, C.: Combining Kelley's and conjugate gradient methods. In: Abstracts, IX. Intern. Symp. on Math. Prog., Budapest (1976).
103. Lemaréchal, C.: Nonsmooth optimization and descent methods. Research Report **78**,4 (1978) IIASA, 2361 Laxenburg, Austria.
104. Lemaréchal, C.: Nonlinear programming and nonsmooth optimization: a unification. Rapport Laboria **332** (1978) INRIA.
105. Lemaréchal, C.: A view of line-searches. In: *Optimization and Optimal Control* (A. Auslender, W. Oettli, J. Stoer, eds.). Lecture Notes in Control and Information Sciences, vol. 30. Springer, Berlin Heidelberg (1981), pp. 59–78.
106. Lemaréchal, C.: Constructing bundle methods for convex optimization. In: *Fermat Days 85: Mathematics for Optimization* (J.-B. Hiriart-Urruty, ed.). North-Holland Mathematics Studies **129** (1986) 201–240.
107. Lemaréchal, C.: An introduction to the theory of nonsmooth optimization. Optimization **17** (1986) 827–858.
108. Lemaréchal, C.: Nondifferentiable optimization. In: *Handbook in OR & MS,* Vol. 1 (G.L. Nemhauser et al., eds.). Elsevier, North-Holland (1989), pp. 529–572.
109. Lemaréchal, C., Mifflin, R.: Global and superlinear convergence of an algorithm for one-dimensional minimization of convex functions. Math. Prog. **24**,3 (1982) 241–256.
110. Lemaréchal, C., Nemirovskij, A.S., Nesterov, Yu.E.: New variants of bundle methods. Math. Prog. (to appear).
111. Lemaréchal, C., Zowe, J.: Some remarks on the construction of higher order algorithms in convex optimization. Appl. Math. Optimization **10** (1983) 51–68.
112. Lion, G.: Un savoir en voie de disparition: la convexité. Singularité **2**,10 (1991) 5–12.
113. Liu, D.C., Nocedal, J.: On the limited memory BFGS method for large-scale optimization. Math. Prog. **45** (1989) 503–528.
114. Luenberger, D.G.: *Optimization by Vector Space Methods*. Wiley & Sons (1969).
115. Magnanti, T.L.: Fenchel and Lagrange duality are equivalent. Math. Prog. **7** (1974) 253–258.
116. Marcotte, P., Dussault, J.P.: A sequential linear programming algorithm for solving monotone variational inequalities. SIAM J. Control Opt. **27** (1989) 1260–1278.
117. Marsten, R.E.: The use of the boxstep method in discrete optimization. In: *Nondifferentiable Optimization* (M.L. Balinski, P. Wolfe, eds.). Math. Prog. Study **3** (1975) 127–144.
118. Marti, J.: *Konvexe Analysis*. Birkhäuser, Basel (1977).
119. Martinet, B.: Régularisation d'inéquations variationnelles par approximations successives. Revue Franc. Rech. Opér. **R3** (1970) 154–158.
120. Mazure, M.-L.: L'addition parallèle d'opérateurs interprétée comme inf-convolution de formes quadratiques convexes. Modélisation Math. Anal. Numér. **20** (1986) 497–515.

121. Mc Cormick, G.P., Tapia, R.A.: The gradient projection method under mild differentiability conditions. SIAM J. Control **10**,1 (1972) 93–98.
122. Mifflin, R.: Semi-smooth and semi-convex functions in constrained optimization. SIAM J. Control Opt. **15**,6 (1977) 959–972.
123. Mifflin, R.: An algorithm for constrained optimization with semi-smooth functions. Math. Oper. Res. **2**,2 (1977) 191–207.
124. Mifflin, R.: A modification and an extension of Lemaréchal's algorithm for nonsmooth minimization. In: *Nondifferential and Variational Techniques in Optimization* (D.C. Sorensen, J.B. Wets, eds.). Math. Prog. Study **17** (1982) 77–90.
125. Minoux, M.: *Programmation Mathématique: Théorie et Algoritmes* I, II. Dunod, Paris (1983).
126. Moré, J.J.: Implementation and testing of optimization software. In: *Performance Evaluation of Numerical Software* (L.D. Fosdick, ed.). North-Holland (1979).
127. Moré, J.J.: Recent developments in algorithms and software for trust region methods. In: *Mathematical Programming, the State of the Art* (A. Bachem, M. Grötschel, B. Korte, eds.). Springer, Berlin Heidelberg (1983), pp. 258–287.
128. Moré, J.J., Thuente, D.J.: Line search algorithms with guaranteed sufficient decrease. ACM Transactions on Math. Software; Assoc. for Comp. Machinery (to appear).
129. Moreau, J.-J.: Décomposition orthogonale d'un espace hilbertien selon deux cônes mutuellement polaires. C.R. Acad. Sci. Paris **255** (1962) 238–240.
130. Moreau, J.-J.: Proximité et dualité dans un espace hilbertien. Bull. Soc. Math. France **93** (1965) 273–299.
131. Moreau, J.-J.: *Fonctionnelles Convexes*. Lecture notes, Séminaire "Equations aux dérivées partielles", Collège de France, Paris (1966).
132. Moulin, H., Fogelman-Soulié, F.: *La Convexité dans les Mathématiques de la Décision*. Hermann, Paris (1979).
133. Nemirovskij, A.S., Yudin, D.B.: *Problem Complexity and Method Efficiency in Optimization*. Wiley-Interscience (1983).
134. Nesterov, Yu.E.: Minimization methods for nonsmooth convex and quasiconvex functions. Matekon **20** (1984) 519–531.
135. Niven, I.: *Maxima and Minima Without Calculus*. Dolciani Mathematical Expositions **6** (1981).
136. Nurminskii, E.A.: On ε-subgradient mappings and their applications in nondifferentiable optimization. Working paper 78,58 (1978) IIASA, 2361 Laxenburg, Austria.
137. Nurminskii, E.A.: ε-subgradient mapping and the problem of convex optimization. Cybernetics **21**,6 (1986) 796–800.
138. Nurminskii, E.A.: Convex optimization problems with constraints. Cybernetics **23**,4 (1988) 470–474.
139. Nurminskii, E.A.: A class of convex programming methods. USSR Comput. Maths Math. Phys. **26**,4 (1988) 122–128.
140. Overton, M.L., Womersley, R.S.: Optimality conditions and duality theory for minimizing sums of the largest eigenvalues of symmetric matrices. Math. Prog. (to appear).
141. Penot, J.-P.: Subhessians, superhessians and conjugation. Nonlinear Analysis: Theory, Methods and Appl. (to appear).
142. Peressini, A.L., Sullivan, F.E., Uhl, J.J.: *The Mathematics of Nonlinear Programming*. Springer, New York (1988).

143. Phelps, R.R.: *Convex Functions, Monotone Operators and Differentiability.* Lecture Notes in Mathematics, vol. 1364. Springer, Berlin Heidelberg (1989, new edition in 1993).
144. Polak, E.: *Computational Methods in Optimization.* Academic Press, New York (1971).
145. Poljak, B.T.: A general method for solving extremum problems. Soviet Math. Dokl. **174**,8 (1966) 33–36.
146. Poljak, B.T.: Minimization of unsmooth functionals. USSR Comput. Maths Math. Phys. **9** (1969) 14–29.
147. Popova, N.K., Tarasov, V.N.: A modification of the cutting-plane method with accelerated convergence. In: *Nondifferentiable Optimization: Motivations and Aplications* (V.F. Demjanov, D. Pallaschke, eds.). Lecture Notes in Economics and Mathematical Systems, vol. 255. Springer, Berlin Heidelberg (1984), pp. 284–190.
148. Ponstein, J.: Applying some modern developments to choosing your own Lagrange multipliers. SIAM Review **25**,2 (1983) 183–199.
149. Pourciau, B.H.: Modern multiplier rules. Amer. Math. Monthly **87** (1980), 433–452.
150. Powell, M.J.D.: Nonconvex minimization calculations and the conjugate gradient method. In: *Numerical Analysis* (D.F. Griffiths ed.). Lecture Notes in Mathematics, vol. 1066. Springer, Berlin Heidelberg (1984), pp. 122–141.
151. Prekopa, A.: On the development of optimization theory. Amer. Math. Monthly **87** (1980) 527–542.
152. Pshenichnyi, B.N.: *Necessary Conditions for an Extremum.* Marcel Dekker (1971).
153. Pshenichnyi, B.N.: Nonsmooth optimization and nonlinear programming. In: *Nonsmooth Optimization* (C. Lemaréchal, R. Mifflin, eds.), IIASA Proceedings Series 3, Pergamon Press (1978), pp. 71–78.
154. Pshenichnyi, B.N.: *Methods of Linearization.* Springer, Berlin Heidelberg (1993).
155. Pshenichnyi, B.N., Danilin, Yu.M.: *Numerical Methods for Extremal Problems.* Mir, Moscow (1978).
156. Quadrat, J.-P.: Théorèmes asymptotiques en programmation dynamique. C.R. Acad. Sci. Paris, **311**, Série I (1990) 745–748.
157. Roberts, A.W., Varberg, D.E.: *Convex Functions.* Academic Press (1973).
158. Rockafellar, R.T.: Convex programming and systems of elementary monotonic relations. J. Math. Anal. Appl. **19** (1967) 543–564.
159. Rockafellar, R.T.: *Convex Analysis.* Princeton University Press (1970).
160. Rockafellar, R.T.: *Convex Duality and Optimization.* SIAM regional conference series in applied mathematics (1974).
161. Rockafellar, R.T.: Augmented Lagrangians and applications of the proximal point algorithm in convex programming. Math. Oper. Res. **1**,2 (1976) 97–116.
162. Rockafellar, R.T.: Lagrange multipliers in optimization. SIAM-AMS Proceedings **9** (1976) 145–168.
163. Rockafellar, R.T.: Solving a nonlinear programming problem by way of a dual problem. Symposia Mathematica **XIX** (1976) 135–160.
164. Rockafellar, R.T.: *The Theory of Subgradients ad its Applications to Problems of Optimization: Convex and Nonconvex Functions.* Heldermann, West-Berlin (1981).
165. Rockafellar, R.T.: Lagrange multipliers and optimality. SIAM Review (to appear, 1993).
166. Rockafellar, R.T., Wets, R.J.-B.: *Variational Analysis* (in preparation).
167. Rosen, J.B.: The gradient projection method for nonlinear programming; part I: linear constraints. J. SIAM **8** (1960) 181–217.

168. Schramm, H., Zowe, J.: A version of the bundle idea for minimizing a nonsmooth function: conceptual idea, convergence analysis, numerical results. SIAM J. Opt. **2** (1992) 121–152.
169. Schrijver, A.: *Theory of Linear and Integer Programming*. Wiley-Interscience (1986).
170. Seeger, A.: Second derivatives of a convex function and of its Legendre-Fenchel transformate. SIAM J. Opt. **2**,3 (1992) 405–424.
171. Shor, N.Z.: *Minimization Methods for Nondifferentiable Functions*. Springer, Berlin Heidelberg (1985).
172. Smith, K.T.: *Primer of Modern Analysis*. Springer, New York (1983).
173. Stoer, J., Witzgall, C.: *Convexity and Optimization in Finite Dimension I*. Springer, Berlin Heidelberg (1970).
174. Strang, G.: *Introduction to Applied Mathematics*. Wellesley – Cambridge Press (1986).
175. Strodiot, J.-J., Nguyen, V.H., Heukemes, N.: ε-optimal solutions in nondifferentiable convex programming and some related questions. Math. Prog. **25** (1983) 307–328.
176. Tikhomirov, V.M.: Stories about maxima and minima. In: *Mathematical World* **1**, Amer. Math. Society, Math. Association of America (1990).
177. Troutman, J.L.: *Variational Calculus with Elementary Convexity*. Springer, New York (1983).
178. Valadier, M.: Sous-différentiels d'une borne supérieure et d'une somme continue de fonctions convexes. Note aux C. R. Acad. Sci. Paris, Série A **268** (1969) 39–42.
179. Valadier, M.: *Contribution à l'Analyse Convexe*. Thèse de doctorat ès sciences mathématiques, Paris (1970).
180. Valadier, M.: Intégration de convexes fermés notamment d'épigraphes. Inf-convolution continue. Revue d'Informatique et de Recherche Opérationnelle (1970) 47–53.
181. Van Rooij, A.C.M., Schikhof, W.H.: *A Second Course on Real Functions*. Cambridge University Press (1982).
182. Van Tiel, J.: *Convex Analysis. An Introductory Text*. Wiley & Sons (1984).
183. Wets, R. J.-B.: *Grundlagen konvexer Optimierung*. Lecture Notes in Economics and Mathematical Systems, vol. 137. Springer, Berlin Heidelberg (1976).
184. Willem, M.: *Analyse Convexe et Optimisation*, 3rd edn. Editions CIACO Louvain-La-Neuve (1989).
185. Wolfe, P.: Accelerating the cutting plane method for nonlinear programming. J. SIAM **9**,3 (1961) 481–488.
186. Wolfe, P.: Convergence conditions for ascent methods. SIAM Review **11** (1968) 226–235.
187. Wolfe, P.: A method of conjugate subgradients for minimizing nondifferentiable functions. In: Proceedings, XII. Annual Allerton conference on Circuit and System Theory (P.V. Kokotovic, E.S. Davidson, eds.). Univ. Illinois at Urbana-Champaign (1974), pp. 8–15.
188. P. Wolfe: A method of conjugate subgradients for minimizing nondifferentiable functions. In: *Nondifferentiable Optimization* (M.L. Balinski, P. Wolfe, eds.). Math. Prog. Study **3** (1975) 145–173.
189. Zarantonello, E.H.: Projections on convex sets in Hilbert spaces and spectral theory. In: *Contributions to Nonlinear Functional Analysis*. Academic Press (1971), pp. 237–424.
190. Zeidler, E.: *Nonlinear Functional Analysis and its Applications III. Variational Methods and Optimization*. Springer, New York (1985).

Index

asymptotic function, **XVII**, 111

breadth, **XVII**
bundle, bundling, 7, 157, 196, 228, 244, 327
– (compression, aggregation of), **14**, 177, 228, 230, 232, 247, 256, 301
– (of information), 14, 304

Clarke, 331
closed convex
– cone, 186
– function, **XVI**, 38, 151, 161
– polyhedron, 115
closure of a function, **XVI**, 45
coercive, 46
– (1-), 50, 82, 89, 318, 333
coincidence set, **122**
complexity, 20, 157
computer, computing, 2, 211, 256, 263
conjugate function, 98, 132, 179, 298, 320
conjugate gradient, 229
constraint, 137
convergence, *see* speed of –
convex combination, **XVI**, 156, 211
convex hull, 159
convex multiplier, **XVI**, 26, 216
critical point, *see* stationary
curvature, 106, 208
curved-search, 284, 316
cutting plane, 77, 229, 275, 330

decomposition, 43, 116, 137, 141, 184, 313
degenerate, 38, 285
derivative, 317
– (second), 106
descent direction, 156
descent-step, 204, 231, 248, 283, 305
difference quotient, **67**

– (approximate), **106**, 200
dilation, 334
directional derivative, **XVI**, 66, 196
– (approximate), **102**
directionally quadratic function, **84**
distance, 65, 173
– of bounded sets, 129
divergent series, 198, 325
domain, **XVII**
dot-product, **XVI**, 138
dual, duality, 22, 238, 306
– function, **148**
– gap, **153**, 155, 179

eigenvalue, 137, 323
ellipsoid, 1, 80, 169
entropy, 151, 153, 156, 160
epigraph, **XVII**
Everett, 147, 163
exposed (face, point), 97, 217

Fenchel
– duality theorem, **63**
– inequality, **37**
– transformation, **37**
filling property, **154**, 155, 164
fixed point, 322

Gâteaux, 49, 54
gauge, **XVII**, 71
Gauss-Newton, 290
gradient, 96, 320
– method, 28
graph theory, 22

half-space, **XVII**, 45
hyperplane, **XVII**, 199

image-function, **54**, 72

indicator function, **XVII**, 39, 93
inf-convolution, 55, 187
– (exact), 62, 119, 120
interior points, 335

Lagrange, Lagrangian, **138**, 237, 307
Legendre transform, **35**, 43, 81
Levenberg-Marquardt, 290
line-search, 4, 196, 283
linearization error, **131**, 201, **232**
Lipschitz, 122, 128
local problem, 140
locally bounded, 127

marginal function, 55, 320
marginal price, 151
master problem, 140
mean-value theorem, 112
minimality conditions (approximate), 115
minimizing sequence, **XVII**, 218, 288, 309
minimum, minimum point, 49
– (global), 333
minorize, minorization, **XVII**
model, 101, 279
Moreau-Yosida, **121**, 183, 187
multi-valued, *see* multifunction
multifunction, **XVI**, 99, 112

Newton, quasi-Newton, 228, 283, 293, 326
nonnegative, **XVII**
normal set (approximate), **93**, 115
normalization, norming, 279, 280
null-step, **7**, 156, 204, 231, 248, 283, 295, 305

objective function, 137, 152
orthant, **XVII**
outer semi-continuous, **XVII**, 128

penalty, 142
– (exact), 185
perspective-function, **XVII**, 41, 99
piecewise affine, **76**, 125, 156, 245
polar cone, **XVII**, 45, 186
polyhedral function, **77**, 307
positively homogeneous, 84, 280
primal problem, 137
programming problem
– (integer), 142, 181
– (linear), 116, 145, 181, 276

– (quadratic), 234, 272, 299, 304, 332
– (semi-infinite), 174
proximal point, 318, 322

qualification, 58, 62, 72, 125, 191
quasi-convex, 201

rate of convergence, *see* speed –
recession(cone, function), *see* asymptotic
relative interior, **XVII**, 62
relaxation
– (Lagrangian), 181, 216
– (convex), 157, 181
– (method), 174, 293

saddle-point, 188
safeguard-reduction property, **208**, 251
semi-smooth, 331
separation, 10, 195, 199, 222, 226, 254
set-valued, *see* multifunction
slack variable, 142
Slater, 165, 188
speed of convergence, 33, 288, 310, 312
– (fast), 208
– (linear), 20
– (sublinear), 16, 18, 271
stability center, 279
stationary point, 50
steepest-descent direction, **1**, 196, 225, 262, 315
strictly convex, 79, 81, 174, 181
strongly convex, 82, 83, 318
subdifferential, subgradient, 47, 151
subgradient algorithm, 171, 315
sublevel-set, **XVII**
support function, **XVII**, 30, 40, 66, 97, 200, 299

transportation formula, **131**, 211
transportation problem, 21
travelling salesman problem, 22
trust-region, 284

unit ball, **XVI**
unit simplex, **XVI**
unit sphere, **XVI**

vertex, *see* exposed (point)

Wolfe, 177, 249, 284

zigzag, 1, 7, 28

Grundlehren der mathematischen Wissenschaften
A Series of Comprehensive Studies in Mathematics

A Selection

200. Dold: Lectures on Algebraic Topology
201. Beck: Continuous Flows in the Plane
202. Schmetterer: Introduction to Mathematical Statistics
203. Schoeneberg: Elliptic Modular Functions
204. Popov: Hyperstability of Control Systems
205. Nikol'skiĭ: Approximation of Functions of Several Variables and Imbedding Theorems
206. André: Homologie des Algébres Commutatives
207. Donoghue: Monotone Matrix Functions and Analytic Continuation
208. Lacey: The Isometric Theory of Classical Banach Spaces
209. Ringel: Map Color Theorem
210. Gihman/Skorohod: The Theory of Stochastic Processes I
211. Comfort/Negrepontis: The Theory of Ultrafilters
212. Switzer: Algebraic Topology – Homotopy and Homology
215. Schaefer: Banach Lattices and Positive Operators
217. Stenström: Rings of Quotients
218. Gihman/Skorohod: The Theory of Stochastic Procrsses II
219. Duvant/Lions: Inequalities in Mechanics and Physics
220. Kirillov: Elements of the Theory of Representations
221. Mumford: Algebraic Geometry I: Complex Projective Varieties
222. Lang: Introduction to Modular Forms
223. Bergh/Löfström: Interpolation Spaces. An Introduction
224. Gilbarg/Trudinger: Elliptic Partial Differential Equations of Second Order
225. Schütte: Proof Theory
226. Karoubi: K-Theory. An Introduction
227. Grauert/Remmert: Theorie der Steinschen Räume
228. Segal/Kunze: Integrals and Operators
229. Hasse: Number Theory
230. Klingenberg: Lectures on Closed Geodesics
231. Lang: Elliptic Curves: Diophantine Analysis
232. Gihman/Skorohod: The Theory of Stochastic Processes III
233. Stroock/Varadhan: Multidimensional Diffusion Processes
234. Aigner: Combinatorial Theory
235. Dynkin/Yushkevich: Controlled Markov Processes
236. Grauert/Remmert: Theory of Stein Spaces
237. Köthe: Topological Vector Spaces II
238. Graham/McGehee: Essays in Commutative Harmonic Analysis
239. Elliott: Proabilistic Number Theory I
240. Elliott: Proabilistic Number Theory II
241. Rudin: Function Theory in the Unit Ball of C^n
242. Huppert/Blackburn: Finite Groups II
243. Huppert/Blackburn: Finite Groups III
244. Kubert/Lang: Modular Units
245. Cornfeld/Fomin/Sinai: Ergodic Theory
246. Naimark/Stern: Theory of Group Representations
247. Suzuki: Group Theory I
248. Suzuki: Group Theory II
249. Chung: Lectures from Markov Processes to Brownian Motion
250. Arnold: Geometrical Methods in the Theory of Ordinary Differential Equations
251. Chow/Hale: Methods of Bifurcation Theory
252. Aubin: Nonlinear Analysis on Manifolds. Monge-Ampère Equations

253. Dwork: Lectures on p-adic Differential Equations
254. Freitag: Siegelsche Modulfunktionen
255. Lang: Complex Multiplication
256. Hörmander: The Analysis of Linear Partial Differential Operators I
257. Hörmander: The Analysis of Linear Partial Differential Operators II
258. Smoller: Shock Waves and Reaction-Diffusion Equations
259. Duren: Univalent Functions
260. Freidlin/Wentzell: Random Perturbations of Dynamical Systems
261. Bosch/Güntzer/Remmert: Non Archimedian Analysis – A System Approach to Rigid Analytic Geometry
262. Doob: Classical Potential Theory and Its Probabilistic Counterpart
263. Krasnosel'skiĭ/Zabreĭko: Geometrical Methods of Nonlinear Analysis
264. Aubin/Cellina: Differential Inclusions
265. Grauert/Remmert: Coherent Analytic Sheaves
266. de Rham: Differentiable Manifolds
267. Arbarello/Cornalba/Griffiths/Harris: Geometry of Algebraic Curves, Vol. I
268. Arbarello/Cornalba/Griffiths/Harris: Geometry of Algebraic Curves, Vol. II
269. Schapira: Microdifferential Systems in the Complex Domain
270. Scharlau: Quadratic and Hermitian Forms
271. Ellis: Entropy, Large Deviations, and Statistical Mechanics
272. Elliott: Arithmetic Functions and Integer Products
273. Nikol'skiĭ: Treatise on the Shift Operator
274. Hörmander: The Analysis of Linear Partial Differential Operators III
275. Hörmander: The Analysis of Linear Partial Differential Operators IV
276. Liggett: Interacting Particle Systems
277. Fulton/Lang: Riemann-Roch Algebra
278. Barr/Wells: Toposes, Triples and Theories
279. Bishop/Bridges: Constructive Analysis
280. Neukirch: Class Field Theory
281. Chandrasekharan: Elliptic Functions
282. Lelong/Gruman: Entire Functions of Several Complex Variables
283. Kodaira: Complex Manifolds and Deformation of Complex Structures
284. Finn: Equilibrium Capillary Surfaces
285. Burago/Zalgaller: Geometric Inequalities
286. Andrianov: Quadratic Forms and Hecke Operators
287. Maskit: Kleinian Groups
288. Jacod/Shiryaev: Limit Theorems for Stochastic Processes
289. Manin: Gauge Field Theory and Complex Geometry
290. Conway/Sloane: Sphere Packings, Lattices and Groups
291. Hahn/O'Meara: The Classical Groups and K-Theory
292. Kashiwara/Schapira: Sheaves on Manifolds
293. Revuz/Yor: Continuous Martingales and Brownian Motion
294. Knus: Quadratic and Hermitian Forms over Rings
295. Dierkes/Hildebrandt/Küster/Wohlrab: Minimal Surfaces I
296. Dierkes/Hildebrandt/Küster/Wohlrab: Minimal Surfaces II
297. Pastur/Figotin: Spectra of Random and Almost-Periodic Operators
298. Berline/Getzler/Vergne: Heat Kernels and Dirac Operators
299. Pommerenke: Boundary Behaviour of Conformal Maps
300. Orlik/Terao: Arrangements of Hyperplanes
301. Loday: Cyclic Homology
303. Lange/Birkenhake: Complex Abelian Varieties
303. DeVore/Lorentz: Constructive Approximation
304. Lorentz/v. Golitschek/Makovoz: Constructive Approximation. Advanced Problems
305. Hiriart-Urruty/Lemaréchal: Convex Analysis and Minimization Algorithms I. Fundamentals
306. Hiriart-Urruty/Lemaréchal: Convex Analysis and Minimization Algorithms II. Advanced Theory and Bundle Methods
307. Schwarz: Quantum Field Theory and Topology